WITHDRAWN

PROGRESS IN

Nucleic Acid Research and Molecular Biology

Volume 31

PROGRESS IN
Nucleic Acid Research and Molecular Biology

edited by

WALDO E. COHN
Biology Division
Oak Ridge National Laboratory
Oak Ridge, Tennessee

KIVIE MOLDAVE
Department of Biological Chemistry
College of Medicine
University of California
Irvine, California

Volume 31

1984

ACADEMIC PRESS, INC.
(Harcourt Brace Jovanovich, Publishers)

Orlando San Diego New York London
Toronto Montreal Sydney Tokyo

COPYRIGHT © 1984, BY ACADEMIC PRESS, INC.
ALL RIGHTS RESERVED.
NO PART OF THIS PUBLICATION MAY BE REPRODUCED OR
TRANSMITTED IN ANY FORM OR BY ANY MEANS, ELECTRONIC
OR MECHANICAL, INCLUDING PHOTOCOPY, RECORDING, OR ANY
INFORMATION STORAGE AND RETRIEVAL SYSTEM, WITHOUT
PERMISSION IN WRITING FROM THE PUBLISHER.

ACADEMIC PRESS, INC.
Orlando, Florida 32887

United Kingdom Edition published by
ACADEMIC PRESS, INC. (LONDON) LTD.
24/28 Oval Road, London NW1 7DX

LIBRARY OF CONGRESS CATALOG CARD NUMBER: 63-15847
ISBN 0-12-540031-4

PRINTED IN THE UNITED STATES OF AMERICA

84 85 86 87   9 8 7 6 5 4 3 2 1

Contents

CONTRIBUTORS	ix
ABBREVIATIONS AND SYMBOLS	xi
SOME ARTICLES PLANNED FOR FUTURE VOLUMES	xv

Immunoassay of Carcinogen-Modified DNA

Paul T. Strickland and John M. Boyle

I.	Methodology	2
II.	Undamaged Nucleic Acid Components	7
III.	Ultraviolet Radiation-Damaged DNA	12
IV.	Ionizing Radiation-Damaged DNA	18
V.	Chemical Adducts in DNA	25
VI.	Perspectives	47
	References	51

On the Biological Significance of Modified Nucleosides in tRNA

Helga Kersten

I.	Evolutionary Aspects	62
II.	Role of Ribosylthymine in tRNA Function of Eubacteria	65
III.	Involvement of Ribosylthymine in Regulatory Mechanisms	71
IV.	Ribosylthymine in Eukaryotic Elongator tRNA	79
V.	N^6-(2-Isopentenyl)adenosine and Derivatives in tRNA	83
VI.	Central Role of N^6-(2-Isopentenyl)adenosine and Derivatives in Growth and Development	86
VII.	On the Function of Queuine in tRNA	95
VIII.	Summary and Outlook	103
	References	107

The Organization and Transcription of Eukaryotic Ribosomal RNA Genes

Radha K. Mandal

I.	Ribosomal RNA Genes	117
II.	Organization of rRNA Genes	120

III.	Transcription of rRNA Genes	138
IV.	Concluding Remarks	150
	References	151

Structure, Function, and Evolution of 5-S Ribosomal RNAs

Nicholas Delihas, Janet Andersen, and Ram P. Singhal

I.	Generalized Structures of the 5-S Ribosomal RNAs	161
II.	"Hot Spots" of Insertions and Deletions in 5-S RNAs	169
III.	Conformation of 5-S RNA Derived from Enzymatic, Chemical, and Physical Studies	174
IV.	5-S RNA Interactions with Protein	177
V.	Phylogeny and Endosymbiosis	181
VI.	Summary	187
	References	188

Optimization of Translational Accuracy

C. G. Kurland and Måns Ehrenberg

I.	Gene Expression by Ambiguous Translation	192
II.	Correlations of Speed and Accuracy in Translation	194
III.	Codon–Anticodon Selectivity on the Ribosome	197
IV.	Kinetic Proofreading: One More Time	200
V.	The Problem Solved by Kinetic Proofreading: Rate or Accuracy?	205
VI.	Translational Accuracy and Exponential Growth	208
VII.	Kinetic Options for Inexpensive Proofreading and Error Regulation	212
VIII.	Error Feedback Increases Accuracy: End of Error Catastrophe	216
	References	217

Molecular Aspects of Development in the Brine Shrimp *Artemia*

Albert J. Wahba and Charles L. Woodley

I.	Polypeptide-Chain Initiation	224
II.	Status of Message in *Artemia* Embryos	245
III.	Questions for the Future	255
	References	260

Translational Control Involving a Novel Cytoplasmic RNA and Ribonucleoprotein

Satyapriya Sarkar

I.	Eukaryotic RNAs with Translation Modulator Activities	269
II.	The Translation-Inhibiting Cytoplasmic 10-S Ribonucleoprotein of Chick Embryo Muscle	273
III.	The Secondary Structure of iRNA	277
IV.	Dissociation and Reassociation of iRNP	278
V.	Differential Effects of iRNA and iRNP on Endogenous and Exogenous mRNA Translation in Reticulocyte Lysate	281
VI.	Effect of iRNA on the Different Intermediates in Polypeptide Chain Initiation	283
VII.	Concluding Remarks	290
	References	291

The Hypoxanthine Phosphoribosyltransferase Gene: A Model for the Study of Mutation in Mammalian Cells

A. Craig Chinault and C. Thomas Caskey

I.	General Aspects: Enzyme and Genetic Properties	296
II.	Comparison of Normal and Mutant Hypoxanthine Phosphoribosyltransferase Proteins	298
III.	Molecular Cloning and Analysis of Hypoxanthine Phosphoribosyltransferase Sequences	301
IV.	Normal and Mutant Gene Structures	305
V.	Future Prospects	309
	References	310

The Molecular Genetics of Human Hemoglobin

Francis S. Collins and Sherman M. Weissman

I.	Structure of Human Globin Genes	317
II.	Approaches for Studying Expression of Globin Genes	325
III.	Physiological Globin Gene Expression	330
IV.	Structure of Human Globin Gene Clusters	359
V.	Evolutionary Considerations	379
VI.	Naturally Occurring Mutations	390
VII.	Summary and Prospects	419
	References	421

INDEX	467
CONTENTS OF PREVIOUS VOLUMES	473

Contributors

Numbers in parentheses indicate the pages on which the authors' contributions begin.

JANET ANDERSEN (161), *Department of Microbiology, School of Medicine, SUNY at Stony Brook, Stony Brook, New York 11794*

JOHN M. BOYLE (1), *Paterson Laboratories, Manchester M20 9BX, England*

C. THOMAS CASKEY (295), *Howard Hughes Medical Institute Laboratories and Departments of Biochemistry and Medicine, Baylor College of Medicine, Houston, Texas 77030*

A. CRAIG CHINAULT (295), *Howard Hughes Medical Institute Laboratories and Departments of Biochemistry and Medicine, Baylor College of Medicine, Houston, Texas 77030*

FRANCIS S. COLLINS (315), *Departments of Human Genetics, Medicine, and Molecular Biophysics and Biochemistry, Yale University School of Medicine, New Haven, Connecticut 06510*

NICHOLAS DELIHAS (161), *Department of Microbiology, School of Medicine, SUNY at Stony Brook, Stony Brook, New York 11794*

MÅNS EHRENBERG (191), *Department of Molecular Biology, Biomedical Center, S-751 24 Uppsala, Sweden*

HELGA KERSTEN (59), *Institute of Physiological Chemistry, The University of Erlangen-Nürnberg, Nürnberg, Federal Republic of Germany*

C. G. KURLAND (191), *Department of Molecular Biology, Biomedical Center, S-751 24 Uppsala, Sweden*

RADHA K. MANDAL (115), *Department of Biochemistry, Bose Institute, Calcutta 700 009, India*

SATYAPRIYA SARKAR (267), *Department of Muscle Research, Boston Biomedical Research Institute, and Department of Neurology, Harvard Medical School, Boston, Massachusetts 02114*

RAM P. SINGHAL (161), *Chemistry Department, Wichita State University, Wichita, Kansas 67208*

PAUL T. STRICKLAND (1), *NCI Frederick Cancer Research Facility, Frederick, Maryland 21701*

ALBERT J. WAHBA (221), *Department of Biochemistry, University of Mississippi Medical Center, Jackson, Mississippi 39216*

SHERMAN M. WEISSMAN (315), *Departments of Human Genetics, Medicine, and Molecular Biophysics and Biochemistry, Yale University School of Medicine, New Haven, Connecticut 06510*

CHARLES L. WOODLEY (221), *Department of Biochemistry, University of Mississippi Medical Center, Jackson, Mississippi 39216*

Abbreviations and Symbols

All contributors to this Series are asked to use the terminology (abbreviations and symbols) recommended by the IUPAC-IUB Commission on Biochemical Nomenclature (CBN) and approved by IUPAC and IUB, and the Editor endeavors to assure conformity. These Recommendations have been published in many journals (1, 2) and compendia (3) in four languages and are available in reprint form from the Office of Biochemical Nomenclature (OBN), as stated in each publication, and are therefore considered to be generally known. Those used in nucleic acid work, originally set out in section 5 of the first Recommendations (1) and subsequently revised and expanded (2, 3), are given in condensed form (I–V) below for the convenience of the reader. Authors may use them without definition, when necessary.

I. Bases, Nucleosides, Mononucleotides

1. *Bases* (in tables, figures, equations, or chromatograms) are symbolized by Ade, Gua, Hyp, Xan, Cyt, Thy, Oro, Ura; Pur = any purine, Pyr = any pyrimidine, Base = any base. The prefixes S-, H_2, F-, Br, Me, etc., may be used for modifications of these.

2. *Ribonucleosides* (in tables, figures, equations, or chromatograms) are symbolized, in the same order, by Ado, Guo, Ino, Xao, Cyd, Thd, Ord, Urd (Ψrd), Puo, Pyd, Nuc. Modifications may be expressed as indicated in (1) above. Sugar residues may be specified by the prefixes r (optional), d (=deoxyribo), a, x, l, etc., to these, or by two three-letter symbols, as in Ara-Cyt (for aCyd) or dRib-Ade (for dAdo).

3. *Mono-, di-, and triphosphates of nucleosides* (5') are designated by NMP, NDP, NTP. The N (for "nucleoside") may be replaced by any one of the nucleoside symbols given in II-1 below. 2'-, 3'-, and 5'- are used as prefixes when necessary. The prefix d signifies "deoxy." [Alternatively, nucleotides may be expressed by attaching P to the symbols in (2) above. Thus: P-Ado = AMP; Ado-P = 3'-AMP] cNMP = cyclic 3':5'-NMP; Bt_2cAMP = dibutyryl cAMP, etc.

II. Oligonucleotides and Polynucleotides

1. Ribonucleoside Residues

(a) Common: A, G, I, X, C, T, O, U, Ψ, R, Y, N (in the order of I-2 above).

(b) Base-modified: sI or M for thioinosine = 6-mercaptopurine ribonucleoside; sU or S for thiouridine; brU or B for 5-bromouridine; hU or D for 5,6-dihydrouridine; i for isopentenyl; f for formyl. Other modifications are similarly indicated by appropriate *lower-case* prefixes (in contrast to I-1 above) (2, 3).

(c) Sugar-modified: prefixes are d, a, x, or l as in I-2 above; alternatively, by *italics* or **boldface** type (with definition) unless the entire chain is specified by an appropriate prefix. The 2'-O-methyl group is indicated by *suffix* m (e.g., -Am- for 2'-O-methyladenosine, but -mA- for 6-methyladenosine).

(d) Locants and multipliers, when necessary, are indicated by superscripts and subscripts, respectively, e.g., -m_2^6A- = 6-dimethyladenosine; -s^4U- or -^{4}S- = 4-thiouridine; -ac^4Cm- = 2'-O-methyl-4-acetylcytidine.

(e) When space is limited, as in two-dimensional arrays or in aligning homologous sequences, the prefixes may be placed *over the capital letter*, the suffixes *over the phosphodiester symbol*.

2. Phosphoric Residues [left side = 5', right side = 3' (or 2')]

(a) Terminal: p; e.g., pppN . . . is a polynucleotide with a 5'-triphosphate at one end; Ap is adenosine 3'-phosphate; C > p is cytidine 2':3'-cyclic phosphate (1, 2, 3); p < A is adenosine 3':5'-cyclic phosphate.

(b) Internal: hyphen (for known sequence), comma (for unknown sequence); unknown sequences are enclosed in parentheses. E.g., pA-G-A-C(C_2,A,U)A-U-G-C > p is a sequence with a (5') phosphate at one end, a 2':3'-cyclic phosphate at the other, and a tetranucleotide of unknown sequence in the middle. (**Only codon triplets should be written without some punctuation separating the residues.**)

3. Polarity, or Direction of Chain

The symbol for the phosphodiester group (whether hyphen or comma or parentheses, as in 2b) represents a 3'-5' link (i.e., a 5'... 3' chain) unless otherwise indicated by appropriate numbers. "Reverse polarity" (a chain proceeding from a 3' terminus at left to a 5' terminus at right) may be shown by numerals or by right-to-left arrows. Polarity in any direction, as in a two-dimensional array, may be shown by appropriate rotation of the (capital) letters so that 5' is at left, 3' at right when the letter is viewed right-side-up.

4. Synthetic Polymers

The complete name or the appropriate group of symbols (see II-1 above) of the repeating unit, **enclosed in parentheses if complex or a symbol,** is either (a) preceded by "poly," or (b) followed by a subscript "n" or appropriate number. **No space follows "poly"** (2, 5).

The conventions of II-2b are used to specify known or unknown (random) sequence, e.g.,

polyadenylate = poly(A) or A_n, a simple homopolymer;

poly(3 adenylate, 2 cytidylate) = poly(A_3C_2) or $(A_3,C_2)_n$, an *irregular* copolymer of A and C in 3:2 proportions;

poly(deoxyadenylate-deoxythymidylate) = poly[d(A-T)] or poly(dA-dT) or $(dA-dT)_n$ or $d(A-T)_n$, an *alternating* copolymer of dA and dT;

poly(adenylate,guanylate,cytidylate,uridylate) = poly(A,G,C,U) or $(A,G,C,U)_n$, a random assortment of **A, G, C,** and **U** residues, proportions unspecified.

The prefix copoly or oligo may replace poly, if desired. The subscript "n" may be replaced by numerals indicating actual size, e.g., $A_n \cdot dT_{12-18}$.

III. Association of Polynucleotide Chains

1. *Associated* (e.g., H-bonded) chains, or bases within chains, are indicated by a *center dot* (not a hyphen or a plus sign) separating the *complete* names or symbols, e.g.:

poly(A) · poly(U)   or $A_n \cdot U_m$
poly(A) · 2 poly(U) or $A_n \cdot 2U_m$
poly(dA-dC) · poly(dG-dT)   or $(dA-dC)_n \cdot (dG-dT)_m$.

2. *Nonassociated* chains are separated by the plus sign, e.g.:

2[poly(A) · poly(U)] → poly(A) · 2 poly(U) + poly(A)
or $2[A_n \cdot U_m] \rightarrow A_n \cdot 2U_m + A_n$.

3. Unspecified or unknown association is expressed by a comma (again meaning "unknown") between the completely specified chains.

Note: In all cases, each chain is completely specified in one or the other of the two systems described in II-4 above.

IV. Natural Nucleic Acids

RNA	ribonucleic acid or ribonucleate
DNA	deoxyribonucleic acid or deoxyribonucleate
mRNA; rRNA; nRNA	messenger RNA; ribosomal RNA; nuclear RNA
hnRNA	heterogeneous nuclear RNA
D-RNA; cRNA	"DNA-like" RNA; complementary RNA

ABBREVIATIONS AND SYMBOLS

mtDNA	mitochondrial DNA
tRNA	transfer (or acceptor or amino-acid-accepting) RNA; replaces sRNA, which is not to be used for any purpose
aminoacyl-tRNA	"charged" tRNA (i.e., tRNA's carrying aminoacyl residues); may be abbreviated to AA-tRNA
alanine tRNA or tRNA^{Ala}, etc.	tRNA normally capable of accepting alanine, to form alanyl-tRNA, etc.
alanyl-tRNA or alanyl-tRNA^{Ala}	The same, with alanyl residue covalently attached. [*Note:* fMet = formylmethionyl; hence tRNA^{fMet}, identical with tRNA$_f^{Met}$]

Isoacceptors are indicated by appropriate subscripts, i.e., tRNA$_1^{Ala}$, tRNA$_2^{Ala}$, etc.

V. Miscellaneous Abbreviations

P_i, PP_i	inorganic orthophosphate, pyrophosphate
RNase, DNase	ribonuclease, deoxyribonuclease
t_m (not T_m)	melting temperature (°C)

Others listed in Table II of Reference 1 may also be used without definition. No others, with or without definition, are used unless, in the opinion of the editor, they increase the ease of reading.

Enzymes

In naming enzymes, the 1978 recommendations of the IUB Commission on Biochemical Nomenclature (4) are followed as far as possible. At first mention, each enzyme is described *either* by its systematic name *or* by the equation for the reaction catalyzed *or* by the recommended trivial name, followed by its EC number in parentheses. Thereafter, a trivial name may be used. Enzyme names are not to be abbreviated except when the substrate has an approved abbreviation (e.g., ATPase, but not LDH, is acceptable).

REFERENCES*

1. *JBC* **241**, 527 (1966); *Bchem* **5**, 1445 (1966); *BJ* **101**, 1 (1966); *ABB* **115**, 1 (1966), **129**, 1 (1969); and elsewhere.†
2. *EJB* **15**, 203 (1970); *JBC* **245**, 5171 (1970); *JMB* **55**, 299 (1971); and elsewhere.†
3. "Handbook of Biochemistry" (G. Fasman, ed.), 3rd ed. Chemical Rubber Co., Cleveland, Ohio, 1970, 1975, Nucleic Acids, Vols. I and II, pp. 3–59.
4. "Enzyme Nomenclature" [Recommendations (1978) of the Nomenclature Committee of the IUB]. Academic Press, New York, 1979.
5. "Nomenclature of Synthetic Polypeptides," *JBC* **247**, 323 (1972); *Biopolymers* **11**, 321 (1972); and elsewhere.†

Abbreviations of Journal Titles

Journals	Abbreviations used
Annu. Rev. Biochem.	ARB
Annu. Rev. Genet.	ARGen
Arch. Biochem. Biophys.	ABB
Biochem. Biophys. Res. Commun.	BBRC

*Contractions for names of journals follow.

†Reprints of all CBN Recommendations are available from the Office of Biochemical Nomenclature (W. E. Cohn, Director), Biology Division, Oak Ridge National Laboratory, Box Y, Oak Ridge, Tennessee 37830, USA.

Biochemistry	Bchem
Biochem. J.	BJ
Biochim. Biophys. Acta	BBA
Cold Spring Harbor	CSH
Cold Spring Harbor Lab.	CSHLab
Cold Spring Harbor Symp. Quant. Biol.	CSHSQB
Eur. J. Biochem.	EJB
Fed. Proc.	FP
Hoppe-Seyler's Z. physiol. Chem.	ZpChem
J. Amer. Chem. Soc.	JACS
J. Bacteriol.	J. Bact.
J. Biol. Chem.	JBC
J. Chem. Soc.	JCS
J. Mol. Biol.	JMB
J. Nat. Cancer Inst.	JNCI
Mol. Cell. Biol.	MCBiol
Mol. Cell. Biochem.	MCBchem
Mol. Gen. Genet.	MGG
Nature, New Biology	Nature NB
Nucleic Acid Research	NARes
Proc. Nat. Acad. Sci. U.S.	PNAS
Proc. Soc. Exp. Biol. Med.	PSEBM
Progr. Nucl. Acid. Res. Mol. Biol.	This Series

Some Articles Planned for Future Volumes

Role of Editing in DNA Fidelity
  MAURICE J. BESSMAN

Control of Transcription in SV40
  GOKUL C. DAS, S. K. NIYOGI, AND N. SALZMAN

ATP-Ubiquitin-Dependent Degradation of Intracellular Proteins
  AVRAM HERSHKO

Chemical Changes Induced in DNA by Gamma Rays
  FRANKLIN HUTCHINSON

The Role of tRNA Anticodon Triplet and Loop in the Specific Interaction with Aminoacyl-tRNA Synthetases
  LEV L. KISSELEV

Interactive Three-Dimensional Computer Graphics in Protein and Genetic Engineering
  ROBERT LANGRIDGE

The Photobiology and Radiobiology of *Micrococcus radiodurans*
  R. E. MOSELEY

Ribosomal Translocation: Facts and Models
  ALEXANDER S. SPIRIN

Sequence-Specific Affinity Modification of Nucleic Acids or Reactive Derivatives of Oligo- and Polynucleotides in Molecular Biological Investigations
  VALENTIN V. VLASSOV AND D. G. KNORRE

PROGRESS IN

Nucleic Acid Research and Molecular Biology

Volume 31

Immunoassay of Carcinogen-Modified DNA

PAUL T. STRICKLAND

NCI-Frederick Cancer Research
  Facility
Frederick, Maryland

JOHN M. BOYLE

Paterson Laboratories
Manchester, England

I. Methodology	2
II. Undamaged Nucleic Acid Components	7
III. Ultraviolet Radiation-Damaged DNA	12
IV. Ionizing Radiation-Damaged DNA	18
V. Chemical Adducts in DNA	25
A. Alkylation Damage	25
B. 2-Acetamidofluorene	31
C. Benzo[a]pyrene	38
D. Aflatoxin B_1	41
E. Cisplatin	43
F. Z-DNA	45
VI. Perspectives	47
References	51

The development of immunological assays for the detection of damaged and undamaged nucleic acids has advanced rapidly in recent years. This approach, which complements existing biochemical and physical approaches, provides a valuable tool for investigators in the field of nucleic acid research. Early progress in the production of antisera specific for photodamaged and undamaged DNA and for normal and rare nucleotides has been summarized in a symposium (1) and in general reviews (2–4). More recent reviews discuss immunological detection of radiation (UV^1 and ionizing) damage in DNA (5), of chemical damage in DNA (6, 7), and of naturally occurring modified nucleosides (8).

[1] Abbreviations: AAF, acetamidofluorene; AFB_1, aflatoxin B_1; BPDE, benzo[a]pyrene-7,8-diol 9,10-epoxide; BSA, bovine serum albumin; ELISA, enzyme-linked immunoadsorbent assay; FA, Freund's adjuvant; FI, fluorescein isothiocyanate; Ig, immunoglobulin; KLH, keyhole limpet hemocyanin; RIA, radioimmunoassay; RIST, radioimmunosorbent technique; SLE, systemic lupus erythematosus; USERIA, ultrasensitive enzymatic RIA; UV, ultraviolet (radiation).

A major advantage of quantitative immunoassays for DNA damage is that the DNA or chemical adduct under investigation does not have to be labeled. This is particularly useful for analysis of DNA from tissue in which the cells are slowly proliferating and therefore cannot be labeled at high specific activity and also in cases where a radioactive tag is required on a compound (e.g., a carcinogen) to detect its binding to DNA. In the former case, the sensitivity (level of detection) of some assays depends on the specific activity of the DNA; in the latter case, the particular compound may be difficult and/or expensive to label. In addition, several methods (immunofluorescence, immunoperoxidase, immunoautoradiography) allow the observation of DNA damage and repair in cells and tissue sections *in situ* (9–14). This is particularly useful when heterogeneous cell populations are being studied, and chemical extraction and subsequent analysis of DNA allow only the average amount of damage sustained by the whole cell population to be measured. Immunological assay of damage in individual cells and/or cell types can be used to determine how DNA damage is distributed within a cell population (or tissue), and may provide information unobtainable by other means.

In this review we concentrate on the application of immunological methods for determining the quantity and distribution of various modified DNA structures, particularly those that are potentially mutagenic or carcinogenic. Many of these methods are similar to those used for detecting specific structures in undamaged DNA, and they have been well described in an earlier review (8), which may be read as a companion to the present paper. In Section I, we describe methods of special relevance to antibodies against DNA adducts and mention newer techniques, such as hybridoma technology, introduced since the earlier review. Similarly, Section II is intended, in part, to update this work and provide a context for discussing antibodies to DNA adducts.

I. Methodology

A. Immunogens

In general, the methods used to produce antibodies specific for carcinogen-altered nucleosides or polynucleotides are adapted from studies with unaltered or naturally occurring modified nucleic acid components (8, 15, 16). Like other haptens, nucleosides and polynucleotides must be linked to carrier proteins to render them immunogenic in animals, and this is accomplished either by covalent coupling in the case of nucleosides or by electrostatic coupling in the case of polynucleotides.

The most widely used procedure for covalent coupling is that in which the ribose group of the nucleoside is oxidized at positions C-2 and C-3 by periodate; the resulting carbonyl groups react spontaneously with free amino groups of the protein (17, 18). The bond is then stabilized with sodium borohydride. This coupling procedure has been used successfully to raise antibodies against a number of carcinogen-modified ribonucleosides (Sections III–V). Although the procedure can be used only with ribonucleosides, this has not been a serious problem since antibodies raised against ribonucleoside–protein conjugates react equally well (or better) with the analogous deoxyribonucleosides.

Several methods exist for coupling deoxyribonucleosides to carrier proteins. They can be oxidized in the presence of a platinum catalyst (19) to 5'-carboxylic acids, which can then be coupled to the carrier protein by means of a carbodiimide (20). Alternatively, deoxyribonucleoside 5'-monophosphates or -diphosphates can be linked to proteins by means of water-soluble carbodiimides (21, 22).

Polynucleotides, on the other hand, can easily be complexed with methylated bovine serum albumin (BSA). The original rationale for using methylated BSA·polynucleotide complexes for immunization was quite simple (1). Albumin had often been used as a protein carrier for simple haptens (and for nucleosides), and it was possible to produce a basic form of this protein by methylation of the negatively charged exposed carboxyl groups. The mBSA binds electrostatically to the acidic DNA polymer and forms stable complexes. Nucleic-acid·mBSA immunizing complexes have been used to raise antibodies to a variety of damaged polynucleotides, including UV-irradiated DNA (23), photooxidized DNA (24), acetamidofluorene-damaged DNA (25), benzo[a]pyrene-damaged DNA (26), and aflatoxin B-damaged DNA (27).

Although relatively pure samples of carcinogen-modified nucleosides can be prepared for linkage to proteins, polynucleotides treated with carcinogens always contain a mixture of modified and unmodified nucleotides. The level of modification of the polynucleotide can influence the specificity of the immunogenic response (28). Modification in the range of 1 to 10% has been used successfully with various carcinogens (7, 27, 29).

The choice of an appropriate immunogen (nucleoside or polynucleotide), depends on how the DNA is to be treated (30). If the modified DNA moiety of interest is to be detected in hydrolyzed DNA samples (degraded to nucleosides), the preferred immunogen would be the specific damaged nucleoside conjugated with a carrier protein. However, if the damaged moiety is to be detected in intact DNA, the preferred immunogen may be a protein conjugate of DNA treated with an agent capable of inducing the required modification. The latter method will, in most cases, produce a heterogeneous range of damage characteristic of the damaging agent that will give rise to antisera containing antibodies of corresponding specificities. As a result, the antibodies may have to be purified for specific forms of damage (by affinity chromatography) or produced by hybridoma methods to ensure monoclonal origin (and monospecificity).

Both methylated and unmethylated BSA are widely used carrier proteins for nucleic acid haptens and produce quite satisfactory results. A few studies have been undertaken to find more effective carrier proteins for carcinogen-damaged DNA. One of these studies compared BSA, rabbit serum albumin, and keyhole limpet hemocyanin (KLH) for their ability to render O^6-methyldeoxyguanosine (O^6-MedGuo) immunogenic in rabbits (31). KLH proved to be the best carrier of the three on the basis of highest antibody affinity constant and lowest degree of cross-reactivity with unaltered dGuo. This finding was attributed to the high molecular weight (800,000) of KLH and its dissimilarity to the proteins of the immunized animal. Similarly, methylated KLH was found to be superior to mBSA as a carrier for aflatoxin B_1-damaged DNA (28). However, KLH is not universally superior to BSA (31a), and the choice of carrier may be affected also by the nature of the hapten and, possibly, the strain of animal immunized.

B. Immunization Schedules

Injection schedules for immunizing rabbits with nucleic acid haptens vary considerably among laboratories; however, a general approach that has been described as "fast-release" (6) is fairly common (15). Nucleoside-protein conjugate (0.5–5.0 mg per rabbit) emulsified in complete Freund's adjuvant (FA) is injected at subcutaneous, intradermal, or intramuscular sites or in hind footpads. Additional weekly or biweekly injections of conjugate emulsified in incomplete FA are given at the same sites or intravenously, and animals are bled 5–10 days after final injection.

A second approach, described as slow-release (6, 31), uses nucleoside–protein conjugate immobilized by adsorption onto aluminum hydroxide. This complex is emulsified with complete FA and injected

(0.5 mg per rabbit) into hind footpads and many (~50) intradermal sites. Eight weeks later the animals are given booster injections by the same procedure. After another 8 weeks, a second booster is administered by intramuscular injection of conjugate (0.5 mg per rabbit) emulsified in incomplete FA. This schedule was found to be superior to the fast-release schedule on the basis of higher titer and higher affinity constant of the antibodies induced (31).

Immunization of mice for the production of monoclonal antibodies requires the same immunogens as those used in rabbits; however, different doses of immunogen and, in some cases, different injection sites are used. When quantities of immunogen are limited, a slow-release approach can be useful. Intradermal injections of conjugate (20 μg) adsorbed onto aluminum hydroxide and emulsified in complete FA are administered and repeated 3 and 6 months later, followed 2 weeks later by intraperitoneal injection (30). When immunogen is plentiful, higher doses of conjugate (50–200 μg) emulsified in complete FA can be so administered at 2–3-week intervals (27, 29). At this stage, serum samples are assayed for antibody titer to eliminate animals with low responses, and a final intravenous booster of conjugate without FA is given 3–4 days before fusion of spleen cells.

A typical immunization schedule for raising antibodies against thymine photoproducts (mainly thymine dimers) is shown in Table I. Generally BALB/c mice are used for immunization because the myeloma cell lines commonly used in hybridoma production are derived from this strain, but (C57B1/6 × BALB/c)F_1 mice have also been used and may produce higher serum titers than BALB/c mice (28).

C. Monoclonal Antibodies

The methods used to produce monoclonal antibodies specific for carcinogen-modified haptens are identical to those used for other hap-

TABLE I
IMMUNIZATION PROCEDURE

Week	Treatment[a]	Injection site[b]
1	UVssDNA/mBSA (100 μg) + CFA	i.p.
3	UVssDNA/mBSA (200 μg) + ICFA	i.p.
5	UVpoly(dT)/mBSA (75 μg) + ICFA	i.p.
7	UVpoly(dT)/mBSA (50 μg)	i.v.
7 + 4 days	Fuse spleen cells	

[a] mBSA, methylated bovine serum albumin; CFA, Freund's adjuvant; ICFA, incomplete CFA.

[b] i.p., intraperitoneal; i.v., intravenous (tail vein).

tens. Polyethylene glycol is used to fuse spleen cells from immunized mice with a myeloma cell line that is deficient in hypoxanthine phosphoribosyltransferase (EC 2.4.2.8) (e.g., P3-NS1-1-Ag4 or P3-X63-Ag8) (27, 29, 30, 32). Complementation of this deficiency permits hybrid clones to be selected in media supplemented with hypoxanthine, aminopterin and thymidine. Feeder cells can be used to increase the yield of hybridomas (32, 33) if necessary; however, feeder cells are not used routinely in most laboratories at this stage. Supernatants from individual culture wells containing hybrid colonies are screened for hapten-specific antibodies, usually by either hapten-binding assay [e.g., $(NH_4)_2SO_4$ precipitation] or enzyme-linked immunoadsorbent assay (ELISA), which are rapid and can easily accommodate a large number of samples. Cells from positive wells are cloned in soft agar or by limiting dilution. The latter method may require the use of spleen cells (30), thymocytes (27), or peritoneal macrophages (28) as feeder cells. Monoclonal antibody is obtained from the supernatant of hybridoma cells grown *in vitro* or from ascites fluid and sera of mice carrying the hybridoma cells as an ascites tumor.

D. Antibody Specificity and Affinity

Older procedures of immunodiffusion, phage neutralization, complement fixation, and quantitative precipitin have been superseded in large part by more accurate and rapid quantitative methods including RIA and solid-phase assays: ELISA; ultrasensitive enzymatic RIA (USERIA); radioimmunosorbent technique (RIST); high-sensitive ELISA (HS-ELISA). RIAs measure the binding in solution of antibody to the appropriate radiolabeled nucleoside or polynucleotide (tracer) by precipitation of immune complexes with anti-Ig (26, 34), protein A (35), ammonium sulfate (10, 29, 31) or ethylchloroformate (13). Solid-phase assays measure antibody bound to unlabeled antigen that is immobilized on plastic Microtiter plates. The amount of antibody bound is determined by a second antibody that is either labeled with ^{125}I (RIST) (31) or linked to an enzyme (e.g., alkaline phosphatase) that cleaves a chromogenic substrate (ELISA) (31, 36), a fluorogenic substrate (HS-ELISA) (37), or a radiolabeled substrate (USERIA) (36).

In both RIA and solid-phase assays, antibody specificity is examined by competitive inhibition of antibody binding to antigen (hapten) by compounds structurally related to the antigen. The degree of inhibition is determined as a function of inhibitor concentration and indicates the degree of cross-reactivity with the inhibitor (see Table IV in Section V). The affinity constant (K) of the antibodies for a particular inhibitor can be calculated by a convenient formula using data obtained in a competitive RIA (38):

$$K \text{ (liters/mole)} = ([I] - [T])^{-1}(1 - 1.5b + 0.5b^2)^{-1}$$

where [I] is the concentration of inhibitor required for 50% inhibition of antibody-tracer binding, [T] is the tracer concentration, and b is the fraction of tracer bound in the absence of inhibitor.

Quantitative assay of carcinogen-modified DNA components is performed by using the test samples as inhibitors in competitive RIA or solid-phase assays that have been calibrated with DNA samples containing known amounts of modification. ELISA and USERIA are reportedly more sensitive (10 to 80-fold) than RIA in some cases [dGuo-C8-AAF, BPDE-DNA (36, 39)], but not others [O^6-EtdGuo (31)]. Although solid-phase assays are more rapid for processing large numbers of samples, the RIA is preferred by some for its simplicity and reproducibility.

In many cases, unpurified antisera are used in assays for modified DNA components; however, some investigators prefer to use enriched or partially purified immunoglobulin (Ig) samples for RIA (40) or immunoflurorescence assay (41). Purified IgG (or IgM) is necessary for certain specialized applications, such as electron microscopy or immunoadsorbent preparation. Standard procedures for isolating Ig include ammonium sulfate precipitation (42) and DEAE-cellulose chromatography (43). A specific Ig can be purified by affinity chromatography using specific antigen immobilized on Sepharose (25, 44). Alternatively, nonspecific antibodies can be removed from a polyclonal antiserum by absorption with nonspecific antigen (e.g., undamaged DNA) (34). However, care must be exercised to avoid unwanted removal of specific antibodies that may cross-react weekly with nonspecific antigen (see Section III,B).

II. Undamaged Nucleic Acid Components

A. Normal Nucleosides

The first demonstration of antibodies specific for individual nitrogenous bases utilized the purine and pyrimidine bases or ribonucleosides as haptens covalently coupled to BSA (17, 45, 46). These antibodies reacted with single-stranded, denatured, or partially denatured DNA, but not with native DNA, suggesting that the antigenic determinant was the base component in the DNA, and had to be exposed for recognition. Deoxyribonucleotides, deoxyribonucleosides, and cyclic nucleotides coupled to proteins have also been used for immunization, and, in general, the antibodies produced show specificity for the purine or pyrimidine base component (21, 47, 48).

The ability of nucleoside-specific antibodies to distinguish between partially denatured DNA and native DNA has been exploited for the detection of DNA damaged by UV (49), ionizing radiation (50), chemical carcinogens (12, 51), and photochemicals (52, 53). UV is known to denature DNA by creating pyrimidine dimers with resultant breaking of hydrogen bonds between base-pairs. On the other hand, with ionizing radiation and chemical carcinogens, it is not clear whether the DNA denaturation detected by the antibodies is caused by direct DNA strand breakage and unwinding or by DNA repair processes.

B. DNA and RNA

The first report of a thoroughly characterized antiserum to DNA appeared in 1960 (54). Rabbits were immunized with T4 phage lysate containing denatured DNA, and the resulting antisera bound purified denatured DNA only from T-even phages (T2, T4, T6). Denatured DNA from other organisms that did not contain glucosylated 5-hydroxymethyl cytosine was not cross-reactive with the T4 antiserum. This modified form of cytosine was shown to be the antigenic determinant of the T4 antiserum by competitive inhibition of T4 DNA to antibody binding by glucosylated hydroxmethylcytidylic acid but not by cytidylic acid (55).

Attempts to induce antibodies specific for DNA containing no unusual modified bases were unsuccessful until it was demonstrated (56) that insoluble complexes of denatured DNA and mBSA could be used as effective immunogens (Section I). Antibodies produced by this method react with denatured DNA (not native DNA) and with each of the deoxyribonucleotides, dAMP, dTMP, dGMP, dCMP, but not with ribonucleotides, deoxyribose, or free bases.

On the other hand, native DNA and native tRNA are not immunogenic when complexed with mBSA (57). The failure of experimental methods to produce native DNA-specific antibodies is unexpected, since antibodies with this specificity have been detected in the sera of patients with the autoimmune disease systemic lupus erythematosus (SLE) and in the sera of NZB/NZW mice (57, 58). The human antibodies do not react with dsRNA and react weakly with RNA·DNA complexes. Cross-reaction with denatured DNA has been attributed to small regions of secondary helical structure in denatured DNA (59).

Another class of SLE antibody reacts only with denatured DNA, recognizing the purine and pyrimidine bases in a manner analogous to that of the experimentally induced antibodies previously described.

Identification of the antigenic determinants for SLE anti-native DNA sera has been the focus of considerable investigation (for review, see 57). Initial studies showed that binding of the sera to native DNA is not inhibited by mononucleotides (60) or oligonucleotides prepared from DNA digests (57), ruling out the possibility that nucleotides or nucleotide sequences are determinants. In addition to reacting with native DNA from a number of sources (plant, animal, bacterial, viral) (61), SLE anti-native DNA sera also react with several synthetic double-stranded polydeoxyribonucleotides including poly(dA-dT)·poly(dA-dT), poly(dA)·poly(dT), and poly(dG)·poly(dC) (62). In contrast, no cross-reaction is observed with synthetic double-stranded polyribonucleotides, including poly(A)·poly(U), and poly(I)·poly(U), or with reovirus RNA (60, 62). Thus, the antibodies apparently recognize the 2-deoxyribose-phosphate backbone and can distinguish it from a ribose-phosphate backbone, in addition to recognizing nucleic acid structural conformation [i.e., double-stranded (ds) vs single-stranded (ss)].

The use of hybridoma technology has allowed the isolation of monoclonal antibodies specific for ssRNA (63). This was accomplished by fusing spleen cells from an unimmunized NZB/NZW mouse with a drug-resistant plasmacytoma cell line. The results of this study indicate that NZB/NZW mice possess spleen cells that spontaneously produce autoantibodies to RNA. A similar study (64), using another autoimmune mouse strain (MRL/1), showed that hybridoma autoantibodies could be produced with various specificities, including ssDNA, dsDNA, or certain oligonucleotides in DNA digests. Monoclonal autoantibodies specific for DNA, rRNA, or Sm (an antigen residing on small nuclear ribonucleoproteins) isolated from MRL/1 mice have been used as probes in immunoprecipitation and immunofluorescence methods (65).

C. Synthetic Polynucleotides

Antibodies have been raised against a number of double-stranded synthetic polynucleotides complexed with mBSA; in general, they show specificity for the helical conformation of the antigens rather than for specific base sequences or base composition. Antibodies raised against poly(A)·poly(U) will cross-react with poly(I)·poly(C) or viral dsRNA, but not with poly(G)·poly(C) or polydeoxyribonucleotides (57). The absence of cross-reactivity with poly(G)·poly(C) may be due to subtle variations in the orientation of the backbone phosphate and ribose groups (66). Immunization with poly(I)·poly(C) induces two distinct antibody populations, one specific for poly(I)·poly(C) (67) and another specific for single-stranded poly(I) (68). The anti-[poly(I)·poly(C)] apparently binds to the minor groove of the helix, interacting with the sugars of the pyrimidine strand and the phosphate groups (67, 69).

Antibodies raised against the hybrid polynucleotide poly(A)·poly(dT) cross-react with poly(I)·poly(dC) as well as natural DNA·RNA complexes, but not with dsRNA or dsDNA, indicating a specificity for hybrid structure (62, 70). The involvement of both helical strands in the antigenic determinant is suggested by the observation that neither poly(dA)·poly(dT) nor poly(A)·poly(U) cross-react with the antibody. [For a review of the specificity of antibodies to helical nucleic-acid structures, see (57).]

D. Naturally Occurring Modified Nucleotides

Antibodies have been produced against many of the naturally occurring modified bases found in nucleic acids including N^2-dimethylguanosine, N^2-methylguanosine, 1-methyladenosine, 5-methylcytidine, 7-methylguanosine, pseudouridine (71, 72), N^6-(2-isopentenyl)adenosine (73–75), 6-methyladenosine (76, 77), and 2-methylthioadenosine (78). In most cases, immunogens were prepared by conjugation of the nucleoside to serum albumin via the periodate method (17). A complete review of the production and applications of antibodies specific for modified nucleosides may be found in this series (8).

Several investigators have used antibodies specific for 5-methylcytosine to study the localization of these residues in denatured metaphase chromosomes. Immunofluorescent staining of UV-irradiated mouse and human metaphase chromosomes (79) and immunoperoxidase staining of human chromosomes (80) reveal intense staining of the centromeric regions of chromosomes 1, 9, and 16, the short arm of chromosome 15, and the midportion of the long arm of the Y chromosome. The resolution of the immunoperoxidase method has been increased to approximately 20 Å by use of electron microscopic visualization of the stained chromosomes (81). More recently, a method has been described in which anti-5-MeCyt is used to identify 5-MeCyt in restriction fragments of DNA immobilized on nitrocellulose paper (82). Bound antibody is detected by incubation of ^{125}I-labeled anti-IgG followed by autoradiography. With this method, the methylation patterns of DNAs from several sources were studied, including Chinese hamster cells, calf thymus (satellite I), and *Chlamydomonas* chloroplasts.

E. Oligonucleotide Sequences

Early studies with di- and triribonucleotides have demonstrated the immunogenicity of these components when complexed with carrier proteins (83–86), although precise specificity was not always reported. Antibodies specific for three triplet codons [p-d(A-A-A), p-d(A-A-C), p-d(A-U-G)] have been produced (87), as well as three antisera highly specific for the dideoxyribonucleotide p-d(A-T) and two trideoxyribonucleotide sequences, p-d(A-T-A) and p-d(A-A-T) (88, 89). Anti-p-d(A-T) showed some cross-reactivity with oligonucleotides containing this sequence as well as with denatured DNA. Anti-p-d(A-T-A) and anti-p-d(A-A-T) showed some cross-reactivity with oligonucleotides containing all, or portions of, their respective sequences; however, little binding was observed to denatured DNA. Antibodies specific for longer oligonucleotide sequences have been produced by immunization with p-dT_3, p-dT_4, p-dT_6, or p-d()A-A-T-T) complexed with mBSA (90). In the case of p-d(A-A-T-T), a heterogeneous antibody population was induced that contained subpopulations specific for p-dA_2 and for p-d(A-A-T-T). Antibodies specific for each of the four oligonucleotide sequences bound denatured DNA very efficiently, particularly DNAs with a high adenine + thymine content.

Hybridoma autoantibodies have been produced from NZB/NZW mice and exhibit sequence specificity in their interaction with DNA (91). Antibodies from six different hybridoma cell lines were tested (four producing IgGs and two producing IgMs) and found to bind ssDNA very efficiently, dsDNA less efficiently, and RNA not at all. Each monoclonal antibody showed a distinct preference for different base sequences, including p-dT_n in one case, p-d$(T-G)_n$ or p-d$(C-A)_n$ in another case, and p-d$(T-T-C)_n$ in a third case.

F. Summary: Sequence versus Conformation

It is clear from the preceding examples that a wide range of undamaged nucleic acids and their components can be rendered immunogenic by conjugation with appropriate protein carriers, the most common being BSA. In general, the antibodies fall into two main categories depending on the nature of the antigenic determinant. Those antibodies that recognize small sections of a polynucleotide rather than conformational structure may be described as having *sequential* determinants (57, 92). Examples of these antibodies include those specific for single nucleotides (normal or modified) or nucleotide sequences that, in general, are accessible only when the polynucleotide is denatured.

On the other hand, antibodies that recognize a particular structure (e.g., a double-helix) and do not cross-react with altered structures (e.g., denatured nucleic acids) may be described as having *conformational* determinants. Examples of this class of antibody are anti-[poly(I)·poly(C)], anti-[poly(A)·poly(dT)], or anti-Z-form DNA (Section V), which require specific conformational shapes for recognition. Some determinants may be influenced by both sequential and conformational elements. In the case of an antibody recognizing the pentose-phosphate backbone of a polynucleotide, binding may be dependent on conformation (double-helix vs single-stranded) as well as composition (ribose or 2-deoxyribose) of the backbone. This concept of sequential and conformational determinants is useful for analysis of antigenic determinants in chemically and physically damaged DNA, as various changes in conformation and/or chemical composition may be present.

III. Ultraviolet Radiation-Damaged DNA

A. Early Studies

Antibodies to photoproducts in DNA were first produced by immunizing rabbits with denatured calf thymus DNA irradiated with high doses of monochromatic UV (270 nm) and complexed with methylated BSA (23). Serologic activity was detected by complement fixation using native DNA from *Proteus vulgaris* irradiated with various doses of 254-nm UV as the test antigen. Cross-reactivity with unirradiated DNA was detected only if 5–10 times more antiserum was used, and UV-irradiated denatured DNA was a better antigen than native UV-DNA. The antibodies appeared to be directed toward thymine photoproducts (Fig. 1), based on reactivity with UV-DNAs containing 28 to 68% adenine + thymine. In addition, photoproducts detected by the antiserum were photoreversible when 270 nm UV was followed by shorter wavelength (235 nm) UV, strongly implicating pyrimidine dimers as the antigenic determinant. The serologic reactivity was inhibited by addition of UV-irradiated p-dT$_3$ and p-dT$_4$, and to a lesser extent by UV-p-dT$_2$, but not by UV-p-dC$_3$ or UV-p-dC$_4$.

In 1972, an RIA was developed (34) for the detection of thymine dimers in DNA produced by biologically relevant doses of UV. UV-irradiated denatured *P. vulgaris* DNA labeled with ^{125}I was used as the tracer, and photoproduct-specific antibody complexes were pre-

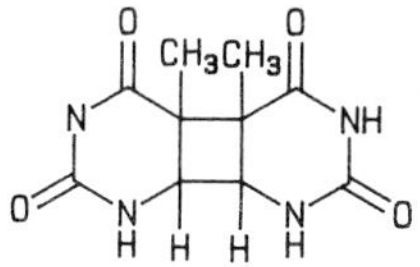

Cyclobutadithymine
(Thymine dimer)

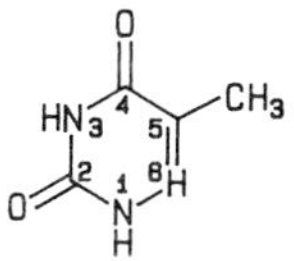

Thymine

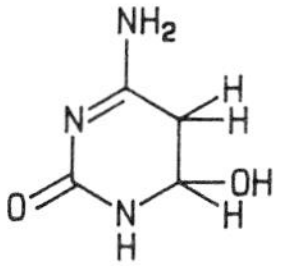

Cytosine photohydrate
5, 6-Dihydro-6-hydroxycytosine

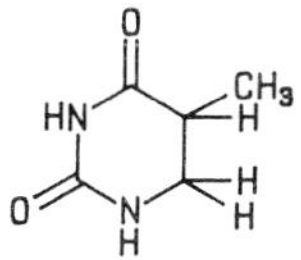

5, 6-Dihydrothymine

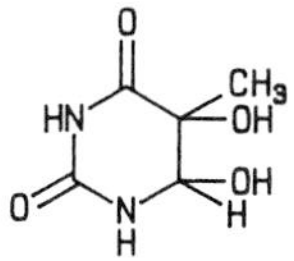

5, 6-Dihydro-5, 6-dihydroxythymine (Thymine glycol)

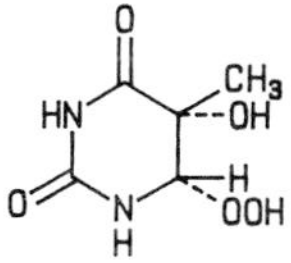

cis -5, 6-Dihydro-6-hydroperoxy-5-hydroxythymine (6-TOOH)

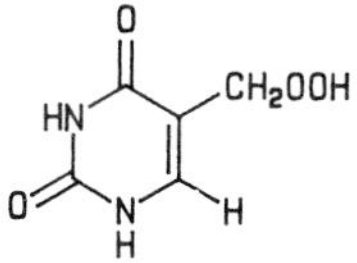

5-Hydroperoxymethyluracil or
α-Hydroperoxythymine (α-TOOH)

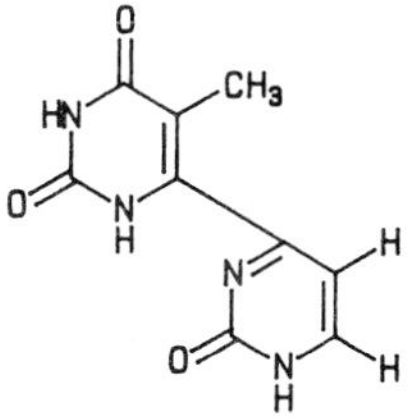

4-(Thymin-6-yl) pyrimidin-2-one
(Thy (6-4) Pyo)

FIG. 1. Major photoproducts in DNA. (See also *92a*.)

cipitated with anti-rabbit-IgG serum. Binding of the tracer was inhibited by a prior incubation of antiserum with various unlabeled test DNAs to estimate the photoproduct content of these samples. The limit of detection in this assay was about 5 J/m² (254 nm), and the antibody used was specific for photoproducts in heat-denatured UV-DNA. The size of the antibody binding site was estimated by inhibition of anti-UV-DNA binding to ^{125}I-UV-DNA by UV-irradiated thymidine oligonucleotides, p-dT$_2$, p-dT$_3$, p-dT$_4$, p-dT$_6$, p-dT$_8$. Antiserum from one rabbit was most effectively inhibited by UV-p-dT$_3$ (or longer) oligonucleotides, whereas antiserum from another rabbit preferred UV-p-dT$_6$ (or longer) oligonucleotides. The RIA was used to analyze DNA extracted from two strains of *B. subtilis* (UV-resistant/UV-sensitive) that had been irradiated. The induction and removal of photoproducts was correlated with the transforming activities of the DNA extracted from these strains before and after repair. Finally, the amount of immunologically detectable damage in HeLa cells decreased in a 24-hour recovery period following UV (10–40 J/m$_2$), whereas the amount of damage in rat glial and pituitary cell lines remained unchanged after 24 hours.

The repair of photoproducts in cultured human cells was also reported (*10*) in a study using an immunofluorescent assay that allowed visual observation of damage in individual cell nuclei. The antiserum was characterized by complement fixation and by a modified RIA using ammonium sulfate precipitation of antibody·DNA complexes (*93*) and was apparently specific for pyrimidine dimers. After incubation of antiserum with UV-irradiated human amnion cells fixed with cold acetone, attached Igs were stained with fluorescein isothiocyanate (FI)-conjugated anti-rabbit Ig serum. Faint fluorescence was detected after a dose of 10 J/m² (254 nm), and clearly fluorescent nuclei were visible after a dose of 30 J/m². Postirradiation incubation of cells in fresh medium for 1–2 hours before staining resulted in a migration of stain (dimers) from nuclei to cytoplasm. By 4 hours postirradiation, no fluorescence was detectable in nuclei and faint fluorescence was visible in the cytoplasm.

An immunofluorescence assay was also used to demonstrate that photoproducts are induced in mouse skin by UV treatment of animals (*9*). The antibodies were induced by immunization with UV-irradiated native DNA complexed with mBSA, in contrast to previous studies that used denatured UV-DNA (*94*). Reactivity with UV-DNA was demonstrated by immunodiffusion, complement fixation, and immunofluorescence. A subsequent report (*95*) showed that the antibodies were specific for a thymine-associated photoproduct. Fluorescence was observed over the nuclei of epidermal cells in cryostat sections of mouse skin exposed to 254 nm UV. The assay was also used after treatment with monochromatic UV (96) to investigate the induction and distribution of photoproducts in skin as a function of wavelength. Similar studies on induction and persistence of photoproducts in mouse skin (41, 97) were conducted by other investigators.

B. Antibody Specificity

Using similar methods, a number of investigators have produced antibodies to UV-DNA (13, 98–102); in each case, the specificity of the antibody has been determined to a different extent (Table II). The antigenic determinant reported in several studies was apparently pyrimidine dimers as suggested by photoreversal (23, 103) or photoreactivation of photoproducts (13, 35, 104), or by correlation of antigenic reactivity with expected and/or chromatographically measured repair kinetics of dimers (34, 104, 105). Several studies have demonstrated antibodies specific for a photoproduct of thymine (23, 34, 95) or a photoproduct of adjacent thymines (29, 102) suggesting that thymine dimers (T[ ]T) are the preferred determinant rather than cytosine dimers (C[ ]C). Cross-reaction with mixed cytosine–thymine dimers C[ ]T was tested in two studies. A monoclonal antibody was reported that is specific for T[ ]T but not C[ ]T (106), whereas a rabbit antiserum was found to bind both T[ ]T and C[ ]T (107). In two additional studies, cross-reaction with cytosine-containing photoproducts was reported (35, 104). Another anti-UV-DNA serum (101), which did

TABLE II
SPECIFICITY OF DNA PHOTOPRODUCT ANTIBODIES[a]

UV-DNA	Pyrimidine dimers			Nondimer photoproducts	References
	T[ ]T	C[ ]T	C[ ]C		
+	+	NT[b]	−	−	(23, 34)
+	+	NT	NT	NT	(10)
+	+?	NT	−	NT	(9, 94–96)
+	NT	NT	NT	NT	(97–99)
+	+	NT	+	−	(13, 104, 105)
+	−	−	−	NT	(101, 108)
+	+	+	−	−	(102, 107)
+	+	−	−	−	(29, 106)
+	−	−	−	+	(113)
−	+	−	−	NT	(112)
+	+	+ and/or +		−	(35)

[a] +, Antibody binds this antigen; −, antibody does not bind this antigen.
[b] NT, not tested.

not bind thymine-containing pyrimidine oligonucleotides but did bind mixed purine and pyrimidine oligonucleotides that had been UV-irradiated, was subsequently shown (*108*) to contain two different antibody populations specific for UV-DNA and thermally denatured DNA.

Attempts to remove nonspecific antiserum that binds to undamaged ssDNA by absorption with ssDNA prior to use produced mixed results. Preabsorption of anti-UV-DNA serum with ssDNA appeared to remove nonspecific binding to ssDNA without reducing specific (UV-ssDNA) binding (*34, 101, 101a*). However, other studies (*104, 109*) indicate that anti-UV-ssDNA activity cannot be separated from cross-reactivity with undamaged ssDNA. These results suggest that, in the former case, two different antibody populations were present in the antiserum, one specific for UVssDNA and the other specific for ssDNA. The latter case, in which cross-reactivity could not be separated, suggests the presence of a single antibody population with high affinity for UV-ssDNA and low affinity for ssDNA.

Several of these antibodies have binding sites between 3 and 6 nucleotides long as determined by competitive inhibition experiments using UV-irradiated oligonucleotides of various lengths (*29, 34, 102*). This is similar in size to the binding sites of antibodies to undamaged DNA (~5 nucleotides) (*110*) and undamaged synthetic polynucleotides (4–6 nucleotides) (*66, 67*), and agrees quite well with the estimated maximum size of an antibody combining site (2.4–3.6 nm) that could theoretically accommodate an oligonucleotide not exceeding 7–10 bases (*111*).

None of the anti-UV-DNA sera discussed above have been shown to bind strongly to isolated pyrimidine (or thymine) dimers. Additional flanking nucleotides are necessary to complete the binding site, suggesting that the antigenic determinant in UV-DNA may be the conformational distortion or constraint associated with the thymidine dimer rather than the dimer itself. One study showed that the antigenic determinants of UV-DNA are located on exposed single-stranded regions of DNA; that is, regions that have been sufficiently distorted by photoproducts to induce local denaturation (*108*). However, most of the anti-UV-DNA sera described above require antigens that have been subjected to further (heat) denaturation in order to exhibit binding. Since most of the test antigens (UV-DNA, UV-oligonucleotides, UV-polynucleotides) are heat-denatured, they would not be expected to exhibit the same conformational distortion that is associated with a thymine dimer in native DNA. Therefore, the constraint (or alignment) imposed on adjacent thymines by the formation of cyclobutane bonds may be the determinant shape that the antibodies recognize. Additional flanking bases, while not rigidly positioned in the unbound antigen, may add stability to the antigen-antibody complex.

Antiserum specific for isolated T[ ]T dimers has been reported in only one case (112). Purified thymine dimers coupled to BSA rather than UV-irradiated DNA complexed with methylated BSA were used for immunization. The antibodies do not cross-react with U[ ]U or C[ ]C dimers, C[ ]T dimers were not tested. Recognition of T[ ]T in UV-irradiated DNA was very poor, but hydrolysis of UV-DNA or treatment with dimer-specific endonuclease from *M. luteus* made them accessible to antibody binding.

Antibodies against nondimer DNA photoproducts have been isolated (113). Anti-UV-DNA serum was raised by immunizing with UV-native DNA (94) and purified on UV-DNA-cellulose columns. A second purification using a photoreactivated UV-DNA-cellulose column bound only those antibodies specific for nonphotoreactivable (i.e., nondimer) photoproducts. Most of the nondimer antibodies were specific for the photoproduct dihydrodihydroxythymine (which is also produced by ionizing radiation or osmium tetroxide treatment of DNA). The remaining nondimer antibodies were not specific for dihydrodihydroxythymine but apparently were bound to other unidentified minor photoproducts. Induction and repair of nondimer UV damage was measured with these antibodies in normal and dimer-repair-deficient (*Xeroderma pigmentosum*) human cell lines. Interestingly, both cell lines were proficient in repair of nondimer photoproducts.

Monoclonal antibodies specific for thymine dimers in a polynucleotide or an oligonucleotide sequence at least four bases long have been characterized (29). Hybrid cell lines were isolated from fusions between mouse myeloma cells and spleen cells from BALB/c mice hyperimmunized with UV-ssDNA and UV-poly(dT) complexed with methylated BSA. In addition, a second antibody that recognized both unirradiated and UV-irradiated polynucleotides (and oligonucleotides) containing tracts of adjacent thymidines, but not polynucleotides devoid of adjacent thymidines, was isolated.

C. Quantitation and *in Situ* Detection

As mentioned earlier, simple radioimmunoassays based on precipitation of immune complexes by anti-Ig (34, 102), ammonium sulfate, (10, 29, 101, 109, 110), or ethyl chloroformate (13) have found widespread use and can detect photodamage caused by UV doses of 5–10

J/m² (254 nm). The sensitivity of the RIA can be increased by using plasmid DNA labeled *in vitro* by nick-translation with deoxyribonucleoside [^{32}P]triphosphates as tracers (*102*). The higher specific activity of this probe (2×10^7 cpm per microgram of DNA) enables the assay to detect photoproducts in cellular DNA at doses of 2.5 J/m² (254 nm). In addition, synthetic polynucleotides labeled with ^{32}P can be used as tracers to gain additional information about the nature of the photoproduct. This assay has been used to study the loss of T[ ]T from mammalian cell DNA (*107*). Interestingly, the loss of antibody binding sites occurred much more rapidly ($T_{1/2} \sim 1$ hour) than the loss of dimer-specific T4 endonuclease V sites ($T_{1/2} \sim 24$ hours). Loss of antibody binding sites was not due to (a) masking of dimers by repair enzymes or (b) deglycosylation of dimers.

The use of radiolabeled antibodies directed against UV-DNA has allowed sensitive, quantitative measurement of UV damage in cells by autoradiography (*13*). This method can detect pyrimidine dimers induced by 2 J/m2 (254 nm) in individual eukaryotic cells and has been used to measure dimer repair kinetics in many cell lines: Chinese hamster lung, toad kidney, human HeLa, African green monkey kidney, Syrian hamster embryo, bat lung, chick embryo, and human diploid fibroblast (*104, 105*). The technical aspects of this assay have been reported in detail (*115*). Immunofluorescent methods have provided a means for investigators to examine *in vivo* the distribution of DNA photoproducts in individual cultured cells (*10*) and in skin cells (*9, 77, 78, 110a*). Immunofluorescent staining for DNA photoproducts and nuclear Feulgen staining (for DNA) have been quantitated simultaneously by means of cytofluorometry to provide a measure of the amount of photoproducts produced per unit amount of DNA (*98*). In addition, immunofluorescent staining has been used to detect photoproducts induced in interphase cells by laser-UV microirradiation of small areas in the nucleus (*114*). By studying the distribution of photoproducts on chromosomes in the subsequent metaphase, it is possible to gain information on the spatial arrangement of the chromosomes at the time of irradiation (interphase).

IV. Ionizing Radiation-Damaged DNA

A. Early Studies

In constrast to 254 nm UV, which primarily induces pyrimidine dimers in DNA without causing strand breakage, ionizing radiation produces many structural modifications in DNA. Table III lists the principal types of base damage resulting from irradiation of adenine and thymine, and shows the conditions and efficiencies of their formation. [For more complete accounts, see reviews (*116i*, *116j*).] Some modifications are labile to most chemical procedures (*116*), and enzymatic hydrolysis of irradiated DNA is incomplete (*117*). Hence rapid assay of these lesions in DNA by serological means is an attractive idea.

The first suggestion that antibodies recognizing damage produced in DNA by ionizing radiation could be produced came in 1974 from a demonstration (*118*) that rabbits immunized with ssDNA that had received 60 krad of ionizing (X-)radiation produced antibodies earlier and in higher titer than rabbits immunized with unirradiated ssDNA. Moreover, passage of antiserum through an affinity column of ssDNA bound to DEAE-cellulose did not remove all antibodies reactive with irradiated DNA, suggesting that irradiated DNA elicited an immune response to radiation-induced modifications.

Another group (*119*) produced a rabbit antiserum by immunization with γ-irradiated thermally denatured coliphage T1 DNA conjugated to mBSA. Reaction with the immunizing antigen was determined by isopycnic density-gradient sedimentation in CsCl in which the buoyancy of the antigen-antibody complex is proportional to the number of antibody molecules bound per DNA molecule. In this study, the specificity of binding was not demonstrated by reaction of the antiserum with unirradiated DNA. However, other investigators (*120*) have produced rabbit antisera following a similar immunization schedule and have showed that some of the antigen-antibody complexes formed with X-irradiated DNA, and observed in CsCl gradients, were inhibited by unirradiated ssDNA. In an ammonium-sulfate-precipitation binding assay, there was only a weak nonlinear increase in binding to DNA exposed to increasing radiation doses, and most of the reactivity was competitively inhibited by unirradiated ssDNA. At doses above 1 krad, they observed a decrease in binding of irradiated tracer DNA, possibly resulting from radiation-induced shearing of the DNA.

This artifact arises from the fact that separation of bound and unbound antigens is achieved by precipitation of the Ig proteins with ammonium sulfate ("Farr" assay). Since binding of a single Ig molecule is sufficient to cause coprecipitation of a bound DNA molecule, one double-strand break will produce two DNA molecules requiring two Ig molecules for precipitation (*121*). Thus, at limiting antibody dilutions, a progressively greater fraction of DNA will remain unbound as increasing radiation exposure produces more strand breaks.

TABLE III
PRINCIPAL ADENINE AND THYMINE RADIATION PRODUCTS

Structure	Substance	Radiation conditions[a]	G value[b]	Reference
(8-hydroxyadenine structure)	8-Hydroxyadenine	Ade (7.4 mM), N_2	0.19–0.20	116a
		Ade (1 mM), air	0.15	116b
		Ade (1 mM), vacuum	0.05	116b
		Ade (0.1–6.5 mM), N_2	0.05–0.32	116c
		Ade (0.1–4.8 mM), N_2O	0.16–0.6	116c
(adenosine 1-oxide structure)	Adenosine 1-oxide	Ado, air	$<2 \times 10^{-5}$	125
(7-8-dihydro-8-hydroxyadenine structure)	7-8-Dihydro-8-hydroxyadenine	Ade (0.1–6.5 mM), N_2	0.12–0.50	116c
		Ade (1.0–6.1 mM), N_2O	0.05–0.33	116c
(6-amino-5-formamido-2-deoxycytosine structure)	6-Amino-5-formamido-2-deoxycytosine	Ade (7.4 mM), N_2	0.14–0.36	116a
		Ade (0.1–6.5 mM), N_2	0.07–0.28	116c
		Ade (1.0–6.1 mM), N_2O	0.22–0.28	116c

	5′ AMP (1.0 mM), N₂O	0.13–0.45 (pH dependent, G_{max} pH 9)	116d
5,6-Dihydro-5,6-dihydroxythymine (thymine glycol)	DNA, air	0.018–0.06	116e
5,6-Dihydro-5-hydroperoxy-6-hydroxythymine	Thy (2 mM), pH 5, O₂	0.83 *trans* 0.31 *cis*	116g
5-Hydroxymethyluracil	Thy (anoxic) Thy (0.2 mM), air Thy (10 mM)	0.1 ? 0.01	116f 125 125
5-Hydroperoxymethyluracil	Thy (2 mM) pH5; O₂	0.047	116h

[a] Aqueous solutions were irradiated.
[b] Number of molecules produced per 100 eV absorbed.

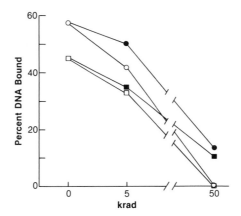

FIG. 2. Antiserum binding to ^{137}Cs-irradiated DNA. Mouse antiserum was raised by immunization with mBSA complexed to calf thymus DNA exposed to 50 krad of ^{137}Cs prior to (○, ●) or after (□, ■) heat denaturation. Antiserum diluted 1 : 2 was used to bind single-stranded [^{3}H]dThd-labeled *Escherichia coli* DNA irradiated prior to (○, □) or after (●, ■) denaturation (J. M. Boyle, unpublished data).

We have made observations similar to those described above. Figure 2 shows binding data obtained with a mouse antibody raised against calf thymus ssDNA irradiated with 50 krad (^{137}Cs) and conjugated to methylated BSA. The antibody binds primarily to ssDNA with an apparent inhibition of binding following irradiation of the ssDNA. A similar effect has been observed with methylnitrosourea-treated DNA. That this inhibition is an artifact of the binding assay, not a property of the antibodies, was shown by the the fact that single-stranded regions induced by ionizing radiation in purified dsDNA could be detected by a rabbit antiserum (serum 28) specific for ssDNA (*122*). Precipitation lines were observed in double-diffusion plates using DNA that had received 0.5 krad X-radiation, and the intensity of reaction increased to the highest dose tested, 30 krad.

B. Antinucleoside Antibodies

Antibodies raised against nucleosides (Section II) have also been used to detect single-stranded regions in DNA (*50*). When synchronized HeLa cells were fixed with acetone and treated with fluorescein-labeled antinucleoside antibodies, the proportion of cells with intense staining was >75% for S-phase cells, but only 10–15% for G1-phase cells. When G1 cells were irradiated and fixed immediately, there was a dose-dependent increase in the proportion of cells showing intense fluorescence over the range tested (0–1 krad). Comparison of cells irradiated in air and under anoxia showed an oxygen enhancement ratio for immunoreactivity of 2.4 (*123*). Incubation of irradiated cells prior to fixation resulted in a reduction of radiation-induced immunoreactivity to control levels within 30 minutes (anoxic cells) or 60–90 minutes (oxic cells) suggesting that single-strand breaks were repaired rapidly. The majority of cells irradiated in air with 0.5 or 1 krad failed to divide and accumulated at later times in an extended G2 phase, defined in synchronous populations as the period between completion of thymidine incorporation and completion of residual population doubling (*124*). In contrast to unirradiated populations in which <10% of cells in G2 phase showed immunoreactivity, a significant proportion (20–60%) of radiation-blocked G2 cells were immunoreactive, suggesting that the single-strand breaks and regions of local denaturation generated as cells passed through S phase were persisting into the G2 phase in the irradiated cells.

Such studies clearly show the usefulness for DNA repair studies of antibodies that recognize denatured regions in irradiated DNA that can arise directly from radiation attack or can be formed enzymatically during repair of radiation-modified nucleotides in DNA.

C. Radiation-Damaged Nucleosides

In addition to the production of antibodies against dihydrodihydroxythymine (Thy glycol) (*116e*), Ward and his co-workers have developed antisera against 5-hydroxymethyldeoxyuridine (hm^{5}dUrd) (*125–127*), adenosine 1-oxide (*125*) and 8,5'-cycloadenosine (*127*). Antisera to each modified structure were produced in rabbits immunized with BSA conjugates and were assayed by phage neutralization assay (*128*).

Antibodies to hm^{5}Urd were detected in antiserum diluted 2×10^5-fold (*126*). Inhibition of phage neutralization by competing haptens showed that antiserum raised against hm^{5}Urd detected 20 pmol of hm^{5}dUrd or 100 pmol of hm^{5}Urd per assay, and other haptens in the order hm^{5}dUrd > IdUrd $\geq$ dThd $\geq$ BrUrd > 6-MeUrd $\geq$ Thy (*127*). However, when conjugated with ovalbumin, 10 fmol of hm^{5}Urd was detected although ovalbumin alone did not cross-react, implying that the antiserum strongly recognized the linkage of nucleoside ribose to protein amino groups. Of the normal components of DNA, only dThd cross-reacted: one hm^{5}Urd residue was detected per 4×10^3 dThd molecules. To overcome this cross-reactivity, hm^{5}Urd was assayed in the presence of excess dThd but at the expense of reduced sensitivity. Affinity purification of antiserum on Sepharose containing ribosylthymine·protein conjugate in the presence of excess hm^{5}dUrd produced antibodies with increased sensitivity and selectivity against dThd (127). The formation of hm⁵dUrd in calf thymus DNA irradiated in O_2 gave a G value similar to that for λ-irradiation of dThd ($G \sim 0.05$) and was detectable at levels of one modified nucleoside for 3×10^6 daltons of DNA. However, approximately 140 krad was required to produce a yield of hm⁵Urd in irradiated heat-denatured DNA of twice the background observed in unirradiated control DNA; hence, no biological studies have been reported using these antibodies.

There are preliminary reports on the use of other antibodies in phage-neutralization assays to demonstrate the formation of adenine 1-oxide from irradiated Ado (G value $< 2 \times 10^{-5}$) (125 and of 8, 5'-cAMP from irradiated solutions of 5'-AMP, Ado, and Ade (127). After irradiation in O_2, virtually no cyclo product was formed from AMP, but a measurable amount was formed from Ado ($G = 0.06$). Under anoxic conditions, yields were increased with G values of 2.36 and 0.043 for Ado and Ade, respectively.

Recently, an RIA of 8-hydroxyadenine (8-hAde) has been described (117) using rabbit antiserum raised against an 8-hAdo-BSA conjugate. The labeled probe was 8-hAdo conjugated to rabbit serum albumin that was then iodinated. In the RIA, 8-hAdo and 8-hAde competed equally and 50% inhibition was observed at 2 fmol of modified nucleoside per assay. The same inhibition required 2000-fold more Ade, which showed the greatest cross-reactivity of all the unmodified nucleic acid components tested. Although free 8-hAde reacted more than 6-fold better than the modified nucleoside in irradiated DNA, it was possible to detect specific damage in DNA irradiated in solution with 1 krad or irradiated *in vivo* in *E. coli* with 10 krad, these being the lowest doses tested. From these experiments the yield of 8-hAde in *E. coli* DNA irradiated *in vivo* was estimated as 0.04 residue per 10^6 daltons per kilorad.

Antibodies specific for thymine glycol were obtained by immunizing rabbits with a methylated BSA conjugate of ssDNA treated with osmium tetroxide (116e). An RIA that could measure 4 fmol of thymine glycol was set up; however, the assay was dependent not only on the total concentration of the glycol, but also on the proportion of thymine coverted to the glycol. The problem was overcome by the finding that the concentration of DNA needed for 50% inhibition of tracer binding was a linear function of the square of the glycol concentration. From this analysis, the limit of detection was one thymine glycol residue per 3×10^4 thymine residues, and the lowest dose of ^{60}Co radiation for which thymine glycols could be measured in *E. coli* DNA irradiated in aqueous solution in air was 2 krad. As with antibodies against thymine dimers (Section III), antibodies against thymine glycols reacted more strongly against thymine glycols in oligothymidylates containing four residues than with the monomer or dimer. Thus, the dependency of binding on the square of the thymine glycol concentration could be explained either by a requirement for more than one such glycol at the antibody binding site or by binding of both arms of the immunoglobulin to the glycols in the same polymer. The yield of thymine glycol induced in *E. coli* DNA by radiation was $G = 0.018$ when measured by RIA compared to $G = 0.06$ when measured chemically. This difference was explained by the probable ability of the RIA to distinguish *cis* and *trans* thymine glycols, and to exclude hydroperoxythymine residues.

V. Chemical Adducts in DNA

A. Alkylation Damage

Monofunctional alkylating agents react with the nitrogen and oxygen atoms of nucleic acid bases and with the oxygen atoms of phosphate groups to form phosphotriesters. The frequency of alkylation at different sites may be characteristic of individual agents (*129, 129a*). Conventional analysis of alkylated DNA involves acid or enzyme hydrolysis to bases or nucleosides followed by separation of the products by chromatography (*130*). The quantity of normal nucleosides is determined by UV absorption, and minor alkylated bases are assayed by measuring the radioactive alkyl residue transferred from a ^{3}H- or ^{14}C-labeled alkylating agent. Alternatively, alkylation of specific nucleosides can be measured using DNA synthesized with a labeled nucleoside precursor. These methods are limited by the specific activity of labeled carcinogen available in the former case, or confined to the analysis of replicating cell populations in the latter case. The use of high specific radioactivity is ethically unacceptable for *in vivo* studies in humans. Immunoassays provide a rapid and convenient means of circumventing these difficulties.

The predominant alkyl adduct in DNA is 7-alkylguanine (*129a*). *In vitro*, 7-methylguanine (7-MeGua) in DNA decays with a half-life in excess of 100 hours. *In vivo*, a faster rate of decay is indicative of enzymatic removal; however, because the rate of removal is very slow in some tissues and species, the quantity of minor alkylated bases is often expressed relative to the amount of 7-MeGua, which acts as an internal standard for the efficiency of DNA extraction and hydrolysis. With one exception (*131*), antibodies have not been used to quantify 7-alkylguanine in DNA, but because it occurs naturally in some tRNAs and in the cap structure of eukaryotic messenger RNAs, several workers have produced antibodies against 7-MeGua for use in affinity purification of these RNA molecules (18, 71, 72, 131–134) or as a probe to inhibit translation of eukaryotic messenger RNA (135).

Methylation of the N-7 atom of guanine makes the deoxynucleoside unstable to alkali (136, 137), resulting in hydrolysis of the glycosidic bond at pHs below ~8.5; at pH 7.4 and 37°C the half-life of 7-MedGuo is about 5 hours, whereas 7-MeGuo is more stable, requiring a lower pH for base release. Above pH 8.5, the 8,9 bond in the imidazole ring hydrolyzes, producing 2-amino-4-hydroxy-5-(N-methylformamido)-6-ribosylaminopyrimidine (7-MeGuo*), an isocytosine derivative; at pH 8.9 and 37°C, the half-lives of 7-MeGuo and 7-MedGuo to this hydrolysis are 6.1 and 9.8 hour, respectively. Conditions producing imidazole-ring opening occur during conjugation of the 7-MeGuo and carrier protein by the commonly applied method of Erlanger and Beiser (17), in which the pH is raised to 9.0–9.5 for 45 minutes at room temperature (Section I). Some studies did not consider this possibility, and the specificity of the antibodies with respect to the integrity of the imidazole-ring configuration was not investigated (71, 72, 131, 135). Using this method, Munns et al. (8, 44, 132) produced an antiserum that reacted strongly with 7-MeGuo and 7-MeGuo*, but whether this property resided in a single immunoglobulin species or in different species was not determined. In two recent reports (18, 134), the investigators recognized the possibility of ring-opening and modified the conjugation procedure to minimize it, either by reducing the time at high pH and lowering the temperature to 4°C (134) or by conjugating the more stable 5'-phosphate of 7-MeGuo at 4°C and keeping the pH as low as possible (18). Both modifications resulted in antisera with two to four orders-of-magnitude preference for the ring-closed form. The sensitivity of RIA (50% inhibition values) for 7-MeGuo was 20 nmol using 7-MeGMP as tracer (18) and 30 nmol using 7-MeGuo as tracer (134). No cross-reaction was observed with 200 nmol of guanosine (134). Using a microprecipitation assay, Sawacki et al. (131) investigated the reactivity between purified anti-7-MeGuo globulin and calf thymus ssDNA alkylated with dimethyl sulfate to give 240, 100, and 10 nmol of 7-MeGuo per milligram of DNA. Tenfold higher concentrations of DNA containing 10 nmol of 7-MeGuo per milligram of DNA were required for maximum precipitation compared to DNA containing 100 nmol of 7-MeGuo per milligram of DNA, as expected. It was possible to detect one 7-MeGuo per 300 nucleosides; in perspective, this level of alkylation is approximately 10 times the lethal dose to mammalian cells.

At present, these antibodies would appear to be insufficiently sensitive to measure 7-MeGuo in alkylated DNA. A mean lethal dose of ~200 μmol of 7-MeGuo per mole of guanine is equivalent to 2.0 pmol of 7-MeGuo per 10 μg of DNA from about 2×10^6 cells. Thus the sensitivity of detection needs to improve by three orders of magnitude. The prospects for achieving this appear to lie mainly with the application of more sensitive assay systems (*138, 139*), combined with monoclonal antibodies to provide increased specificity for the deoxynucleoside.

Some improvements in the choice of hapten may also be possible. It has been reported (*140*) that methylation of DNA by dimethyl sulfate in the presence of tri(*n*-butyl)amine causes 80% of guanine residues to be methylated at N-7 in the absence of methylation of other bases. Although secondary structure was not destroyed, the t_m of the methylated DNA was lowered by 14°C, indicating a reduction in the stability of the duplex. This observation may explain how methylated salmon sperm dsDNA could elicit antibodies specific for methylated DNA, even though dsDNA is not immunogenic (*141*).

Besides the use of antinucleoside antibodies to detect single-stranded regions in cellular DNA following exposure to monofunctional alkylating agents (*12*), antibodies specific for base-adducts are available to quantify alkylated O^6 of guanine and O^2 and O^4 of thymine. Effort has been concentrated on these adducts because of their miscoding properties when incorporated in DNA templates for *in vitro* synthesis directed by DNA or RNA polymerase (*142–147*) and because of the implications for mutagenesis (*148–150*) and carcinogenesis (*151, 152*). A comparison of the sensitivity and specificity of rabbit antisera, affinity-purified immunoglobulin, and monoclonal antibodies to these adducts has been presented recently (*6*).

Antibodies against O^6-MedGuo have been prepared as rabbit polyclonal antisera (*6, 31a, 153, 154, 154a*) and mouse monoclonal antibodies (*154a, 155*). The monoclonal antibody C4 was slightly more cross-reactive with other modified nucleosides than the best of the rabbit antisera (Table IV). Its affinity constant ($K = 3.0 \times 10^9$ liters/mole) was intermediate between those of rabbit antisera R68/4 ($K = 4.0 \times 10^8$) and R45 ($K = 5.9 \times 10^9$). Antibody species specific for O^6-MedGuo ($K = 3.0 \times 10^8$) and O^6-EtGuo ($K = 1.8 \times 10^{10}$) have been affinity-purified from rabbit antiserum raised against O^6-EtdGuo coupled to KLH (*6*).

A different antiserum required a 10-fold higher concentration of O^6-MedGuo in ssDNA to produce 50% inhibition of tracer-antibody binding in RIA, compared to the isolated nucleoside (*153*). In addi-

TABLE IV
Relative Specificities of Antibodies to O^6-Alkylguanine[a,b]

Hapten:	O^6-MeGuo			O^6-EtGuo			O^6-BuGuo	
Reference:	(31a)	(154a)	(154a)	(40)	(31)	(30)	(30)	(158)
Antibody:	R68/4	R45	C4	—	E3	ER-6	E20	α-O^6BudGuo
Nucleoside inhibitor								
O^6-MedGuo	1	1	1	—	48	170	14,000	200,000
O^6-MeGuo	11	—	—	460	240	6500	—	—
O^6-MeGua	270	—	60	—	—	—	—	—
O^6-EtdGuo	—	500	70	1	1	1	200	1,200
O^6-EtGuo	—	—	—	6	3	22	—	—
O^6-BudGuo	—	0.5×10^6	0.03×10^6	—	9	82	1	1
O^6-BuGuo	—	—	—	—	40	140	1	3
dGuo	—	4.0×10^6	$>1.0 \times 10^6$	$>3 \times 10^6$	5×10^6	20×10^6	$>100 \times 10^6$	$>10 \times 10^6$

[a] Values are relative amounts of inhibitor required to inhibit binding to homologous tracer by 50% in RIA.
[b] Me, methyl; Et, ethyl; Bu, butyl.

tion, no reaction was observed with methylated dsDNA, implying decreased accessibility of the adduct to antibodies with increasing secondary structure. The specificity of the reactivity in ssDNA was shown by the observation that reaction occurred with DNA methylated by methylnitrosourea, which produces O^6-MedGuo adducts, but not with DNA methylated with dimethylsulfate, which produces O^6-MedGuo only in small yields. Apurinic sites resulting from hydrolysis of 7-MeGuo and 1-MeGuo were not recognized by the antibody.

Because antibody accessibility is reduced in polynucleotides, the preferred method of analysis is to measure O^6-MedGuo by RIA of enzyme hydrolyzates (154) or after separation of the modified nucleoside, e.g., by chromatography on Aminex-A6 with 10 mM NH_4HCO_3, at 50°C (154a, 155). The latter procedure has the merit of allowing concentration of O^6-MedGuo by lyophilization. Validation of both systems has been made by analysis of DNA samples treated with ^{14}C-labeled carcinogen simultaneously by RIA and by chromatography. After administration of a single dose of dimethylnitrosamine, the production and persistence of O^6-MedGuo in mainband and satellite DNA from mouse (154) and rat (156) liver was measured by RIA. Adduct yield was 20–30% less in mouse satellite DNA than in main-band DNA, and its rate of removal was identical in both DNAs. In rats, both adduct production and removal were identical in satellite and main-band DNA. The production and persistence of O^6-MedGuo in human and rodent cells following treatment with methylnitrosourea have been measured (154a, 155).

Mouse monoclonal antibodies have also been prepared against O^2-MedThd and O^4-MedThd (157). In three attempts to produce stable hybridomas secreting anti-O^4-MedThd, all clones lost the capacity to produce high titer antibodies, and none grew successfully as ascites. Consequently, we have been able to characterize only polyclonal mouse serum and some uncloned hybridoma culture supernatants. No problems were encountered in producing monoclonal antibodies against O^2-MedThd. The properties of these antibodies are shown in Table V.

Rabbit antisera (31, 37, 40) and monoclonal antibodies from mouse × mouse and rat × rat fusions (30) have been produced with high specificity for O^6-EtdGuo. The influence of the carrier protein and the immunization schedule on antibody titers and specificity has been studied systematically (30), favoring the use of KLH as carrier and an immunization schedule using low doses (0.5 mg per rabbit) of conjugate absorbed to aluminum hydroxide to permit slow release and distributed over multiple injection sites with injections repeated several

TABLE V
Radioimmunoassay of Alkylated Nucleosides Using Mouse Monoclonal and Polyclonal Antibodies[a]

Inhibitor	Monoclonal antibody against O^2-MedThd	Polyclonal antibody against	
		O^4-MedThd	O^2-MedThd
O^6-MedGuo	$>10^6$	1×10^5	—
O^6-EtdGuo	$>10^6$	$>2 \times 10^5$	0.67×10^5
O^6-BudGuo	$>10^6$	$>2 \times 10^5$	—
O^2-MedThd	1*	0.8×10^5	1[†]
O^2-EtdThd	154	—	1.8
O^2-PrdThd	89	—	3.6
O^2-BudThd	625	—	11.8
O^4-MedThd	20×10^3	1[‡]	4×10^3
3-MedThd	60×10^3	8×10^3	1.5×10^3
7-MeGuo	80×10^3	—	—
dAdo	1.6×10^5	$>2 \times 10^5$	0.16×10^5
dCyd	$>>10^5$	$>2 \times 10^5$	$>0.67 \times 10^5$
dGuo	$>>10^5$	$>2 \times 10^5$	$>0.67 \times 10^5$
dThd	7×10^5	$>2 \times 10^5$	0.16×10^5

[a] Values are picomoles of unlabeled inhibitor per 150-μl assay required to inhibit binding of tracer by 50%. Absolute amounts are indicated as follows: *, 0.56; [†], 15.0; [‡], 4.5.

months apart (Section I) (6). When present in ssDNA and dsDNA, 13- and 250-fold more O^6-EtdGuo were required to produce 50% inhibition of tracer binding than when present as the nucleoside (31), thus confirming the observation that O^6-alkyl sites are less accessible in DNA (153).

The sensitivities of different types of immunoassay have been compared using antiserum E2 with affinity constant for O^6-EtdGuo of ~1×10^{10} as determined by RIA (31). The limits of useful response (~10–90% inhibition of antibody binding) were found over the ranges 5–500, 20–2500, and 50–20,000 fmol for RIA, ELISA, and RIST, respectively. The ability to quantify O^6-EtdGuo in DNA has been improved by the application of the high-sensitive enzymatic linked immunosorbent assay (HS-ELISA) (37, 139). Individual wells of 96-well polystyrene Microtiter plates were coated with 2.8 μg of ethylated ssDNA by evaporation at 37°C overnight. Using a rabbit antiserum with an affinity constant for O^6-EtdGuo of 9.0×10^9 liters/mole and 50% inhibition of antibody binding by 92 fmol measured by RIA, a linear response over the range 0.5 to 50 fmol O^6-EtdGuo per 2.8 μg of DNA was obtained, permitting measurement of 0.1 μmol of adduct per mole of guanine without the need to hydrolyze DNA samples. This level of sensitivity should allow measurement of industrial and environmental exposure to alkylating agents on small biopsy or blood samples.

Preliminary characteristics of affinity-purified rabbit antiserum and monoclonal antibodies specific for O^4-EtdThd have been reported (6, 30).

RIA of O^6-BudGuo using mouse monoclonal antibody had a sensitivity of 44 fmol (158), similar to that of affinity-purified rabbit antiserum and rat monoclonal antibodies (6, 30). Mouse monoclonal antibodies to O^4-BudThd and O^2-BudThd were slightly less sensitive (450 and 69 fmol, respectively). The application of all three mouse monoclonal antibodies to the analysis of DNA samples was validated by correlation of RIA and high-pressure liquid-chromatography measurements.

An interesting feature of antibodies to alkylated nucleosides is that, although they result from immunizations with protein conjugates linked through the ribose moiety (17), there is a tendency to favor specific interaction with deoxyribonucleosides rather than ribonucleosides (Table IV). As the alkyl chain length increases, however, this specificity decreases. Specificity in terms of the alkyl chain length is greater for anti-methyl and anti-butyl antibodies than for anti-ethyl antibodies (Table IV).

B. 2-Acetamidofluorene

Two ultimate metabolites of 2-acetamidofluorene (2-acetylaminofluorene, 2-AAF or AAF) are produced *in vivo:* acetamidofluorene sulfate ester, of which 2-(N-acetoxy)acetamidofluorene (N-AcO-AAF) is an active analog, and 2-(hydroxylamino)fluorene (N-OH-AF) (Fig. 3). Reaction of N-AcO-AAF with DNA *in vitro* produces a major arylamidation product at the C8 of guanine, 2-[N-(2′-deoxyguanosin-8-yl)]acetamidofluorene (dGuo-C8-AAF), and a minor arylation product at the 2-NH_2 group of guanine, 3-[N-(2′-deoxyguanosin-N^2-yl)]acetamidofluorene (dGuo-N^2-AAF) in a ratio of about 6:1 (159–161). A third product found *in vivo* is the deacylated adduct 2-(2′-deoxyguanosin-8-yl)aminofluorene (dGuo-C8-AF) (162). A fourth adduct, 1-[6-(2,5 diamino-4-oxo)pyrimidinyl]-1-(2′-deoxyribosyl)-3-(2-fluorenyl)urea (ro-dGuo-C8-AF), can be obtained *in vitro* by alkaline hydrolysis of the imidazole ring of dGuo-C8-AF at the N7–C8 bond; it has been identified in rat liver DNA exposed to N-OH-AF (163–165).

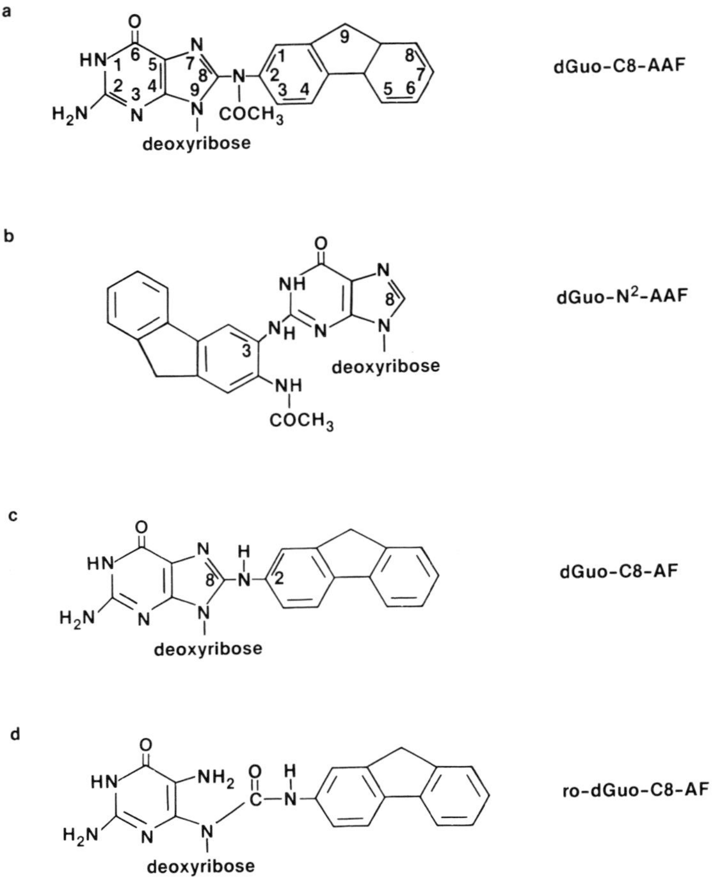

FIG. 3. Reaction products of DNA with N-AcO-AAF or N-OH-AF.

In preparing rabbit antibodies against acylated guanosine (C8) adducts and against native DNA containing this lesion, care must be taken to avoid unwanted deacylation of Guo-C8-AAF during conjugation with BSA (166). This can be reduced by omitting the sodium borohydride reduction step (17), thus avoiding prolonged alkaline conditions (37). Similarly, care must be taken to avoid inducing imidazole ring-opening during isolation of modified cellular DNA (167).

The sensitivities of different immunoassays for AAF-modified DNA using two rabbit antibodies that have similar affinity constants for dGuo-C8-AAF (7) are compared in Table VI. With dGuo-C8-AAF as a competing ligand (39), ELISA and USERIA were 6 and 60 times

TABLE VI
COMPARISON OF SENSITIVITIES OF IMMUNOASSAYS FOR dGuo-C8-AAF IN DNA

	Competitive assays (fmol dGuo-C8-AAF) for		Noncompetitive assays Linear range of detection of dGuo-C8-AAF in DNA (fmol)
	50% inhibition modified-DNA binding	Limit of detection	
Hsu et al. (39)			
RIA	250	~100	—
ELISA	40	~10	—
USERIA	4	~1	2–1000[a]
Van der Laken et al. (37)			
RIA	150	—	—
HS-ELISA	—	—	0.2–5[b]
USERIA	—	—	0.2–25[b]

[a] Per 10 ng of modified DNA per well.
[b] Per 800 ng of modified DNA per well.

more sensitive than RIA. On the other hand, when modified DNA was the competitor, the competitive assays were found to be less sensitive and less reproducible than noncompetitive assays (37). However, under the noncompetitive conditions used in the latter study the range of detection of dGuo-C8-AAF in DNA was approximately one-tenth that reported by Hsu et al. (39), and the sensitivity of HS-ELISA was the same as that of USERIA, the limit of detection being 0.2 fmol at a modification level of one dGuo-C8-AAF per 10^7 nucleotides in DNA.

Examination of Table VII shows that antibodies prepared against modified nucleosides also bind to AAF-modified DNA. The extent of binding is not dependent on the extent of DNA modification over the range tested (0.62 to 28%). Similarly, antibodies prepared against modified DNA react with modified nucleosides, particularly if these are phosphorylated. The observation that antibodies against Guo-C8-AAF have a lower affinity for AAF-modified DNA than for the modified nucleoside was interpreted as indicating that adducts are less readily accessible to interaction with antibody when part of a polynucleotide (168).

Carrying the concept of accessibility further, the conformation of adducts in DNA was investigated with particular reference to the base-displacement and insertion-denaturation models (169–171), which propose that the covalent binding of AAF to guanine in DNA results in a local distortion with the AAF residue lying internal and the guanine lying external to the double helix. A prediction of this

TABLE VII
Relative Inhibition of Antibody Binding in RIA by Modified Nucleosides and DNA[a]

Antibody prepared against	Tracer	Inhibitors compared: A/B	Relative concentration of inhibitors (A/B) at 50% inhibition	Reference
Guo-C8-AAF	[³H]dGuo-C8-AAF	ssDNA-AAF (28%)/Guo-C8-AAF	6.3	(168)
		dsDNA-AAF (28%)/Guo-C8-AAF	13	
Guo-C8-AAF	[³H]Guo-C8-AAF	GMP-AAF/Guo-C8-AAF	2.5	(166)
		GMP-AF/Guo-AAF	20	
		N-OH-AAF/Guo-AAF	100	
		ssDNA-AAF (6.7%)/Guo-AAF	125	
		dsDNA-AAF (6.7%)/Guo-AAF	125	
Guo-C8-AAF	[³H]Guo-C8-AAF	ssDNA-AAF (0.62%)/dGuo-C8-AAF	7.6	(37)
		dsDNA-AAF (0.62%)/dGuo-C8-AAF	46.6	
		Guo-C8-AAF/dGuo-C8-AAF	28.0	
		dGuo-C8-AF/dGuo-C8-AAF	5.8	
Guo-C8-AF	[³H]Guo-AF	Guo-AAF/Guo-AF	20	(182)
Guo-C8-AAF	[³H]Guo-AAF	Guo-AF/Guo-AAF	18	
Guo-C8-AAF	[³H]Guo-C8-AF	ssDNA-AF (1.5%)/GMP-AF	20	(172)
		dsDNA-AF (4.5%)/GMP-AF	200	
dsDNA-AAF	[³H]Guo-C8-AF	ssDNA-AF (1.5%)/GMP-AF	16	(172)
		dsDNA-AF (4.5%)/GMP-AF	160	
dsDNA-AAF	[³H]Guo-AAF	ssDNA-AAF (5.3%)/GMP-AAF	2.0	(173)
	[³H]ssDNA-AAF	dsDNA-AAF (5.5%)/GMP-AAF	2.0	
		GMP-AF/GMP-AAF	8.6	
dsDNA-AF	[³H]ssDNA-AF (1%)	dsDNA-AF (5.5%)/ssDNA-AF (1.4%)	17	(174)
		ssDNA-AAF (8.0%)/ssDNA-AF (1.4%)	85	
		dsDNA-AAF (5.5%)/ssDNA-AF (1.4%)	150	
		dGuo-AF/ssDNA-AF (1.4%)	370	
		dGMP-AF/ssDNA-AF (1.4%)	80	
ro-dGuo-C8-AF	[³H]ro-Guo-C8-AF	ssDNA-ro-dGuo-C8-AF/ro-dGuo-C8-AF	16	(167)

[a] Values in parentheses indicate percentage of bases modified.

model is that AAF residues in ssDNA should be more accessible to antibodies than AAF residues in dsDNA. However, when affinity-purified anti-Guo-C8-AAF was tested for binding to AAF-modified DNA, the affinity for dsDNA and ssDNA was identical (Table VII) (*166*), but when tested against AF-modified dsDNA, the same antibody showed one-tenth the affinity for AF-modified dsDNA compared to AF-modified ssDNA (*172*).

These results suggest that AF residues are buried deeper within the double helix than are AAF residues. Consistent with this interpretation are parallel observations that antibodies prepared against dsDNA-AAF bind equally to dsDNA-AAF and ssDNA-AAF (*173*), whereas antibodies against dsDNA-AF bind better to ssDNA-AF than to dsDNA-AF (*174*). On the other hand, when antibodies raised against Guo-C8-AAF are used without purification, some preference for binding to ssDNA-AAF (relative to dsDNA-AAF) is observed (*37, 168*) suggesting that some antibodies may recognize more than C8-modified dGuo.

Studies with anti-cytidine and anti-adenosine antibodies have supported the interpretation that AF residues are more internal than are AAF residues in dsDNA. The relative destabilization of DNA as assessed by susceptibility to S1 nuclease showed that, although AAF destabilizes the helix more than AF, the degree of destabilization is not sufficient to allow binding of anti-adenosine (*175*). Destabilization of dsDNA by AF modification results in a lowering of t_m, and a strong influence of base stacking was inferred from a comparison of circular dichroism spectra of modified DNA with those of modified dGMP (*166, 172*). Taken together, the physical, enzymological, and immunological data suggest a more dynamic conformation of bound residues than was previously considered.

Differences in the kinetics and extent of binding of anti-Guo-C8-AAF to dGuo-C8-AAF and dGuo-C8-AF have been used to estimate the extent of deacetylated vs acetylated nucleosides in enzyme hydrolyzates of DNA from cells exposed to N-AcO-AAF (*176*). When more than 80% of the adducts in standard mixtures of dGuo-C8-AAF and dGuo-C8-AF were deacetylated, RIA profiles showed decreasing slopes and lower levels of maximum inhibition (Fig. 4). Comparison of RIA profiles of hydrolyzed DNA samples with standard curves of appropriate mixtures of deacetylated and acetylated dGuo has shown that more than 90% of the total C8 adducts in primary epidermal and fibroblast cell cultures of BALB/c and Sencar mice, rats, and human foreskin are deacetylated after exposure to 10^{-5} M N-AcO-AAF for 1 hour (*176, 177*). In contrast, primary rat hepatocytes treated with ei-

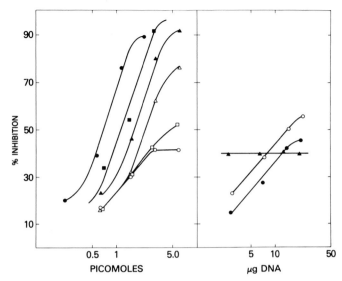

FIG. 4. (A) RIA standard curves for dGuo-C8-AF and dGuo-C8-AAF. Tracer is [³H]dGuo-C8-AAF. Percentage of modified dGuo (dGuo-C8-AF : dGuo-C8-AAF) used as inhibitor is ○, 100 : 0; □, 95 : 5; △, 89 : 11; ▲, 78 : 22; ■, 52 : 48; ●, 0 : 100. (B) RIA of DNA from N-AcO-AAF-treated cells. Tracer is [³H]dGuo-C8-AAF. ○, DNA from YDF cells; ●, DNA from BALB/c cells; ▲, untreated calf thymus DNA to which has been added a saturating level (6–12 pmol) of dGuo-C8-AF. Adapted from Poirier et al. (176).

ther N-AcO-AAF or 2-AAF have more than 80% of C8 adducts in the acetylated form. After incubation of treated cells for 24 hours, mouse epidermal cells removed 40% of the C8 adducts and human fibroblasts removed 50% (176).

The feasibility of similar studies with intact animals has also been demonstrated (178). During continuous feeding of male Wistar–Furth rats with 0.02 or 0.04% 2-AAF, 80% of all C8 adducts in the liver were found to be deacetylated during the first week, increasing to 97–100% by 30 days. The level of C8 adducts decreased if feeding was discontinued after 3 or 7 days (65–90% decrease by 28 days), but persisted if feeding was discontinued after 28 days (0–30% decrease by 28 days). In another study, no loss of dGuo-C8-AAF, dGuo-C8-AF, and ro-dGuo-C8-AF was observed in rat liver up to 45 hours after i.p. injection of N-OH-AAF (167). Using antiserum against ro-Guo-C8-AF (with [³H]ro-dGuo-C8-AF as tracer), ro-dGuo-C8-AF was assayed in enzyme digests of rat liver DNA, and total C8 adducts were measured in the same digests after alkali treatment to convert all C8 adducts to the ring-opened form. Using antiserum against Guo-C8-AAF (with

[³H]Guo-C8-AAF as tracer), the amount of dGuo-C8-AAF in the digests was estimated, and, by difference, the amount of dGuo-C8-AF was calculated.

With antibodies prepared against AAF- or AF-modified DNA, the immunodeterminant groups are dGMP-AAF or dGMP-AF ((*173, 174*). The association constant for binding of antibody-F_{ab} fragments to AAF-modified DNA was linearly and inversely dependent on ionic strength, since binding is related to the ionization of the nucleotide phosphate group. Difference UV spectra for F_{ab} fragments with the ligands GMP-AAF, dsDNA-AAF, and ssDNA-AAF all showed a positive band at 312 nm, at which wavelength only the fluorene ring absorbs, thus suggesting that the fluorene ring is located in the binding site of the antibody.

Antibodies to both DNA-AAF and to Guo-C8-AAF have been used to stain the nuclei of HeLa cells treated with 1–5 µg of N-AcO-AAF per milliliter for 20 min using fluorescein-isothiocyanate-labeled sheep anti-rabbit immunoglobulin as the second antibody (*179*). Structural fluorescence of specifically stained nuclei was observed against a background of homogeneous fluorescence in nonspecifically stained nuclei. About 15% of cells not exposed to AAF showed specific fluorescence that the investigators ascribed to the presence of autoantibodies to DNA in the rabbit antiserum. The fraction of immunoreactive cells in N-AcO-AAF-treated populations was proportional to carcinogen dose. After doses of 0, 1, and 5 µg of N-AcO-AAF per milliliter, cells were incubated for 0, 1, 3, and 21 hours, at which times samples were fixed and stained with antibodies to DNA-AAF, Guo-C8-AAF, and Guo. Immunoreactivity with anti-DNA-AAF decreased rapidly at 1 and 3 hours in cells exposed to either dose of carcinogen, but with anti-Guo-C8-AAF, immunoreactivity decreased rapidly only at 1 hour after 1 µg of N-AcO-AAF per milliliter. After 5 µg/ml, there was little decrease at early times; however, this analysis was complicated by an increase in immunoreactivity of control cells from 10% at 0 hour to more than 20% at 3 hours.

The authors suggest that the different kinetics seen with these antibodies may reflect differences in affinity for DNA-AAF. However, while it is true that anti-Guo-AAF has a higher affinity for the free modified nucleoside (-tide) than anti-DNA-AAF, the converse is true of its reactivity with DNA-AAF, which is more pertinent to the present situation. The disappearance of radioactivity from cells exposed to ³H labeled N-AcO-AAF was monitored for 6 hours; during this time there was little removal of label, a result consistent with the persistence of immunoreactivity by anti-Guo-C8-AAF. By 21 hours, the immunoreactivity to both specific antibodies had declined to that of control cells. The significance of these results is thus unclear, and the study demonstrates the need to use more specific (or purified) antibodies to minimize background fluorescence and the need for more quantitative methods for measuring specific immunofluorescence.

Affinity-purified anti-Guo-C8-AAF has been used with electron microscopy to visualize adducts in Col-E1 DNA, erythrocyte core particles, and trinucleosomes treated with N-AcO-AAF (*180, 181*). The structural integrity of the carcinogen-treated chromatin subunits was checked by circular dichroism and electron microscopy before modified DNA was extracted. All the DNA samples were incubated with a 10:1 ratio of antibody:antigen, and unbound antibody was removed by gel filtration prior to preparation for electron microscopy (Fig. 5). With linear Col-E1 DNA, the average number of antibody molecules bound per DNA molecule plotted against the number of dGuo-C8-AAF residues per DNA molecule measured radiochemically gave a linear response with a slope of 0.4. A slope of 0.5 would be expected from the bivalent nature of IgG binding, and cross-linking of DNA by a large proportion of antibodies was observed. With nucleosome core particles, antibodies bound only at the extremities of the 145-basepair DNA, whereas with DNA from trinucleosomes, antibodies bound both at the extremities and internally with an average distance between two IgGs of 737 ± 200 Å which compares with the average repeat length of erythrocyte chromatin of 734 ± 17 Å. The inference is, therefore, that N-AcO-AAF reacts mainly with linker DNA.

C. Benzo[*a*]pyrene

The aromatic hydrocarbon benzo[*a*]pyrene (BP) is a potent carcinogen in laboratory animals and is associated with lung cancer in humans. Smoking produces exposures of 20–50 ng of BP per cigarette, and the incomplete combustion of fossil fuels makes airborne BP ubiquitous in industrial societies, with occupational exposure of asphalt, coke oven, and gas works personnel as high as 2 mg of BP per cubic meter air (*183, 184*).

Metabolism of BP yields four stereoisomeric 9,10-epoxides of the *trans*-7,8-dihydrodiol (Fig. 6) (*184a*). The two diastereomeric forms, BPDE-I (*anti*) and BPDE-II (*syn*), are each formed as either (+) and (−) or (*R*) and (*S*) enantiomers. *In vitro*, all four enantiomers can form adducts with guanine and adenine residues, but *in vivo* the main reaction with DNA is at the N-2 of guanine with (7*R*)BPDE-I-dGuo as the major adduct. BPDE-I is more mutagenic (*185*) and carcinogenic (*186*) than BPDE-II.

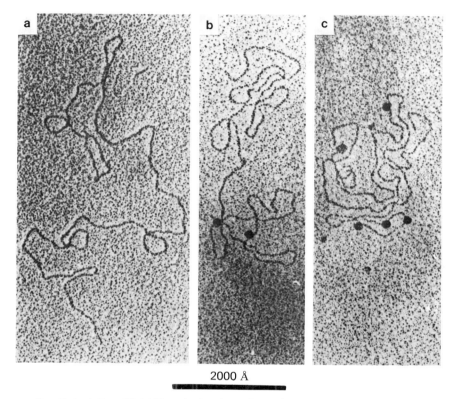

FIG. 5. Anti-Guo-C8-AAF antibodies bound to Col-E1 DNA visualized by electron microscopy. AAF modification is 0.07% (a and b) or 0.15% (c). DNA samples were incubated with purified anti-Guo-C8-AAF antibodies, and then with ferritin-labeled goat anti-rabbit IgG, and purified by filtration (b and c). In one sample (a), the first step, incubation with anti-Guo-C8-AAF antibodies, was omitted. From DeMurcia *et al.* (*180*), with permission.

Two groups have reported the production of rabbit antisera specific for BP-modified DNA following immunization with calf thymus DNA modified by *in vitro* reaction with (±)BPDE-I and complexed with carrier mBSA (*11, 26*). The specificity of one antiserum was determined by RIA. A 50% inhibition of binding of BPDE-I-dGuo was obtained with 5 pmol of calf thymus DNA modified 1.4% with BPDE-I or with about 40 pmol of either (7R)BPDE-I-dGuo or BPDE-II-dGuo, suggesting that configurational elements of the DNA contribute to the antigenic determinant. The antiserum was twice as reactive with BP-modified ssDNA as with BP-modified dsDNA (*187*), and it also recognized BP-modified RNA (*14*).

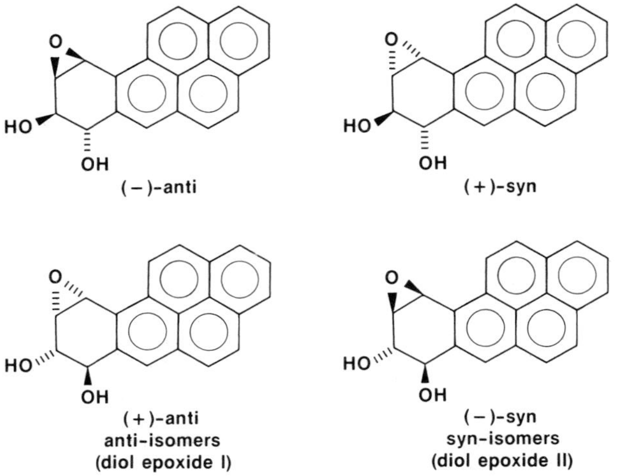

FIG. 6. Structures of four stereoisomeric benzo[a]pyrene-7,8-diol 9-10-epoxides. Alternative nomenclature for *anti* isomer:
r-7,t-8-dihydroxy-t-9,10-oxy-7,8,9,10-tetrahydrobenzo[a]pyrene or
(±)-7α,8β-dihydroxy-9β,10β-epoxy-7,8,9,10-tetrahydrobenzo[a]pyrene
Alternative nomenclature for *syn* isomer:
r-7,t-8-dihydroxy-c-9,10-oxy-7,8,9,10-tetrahydrobenzo[a]pyrene or
(±)-7α,8β-dihydroxy-9α,10α-epoxy-7,8,9,10-tetrahydrobenzo[a]pyrene

Sensitivity was improved about 500-fold and 100-fold in competitive USERIA and ELISA, respectively, using 1 µg of DNA in antigen–antibody mixtures (36). Subsequent studies showed that these assays were unaffected by up to 50 µg of unmodified DNA (186) and easily measured 25-fmol adducts in 5 µg of DNA from mouse epidermal cells exposed to 10^{-7} M (±)BPDE-I for 1 hour (36). Although useful in permitting amplification of the antibody–antigen reaction signal, solid-phase assays include steps to ensure blocking of nonspecific binding and extensive washing to remove reagents applied in the preceding steps. These operations and other factors, such as the conformation of the DNA antigen, are variables that must be considered when working at the practical limits of assay sensitivity.

In a pilot study to test the feasibility of using ELISA to measure BP adducts in lung cancer patients, variability was 15–20%, on samples assayed in quadruplicate and repeated on 2 or 3 different days (187). Therefore, the lowest value in the ELISA considered significant was 30% inhibition (corresponding to 0.08–0.10 fmol of adduct per microgram of DNA), but this value was accepted only if the mean value of quadruplicate wells was more than two standard deviations below that of quadruplicate wells containing antibody but no competitor.

Analysis of lung DNA samples from 27 patients gave positive results (31–38% inhibition corresponding to 0.14–0.18 fmol of adduct per microgram of DNA) in samples from 4 of 19 patients with lung cancer. In addition, no adducts were detected in blood lymphocytes of dogs that received up to 252 ng of BP per day by smoke inhalation for at least 20 months, but adducts were observed in lungs from mice that received as little as 0.1 mg of BP i.p.

One implication of these results is that exposure to BP from smoking may be too low to be detectable consistently, even with the sensitivity of the present techniques. Clearly, detoxification processes, including DNA repair, help to determine the adduct load carried by individuals, and further studies may reveal individuals with high adduct levels resulting from genetic deficiencies in these processes, which would be a contraindication for their employment in high-risk industries.

The localization of BP adducts in cell nuclei in DNA has been visualized with indirect immunofluorescence (11, 14). The intensity of nuclear fluorescence was proportional to BP dose, and all cells receiving a nontoxic dose showed similar levels of fluorescence, which declined with time owing to repair of BP adducts and dilution by replication. After toxic BP doses, some cells presumed to be dead showed strong staining that persisted for long times. Fluorescence was brightest in RNase-sensitive spots probably corresponding to nucleoli and also in the perinuclear region, owing possibly to the presence of mtDNA that is preferentially modified by BP.

Visualization of specific antibody bound to DNA molecules was achieved by using a second antibody labeled with ferritin. Over 60% of BPDE-1-DNA adducts measured by RIA were observed by electron microscopy of calf thymus DNA modified *in vitro* to 0.005% or less, but higher levels of modification caused aggregation of DNA and antibodies (188). To overcome the problem of adherence of excess antibodies, labeled DNA was separated by preparative sucrose gradient centrifugation prior to preparation for electron microscopy. In this way, BP adducts in Col-E1 DNA were shown to be randomly distributed (11).

D. Aflatoxin B_1

Aflatoxin B_1 (AFB_1) is the most potent member of a series of metabolites produced by *Aspergillus flavus* (184a). It is toxic, mutagenic, and carcinogenic in many species, is a frequent contaminant of food, including cereals and peanuts, and is statistically associated with human liver cancer in Africa and Asia. The major adduct in DNA (Fig. 7)

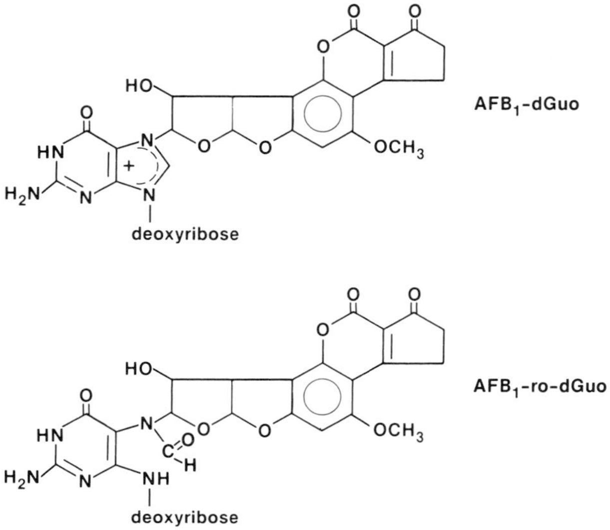

FIG. 7. Aflatoxin B_1 adducts of guanosine.

is 2-(7-guanyl)-2,3-dihydro-3-hydroxyaflatoxin B_1 (AFB_1-Guo) formed by nucleophilic attack on the N-7 guanine by the 2,3-epoxide of AFB_1, which results from monooxygenase activity on AFB_1. The adduct is unstable owing to a localized positive charge on the imidazole ring and breaks down to the ring-opened (ro) derivative 2-[2-amino-6-(2'-deoxyribosylamino)-5-formamido-4-oxopyrimidin-N^5-yl]-3-hydroxy-AFB_1 (ro-AFB_1-Guo), which is the persistent adduct *in vivo*.

Reaction of AFB_1 with nucleotides does not give large yields, but reaction with dsDNA mediated either by *m*-chloroperoxybenzoic acid or by phenobarbital-induced rat liver microsomes can yield up to 5% modified nucleotides (27, 28). AFB_1-modified DNA complexed with mBSA provides an antigen, either directly (27, 189) or after incubation at 37°C for 2 hours at pH 9.5, resulting in the formation of ro-AFB_1-DNA (32). Mouse monoclonal antibodies have been selected by screening hybridoma supernatants by ELISA using either AFB_1-DNA or ro-AFB_1-DNA bound to Microtiter plates. These antibodies are very sensitive in competitive and noncompetitive ELISA and USE-RIA assays (27), with levels of one AFB_1 residue per 1.4×10^6 nucleotides being assayed in rat liver DNA by USERIA (189) and one in 3×10^5 nucleotides by ELISA (32).

The antibodies produced against AFB_1-DNA recognize neither free aflatoxins nor the free modified bases, AFB_1-Gua and its ring-

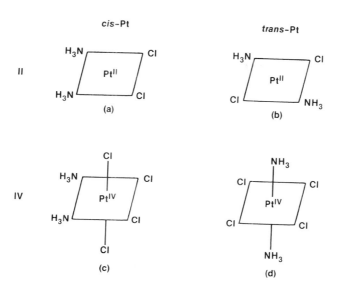

FIG. 8. Isomeric platinum coordination compounds:
(a) *cis*-Diamminedichloroplatinum(II); cisplatin (*cis*-Pt^{II})
(b) *trans*-Diamminedichloroplatinum(II); (*trans*-Pt^{II})
(c) *cis*-Diamminetetrachloroplatinum(IV); (*cis*-Pt^{IV})
(d) *trans*-Diamminetetrachloroplatinum(IV); (*trans*-Pt^{IV}).

opened form (27). In contrast, a wide spectrum of specificities was observed with five antibodies raised against ro-AFB$_1$-DNA (28, 32). Two of these bind to AFB$_1$-Gua and AFB$_1$-DNA with about the same affinity, and some react with unmodified ssDNA (28). It is not clear from the small number of hybridomas characterized whether the differences in specificity observed are merely a sample artifact or represent real differences in responses to the different immunogens. Clearly the range of specificities produced provides potential probes for the quantitation of AFB$_1$ adducts in DNA, the quantitation of AFB$_1$-Gua in urine or serum, and the localization of AFB$_1$-DNA in tissues by immunohistological staining.

E. Cisplatin

In contrast to the *trans* isomers, cisplatin [*cis*-diamminedichloroplatinum(II), formerly called (190) *cis*-platinum(II) and *cis*-Pt(II)] and similar coordination compounds of platinum are highly cytotoxic and effective antitumor agents, being primary agents for the treatment of testicular cancers and ovarian carcinoma (for reviews, see 190, 191). Both geometric isomers (Fig. 8) react preferentially with the N-7 of guanine in DNA as well as with adenine and cytosine, and to a much lesser extent with thymidine. Monofunctional binding is probably not the reason for the higher toxicity of the *cis* compounds, because at equitoxic doses there is less reaction with the *cis* than with the *trans* isomers (*192*). On the other hand, at equimolar doses, more interstrand cross-links are observed with the toxic *cis* compounds than with the nontoxic *trans* isomers, suggesting that bifunctional reactions are important for toxicity (*193, 194*). Several bifunctional reactions have been postulated, including (a) interstrand cross-linking of guanines on opposite sides of the duplex (*192, 193*); (b) intrastrand cross-links between adjacent guanines in the same strand (*195, 196*), or between the guanines of GAG and GCG codons giving rise to amber and ochre mutational "hot spots" (*197*); and (c) intrabase chelation involving the N-7 and O-6 of guanine (*198*).

In an attempt to resolve more clearly the differences in *cis* and *trans* reactions with DNA, two groups have adopted immunological approaches (*199, 200*). Rabbits were immunized with a complex of methylated BSA and deproteinized calf-thymus DNA treated with cisplatin to the extent that the molar ratio of Pt to nucleotide was 0.04–0.05 (cisplatin/DNA-P = 4–5%). In the first study (*199*), in which Pt adducts in DNA were assayed by competitive RIA, the antibodies appeared to recognize a distortion of the double helix produced by the three *cis* compounds, cisplatin, *cis*-dichloro(ethylenediammine)platinum(II), and (1,2-cyclohexanediamine)sulfatoplatinum(II). No cross-reaction was seen with native DNA modified with *trans*-diamminedichloroplatinum(II), *cis*-diamminotetrachloroplatinum(IV), or chloro(diethylenetriamine)platinum(II) chloride (Cl_2dienPt), nor with cisplatin-modified RNA or polyribonucleotides. Although cisplatin-ssDNA was able to compete for antibody, it did so poorly, owing perhaps to the presence of short regions of modified dsDNA. Quantitative precipitation analysis indicated that at low levels of modification (cisplatin/DNA-P = 5%), the molar ratio of antibody to bound Pt was about one, indicating that all Pt adducts were accessible to antibody. At high levels of modification (cisplatin/DNA-P = 20%), fewer antibodies were bound, suggesting that the rabbit antiserum contained at least two classes of antibody, only one of which binds to highly modified DNA.

In the second study (*200*), using competitive ELISA, 50% inhibition of binding was obtained with 50 (±10) fmol of cisplatin-DNA compared to about 40 pmol by RIA (*199*). The higher sensitivity of the ELISA permitted measurement of adducts in DNA of L1210 cells exposed to cisplatin either in cell culture or in ascites. Cells exposed *in vitro* to 100 μM cisplatin for 1 hour or exposed *in vivo* to 10 mg of cisplatin per kilogram were modified to similar extents (1.4–2.1 fmol of Pt per microgram of DNA). However, no adducts were detected by RIA of rat liver DNA after i.p. injection of 3.5 mg of cisplatin per kilogram. Thus at least some, but probably not all, adducts formed *in vivo* are recognized by antiserum and are the same as those formed *in vitro* (200). The consensus of both studies was that neither antiserum recognized interstrand cross-links in DNA, and at present, the precise nature of the binding site(s) is not known.

F. Z-DNA

At this point it may be pertinent to mention the role of antibodies in characterizing Z-DNA. The left-handed or Z conformation was first described by X-ray crystallography of the hexanucleotide p-d(C-G-C-G-C-G) (201). In solution, poly(dG-dC)·poly(dG-dC) undergoes a salt-induced cooperative transition from the B form in low salt to the Z form in high salt with the midpoint at 2.5 M NaCl, a transition that can be followed by inversion of the circular dichroism spectrum, by changes in optical absorbance and nuclear magnetic resonance spectra, and by altered rate of proton exchange with solvent (202–204).

Poly(dG-dC)·poly(dG-dC), stabilized in the Z form by bromination at the C-8 of guanine or the C-5 of cystosine (205, 206) or by Cl_2dienPt at the N-7 of guanine (206, 207), has been used to generate polyclonal (205, 207, 208) and monoclonal (208) antibodies specific for Z-DNA. In the absence of stabilizing modifications, poly(dG-dC)·poly(dG-dC) may be partially in the Z conformation when complexed with methylated BSA, because this antigen also elicits some anti-Z-DNA antibodies. It is of interest that these antibodies are also present in the sera of some MRL/lpr autoimmune mice (205).

The specificities of anti-Z sera have been carefully characterized by binding to poly(dG-dC)·poly(dG-dC) under conditions that produced Z conformation as determined by physical measurements. None of the antibodies recognized the bromine or Cl_2dienPt moieties present in the immunogen. In RIA, no inhibition of binding of brominated poly(dG-dC)·poly(dG-dC) was observed with 10 μg of brominated poly(dG), brominated poly(dG)·poly(dC), poly(dG), or poly(dG-dC)·poly(dG-dC) when assayed in 0.2 M NaCl buffer under conditions in which 45 ng of brominated poly(dG-dC)·poly(dG-dC) caused 50% inhibition of tracer binding (205). However, antiserum drawn 5 weeks later from the same rabbit immunized with Z-form polymer contained different antibodies that bound to B-form poly(dG-dC)·poly(dG-dC) and that recognized both B- and Z-form polymer in RIA. Similar results were obtained with antisera raised against Cl_2dienPt-modified polymer (206). None of the anti-Z sera were inhibited in RIA by native or denatured DNA from a variety of sources with the exception of form V DNA (208a), which is either a poor competitor or contains small Z regions (208).

Modification of more than 20% of the guanine in poly(dG-dC)· poly(dG-dC) by AAF also results in Z form. Unlike highly modified DNA, this Z-form polymer is not susceptible to S_1 endonuclease (EC 3.1.30.1) digestion, and anticytosine antibodies will not bind, implying that Z-form polymer is totally duplexed (209). However, anti-Z antibodies induced by Cl_2dienPt-modified polymer will bind to the AAF-modified polymer (206). CD-spectra for AF-modified polymer indicate B conformation (210), but the predicted lack of reaction with anti-Z antibodies has not yet been reported. The results to date are consistent with AAF acting as proposed by the insertion-displacement hypothesis (see Section V,B), resulting in displacement of the C-8 of guanine external to the helix with guanine in the syn conformation, while AF modification causes guanine to remain in the anti conformation with C-8 internal to the helix.

Immunochemical staining by these antibodies has demonstrated the presence of Z-DNA in interbands of *Drosophila* polytene chromosomes (211) and in what is probably highly repetitive (G+C)-rich heterochromatin of *Gerbillus nigeriae* metaphase chromosomes (212). Immunofluorescence also demonstrated Z-DNA in the transcriptionally active macronucleus of the ciliate *Stylonychia mytilus*, but not in the transcriptionally inert micronucleus (213). Immunoreactivity of Z-DNA has been observed in most, but not all, nuclei of rat cerebellar cortex, kidney, liver, and testis (213a). The criticism that the Z conformation may arise during the preparation of samples for immunostaining was partially answered by showing that the staining of spermatogonia nuclei is independent of the method of fixation.

Because of the reversible nature of the B to Z transition, it has been suggested that Z-DNA may be of regulatory importance, but what this is has yet to be defined (214). The finding that poly(dG-dC)·poly(dG-dC) is stabilized in the Z conformation under physiological conditions by methylation at the N-7 of guanine and the C-5 of cytosine (215–217), and by modification with mitomycin C, believed to be at the O-6 of guanine (218) as well as by AAF and Cl_2dienPt, suggests that the Z conformation might play a role in regulating DNA repair.

VI. Perspectives

In this review, we emphasize a number of parameters that affect the specificity and affinity of antibodies to carcinogen-induced adducts in DNA. It is difficult to provide precise formulas for the successful production of antibodies with desired properties because the interplay of parameters is so large. Probably for this reason, as much as for restrictions on time and effort, there have been few systematic studies of the production of antibodies against carcinogen damage, and much of our information derives from imprecise impressions obtained by comparing the experiences of different investigators.

Clearly, the most important consideration is the chemical structure of the adducts. Whether a particular structure is immunogenic may depend on whether it is presented as a nucleoside or within DNA, where it may be less accessible. If presented as part of a DNA molecule, can a sufficient number of adducts be formed? Where modified DNA has been used successfully, the level of modification has usually been in the range of 1 to 10% of the parent base. Is it better to use modified ssDNA or modified dsDNA? The answer may depend on whether the modification lies internal or external to the helix. Regardless of whether ssDNA or dsDNA is used, antibodies against unmodified DNA will usually be formed and may have to be removed by absorption. Apart from cisplatin, there have been no reports of antibodies raised against the adducts of bifunctional chemotherapeutic agents. One reason for this may be that the presence of interstrand cross-links increases the stability of DNA against thermal denaturation. Thus, the adduct may remain hidden unless its presence causes distortion of the helix to reveal the adduct, or to provide a topological structure that is itself immunogenic.

The eventual use to which an antibody is to be put can influence decisions concerning immunization and also the method used to determine antibody titer. This is particularly important when screening hybridomas, where the best method for screening large numbers of clones need not select for optimum specificity. Direct binding assay of antibodies produced after minimum antigen exposure may select antibodies with higher affinities than desirable if they are to be used for affinity chromatography. Antibodies to be used for immunofluorescence may require screening for nonspecific binding on the material to be studied so as to reduce background fluorescence.

In many instances antibodies raised against isolated chemical adducts (Section V) or nucleotides (Section II) will bind to their respective antigens in a polynucleotide, although with lower affinity. This finding emphasizes the importance of flanking nucleotides (and possibly the ribose-phosphate backbone) on antibody–antigen binding. In general, the experience in a number of laboratories indicates that antibodies raised against modified DNA bind most efficiently to modified polynucleotides (and are therefore ideally suited for detecting modified DNA *in situ*), whereas detection of isolated nucleoside adducts in modified DNA hydrolyzates is best accomplished with antibodies raised against nucleoside adducts.

Antibodies raised against UV-damaged ssDNA often show a preference for pyrimidine dimers contained in oligonucleotides at least three to six nucleotides in length (Section III). The role of the flanking nucleotides in the antibody–antigen binding reaction is not clear. Although required for high-affinity binding, they probably do not function as compositional antigenic determinants, because specific nucleotides in the flanking positions are not required. Substitution of different nucleotides in the flanking position(s) can alter binding (106) but does not eliminate it completely. This observation and the fact that it is possible to produce antibodies that recognize triplet (87–89) or longer (90) sequences suggest this possibility of obtaining antibodies capable of recognizing adducts with specific adjacent base sequences that may be useful in defining mutational "hot spots."

A future source of antibodies may be antisera from SLE patients or hybridomas generated from fusions with lymphoid cells from animal strains exhibiting lupus-like symptoms. As discussed in Section II, this disease is characterized by autoantibodies to a range of nucleic acids and nucleoproteins including some that react with dsDNA. SLE patients are frequently photosensitive, an observation that led to the idea that UV may play a role in exacerbating the disease, possibly by causing local tissue damage with the release of cellular DNA, some of which may well contain immunogenic photoproducts. However, the observation that SLE sera can react with UV-irradiated DNA does not imply photoproduct specificity, since these sera react also with unirradiated DNA (9, 94). A study of 70 sera from SLE patients showed many that reacted to both dsDNA and UV-dsDNA and a few that reacted well with one antigen but not the other, although it is difficult to assess from the data presented whether any photoproduct-specific antibodies were present (220). Sera from other photodermatoses failed to show antibodies to UV-DNA (221).

In addition to a genetic hyperresponsiveness to nucleic acid antigens, it has been suggested that lymphocytes of SLE patients are defective in DNA repair (222). Using peripheral blood lymphocytes of these patients, defective unscheduled DNA synthesis after UV radiation and reduced repair of O^6-MedGuo after methylnitrosourea have been observed (222, 223). Defective repair may result in increased toxicity of cells exposed to carcinogens *in vivo* (223) and the release of carcinogen-modified DNA. It would seem worthwhile, therefore, to test stringently for the presence of adduct-specific antibodies in the sera of autoimmune animals that have been exposed to carcinogens. If such antibodies are found, they may be immortalized as monoclonal antibodies, as has already been done for antibodies specific for dsDNA, ssDNA, and rDNA, some or which show interesting sequence or conformational specificities (63–65, 224–228).

An important application of antibodies is their use in visualizing damaged DNA *in situ*. Enzyme-linked secondary antibodies are often used to amplify the primary antibody reaction, and reagents of high quality are now readily available commercially for this purpose. To date, most observations of damage in cells and tissues *in situ* have been qualitative with the notable exception of a series of autoradiographic observations of UV damage quantified by silver grain density (13, 104, 105). These studies were significant in demonstrating that *in situ* methods could yield meaningful data on DNA repair without absolute quantitation of damage, which would require a means of calibrating the accessibility of adducts in chromatin. However, autoradiography is a lengthy procedure and is limited to fixed cells or tissues. An important development would be the quantitation of damage by immunofluorescence using integrated densitometry of nuclei observed microscopically or by automated flow cytometry (98), which has the additional advantage of multiparameter analysis of cellular subpopulations.

Immunofluorescent and immunoperoxidase staining of metaphase chromsomes with anti-5MeCyt and anti-ZDNA antibodies has proved useful in studying the distribution of these antigens in normal chromosomes (79–81, 211, 212). Analysis of chromosomal distribution of DNA damage has included the use of nucleoside-specific antibodies to detect denatured (A+T)-rich regions after UV treatment (49) and (G+C)-rich regions following selective destruction of guanine residues by photooxidation in the presence of methylene blue (53). Other investigators (114) have demonstrated the use of DNA photoproduct-specific antibodies in studies of chromosomal damage following UV microirradiation of small areas of the nucleus. With improvements in amplification methods, it should be possible to determine the chromosomal distribution of other (less frequent) forms of DNA damage.

At the molecular level, electron microscopic localization of damaged sites in DNA spreads has been achieved using antibodies specific for AAF-DNA, Guo-8-AAF (*180*), and BPDE-DNA (*11, 188*). This technique is a useful means of determining whether altered sites in DNA are distributed in a random or a nonrandom fashion. Also of interest is the potential use of antibodies as probes for damage in DNA that has been fractionated electrophoretically and transferred to nitrocellulose or diazobenzyloxymethylcellulose paper. This method has been employed to study natural methylation patterns of DNA using an antibody specific for 5-MeCyt (*82*), and to study the distribution of carcinogen damage in cellular DNA and in the coding and noncoding strands of simian-virus-40 DNA, using an antibody to BPDE-1-modified DNA (*228a*).

Antibodies have also been used to isolate specific nucleic acid fractions. Immunoprecipitation of antigenic small nuclear RNAs[2] from nuclear extracts has enabled investigators to analyze the specificity of human SLE antibodies (*219*) and mouse MRL/1 antibodies (*65*). Antibodies immobilized on chromatographic columns have been used extensively as a means of fractionating oligonucletides and RNAs possessing specific modified nucleotides (*8*). In principle, these approaches should allow the partial separation of damaged DNA from whole-cell DNA when appropriate antibodies are employed.

The immunological recognition of altered and unaltered nucleic acid components provides a means by which to study protein/nucleic acid recognition in general. The determinants on nucleic acids recognized by antibodies may, in some cases, be similar to those recognized by other site-specific binding proteins such as repair or restriction enzymes. The extent of these similarities can be tested by comparing the binding characteristics and kinetics of the two proteins in question. Simultaneous monitoring of alterations in DNA with specific antibodies and specific repair enzymes is one method by which to elucidate differences and similarities in recognition. For example, the repair kinetics of pyrimidine dimers in mammalian cells has been followed by both dimer-specific antibodies and dimer-specific T4 endonuclease (*107*). Surprisingly, the loss of antibody-binding sites occurred much more rapidly than the loss of endonuclease-cleavage sites. The resolution of this paradoxical finding will depend upon the precise determination of both antibody and endonuclease recognition sites, to which end it may be of interest to investigate the idiotypes of adduct-specific antibodies.

---

[2] See review of small nuclear RNAs by Reddy and Busch in Vol. 30 of this series. [Eds.]

TABLE VIII
Antibodies for the Detection of Carcinogens

Carcinogen	References
Aflatoxin B_1	(229, 230, 233, 236)
2-Acetamidofluorene	(231, 232)
1,2,5,6-Dibenzanthracene	(242, 243)
2-Anthrylamine	(235, 245)
7,12-Dimethylbenz[a]anthracene	(239)
5-Fluoro-12-methylbenzanthryl-7-acetic acid	(235)
Benzo[a]pyrene	(235, 237, 238, 240, 241, 244)
Benzo[rst]pentaphene	(235)
Actinomycin D	(234)
Mitomycin C	(246)

Finally, the use of immunological assays overcomes the ethical difficulties of administering radioactivity to humans, which would otherwise be required for standard biochemical analysis of carcinogen exposure. Already, the first epidemiological study has been performed (187), pioneering the use of antibodies for monitoring human environmental exposure to carcinogens. Antibodies to several known carcinogens and mutagens are available also for measuring blood and urine levels (229–234), and, although beyond the scope of this review, a bibliography is listed in Table VIII. In addition, some information is available concerning the effects of immunizing with carcinogens on cytotoxicity (235, 236), mutagenicity (237), and carcinogenicity (235, 238–241) of these agents.

Acknowledgments

We are grateful to M. C. Poirier and G. DeMurcia for permission to reproduce Figs. 4 and 5 and to colleagues who showed us data before publication. Research originating in the authors' laboratories was supported by grants from the Cancer Research Campaign and the Medical Research Council, and the National Cancer Institute, DHHS, under contract No. N01-C0-23909 with Litton Bionetics, Inc. The contents of this publication do not necessarily reflect the views or policies of the Department of Health and Human Services, nor does mention of trade names, commercial products, or organizations imply endorsement by the United States Government.

References

1. O. J. Plescia and W. Braun, "Nucleic Acids in Immunology." Springer-Verlag, Berlin and New York, 1968.
2. L. Levine and H. Van Vunakis, in "Antibodies to Biologically Active Molecules" (B. Cinader, ed.), p. 25. Pergamon, Oxford, 1966.

3. O. J. Plescia and W. Braun, *Adv. Immunol.* **6**, 231 (1967).
4. B. D. Stollar, in "The Antigens" (M. Sela, ed), Vol. 1, p. 2. Academic Press, New York, 1973.
5. H. Van Vunakis, *Photochem. Photobiol. Rev.* **5**, 293 (1980).
6. R. Müller and M. F. Rajewsky, *J. Cancer Res. Clin. Oncol.* **102**, 99 (1981).
7. M. C. Poirier, *JNCI* **67**, 515 (1981).
8. T. W. Munns and M. K. Liszewski, This Series **24**, 109 (1980).
9. E. M. Tan and R. B. Stoughton, *PNAS* **62**, 708 (1969).
10. C. J. Lucas, *Exp. Cell Res.* **74**, 480 (1972).
11. H. Slor, H. Mizusawa, N. Neihart, T. Kakefuda, R. S. Day, and M. Bustin, *Cancer Res.* **41**, 3111 (1981).
12. R. Bases, A. Rubinstein, A. Kadish, F. Mendez, D. Wittner, F. Elequin, and D. Liebeskind, *Cancer Res.* **39**, 3524 (1979).
13. J. J. Cornelis, J. Rommelaere, J. Urbain, and M. Errera, *Photochem. Photobiol.* **26**, 241 (1977).
14. M. C. Poirier, J. R. Stanley, J. B. Beckwith, J. B. Weinstein, and S. H. Yuspa, *Carcinogenesis* **3**, 345 (1982).
15. B. D. Stollar, *Methods Enzymol.* **70**, 70 (1980).
16. B. F. Erlanger, *Methods Enzymol.* **70**, 85 (1980).
17. B. F. Erlanger and S. M. Beiser, *PNAS* **52**, 68 (1964).
18. R. D. Meredith and B. F. Erlanger, *NARes* **6**, 2179 (1979).
19. S. M. Beiser, S. W. Tanenbaum, and B. F. Erlanger, *Methods Enzymol.* **12**, 889 (1968).
20. B. F. Erlanger, *Pharmacol. Rev.* **25**, 271 (1973).
21. M. J. Halloran and C. W. Parker, *J. Immunol.* **96**, 379 (1966).
22. R. Müller, W. Drosdziok, and M. F. Rajewsky, *Carcinogenesis* **2**, 321 (1981).
23. L. Levine, E. Seaman, E.Hammerschlag, and H. Van Vunakis, *Science* **153**, 1666 (1966).
24. E. Seaman, L. Levine, and H. Van Vunakis, *Bchem* **5**, 1216 (1966).
25. M. Leng and E. Sage, *FEBS Lett.* **92**, 207 (1978).
26. M. C. Poirier, R. Santella, I. B. Weinstein, D. Grunberger, and S. H. Yuspa, *Cancer Res.* **40**, 412 (1980).
27. A. Haugen, J. D. Groopman, I.-C. Hsu, G. R. Goodrich, G. N. Wogan, and C. C. Harris, *PNAS* **78**, 4124 (1981).
28. P. J. Hertzog, A. Shaw, and R. C. Garner, *J. Immunol. Methods* (1984) in press.
29. P. T. Strickland and J. M. Boyle, *Photochem. Photobiol.* **34**, 595 (1981).
30. M. Rajewsky, R. Müller, J. Adamkiewicz, and W. Drosdziok, in "Carcinogenesis: Fundamental Mechanisms and Environmental Effects" (B. Pullman, P. O. P. Ts'o, and H. V. Gelboin, eds.), p. 207. Reidel, Dordrecht, 1980.
31. R. Müller and M. F. Rajewsky, *Cancer Res.* **40**, 887 (1980).
31a. H. Salih and P. F. Swann, *Chem.-Biol. Interact.* **41**, 169 (1982).
32. P. J. Hertzog, J. Lindsay-Smith, and R. C. Garner, *Carcinogenesis* **3**, (1982).
33. J. M. Boyle, unpublished.
34. E. Seaman, H. Van Vunakis, and L. Levine, *JBC* **247**, 5709 (1972).
35. R. D. Ley, *Cancer Res.* **43**, 41 (1983).
36. I.-C. Hsu, M. C. Poirier, S. H. Yuspa, D. Grunberger, I. B. Weinstein, R. H. Yolken, and C. C. Harris, *Cancer Res.* **41**, 1091 (1981).
37. C. J. Van der Laken, A. M. Hagenaars, G. Hermsen, E. Kriek, A. J. Kuipers, J. Nagel, E. Scherer, and M. Welling, *Carcinogenesis* **3**, 569 (1982).
38. R. Müller, *J. Immunol. Methods* **34**, 345 (1980).

39. I.-C. Hsu, M. C. Poirier, S. H. Yuspa, R. H. Yolken, and C. C. Harris, *Carcinogenesis* **1**, 455 (1980).
40. R. Müller and M. F. Rajewksy, *Z. Naturforsch. C. Biosci.* **33C**, 897 (1978).
41. P. T. Strickland, unpublished.
42. K. Heide and H. G. Schwick, *in* "Handbook of Experimental Immunology" (D. M. Weir, ed.), pp. 7.2–7.5. Blackwell, Oxford, 1978.
43. J. L. Fahey and E. W. Terry, *in* "Handbook of Experimental Immunology" (D. M. Weir, ed.), pp. 8.3–8.10. Blackwell, Oxford, 1978.
44. T. W. Munns, M. K. Liszewski, and H. F. Sims, *Bchem* **16**, 2163 (1977).
45. V. P. Butler, S. Beiser, B. F. Erlanger, S. W. Tanenbaum, S. Cohen, and A. Bendich, *PNAS* **48**, 1597 (1962).
46. S. W. Tanenbaum and S. M. Beiser, *PNAS* **49**, 662 (1963).
47. A. J. Steiner, D. M. Kipnis, R. Utiger, and C. Parker, *PNAS* **64**, 367 (1967).
48. M. Sela, H. Ungar-Waron, and Y. Schechter, *PNAS* **52**, 285 (1964).
49. R. R. Schreck, B. F. Erlanger, and O. J. Miller, *Exp. Cell Res.* **88**, 31 (1974).
50. D. Liebeskind, K. C. Hsu, B. F. Erlanger, and R. Bases, *Exp. Cell Res.* **83**, 399 (1974).
51. R. Bases, F. Mendez, S. Neubort, D. Liebeskind, and K. C. Hsu. *Exp. Cell Res.* **103**, 175 (1976).
52. A. J. Garro, B. F. Erlanger, and S. M. Beiser, *in* "Nucleic Acids in Immunology" (O. J. Plescia and W. Braun, eds), p. 47. Springer-Verlag, Berlin and New York, 1968.
53. R. R. Schreck, D. Warburton, O. J. Miller, S. M. Beiser, and B. F. Erlanger, *PNAS* **70**, 804 (1973).
54. L. Levine, W. T. Murakami, H. Van Vunakis, and L. Grossman, *PNAS* **46**, 1038 (1960).
55. E. E. Townsend, H. Van Vunakis, and L. Levine, *Bchem* **4**, 943 (1965).
56. O. J. Plescia, W. Braun, and W. C. Palczuk, *PNAS* **52**, 279 (1964).
57. B. D. Stollar, *CRC Crit. Rev. Biochem.* **3**, 45 (1975).
58. A. D. Steinberg, T. Pincus, and N. Talal, *J. Immunol.* **102**, 788 (1969).
59. B. D. Stollar and M. Papalian, *J. Clin. Invest.* **66**, 210 (1980).
60. B. Ginsberg and H. Keiser, *Arthritis Rheum.* **16**, 199 (1973).
61. B. D. Stollar, L. Levine, and J. Marmur, *BBA* **61**, 7 (1962).
62. B. D. Stollar, *Science* **169**, 609 (1970).
63. D. Eilat, R. Asofsky, and R. Laskov, *J. Immunol.* **124**, 766 (1980).
64. C. Andrzejewski, B. D. Stollar, T. M. Lalor, and R. S. Schwartz, *J. Immunol.* **124**, 1499 (1980).
65. E. A. Lerner, M. R. Lerner, C. A. Janeway, and J. A. Steitz, *PNAS* **78**, 2737 (1981).
66. M. I. Johnston and B. D. Stollar, *Bchem* **17**, 1959 (1978).
67. M. Guigues and M. Leng, *EJB* **69**, 615 (1976).
68. E. Sage and M. Leng, *Bchem* **16**, 4283 (1977).
69. M. Leng, M. Guigues, and D. Genest, *Bchem* **17**, 3215 (1978).
70. C. Colby, B. D. Stollar, and M. I. Simon, *Nature NB* **229**, 172 (1971).
71. L. Levine, H. Van Vunakis, and R. C. Gallo, *Bchem* **10**, 2009 (1971).
72. L. Levine and H. Gjika, *ABB* **164**, 583 (1974).
73. B. Hacker, H. Van Vunakis, and L. Levine, *J. Immunol.* **108**, 1726 (1970).
74. M. Z. Humayun and T. M. Jacob, *BBA* **349**, 84 (1974).
75. D. S. Milstone, B. S. Vold, D. G. Glitz, and N. Shutt, *NARes* **5**, 3439 (1978).
76. D. L. Sawicki, B. F. Erlanger, and S. M. Beiser, *Science* **174**, 70 (1971).
77. H. J. Storl, H. Simon, and H. Barthelmes, *BBA* **564**, 23 (1979).

78. B. S. Vold, *NARes* **7**, 193 (1979).
79. O. J. Miller, W. Schnedl, J. Allen, and B. F. Erlanger, *Nature (London)*, **251**, 636 (1974).
80. B. W. Lubit, R. R. Schreck, O. J. Miller, and B. F. Erlanger, *Exp. Cell Res.* **89**, 426 (1974).
81. B. W. Lubit, T. D. Pham, O. J. Miller, and B. F. Erlanger, *Cell* **9**, 503 (1976).
82. H. Sano, H.-D. Royer, and R. Sager, *PNAS* **77**, 3581 (1980).
83. S. M. Beiser and B. F. Erlanger, *Cancer Res.* **26**, 2012 (1966).
84. H. Van Vunakis, E. Seaman, and L. Levine, in "Nucleic Acids in Immunology" (O. J. Plescia and W. Braun, eds.), p. 58. Springer-Verlag, Berlin and New York, 1968.
85. S. P. Wallace, B. F. Erlanger, and S. M. Beiser, *Bchem* **10**, 679 (1971).
86. R. M. D'Alisa and B. F. Erlanger, *Bchem* **13**, 3575 (1974).
87. R. M. D'Alisa and B. F. Erlanger, *J. Immunol.* **116**, 1629 (1976).
88. S. A. Khan, M. Z. Humayum, and T. M. Jacob, *NARes* **4**, 2997 (1977).
89. S. A. Khan and T. M. Jacob, *NARes* **4**, 3007 (1977).
90. H. J. Storl, H. Simon, and H. Barthelmes, *NARes* **5**, 4919 (1978).
91. J. S. Lee, J. R. Lewis, A. R. Morgem, T. R. Mosmann, and B. Singh, *NARes* **9**, 1707 (1981).
92. E. A. Kabat, *J. Immunol.* **97**, 1 (1966).
92a. W. E. Cohn, N. J. Leonard, and S. Y. Wang, *Photochem. Photobiol.* **19**, 89 (1974).
93. R. T. Wold, F. E. Young, E. M. Tan, and R. S. Farr, *Science* **161**, 806 (1968).
94. E. M. Tan, *Science* **161**, 1353 (1968).
95. P. G. Natali and E. M. Tan, *Radiat. Res.* **46**, 506 (1971).
96. E. M. Tan, R. B. Freeman, and R. B. Stoughton, *J. Invest. Dermatol.* **55**, 439 (1970).
97. M. Jarzabek-Chorzelska, Z. Zarebska, H. Wolska, and G. Rzesa, *Acta Dermatol. (Stockholm)* **56**, 15 (1976).
98. M. Fukuda, K. Nakanishi, T. Mukainaka, A. Shima, and S. Fujita, *Acta Histochem. Cytochem.* **9**, 180 (1976).
99. A. S. Saenko, T. P. Ilyina, V. K. Podgorodnichenko, and A. M. Poverenny, *Immunochemistry* **13**, 779 (1976).
100. A. Fink and G. Hotz, *Z. Naturforsch. C: Biosci.* **32C**, 544 (1977).
101. A. Wakizaka and E. Okuhara, *Photochem. Photobiol.* **30**, 573 (1979).
102. D. L. Mitchell and J. M. Clarkson, *BBA* **655**, 54 (1981).
103. P. T. Strickland, unpublished.
104. J. J. Cornelis, *NARes* **5**, 4273 (1978).
105. J. J. Cornelis, *BBA* **521**, 134 (1978).
106. P. T. Strickland, in "Application of Biological Markers to Carcinogen Testing" (H. A. Milman and S. Sell, eds.), p. 337. Plenum, New York, 1983.
107. D. L. Mitchell, R. S. Nairu, J. A. Alvillar, and J. M. Clarkson, *BBA* **697**, 270 (1982).
108. A. Wakizaka and E. Okuhara, *J. Biochem.* **86**, 1469 (1979).
109. P. T. Strickland and J. M. Boyle, *J. Immunol. Methods* **41**, 115 (1981).
110. A. Wakizaka and E. Okuhara, *Immunochemistry* **12**, 843 (1975).
110a. G. Eggset, G. Volden, and H. Krokan, *Carcinogenesis* **4**, 745 (1983).
111. E. Kabat, "Structural Concepts in Immunology and Immunochemistry," p. 127. Holt, New York, 1976.
112. H. Klocker, B. Auer, H. J. Burtscher, J. Hofmann, M. Hirsch-Kaufmann, and M. Schweiger, *MGG* **186**, 475 (1982).
113. H. Slor, in "DNA-Repair and Late Effects" (H. Altmann, E. Riklis, and H. Slor, eds.), p. 103. Nuclear Research Center, Negev, Israel, 1980.
114. C. Cremer, T. Cremer, M. Fukuda, and K. Nakanishi, *Hum. Genet.* **54**, 107 (1980).
115. J. J. Cornelis and M. Errera, in "DNA Repair: A Laboratory Manual of Research

Procedures (E. C. Freidberg and P. C. Hanawalt, eds), Vol. 1, p. 31. Dekker, New York, 1981.
116. R. Latarjet, B. Eckert, S. Apelgot, and N. Rebeyrotte, *J. Chim. Phys.* **58**, 1046 (1961).
116a. C. Ponnamperuma, R. M. Lemmon, and M. Calvin, *Radiat. Res.* **18**, 540 (1963).
116b. J. J. Conlay, *Nature (London)* **197**, 555 (1963).
116c. J. J. van Hemmon and J. F. Bleichrodt, *Radiat. Res.* **46**, 444 (1971).
116d. J. A. Raleigh, W. Kremers, and R. Whitehouse, *Radiat. Res.* **65**, 414 (1976).
116e. G. J. West, I. W. L. West, and J. F. Ward, *Radiat. Res.* **90**, 595 (1982).
116f. R. Latarjet, B. Ekert, and P. Demersemian. *Radiat. Res. Suppl.* **3**, 247 (1963).
116g. L. S. Myers, J. F. Ward W. T. Tsukamoto, D. E. Holmes, and J. R. Luca. *Science* **148**, 1234 (1965).
116h. J. Cadet and R. Teoule, *BBA* **238**, 8 (1971).
116i. J. F. Ward, *Adv. Radiat. Biol.* **5**, 181 (1975).
116j. P. A. Cerutti, in "Photochemistry and Photobiology of Nucleic Acids" (S. Y. Wang, ed.), Vol. 2, p. 375. Academic Press, New York, 1976.
117. G. J. West, I. W.-L. West, and J. F. Ward, *Int. J. Radiat. Biol.* **42**, 481 (1982).
118. I. P. Moskalenko, N. A. Mitriaeva, and V. N. Aleshina, *Radiobiologiya* **14**, 816, (1974).
119. G. Hotz, *Radiat. Environ. Biophys.* **12**, 41 (1975).
120. H. Waller, E. Friess, and J. Kiefer, *Radiat. Environ. Biophys.* **19**, 259 (1981).
121. L. A. Aarden, F. Lakmaker, and T. E. W. Feltkamp, *J. Immunol. Methods* **10**, 39 (1976).
122. P. G. Natali, *Eur. J. Immunol.* **5**, 53 (1975).
123. R. Bases, A. Leifer, F. Mendez, S. Neubort, D. Liebeskind, and K. C. Hsu, *Exp. Cell Res.* **101**, 244 (1976).
124. R. Bases, F. Mendez, K. C. Hsu, and D. Liebeskind, *Exp. Cell Res.* **92**, 505 (1975).
125. H. L. Lewis, D. R. Muhleman, and J. F. Ward, *Radiat. Res.* **67**, 635 (1976).
126. H. L. Lewis, D. R. Muhleman, and J. F. Ward, *Radiat. Res.* **75**, 305 (1978).
127. H. L. Lewis and J. F. Ward, in "DNA Repair Mechanisms" (P. C. Hanawalt, E. C. Friedberg, and C. F. Fox, eds.), p. 35. Academic Press, New York, 1978.
128. B. Bonavida, S. Fuchs, H. Inouye, and M. Sela, *BBA* **240**, 604 (1971).
129. B. Singer, *JNCI* **62**, 1329 (1979).
129a. B. Singer and M. Kröger, This Series **23**, 151 (1979).
130. P. D. Lawley and W. Warren, in "DNA Repair: A Laboratory Manual of Research Procedures" (E. C. Friedberg and P. C. Hanawalt, eds.), Vol. 1, p. 129. Dekker, New York, 1981.
131. D. L. Sawicki, S. M. Beiser, D. Srinivasan, and P. R. Srinivasan, *ABB* **176**, 457 (1976).
132.. T. W. Munns, H. F. Sims, and M. K. Liszewski, *JBC* **252**, 3102 (1977).
134. L. Rainen and B. D. Stollar, *NARes* **5**, 4877 (1978).
135. R. Dante and A. Niveleau, *EJB* **110**, 539 (1980).
136. P. D. Lawley and P. Brookes, *BJ* **89**, 127 (1963).
137. S. Hendler, E. Furer, and P. R. Srinivasan, *Bchem* **9**, 4141 (1970).
138. C. C. Harris, R. H. Yolken, H. Krokan, and I.-C. Hus, *PNAS* **76**, 5336 (1979).
139. A. Shalev, A. H. Greenberg, and P. McAlpine, *J. Immunol. Methods* **38**, 125 (1980).
140. F. Pochon and A. M. Michelson, *BBA* **149**, 99 (1967).
141. J. Igarashi and E. Okuhara, *Tohoku J. Exp. Med.* **120**, 25 (1976).
142. A. Loveless, *Nature (London)* **223**, 206 (1969).
143. A. E. Pegg, *Adv. Cancer Res.* **25**, 195 (1977).

144. P. J. Abbott and R. Saffhill, *BBA* **562**, 51 (1979).
145. P. J. Abbott and R. Saffhill, *NARes* **4**, 761 (1977).
146. B. Singer, H. Fraenkel-Conrat, and J. T. Kusmierek, *PNAS* **75**, 1722 (1978).
147. R. Saffhill and P. J. Abbott, *NARes* **5**, 1971 (1978).
148. R. F. Newbold, W. Warren, A. S. C. Metcalf, and J. Amos, *Nature (London)* **283**, 596 (1980).
149. M. Fox and J. Brennand, *Carcinogenesis* **1**, 795 (1980).
150. W. J. Suter, J. Brennand, S. McMillan, and M. Fox, *Mutat. Res.* **73**, 171 (1980).
151. R. Goth and M. F. Rajewsky, *Z. Krebsforsch. Klin. Onkol.* **82**, 37 (1974).
152. P. J. O'Connor, R. Saffhill, and G. P. Margison, in "Environmental Carcinogenesis, Occurrence, Risk, Evaluation and Mechanism" (P. Emmelot, ed.), p. 73, Elsevier, Amsterdam, 1979.
153. W. T. Briscoe, J. Spizizen, and E. M. Tan, *Bchem* **17**, 1896 (1978).
154. S. A. Kyrtopoulos and P. F. Swann, *J. Cancer Res. Clin. Oncol.* **98**, 127 (1980).
154a. C. P. Wild, G. Smart, P. Saffhill, and J. M. Boyle, *Carcinogenesis* **4**, 1605 (1983).
155. C. P. Wild, R. Saffhill, and J. M. Boyle, *Br. J. Cancer* **48**, 130 (1983).
156. S. A. Kyrtopoulos and B. Vrotsou, *Chem.-Biol. Interact.* **37**, 191 (1981).
157. C. P. Wild, R. Saffhill, and J. M. Boyle, unpublished.
158. R. Saffhill, P. T. Strickland, and J. M. Boyle, *Carcinogenesis* **3**, 547 (1982).
159. E. Kriek, *Cancer Res.* **32**, 2042 (1972).
160. R. P. P. Fuchs, *Anal. Biochem.* **91**, 663 (1978).
161. H. Yamasaki, P. Pulkrabek, D. Grunberger, and I. B. Weinstein. *Cancer Res.* **37**, 3756 (1977).
162. E. Kriek, *Chem.-Biol. Interact.* **1**, 3 (1969).
163. E. Kriek, in "Environmental Carcinogenesis" (P. Emmelot and E. Kriek, eds.), p. 143. Elsevier/North-Holland, Amsterdam, 1979.
164. E. Kriek, in "Carcinogenesis: Fundamental Mechanisms and Environmental Effects" (B. Pullman, P. O. P. Ts'o, and H. Gelboin, eds.), p. 103. Reidel, Dordrecht, 1980.
165. E. Kriek and J. C. Westra, *Carcinogenesis* **1**, 459 (1980).
166. M. Guigues and M. Leng, *NARes* **6**, 733 (1979).
167. P. Rio, S. Bazgar, and M. Leng, *Carcinogenesis* **3**, 225 (1982).
168. M. C. Poirier, S. H. Yuspa, I. B. Weinstein, and S. Blobstein, *Nature (London)* **270**, 186 (1977).
169. J. H. Nelson, D. Grunberger, R. C. Cantor, and I. B. Weinstein, *JMB* **62**, 331 (1971).
170. R. P. P. Fuchs, and M. P. Duane, *FEBS Lett.* **92**, 207 (1971).
171. R. P. P. Fuchs, and M. P. Duane, *Bchem* **11**, 2659 (1972).
172. M. Spodheim-Maurizot, G. Saint-Ruf, and M. Leng, *NARes* **6**, 1638 (1977).
173. E. Sage, R. P. Fuchs, and M. Leng, *Bchem* **18**, 1328 (1979).
174. M. Spodheim-Maurizot, G. Saint-Ruf, and M. Leng, *Carcinogenesis* **1**, 807 (1980).
175. E. Sage, M. Spodheim-Maurizot, P. Rio, M. Leng, and R. P. P. Fuchs, *FEBS Lett.* **108**, 66 (1979).
176. M. C. Poirier, M. A. Dubin, and S. H. Yuspa, *Cancer Res.* **39**, 1377 (1979).
177. M. C. Poirier, G. M. Williams, and S. H. Yuspa, *Mol. Pharmacol.* **18**, 581 (1980).
178. M. C. Poirier, B. True, and B. A. Laishes, *Cancer Res.* **42**, 1317 (1982).
179. E. Sage, N. Gabelman, F. Mendez, and R. Bases, *Cancer Lett.* **14**, 193 (1981).
180. G. DeMurcia, M. C. E. Lang, A. M. Freund, R. P. P. Fuchs, M. P. Duane, E. Sage, and M. Leng, *PNAS* **76**, 6076 (1979).
181. G. DeMurcia, M. C. Lang, A. Mazen, R. P. P. Fuchs, M. Duane, E. Sage, and

M. Leng, in "Electron Microscopy 1980" (P. Brederoo and W. dePriester, eds.), Vol. 2, p. 552 (Proc. 7th Eur. Congr. Electron Microsc.). North Holland, Amsterdam, 1980.
182. P. Rio and M. Leng, Biochimie **62**, 487 (1980).
183. H. V. Gelboin and P. O. P. Ts'o (eds.), "Polycyclic Hydrocarbons and Cancer," Vols. 1 and 2. Academic Press, New York, 1978.
184. K. Bridbord and J. G. French, in "Carcinogenesis" (P. W. Jones and R. I. Freudenthal, eds.), Vol. 3, p. 451. Raven, New York, 1978.
184a. D. Grunberger and I. B. Weinstein, This Series **23**, 105 (1979).
185. E. Huberman, L. Sachs, S. K. Yang, and H. V. Gelboin, PNAS **73**, 607 (1976).
186. T. J. Slaga, W. J. Bracken, G. Gleason, W. Levin, H. Yagi, D. M. Jerina, and A. H. Conney, Cancer Res. **39**, 67 (1979).
187. F. P. Perera, M. C. Poirier, S. H. Yuspa, J. Nakayama, A. Jaretski, M. M. Curnen, D. M. Knowles, and I. B. Weinstein, Carcinogenesis **3**, 1405 (1982).
188. R. S. Paules, M. C. Poirier, M. J. Mass, S. H. Yuspa, and D. G. Kaufman, Carcinogenesis (in press).
189. J. D. Groopman, A. Haugen, G. R. Goodrich, G. N. Wogan, and C. C. Harris, Cancer Res. **42**, 3120 (1982).
190. J. J. Roberts and A. J. Thomson, This Series **22**, 71 (1979).
191. A. W. Prestayko, S. T. Crooke, and S. K. Carter (eds.), "Cisplatin: Current Status and New Developments." Academic Press, New York, 1980.
192. J. M. Pascoe and J. J. Roberts, Biochem. Pharmacol. **23**, 1345 (1974).
193. L. A. Zwelling, T. Anderson, and K. W. Kohn, Cancer Res. **39**, 365 (1979).
194. L. A. Zwelling, M. O. Bradley, N. A. Sharkey, T. Anderson, and K. Kohn, Mutat. Res. **67**, 271 (1979).
195. P. J. Stone, A. D. Kelman, and F. M. Sinex, JMB **104**, 793 (1976).
196. A. D. Kelman, H. J. Peresie, and P. J. Stone, J. Clin. Hematol. Oncol. **7**, 440 (1977).
197. J. Brouwer, P. van de Putte, A. M. J. Fichtinger-Schepman, and J. Reedijk, PNAS **78**, 7010 (1981).
198. J. P. Macquet and J. L. Butour, Biochimie **60**, 901 (1978).
199. B. Malfoy, B. Hartmann, J. P. Macquet, and M. Leng, Cancer Res. **41**, 4127 (1981).
200. M. C. Poirier, S. J. Lippard, L. A. Zwelling, H. M. Ushay, D. Kerrigan, C. C. Thill, R. M. Santella, D. Grunberger, and S. H. Yuspa, PNAS **79**, 6443 (1982).
201. A. H.-J. Wang, G. C. Quigley, F. J. Kolpak, J. L. Crawford, J. H. van Boom, G. van der Marel, and A. Rich, Nature (London) **282**, 680 (1979).
202. F. M. Pohl and T. M. Jovin, JMB **67**, 375 (1972).
203. D. J. Patel, L. L. Canuel, and F. M. Pohl, PNAS **76**, 2508 (1979).
204. J. Ramstein and M. Leng, Nature (London) **288**, 413 (1980).
205. E. M. Lafer, A. Moller, A. Nordheim, B. D. Stollar, and A. Rich, PNAS **78**, 3546 (1981).
206. B. Malfoy, N. Rousseau, and M. Leng, Bchem **21**, 5463 (1982).
207. B. Malfoy and M. Leng, FEBS Lett. **132**, 45 (1981).
208. F. M. Pohl, R. Thomae, and E. DiPapua, Nature (London) **300**, 545 (1982).
208a. V. Stettler, H. Weber, T. Koller, and C. Weissman, JMB **131**, 21 (1979).
209. R. M. Santella, D. Grunberger, S. Broyde, and B. E. Hingerty, NARes **9**, 5459 (1981).
210. E. Sage and M. Leng, PNAS **76**, 6076 (1980).
211. A. Nordheim, M. L. Pardue, E. M. Lafer, A. Moller, B. D. Stollar, and A. Rich, Nature (London) **294**, 417 (1981).

212. E. Viegas-Pequignot, C. Derbin, F. Lemeunier, and E. Tallandier. *Ann. Genet.* **25**, 218 (1982).
213. H. J. Lipps, A. Nordhiem, E. M. Lafer, D. Ammermann, B. D. Stollar, and A. Rich, *Cell* **32**, 435 (1983).
213a. G. Morganegg, M. R. Celio, B. Malfoy, M. Leng, and C. C. Kuenzle, *Nature (London)* **303**, 540 (1983).
214. S. Neidle, *Nature (London)* **302**, 574 (1983).
215. A. Moller, A. Nordheim, S. R. Nichols, and A. Rich, *PNAS* **78**, 4777 (1981).
216. M. Behe and G. Felsenfeld, *PNAS* **78**, 1619 (1981).
217. M. Behe, S. Zimmerman, and G. Felsenfeld, *Nature (London)* **293**, 233 (1981).
218. C. M. Mercado and M. Tomasz, *Bchem* **16**, 2040 (1977).
219. M. R. Lerner and J. A. Steitz, *PNAS* **76**, 5495 (1979).
220. P. Davis, A. S. Russel, and J. S. Percy, *J. Rheumatol.* **3**, 375 (1976).
221. P. Davis, *Br. J. Dermatol.* **97**, 197 (1977).
222. D. J. Beighlie and R. L. Teplitz, *J. Rheumatol.* **2**, 149 (1975).
223. G. Harris, L. J. Asberry, P. D. Lawley, A. M. Denman, and W. Hylton, *Lancet* **2** (8305), 952 (1982).
224. F. Tron, D. Charron, J. F. Bach, and N. Talal, *J. Immunol.* **125**, 2805 (1980).
225. B. Hahn, F. Ebling, S. Freeman, B. Clevinger, and J. Davie, *Arthritis Rheum.* **23**, 942 (1980).
226. J. S. Lee, J. R. Lewis, A. R. Morgem, T. R. Mosmann, and B. Single, *NARes* **9**, 1707 (1981).
227. J. S. Lee, D. F. Dombroski, and T. R. Mosmann, *Bchem* **21**, 4740 (1982).
228. D. S. Pisetsky and S. A. Caster, *Mol. Immunol.* **19**, 645 (1982).
228a. M. Seidman, H. Mizusawa, H. Slor, and M. Bustin, *Cancer Res.* **43**, 743 (1983).
229. P. Sizaret, C. Malaveille, R. Montesano, and C. Frayssinet, *JNCI* **69**, 1375 (1982).
230. J. J. Langone and H. Van Vunakis, *JNCI* **56**, 591 (1976).
231. S. F. Cernosek and R. M. Gutierrez-Cernosek, *FP* **34**, 513 (1975).
232. R. M. Gutierrez-Cernosek and S. F. Cernosek, *Ann. Clin. & Lab. Sci.* **7**, 35 (1977).
233. F. S. Chu and I. Ueno, *Appl. Environ. Microbiol.* **33**, 1125 (1977).
234. A. R. Brothman, T. P. Davis, J. J. Duffy, and T. J. Lindell, *Cancer Res.* **42**, 1184 (1982).
235. F. L. Moolten, N. J. Capparell, E. Boger, and P. Mahathalang, *Nature (London)* **272**, 614 (1978).
236. I. Ueno and F. S. Chu, *Experientia* **34**, 85 (1978).
237. A. Tompa, G. Curtis, W. Ryan, C. Kuszynski, and R. Langenback, *Cancer Lett.* **7**, 163 (1979).
238. F. L. Moolten, B. Schrieber, A. Rizzone, A. J. Weiss, and E. Boger, *Cancer Res.* **41**, 425 (1981).
239. R. M. Peck and E. B. Peck, *Cancer Res.* **31**, 1550 (1971).
240. G. L. Curtis and W. L. Ryan, *FP* **38**, 912 (1979).
241. G. L. Curtis, W. L. Ryan, and F. Stenback, *Cancer Lett.* **4**, 223 (1978).
242. H. J. Creech and W. R. Franks, *Am. J. Cancer* **30**, 555 (1937).
243. H. J. Creech, E. L. Oginsky, and M. Tryon, *Cancer Res.* **7**, 301 (1947).
244. F. Moolten, N. Capparell, and E. Boger, *JNCI* **61**, 1347 (1978).
245. R. Balick and J. Pataki, *BBRC* **82**, 81 (1978).
246. K. Fujiwara, H., Saikusa, M. Yasuna, and T. Kitagawa, *Cancer Res.* **42**, 1487 (1982).

On the Biological Significance of Modified Nucleosides in tRNA

HELGA KERSTEN

Institute of Physiological
  Chemistry
University of Erlangen-Nürnberg
Nürnberg, Federal Republic of
  Germany

I. Evolutionary Aspects	62
II. Role of Ribosylthymine in tRNA Function of Eubacteria	65
A. Initiation of Protein Synthesis	66
B. Ribosomal A-Site Interaction	68
C. Elongation of Peptide Chains	69
D. Fidelity of Translation	70
III. Involvement of Ribosylthymine in Regulatory Mechanisms	71
A. Stringent Response	72
B. Tetrahydrofolate-Independent Initiation	73
C. Translation of Polycistronic mRNA	74
D. Does Undermodified, Unformylated tRNA^{fMet} Function in Regulation?	75
IV. Ribosylthymine in Eukaryotic Elongator tRNA	79
A. Variations of the TψC Sequence	79
B. Influence of Translation *in Vitro*	79
C. Utilization of tRNAs with Variable Thymine Content	80
D. Expression of tRNA Genes and Biosynthesis of Thymine	82
V. N^6-(2-Isopentenyl)adenosine and Derivatives in tRNA	83
A. Evolutionary Aspects	83
B. Codon/Anticodon Recognition	83
C. Transcription Termination Control	84
VI. Central Role of N^6-(2-Isopentenyl)adenosine and Derivatives in Growth and Development	86
A. Sporulation in *Bacillus subtilis*	86
B. Relation to Iron and Amino Acid Metabolism	87
C. The Link between A-37 Modification and Major Routes of Metabolism	88
D. Cytokinin	93
VII. On the Function of Queuine in tRNA	95
A. Queuine in *Dictyostelium discoideum*	96
B. A Possible Role of Queuine and Queuine-Containing tRNAs in Differentiation and Neoplastic Transformation	98
C. Queuine and the Melanophore System of Xiphophorine Fish	101
VIII. Summary and Outlook	103
References	107

During the first two decades of research on modified nucleosides in tRNA, the most important achievements were (a) the discovery of more than 50 modified nucleosides; (b) the elucidation of their structures and positions in specific tRNAs; and (c) the clarification of their biosynthetic pathways. That modified nucleosides play an important role in protein synthesis and cellular regulation is indicated in many reviews (1–2). The principles of the function of modified nucleosides are discussed in this article, although the molecular bases of their functions remain to be solved in the third decade of work in this field.

In 1963 Borek and co-workers showed that the methyl groups in tRNA originate from methionine (13): the tRNA that accumulates during methionine starvation in the *Escherichia coli* mutant K12 58-161 rel^- is devoid of methylated bases. To elucidate the role of methylated nucleosides, the "undermethylated" tRNA was isolated and tested for amino-acid acceptance, recognition of mRNA codons, and association with ribosomes. In all the *in vitro* systems formerly available for the study of each of these functions, methylation of tRNA appears not to be significantly involved. In 1966, Borek and Srinivasen wrote (14): "It must be recalled, before a function of methylation of tRNA is completely ruled out, that the methods used for assaying the three functions are rather crude *in vitro* reconstructions of complex biological systems. Therefore, subtle differences between the methylated and non-methylated species of tRNA may be masked."

The components of the protein synthesizing system have been further purified, and improved cell-free systems of protein synthesis have been developed. These advances have made it possible to study the interactions of tRNAs with each of these components and to investigate the influence of a single modification in a specific tRNA on specific steps of protein synthesis. Moreover, the extraordinary progress in research on the structural organization of genes makes it possible to get some insight into the role of modified nucleosides in mechanisms involved in the control of gene expression.

The aim of this article is to discuss the role of modified nucleosides in tRNA function with respect to two alternative questions: Did the tRNA modifications evolve primarily to improve the precision of the translational apparatus? Did the tRNA modifications evolve primarily as regulatory mechanisms as, for example, in the expression amino-acid biosynthetic operons (15), with the protein synthesizing systems that developed subsequently increasing the flexibility of organisms to adapt to metabolic and environmental changes? I attempt to link early discoveries, observations, and speculations on the role of modified nucleosides with new data and progress in related fields.

The general outlines of the picture are emerging, but many details still remain to be filled in. I hope that this will encourage further research on this subject.

The discussion of this article is concerned mainly with three tRNA modifications: ribosylthymine (T) at position 54; N^6-(2-isopentenyl) adenosine (i^6A) and its derivatives at position 37; queuosine (Q) and its derivatives at position 34 (see Fig. 1 and ref. 12). These modified nucleosides in tRNAs provide cells with a survival advantage under

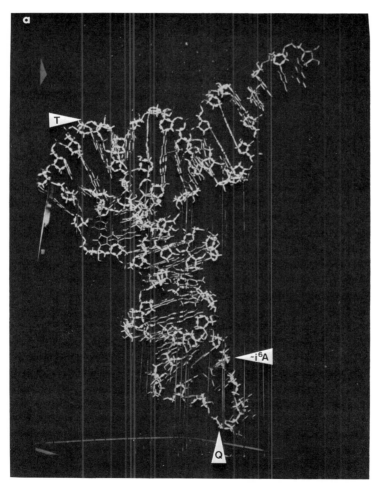

FIG. 1. (a) Three-dimensional structure of tRNA showing position and structures of T, i^6A, ms^2i^6A, Q, and Q derivatives. (b) Structures of T, i^6A, Q, and ms^2i^6A.

b

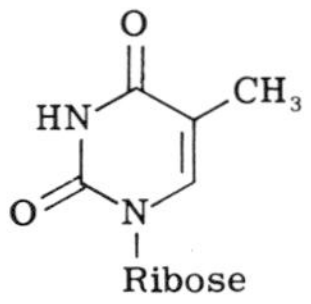

Ribothymidine (T)

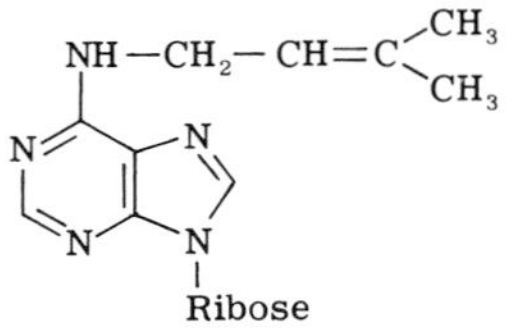

N^6-Isopentenyl-adenosine (i⁶A)

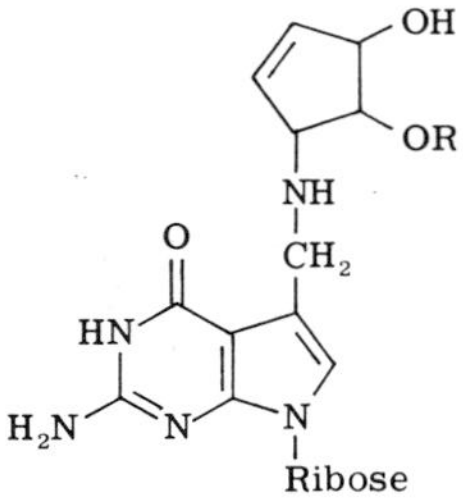

Queuosine structure:
R = H: Q (queuosine)
R = β-D-mannosyl: manQ
R = β-D-galactosyl: galQ

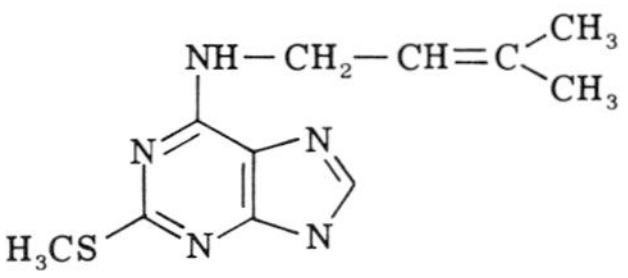

2-Methylthio-N^6-isopentenyladenosine (ms²i⁶A)

FIG. 1. (*Continued*)

unfavorable environmental conditions, thus indicating that these tRNA modifications probably played an important role in the evolution of organisms.

I. Evolutionary Aspects

In the kingdom of living organisms, translational machineries have evolved that differ in primary structure and in modification of the single components. The translational apparatus, including ribosomal RNAs, ribosomal proteins, initiation and elongation factors, tRNAs and their modifications, are structurally related, but exhibit species-specific variations and increasing complexity, as has been shown for mycoplasma, gram-positive and gram-negative eubacteria, archaebacteria, lower and higher eukaryotes and for the components of the protein-synthesizing system of cellular organelles. Therefore, it has been suggested that the translational apparatus has evolved not with the establishment of the mechanism per se, but with the refinement of the mechanism through stages of greater and greater precision.

Eubacterial species were originally classified as gram-positive or gram-negative, according to whether or not cells are decolorized by alcohol after staining with a basic dye and iodine. Interspecies differences in the retention of stain are now considered to reflect differences in the molecular architecture of the cell wall. Gram-positive and gram-negative microorganisms can now be discriminated in that their ribosomes, mRNAs, and tRNAs have different structural and/or functional properties (16, 17). Also, the modification patterns of tRNA show characteristic differences (18) (Fig. 2). Several modifications are common to gram-positive and gram-negative bacteria. However, some are taxon-specific; for example, m^1A at position 22, mo^5U and $cmcm^5U$ at position 34 are specific for gram-positive microorganisms, whereas m^1A at 22 is absent in tRNAs from gram-negative bacteria, and the modifications cmo^5U and mcm^5U occur at position 34 in tRNAs of gram-negative bacteria. The higher an organism stands in the evolutionary scale, the greater the amount of modified nucleosides in its tRNA (Fig. 2).

All tRNAs participating in ribosome-dependent peptide elongation contain a modified sequence in the ribosomal binding region, which is invariant in gram-positive and in gram-negative microorganisms and contain, at positions, 54, 55, and 56, the sequence TψC. In *Thermus thermophilus*, the initiator tRNA and most elongator tRNAs have s^2T at position 54 (12). The thiolation of T appears to be necessary for the structural integrity of tRNAs at the high temperatures at which the organism grows. Eukaryotic tRNAs have been divided into different classes with respect to this sequence (19), one major class of elongator tRNAs with TψC, and minor classes with TmψC, UψC or $\psi\psi$C, $N^+\psi$C or N^+UC (N^+ indicates so-far-unidentified uridine derivative; see 6). The initiator tRNA of eubacteria contains TψC, whereas the initiator tRNA from eukaryotes contains AUC (or AψC, as in wheat germ) in place of the invariant TψC sequence (9). (tRNAs from organelles are not included here.)

These observations show that different modifying enzymes that modify positions 54 and 55 in tRNA have evolved. A unique biosynthetic pathway of T exists in gram-positive bacteria. Usually the methyl groups present in eubacterial tRNAs are derived from methionine and are introduced into tRNA precursor molecules by specific S-adenosylmethionine-dependent tRNA methyltransferases (20). In gram-positive bacteria, however, the methyl group of T is derived from serine (21) and is transferred to tRNA during maturation by a tetrahydrofolate-dependent enzyme (22). This pathway in the biosynthesis of ribosylthymine (T) in tRNA was discovered in *Strep-*

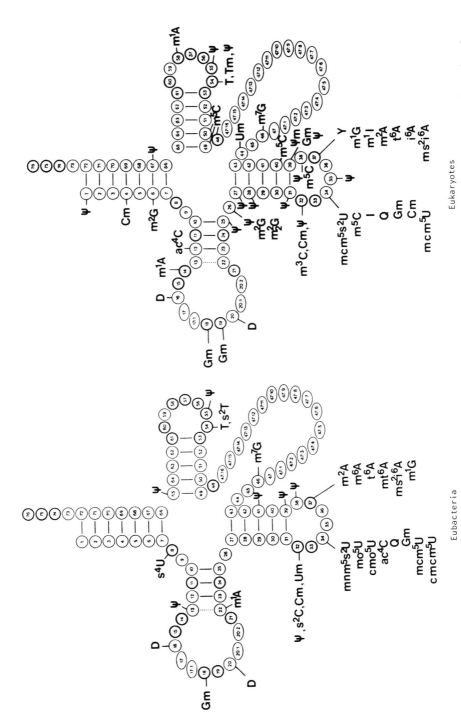

FIG. 2. "Cloverleaf" structure of tRNA and positions of modified nucleosides. Positions and abbreviations as in Sprinzl and Gauss (12).

tococcus faecalis (23), in *Streptococcus faecium, Staphylococcus aureus, Corynebacterium bovis, Arthrobacter albidus*, and in Bacillaceae, but not in *Bacillus stearothermophilus* (24). Trimethoprim, a specific inhibitor of the dihydrofolate reductase, inhibits specifically the tetrahydrofolate-dependent formation of T in tRNA (24). As has been shown for *Streptococcus faecalis*, the methyl carbon of T in tRNA is derived from 5,10-methylenetetrahydrofolate (25).

In over half of the archaebacteria examined, a modified nucleoside (N*) occurs at position 54 that is unique to archaebacterial tRNA, but has not been found in other tRNAs (26). The structural characterization of N* as 1-methyl-5-(β-D-ribofuranosyl)uracil (1-methylpseudouridine), $m^1\psi$, has recently been reported (27); the modified nucleoside was isolated from tRNA of *Halococcus morrhuae*. In all cases in which its location has been determined (i.e., in 15 sequenced tRNAs of *H. volcanii*, in initiator tRNA of *H. morrhuae*, and in a number of tRNAs from various methanogens), $m^1\psi$ always replaces T (see 8 and 8a). The occurrence of $m^1\psi$ in archaebacterial tRNA appears to be confined to some of the methanogens and their specific relatives, the extreme halophiles (28). The nucleoside is not found in tRNAs of either *Sulfolobus acidocaldarius* or of *Thermoplasma acidophilium*, in which position 54 is occupied by ψ or Um (2'-O-methyluridine), respectively (26).

Because the methyl group in $m^1\psi$ occupies the usual position of ribose, the biosynthesis of $m^1\psi$ must first involve rotation of uracil to form ψ, prior to methylation at N-1 (29, 30). The tRNA (pseudouridine-1-)-methyltransferase is thought to be unique to archaebacteria. The methylation step occurs at a site on the uracil ring (N-1), which bears a similar orientation with respect to ribose and the polynucleotide chain, as does the C-5 of uracil in T. The methylated products in each case have similar shapes, which suggests an evolutionary convergence of structures at position 54 in tRNAs that might be closely related to the evolution of the ribosome.

II. Role of Ribosylthymine in tRNA Function of Eubacteria

All tRNAs of eubacteria contain ribosylthymine (T) at position 54. Since T is replaced by U in tRNAs of mycoplasma, in some tRNA species of plants and of mammals, and in tRNAs of eukaryotic cell organelles (1–12), it has been concluded that the methylation of U-54 in those tRNAs participating in the ribosome-dependent peptide elongation, is not essential for protein synthesis.

Matched pairs of transductant strains of *E. coli*, differing by the presence or the absence of T in their tRNAs, have been constructed, and they provide a specific tool for investigating the function of T (*31*). Ordinary measurements of growth rates in different media reveal no effect of the loss of T in tRNA. With the conventional methods used to measure codon recognition, the binding of tRNA to ribosomes, the rate of protein chain elongation, or the macromolecular composition *in vivo*, no disadvantage of the lack of T has been observed. However, in a mixed population, cells possessing T in their tRNAs have a survival advantage over cells lacking the modified nucleoside. This result was the first indication of a biological significance of T (*32*). The survival advantage was thought to be caused by subtle changes in the function of tRNA that are probably not detectable in cell-free systems of protein synthesis.

Early in the history of tRNA research, the TψCG sequence of tRNA was proposed to be the ribosomal binding region (*33*). This idea was further supported by the observed inhibitory effect of added TψCG on the binding of aminoacyl-tRNA to the ribosome (*34, 35*). However, the interaction of the TψCG sequence in tRNA with the ribosome and its involvement in tRNA function during protein synthesis has been a subject of numerous investigations and controversial discussion (for a recent review, see *36*).

A. Initiation of Protein Synthesis

It is generally accepted that initiator Met-tRNA^{fMet} of eubacteria initiates protein synthesis after formylation to fMet-tRNA^{fMet}. The formylation, however, does not seem to be indispensable for the initiation of protein synthesis, as has been shown in *Halobacterium cutirubrum* (*37*). Also, in folate-dependent *Streptococcus faecalis* upon starvation for folate (*38, 39*) and in *B. subtilis* upon inhibition of the formylation reaction by trimethoprim, protein synthesis can be initiated with unformylated tRNA, and peptide bonds can be formed provided that the growth medium is supplemented with all other folate-dependent metabolites (*17, 40*).

In gram-positive eubacteria, the methylation of U-54 depends on a tetrahydrofolate-dependent enzyme. The ability of folate-deficient *Streptococcus faecalis* to initiate protein synthesis with nonformylated tRNA is assumed to be caused solely by the absence of T in initiator tRNA (*38, 39*).

The influence of T on the structure of initiator tRNA, and its function in the formylated and unformylated forms, have been investigated. For this purpose the initiator tRNAs^{fMet} were purified from the *E. coli*

TrmA mutant (IB-5) lacking the T-forming enzyme, or from *B. subtilis* upon inhibition of the tetrahydrofolate-dependent formation of T with trimethoprim.

Sensitive indicators of structural differences that affect the stability of the tertiary structure of tRNAs are the methyl and methylene resonances from the minor nucleosides T, s^2T, h_2U (D), and m^7G in *E. coli* tRNA^{fMet}, as they experience large chemical shifts when the tRNA unfolds. A comparison of NMR melting profiles of the minor nucleosides of tRNA^{fMet} (T-54) and tRNA^{fMet} (U-54) reveals that the tRNA structure is less stable in methyl-deficient tRNA^{fMet} (U-54). The melting temperature, measured by the 270-MHz high-field proton NMR spectra of the tRNA (U-54) is 6°C lower than that of the wild-type tRNA, which has T at position 54 *(41)*. The results demonstrate that the modification of U-54 to T-54 stabilizes the tertiary structure of tRNA^{fMet} (Fig. 3). This structural change does not seem to affect the aminoacylation of the initiator tRNA, since *in vitro* the aminoacylation kinetics for both tRNA^{fMet} (U-54 or T-54) are identical.

The influence of ribosylthymine on the activities of the formylated and the nonformylated initiator tRNAs on the initiation of protein synthesis has been investigated in the *in vitro* protein-synthesizing systems of *E. coli* ribosomes, purified initiation factors IF1, IF2, and IF3, and synthetic AUG. The binding of the tRNAs to 70-S ribosomes

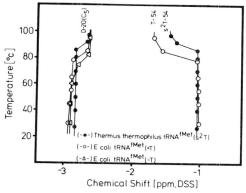

FIG. 3. Thermal melting profiles of three initiator tRNAs with 2-thiothymine (s^2T), thymine (+T) and uracil (−T) at position 54. Proton NMR spectra were measured at various temperatures on a Bruker WH270 MHz spectrometer. The temperature was measured before and after each run, using the chemical shift difference between the methylene and hydroxyl protons of ethylene glycol. The field was locked by D_2O in the solvent. Chemical shifts are reported with respect to 2,2-dimethyl-2-silapentane-5-sulfonate (DSS). From *(41)*.

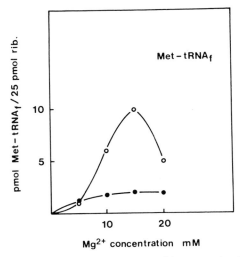

FIG. 4. Initiation with unformylated Met-tRNA^fMet containing U-54 (O——O) or T-54 (●——●). The initiation system from *Escherichie coli* contained 70-S ribosomes (rib.), initiation factors IF1, IF2, and IF3, and synthetic AUG. The binding of the initiator tRNAs to the ribosome was measured, and specific peptidyl-site interactions were calculated from subsequent puromycin release. From (42).

was measured, and specific P-site interaction was calculated from subsequent puromycin release (42).

The activities were highest with the formylated tRNAs and independent of the presence or the absence of the methyl group at U-54. The nonformylated T-containing tRNA poorly supported initiation at low and high concentrations of Mg^{2+}. However, the nonformylated T-lacking tRNA was almost as efficient as its formylated counterpart at Mg^{2+} concentrations between 10 and 50 mM (Fig. 4). These results agree well with those reported for T-lacking initiator tRNA from *Streptococcus faecalis* and support the view that the absence of T in initiator tRNA might compensate for the lack of formylation (43, 44). However, conflicting results have been reported (40).

The relationship between formylation of initiator tRNA and the role of ribosylthymine in initiator and elongator tRNA is presented in detail in Section III.

B. Ribosomal A-Site Interaction

The classical model describing the function of the ribosome includes two tRNA binding sites, the P site as the binding site for peptidyl-tRNA and the A site for aminoacyl-tRNA. This classical A/B-site model offers a satisfactory description of most of the functional experiments. Aminoacyl-tRNA can bind to the A and P sites; it reaches the P site via transient A-site binding; i.e., the A site is the entry site for the aminoacyl-tRNA. Up to two aminoacyl-tRNAs can be bound per ribosome. Peptidyl-tRNA can be bound in one copy per ribosome to either the A or the P site. Further details on the extended three-site functional model of the ribosome and the different steps in protein synthesis have been described (45, 46).

tRNA participates in five reactions during the elongation cycle: (a) formation of the ternary complex (EF-Tu)·GTP·(AA-tRNA); (b) tRNA binding to the A site; (c) the translocation reaction; (d) tRNA binding to the P site; and (e) the peptidyltransferase reaction.

The presence or the absence of T does not influence the formation of the ternary complex, as has been established for Phe-tRNA^{Phe} and Lys-tRNA^{Lys}. The remaining reactions (b–e) were tested under conditions that permit a highly specific binding of AcLys-tRNA^{Lys} to either the A site or the P site on polyadenylate-programmed ribosomes (47, 48). The results showed that the extent of binding of AcLys-tRNA^{Lys} to the P site is not influenced by the absence of T. In contrast, binding to the A site was significantly reduced in the absence of the modified nucleoside (49). The absence of T in AcLys-tRNA^{Lys} enhances the translocation reaction, but has no effect on the peptidyltransferase reaction.

C. Elongation of Peptide Chains

Purified Lys-tRNA^{Lys} and Phe-tRNA^{Phe} containing either U-54 or T-54 were tested in the conventional poly(A)-dependent poly(Lys) and the poly(U)-dependent poly(Phe) *E. coli* cell-free systems; respectively. Lys-tRNA^{Lys} or Phe-tRNA^{Phe} and purified elongation factors EF-Tu and EF-G were used in these experiments. Surprisingly the lack of T enhances poly(A)-dependent poly(Lys) formation, the increase also being evident in the initial rate of poly(Lys) synthesis. However, no significant differences could be found in the conventional poly(U) system when Phe-tRNA^{Phe} either containing or lacking T was used (49). Probably the conventional poly(U) system is not sensitive enough to reveal the subtle effects induced by T. The low sensitivity of this system is evident from the observation that the Phe-tRNA fragment CCA-Phe binds more strongly to the A-site moiety of the peptidyltransferase center than do the corresponding fragments from other tRNAs (50).

A breakthrough was achieved by the development of an optimized poly(U)/poly(Phe) translation system from *E. coli* in which a partial

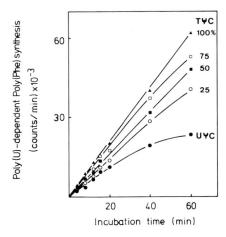

FIG. 5. Kinetics of poly(U)-dependent synthesis of poly(Phe) with tRNAs^{Phe} containing variable amounts of T or U, respectively. Homopeptide synthesis was measured in the polyamine-dependent system (51). tRNAs lacking T were obtained from E. coli TrmA$^-$ and mixed with tRNA from E. coli MRE 600 to give a final composition as indicated at the corresponding curves.

replacement of magnesium, potassium, and ammonium ions by calcium ions and polyamines produces conditions more likely to occur *in vivo* (51). This optimized system was applied, and the kinetics of poly(Phe) formation on poly(U)-programmed ribosomes were studied with mixtures of tRNAs from E. coli TrmA$^-$ and TrmA$^+$ strains with defined ratios of T to U at position 54. Under the conditions described (51), the rate of homopeptide elongation is successively reduced at decreasing ratios of T-54 to U-54 (Fig. 5) (52, 53). The same effect of T on the rate of peptide elongation was observed in an eukaryotic *in vitro* system with tRNA^{Phe} containing variable ratios of U to T. Like tRNA^{Lys} in prokaryotes, tRNA^{Val} of cukaryotes exhibits a higher rate of peptide elongation when it lacks T. But even though some elongator tRNAs seem to be better off without T, the overall rate of protein synthesis will depend on those tRNAs with the lowered V_{max}, when the rate of elongation becomes the rate-limiting step in translation (see Table I in Section III and summarizing references in Section IV).

D. Fidelity of Translation

In the optimized poly(U)-dependent poly(Phe) E. coli system, the relative frequency of misincorporation of leucine in place of phenylalanine is approximately 1×10^{-4} (51). With various preparations of ribosomes and T-54 tRNAS from different E. coli TrmA$^+$ strains, the error frequencies were in the range of 2 to 4×10^{-4}. With a number of preparations of tRNA (U-54) from an isogenic TrmA$^-$ mutant of E. coli, the error frequency increased significantly to 25 to 35×10^{-4} (10, 52, 53).

In order to determine the isoaccepting tRNA^{Leu} species affected by a deficiency of T, the tRNA^{Leu} isoacceptors were separated. tRNA$^{Leu(1-3)}$ and tRNA$^{Leu(5)}$ are incapable of misreading UUU irrespective of the presence or the absence of T. Only tRNA$^{Leu(4)}$ (T) anticodon NAA shows slight codon ambiguity for UUU with a frequency of about 8×10^{-4}. The relative frequency of misincorporation of Leu increases further, to 15 to 20×10^4 when tRNA$^{Leu(4)}$ contains U at position 54 in place of T. The absence of T in both tRNA$^{Leu(4)}$ and in tRNA^{Phe} is associated with the most pronounced loss of accuracy of homopeptide synthesis (54). The high missence error frequency (6×10^{-4}) of tRNA$^{Leu(4)}$ has been traced back to inefficient proofreading of this species (55, 55a).

Methods that permit the study of mistranslation intracellularly in E. coli strain relA$^-$, His$^-$ have been developed. In this strain, starvation for histidine leads to incorporation of glutamine into the newly synthesized proteins in place of histidine (CAU is misread as CAA or CAG; CAC is misread as CAA). This misreading causes stepwise changes ("stuttering") in the isoelectric point of several proteins to more acidic pH values, and thus is observable in the electrophoretic patterns of two-dimensional "O'Farrell gels" (56, 57).

The relA$^-$ mutants E. coli IB5 TrmA$^-$ and CP79 TrmA$^+$ are auxotrophic for histidine. Extensive starvation for listidine causes high levels of mistranslation in both strains.

As has been shown for the authentic elongation factor EF-G and several other proteins, the total number of His/Gln exchanges in both strains increases at definite reduced level of histidine to the same extent (54), indicating that the absence of T in tRNA does not increase the error frequency of the His codons in relA$^-$ strains when the amino acid becomes limiting.

Transductant pairs of E. coli TrmA$^+$ and TrmA$^-$ have been used to explore the possibility of translation ambiguities of terminator codons in cells possessing T-deficient tRNA. A great a number of mutants of phage T4 carrying amber, ochre and opal "stop" codons were tested for their ability to grow on TrmA$^+$ and TrmA$^-$ strains of E. coli. No differences were observed in the suppression patterns between TrmA$^+$ and TrmA$^-$ cells. From this result, it was tentatively assumed that the absence of T in tRNA does not quantitatively affect the translation of the three chain-termination codons (32).

III. Involvement of Ribosylthymine in Regulatory Mechanisms

Regulation of translation in eubacteria differs from that in eukaryotes in that (a) the synthesis of the components of their protein-synthesizing system is stringently controlled by the *relA* gene; (b) formylmethionyl-tRNA is used for the initiation of protein synthesis; (c) initiator tRNAs have T at position 54; and (d) most of the mRNAs are polycistronic. From the available data, reviewed in this section, an important function of T is emerging.

A. Stringent Response

In auxotrophic strains of *E. coli*, deprivation of an essential amino acid initiates a regulatory process termed "stringent response" that is governed by *relA*. This gene regulates the synthesis of tRNA as well as the synthesis of rRNA and some mRNA species relative to protein synthesis (58, 59). Under stringent conditions, two unusual nucleotides, guanosine 3′,5′-bis(diphosphate) (ppGpp) and guanosine 3′-diphosphate, 5′-triphosphate (pppGpp) accumulate. Studies in *in vitro* systems have shown that the synthesis of the guanosine phosphates is catalyzed by an enzyme that requires the ribosome·mRNA complex and uncharged tRNA as activator (60–62). The *relA* gene product has been termed "stringent factor" and is an ATP:GTP (GDP) pyrophosphotransferase.

Gram-positive species such as *Bacillus brevis* (63, 64) and *Bacillus stearothermophilus* (65) contain a (p)ppGpp-synthesizing enzyme apparently different from the *E. coli* stringent factor. The striking difference from the *E. coli* enzyme is that the one from the bacilli functions independently of the ribosome·mRNA·tRNA complex. The *B. brevis* enzyme has been called ppGpp synthetase (64) although others (65, 65a) use the term "stringent factor."[1]

A possible involvement of the TψC region in the tRNA of *E. coli* in stringent response has been suggested and is based on the following observation. The oligonucleotide TψCG can inhibit tRNA^{Phe} binding to poly(U)-instructed *E. coli* ribosomes. This is probably caused by a direct interaction of TψCG with a complementary sequence in ribosomal RNA (66–69). Exposure to TψCG activates the synthesis of ppGpp and pppGpp in an *E. coli* cell-free system by the complex of poly(U), stringent factor, and ribosome (70).

An important interrelationship between the *relA* product and the methylation of U-54 in tRNA has been observed. In *E. coli* the AdoMet-dependent tRNA (uracil-5-)-methyltransferase (EC2.1.1.35), unlike several other methylating enzymes, is stringently controlled (71).

---

[1] The IUPAC—IUB Enzyme Commission has proposed the names GTP pyrophosphokinase (or ATP:GTP 3′-pyrophosphotransferase) and the number EC2.7.6.5 (see EJB **116**, 423, 1981).[Eds.]

This has now been shown conclusively by using two *E. coli* strains differing only in the allelic state of *relA*. Upon partial deprivation of charged tRNA^{Val} (by a temperature shift in strains carrying a temperature-sensitive valyl-tRNA ligase), the rate of tRNA (uracil-5-)-methyltransferase synthesis was found to decrease in a strain with stringent control (*relA*$^+$) while it increased in strains carrying the *relA*$^-$ allele (*72*). This increase of the activity of the T-forming enzyme requires protein synthesis. Thus, when tRNA is partially uncharged in the cell, the *relA* product inhibits the expression of the T-forming enzyme.

B. Tetrahydrofolate-Independent Initiation

In gram-positive microorganism, the formylation of Met-tRNA^{fMet} and the biosynthesis of T are closely linked. Both depend on tetrahydrofolate as coenzyme (*21–24*). When grown in the presence of folic acid, *Streptococcus faecalis* initiates protein synthesis with fMet-tRNA^{fMet}. However, when *S. faecalis* is grown in a folate-free medium, initiation proceeds with nonformylated Met-tRNA^{fMet}. The ability of folate-deficient *S. faecalis* to initiate protein synthesis with nonformylated Met-tRNA^{fMet} *in vitro* is caused solely by the difference in initiator tRNA containing U-54 in place of T-54. It was therefore suggested that the replacement of T by U at position 54 permits initiation of protein synthesis *in vivo* with nonformylated Met-tRNA^{fMet} (*23, 25*) (see also Section II,A).

This view was further supported by the observation that trimethoprim, which inhibits the formylation of Met-tRNA^{fMet} and the biosynthesis of ribosylthymine in gram-positive microorganism, reduces, but does not inhibit, growth, provided that the medium is supplemented with all other folate-dependent metabolites, i.e., with serine, methionine, thymine, a purine base, and panthotenate (*24*).

Initiation of protein synthesis in the presence of trimethoprim and aminopterin has been studied in detail in *Bacillus subtilis* supplemented with the folate-dependent metabolites (*40*).

The amount of *N*-formylmethionine in tRNA and in protein has been determined. The level of formylation of methionyl-tRNA is 70% in control cells and approximately 2% in inhibitor-treated cells; the number of formyl groups in the proteins was also drastically reduced. Trimethoprim or aminopterin did not alter the amount of tRNA^{Met} or the degree of aminoacylation of tRNA^{Met} *in vivo*. From these results, it was concluded that protein synthesis is initiated in trimethoprim- or aminopterin-treated *B. subtilis* by nonformylated methionyl-tRNA^{fMet} in a manner similar to that reported for Lactobacillaceae under folate-independent growth conditions. However, the subsequent addition of

H$_4$folate-dependent metabolites to trimethoprim-inhibited cells completely overcame the inhibition of RNA synthesis, whereas protein synthesis remained inhibited by 10–25% compared with control cells. These results suggest that the synthesis of particular proteins remains dependent on tetrahydrofolate and ribosylthymine (40).

Two mutants from E. coli that initiate protein synthesis without prior formylation have been isolated and characterized. They were obtained by exposing E. coli strains to selective media containing the end products of folate metabolism plus two inhibitors of folate synthesis, sulfathiazole and trimethoprim (72a). Detailed studies with one of these mutants showed that no endogenous formyltetrahydrofolate is present and that protein synthesis, both in vivo and in vitro, occurs without any detectable formylation of Met-tRNA^{fMet}. The tRNAs of these mutants were undermethylated with respect to the parent tRNA, and only T-54 is lacking.

The mutants, lacking T in tRNA, grew significantly better than the isogenic parent strains under conditions forced to take place with unformylated initiator tRNA. It has been suggested that the mutant cells contain additional factors that facilitate initiation without formylation. Some results indicate that the levels of initiation factor IF-2 in the mutants are 3- to 4-fold that in the parent strain.

In addition to the lack of T-54, these mutants are streptomycin-resistant, and ribosomal mutations leading to streptomycin resistance relieve the necessity of formylation for the initiation of protein synthesis (73, 74). Therefore, the observed alterations can be additionally affected by alterations of ribosomal proteins.

C. Translation of Polycistronic mRNA

In eubacteria, RNA transcripts are frequently polycistronic, possessing a number of translational start and stop signals in vivo (for a recent review, see 73).

It has been proposed that two different mechanisms may coexist for initiation of translation of polycistronic mRNA. (a) The first cistron of the messenger is translated using dissociated ribosomes for initiation, which would be rather insensitive to formylation; (b) The following cistrons initiate translation with undissociated 70-S ribosomes, which would certainly require formylation. In this hypothesis, the formyl group would act as a positive effector on a preexisting equilibrium between two conformations of 70 S ribosomes: a major one, inactive and unable to bind fMet-tRNA^{fMet}, and a minor one active in initiation and binding fMet-tRNA^{fMet}. Both forms are presumed to bind the unformylated tRNA equally well, whereas only the active form binds the formylated species, shifting the equilibrium toward activity. (74).

On the basis of this hypothesis, a general function for the regulatory action of the formyl group at the level of translation has been suggested (73, 74) namely, that the formyl group acts as an antidissociating factor in polycistronic messenger translation, so that formylation would protect against premature termination of transcription and thus act as an antipolar effector. The proposed antipolar effect of the formyl group was tested in *E. coli* by measuring the different enzyme activities, encoded by the lactose operon after treatment of the cells, with low levels of the antifolic agent trimethoprim. The results indicated that trimethoprim exhibits a polar effect on the expression of the lactose operon. This has been traced back to the lack of the formyl group in initiator tRNA (for summarizing references, see 74).

D. Does Undermodified, Unformylated tRNA^{fMet} Function in Regulation?

Although bacteriophage RNAs sometimes use different strategies for regulation of polycistronic RNA expression than their hosts, MS2-phage RNA translation with ribosomes, enzymes, and tRNAs from *E. coli* turned out to be useful for the study of the mechanism of operon expression with unformylated Met-tRNA^{fMet} in more detail. The translation was performed in a polyamine-dependent system in which Mg^{2+} was partially replaced by polyamines and Ca^{2+} (51). In these experiments, initiator and elongator tRNAs from *E. coli* MRE 600 and from mutants either containing or lacking the T-forming enzyme (*E. coli* CP79 relA$^-$ TrmA$^+$ and *E. coli* IB5 relA$^-$ TrmA$^-$) have been separated.

Surprisingly, in both mutants, one additional type of initiator tRNA$^{fMet(2)}$ was found (Fig. 6). According to the electrophoretic mobility, the tRNA$^{fMet(2)}$ is not identical with tRNA$^{fMet(1a)}$ and tRNA$^{fMet(1b)}$ from *E. coli* MRE 600. Both types of tRNA$^{fMet(1)}$ are derived from the same gene, but tRNA$^{fMet(1b)}$ lacks s^4U at position 8 (Th. Dingermann, Y. Kushino, R. Praisler, and H. Kersten, unpublished results). Total initiator tRNA from strain CP79 can be methylated with AdoMet and the T-forming enzyme from *E. coli* to an extent of 300 pmol per A_{260} unit of tRNA; the product formed is T. This suggests that tRNA$^{fMet(2)}$ from *E. coli* CP79 is T-deficient. The analysis of tRNA$^{fMet(2)}$ is under way.

In *E. coli* CP79, all elongator tRNAs are fully methylated at U-54, whereas in *E. coli* IB5, initiator and elongator tRNAs totally lack T.

In vitro translation of MS2-phage RNA with bulk tRNAs from the

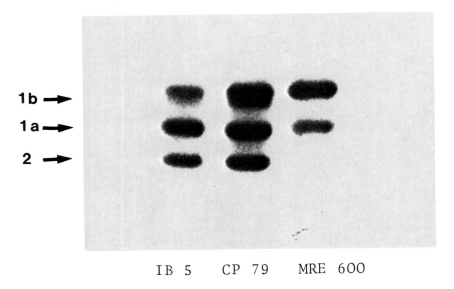

FIG. 6. Electrophoretic patterns of tRNA^{fMet} purified from amino acid auxotrophs of E. coli, IB-5 (TrmA$^-$), CP79 (TrmA$^+$), and the prototrophic strain MRE 600. Eluted tRNAs of the spots indicated by an arrow accepted methionine and were formylatable; they are referred to in the text as tRNA$^{fMet(1a,b)}$ and tRNA$^{fMet(2)}$, respectively.

E. coli MRE 600 and from the two mutants shows that the absence or the presence of a modified uridine at positions 54 in initiator tRNAs does not significantly affect the expression of the coat protein or replicase cistrons if the initiator tRNAs are formylated. Compared with the amount of products, translated with formylated initiator tRNA set at 100%, it is observed that (a) when all initiator tRNA$^{fMet(2)}$ is present and the elongator tRNAs contains T, 68% of the coat protein and 41% of the replicase protein are formed; (b) when both initiator tRNAs are present and are totally lacking T in initiator and elongator tRNA(s), only 32% of the coat protein and 18% of the replicase protein are formed; (c) when tRNA$^{fMet(2)}$ is absent and the usual initiator and elongator tRNAs contain T, 16% of the coat protein and less than 10% of the replicase are produced (Table I) (75).

Assuming that the MS2-phage-RNA translation model applies to the translation of certain cistrons or to a certain class of polycistronic mRNAs in E. coli, it follows that a particular set of proteins can be synthesized under H$_4$folate limitation with specifically undermodified tRNA$^{fMet(2)}$.

TABLE I
OPERON EXPRESSION OF THE MS2-PHAGE RNA: DEPENDENCE ON TETRAHYDROFOLATE AND RIBOSYLTHYMINE[a]

tRNA		Coat protein (cpm)		Replicase (cpm)	
Initiation	Elongation	Formylated	Nonformylated	Formylated	Nonformylated
U(T)	T	4000	2700	300	130
U	U	4400	1400	300	60
T	T	4600	800	400	30

[a] The protein synthesis assay was done as described by Kersten et al (49). (i) The tRNA preparation was obtained from E. coli rel⁻ CP79; (ii) from E. coli Trm⁻ rel⁻ (IB5); iii, from E. coli MRE 600.

The final volume of 50 μl contained a further 9 μl of a mixture of 13 ^{14}C-labeled amino acids (50 mCi/mol) and, in addition, 60 μM of each of the 20 amino acids (unlabeled) used in protein synthesis; also the S-100 supernatant from E. coli MRE 600 corresponding to 144 μg of protein per assay, IF 1–3 each at concentrations between 0.2 and 0.5 μg of protein, and 0.85 A_{260} 70-S ribosomes from E. coli MRE 600. The assay was started with 0.3 A_{260} of MS2-phage RNA. The proteins were isolated and separated by one-dimensional gel electrophoresis. The gels were cut into 1-mm pieces, dissolved in toluene, and transferred into a toluene scintillation "cocktail." The radioactivity was determined; the values are given in total counts per minute for the coat protein and the replicase fractions, respectively.

An interesting regulatory mechanism involving fMet-tRNA^{fMet} has been described (75a). If overall protein synthesis ceases, the initiation rate decreases, yielding a pool of fMet-tRNA^{fMet}. This then binds to the RNA polymerase and induces a biphasic effect: rRNA and r-protein syntheses become reduced, and the synthesis of nonribosomal proteins is stimulated. Since only the formylated form binds to the RNA polymerase, H$_4$folate limitation will prevent this effect, and the translation of rRNA and r-proteins can probably proceed with Met-tRNA$^{fMet(2)}$.

The synthesis of rRNA and r-proteins is also regulated by ppGpp, which interacts with the RNA polymerase and changes its promotor selectivity (75a, 75b, 75c). Although unformylated Met-tRNA^{fMet} is not a signal for the formation of ppGpp in the prototrophic E. coli strain MRE 600 (76), it is possible that tRNA$^{fMet(2)}$ gives rise to the formation of ppGpp in amino-acid auxotrophs of E. coli (77).

A balance between the stimulatory effect of unformylated undermodified tRNA$^{fMet(2)}$ and the inhibitory effect of ppGpp might therefore increase the economy of the cell under starvation conditions and permit cells to synthesize an adapted amount of components of the protein-synthesizing apparatus under H$_4$folate limitation. In this re-

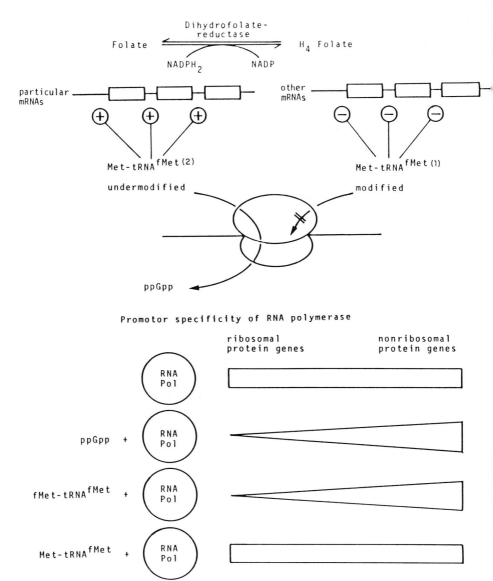

FIG. 7. Schematic representation of the discussed regulatory function, in relation to formylation, of two types of tRNA^{fMet} differing in modification at U-54. *Upper part:* The expression of a particular class of mRNAs is not under metabolic control; that of another class of mRNAs is. One class of mRNAs is initiated under H$_4$ folate restriction with tRNA$^{fMet(2)}$; tRNA$^{fMet(2)}$ might give rise to the formation of ppGpp under H$_4$ folate limitation. tRNA$^{fMet(1)}$ restricts the translation of another class of mRNAs under H$_4$ folate limitation. *Lower part:* Known effect of ppGpp and Met-tRNA^{fMet} (formylated and unformylated) on the specificity of RNA polymerase for ribosomal and nonribosomal protein genes. Bars, no differential specificity; wedges indicate alterations in specificity.

spect, it is of importance to find out whether tRNA^fMet(2) might be involved in the regulation of the synthesis of particular ribosomal proteins (77a–c) or initiation factors considered to be subject to metabolic control (77d,e).

The results presented in this section on the role of alterations in modification and formylation of Met-tRNA^fMet on interactions of transcriptional and translational control mechanisms in gram-negative *E. coli* are summarized in Fig. 7.

IV. Ribosylthymine in Eukaryotic Elongator tRNA

A. Variations of the TψC Sequence

Every tRNA molecule from eubacteria contains one ribosylthymine (T) residue, whereas tRNAs form eukaryotic cells or tissues contain less than molar amounts of T (78). Depending on which nucleoside is present at position 54, mammalian tRNAs have been grouped into four classes: A, those tRNAs in which T is completely replaced by adenosine; B, those tRNAs that contain their full complement of T, pseudouridine, or 2′O-methylribosylthymine; C, those tRNAs showing only a partial conversion of uridine to T; and D, those tRNAs in which only uridine occupies position 54.

B. Influence on Translation *in Vitro*

Class C tRNAs and class D tRNAs have been methylated *in vitro* with the *E. coli* tRNA (uracil-5-)-methyltransferase and S-adenosylmethionine as a methyl donor (79, 80). An analysis of the methylation reaction showed that T is the only product formed and is localized exclusively at position 54 (81). The methylated tRNAs (T-54) were tested with a tRNA-free *in vitro* protein-synthesizing system of wheat germ using tobacco mosaic virion (TMV) RNA, brome mosaic virion (BMV) RNA, or chorion polysomal RNA as mRNAs. Control experiments showed that the extent of protein synthesis with shammethylated or untreated crude tRNA was essentially indistinguishable for all mRNAs tested. The use of an equivalent amount of the *in vitro* methylated tRNA T-54, however, led to a significant reduction in protein synthesis for all mRNAs tested. To identify the specific tRNAs involved in the inhibitory effect of T, pure, 100% methylated tRNA^Gly(1) (class D, U-54) was mixed with untreated crude wheat germ tRNA, and the mixture was examined for protein-synthesizing activity. No influence of the conversion of U to T in tRNA^Gly was observed during BMV RNA translation; however, chorion mRNA translation was inhibited significantly. The inhibition was found to be at the ribosome level and could be abolished by the addition of spermine (80, 81).

tRNAs^{Phe} (class C) purified from rat liver (78% T), from beef liver (72% T), and from human placenta (50% T) were compared with mammalian tRNA^{Phe} methylated to 100% with the T-forming enzyme from *E. coli* in cell-free protein synthesis. The T content of mammalian tRNAs^{Phe} had no effect on their ability to function in the aminoacylation reaction. There was, however, a significant influence of the T content in tRNA to support poly(U)-dependent poly(Phe) synthesis (78, 82). Specifically, an increasing T content in a class C mammalian tRNA^{Phe} increases the V_{max} but does not change the apparent K_m for the poly(Phe) synthesis reaction (82). Since the experiments were carried out in the presence of added initiation and elongation factors required for protein synthesis, it was postulated that the *in vitro* effect of T in mammalian tRNA is to regulate protein synthesis in concert with one or more of these factors.

Further studies (82) showed that the complete conversion of U-54 to T-54 in an *in vitro* partially methylated tRNA (class C) has a positive effect, whereas the conversion of U-54 to T-54 in tRNA^{Val} naturally lacking this modification (class C) has a negative effect, on protein synthesis. Thus the structure of tRNA^{Phe} requires T-54 for optimum function, whereas the structure of tRNA^{Val} requires U at position 54 irrespective of whether homopolynucleotides or natural mRNAs are translated. The results demonstrate that the rate of EF-1-directed ribosomal A-site binding is a major step affected by the T content of class C tRNAs, and suggest that T in the major class of tRNAs has a regulatory effect on those mammalian protein-synthesizing systems in which elongation is rate-limiting.

C. Utilization of tRNAs with Variable Thymine Content

In eukaryotes, variations in the modification level of tRNA occur from organ to organ, in differentiating cells, during embryogenesis and aging, during starvation, in response to hormones, certain drugs, or carcinogens, in malignant cells, and after infection of cells with viruses. Extensive reviews describing these variations have been published (83–89).

Variations in the level of the modified nucleosides in tRNA, especially of T, have been observed during differentiation of *Acetabularia mediterranea* and during development of the slime mold *Dictyoste-*

lium discoideum (*90, 91*). To elucidate whether variations in the degree of U-54 methylation in eukaryotic tRNA might play a regulatory role in protein synthesis, polysomal and overall cellular tRNAs of *D. discoideum* from the vegetative growth stage and the early developmental preaggregation stage were isolated and characterized. In these experiments, axenic strain AX-2 of the slime mold was grown in a well-defined medium (*92*) and synchronous development to the preaggregation stage was induced by starvation of nutrients in shaking suspension (*93*). The nucleoside composition of vegetative and early developmental tRNAs from *D. discoideum* was evaluated, and the amount of tRNAs with T or U at position 54 was determined. The following results were obtained. (i) During vegetative growth, 85% of the tRNAs present on polysomes contain T at position 54 whereas tRNA species containing U there accumulate in nonpolysomal cell fractions. (ii) Overall cellular tRNAs and polysomal tRNAs from the preaggregation stage are less modified with respect to T than those from the vegetative growth stage; in addition pseudouridine and 5-methylcytidine are partially lacking at tRNA positions that have not yet been identified. (iii) The T-containing tRNAs predominate also on polysomes during preaggregation; however, only 65% of the polysomal tRNAs contain T (*93*).

During the early preaggregation stage, the synthesis of tRNAs proceeds almost at the same rate as in vegetatively growing cells, whereas the synthesis of protein is drastically reduced (*94*). A decrease in the amount of T at this developmental stage might therefore be caused by a limitation in the amount of the corresponding T-forming enzyme. Reduced levels of tRNA methyltransferases during the development of *D. discoideum* have been described (*95, 96*). No changes were detected in the base composition of tRNA of an axenic strain of *D. discoideum* between vegetative growth and the late developmental culmination stage (*97*). It has therefore been questioned whether alterations in the base modification of tRNAs from vegetative and from developing *D. discoideum* can be analyzed sufficiently well to detect subtle changes by applying the nucleoside analysis described by Randerath *et al.* (*98*).

In order to clarify the conflicting results, a more sensitive method has been applied to determine the formation of T and other methylated nucleosides in newly synthesized tRNAs of *D. discoideum* during early development. Cells were labeled with L-[^{3}H]methionine as methyl donor and [^{14}C]guanine as tRNA precursor. The methylated nucleosides of the newly synthesized developmental tRNA were assayed by liquid chromatography, and the methyl group incorporation into each nucleoside was referred to the amount of newly synthesized tRNA. The tRNA synthesized early during development was submethylated with regard to T. Also, the methylation of guanine to 7-methylguanine and to N^2,N^2-dimethylguanine was significantly reduced (52). Interestingly, undermodified 4 S RNA, apparently tRNA, accumulates in the nucleus, and shortly thereafter the synthesis of tRNA ceases (52, 99). It is therefore possible that the formation of T is somehow involved in the expression of tRNA genes.

D. Expression of tRNA Genes and Biosynthesis of Thymine

Gene expression in eukaryotes is influenced by a wide variety of mechanisms, including gene loss, gene amplification, gene rearrangement, DNA modification, transcriptional and posttranscriptional control during processing, splicing, and modification of tRNA and of mRNA. Gene expression is controlled also at the translational level, e.g., modulated by one or more of the factors involved in protein synthesis (for recent reviews, see 100, 101).

Transcription of 5 S rRNA and tRNA genes in eukaryotes is catalyzed by RNA polymerase III, which has some unique features. Intragenic promotors have been discovered in 5 S rRNA and tRNA genes. Within the tRNA coding sequence, two control regions occur. These have been identified independently for a tRNA^{Arg} gene from *Drosophila melanogaster* (102), for a tRNA^{Pro} gene from *Cenorabdites elegans* (103), and for a tRNA^{fMet} and a tRNA^{Leu} gene from *Xenopus laevis* (104, 105). One of the intragenic transcriptional control regions resides at those nucleosides of a tRNA gene that encode the D-stem and D-loop region; the other is located within stem and loop IV, which contains the semi-invariant sequence TψC.

Both of these two control regions contain a significant number of invariant or semi-invariant nucleotides in eukaryotic tRNAs. Therefore "general promoter" sequences have been postulated (104, 106). The significance of these general promoters is underscored by the fact that tRNA genes from eubacteria or from eukaryotic cellorganelles can also be transcribed by RNA polymerase III if they fulfill these promoter requirements (107, 108).

Posttranscriptional maturation of the transcribed tRNA involves modification, splicing, and processing. The modified bases of the D loop and the TψC loop are introduced into tRNA before processing and splicing (109, 110). It is therefore possible that these modifications play a role in the transcriptional control of particular tRNA genes.

V. N^6-(2-Isopentenyl)adenosine and Derivatives in tRNA

A. Evolutionary Aspects

Modifications in tRNA frequently occur at position 37 at the 3' end of the anticodon (Fig. 1) Among these, N^6-(Δ^2-isopentenyl)adenosine and derivatives, especially the 2-methylthio derivative ms^{2}i^{6}A, are the only ones being studied extensively with respect to their biological significance. Isopentenyladenosine and the hypermodified ms^{2}i^{6}A are present in eubacterial tRNAs recognizing codons starting with uridine, e.g., tRNA^{Leu}, tRNA^{Phe}, tRNA^{Ser}, tRNA^{Trp}, and tRNA^{Tyr} (for recent reviews, see 9 and 12).

Several strains of yeast contain i^{6}A in tRNA^{Leu}, tRNA^{Ser}, and tRNA^{Tyr}. Hydroxy derivatives, N^6-(cis-4-hydroxyisopentenyl)adenosine, also called cis-zeatosine (c-io^{6}A) and the corresponding methylthio compound (ms^{2}io^{6}A) occur in *Salmonella typhimurium*, in certain Enterobacteriaceae (*111*), and in plants (*112*).

The biosynthesis of the 2-methylthio derivative of isopentenyladenosine involves cysteine and methionine and probably occurs in a sequential manner, i.e., thiolation of isopentenyladenosine followed by methylation (*112a*).

Another family of tRNAs contains the hypermodified adenosine nucleoside N-[(9-β-D-ribofuranosylpurine-6-yl)carbamoyl]threonine (t^{6}A) in position 37. It is also found in tRNAs of fungi, animals, and plants and can be methylated at the N-carbamoyl group to mt^{6}A, or a methylthio group may be incorporated in position 2 of the purine (ms^{2}t^{6}A) (for summarizing references, see *112b*).

In tRNAs of higher eukaryotes, l-methylguanosine and hypermodified nucleosides more or less related to guanosine are discovered at N-37. These modifications are referred to as wyosines. They occur frequently in the tRNA^{Phe} and tRNA^{Lys} of higher eukaryotes.[2] Lysine and methionine seem to be involved in the biosynthesis of these hypermodified nucleosides (112c, d).

B. Codon/Anticodon Recognition

The function of the hypermodified base, ms^{2}i^{6}A, has been studied by several groups. As early as 1969, it was shown that the lack, or even a partial lack, of this normally hypermodified base reduces the efficiency of tRNA^{Tyr} binding to ribosomes in the presence of the appro-

---

[2] Included in the 328 tRNA sequences presented in This Series (refs. 8 and 8a). [Eds.]

priate trinucleotide codon (*113*). Possible physical explanations for this effect have been demonstrated by studying the binding interaction between tRNAs with complementary anticodons (*114*). It was observed that tRNA^{Phe} (anticodon GAA) from yeast or from *E. coli*, each having a hypermodified base next to the anticodon, bind much more strongly to tRNA^{Glu} (anticodon UUC) than does tRNA^{Phe} from mycoplasma lacking the hypermodified base.

Two isoacceptors, tRNA$^{Phe(1)}$ and tRNA$^{Phe(2)}$ have been isolated from *B. subtilis*, and their modification patterns have been established (*115*). tRNA$^{Phe(1)}$ has been shown to be an incompletely modified precursor of tRNA$^{Phe(2)}$: it has i^{6}A at position 37 and lacks a methyl group in the ribose moiety of the guanosine residue at position 34 while tRNA$^{Phe(2)}$ contains Gm at 34 and ms^{2}i^{6}A at 37 (*115*). Both tRNAs behave identically in aminoacylation kinetics. In the factor-dependent AUGU$_3$-directed formation of fMet-Phe, the undermodified tRNA$^{Phe(1)}$ was less efficient than its mature counterpart at Mg^{2+} concentrations from 5 to 15 mM (*42*). The pronounced difference between tRNA$^{Phe(1)}$ and tRNA$^{Phe(2)}$ in this synthesis was thought to be caused by a change in the modification of the hypermodified adenosine residue next to the anticodon. The hypermodification increases the stability of complexes between complementary codon triplets in the presence of ribosomes (*116*). Therefore a lack of the modified nucleosides in the anticodon region of tRNA^{Phe} was thought to interfere with the correct arrangement of the aminoacyl moiety at the "peptidyltransferase center" (*42*).

The role of N^6-(2-isopentenyl)adenosine and its derivatives in tRNA on codon/anticodon recognition has been discussed in previous reports (*4, 6, 7, 9, 11, 12, 116*). The following sections deal mainly with the importance of this modification in control mechanisms of mRNA translation involving transcription termination control.

C. Transcription Termination Control

The three codons UAG (amber), UAA (ochre), and UGA (opal) serve as signals for polypeptide chain termination during messenger RNA translation in various prokaryotic and eukaryotic organisms. When one of these codons occurs internally in a structural gene as a result of mutation, the gene product is an incomplete polypeptide. Such "nonsense" mutations can be reverted by an unlinked "suppressor" mutation that alters one of the components required for protein synthesis and facilitates the formation of a completed polypeptide through the insertion of an amino acid at the position of the nonsense codon. The most extensively studied suppressors arise by mutation of the structural genes for a particular tRNA. Suppressor tRNAs can also result by mutation of genes controlling tRNA modification enzymes or ribosomal proteins (117–119).

1. tRNA^{Trp} IN E. coli

Mutants of E. coli have been isolated in which the structure of tRNA^{Trp} is altered and the ability of the cells to regulate the trp operon is impaired (119). Sequence analyses of wild-type tRNA^{Trp} and UGA suppressor tRNA^{Trp}, both derived from trpX strains, reveal an unmodified A in position 37, which is normally occupied by the hypermodified base ms^{2}i^{6}A. The trpX gene product was necessary for the synthesis of ms^{2}i^{6}A; therefore, the first gene involved in ms^{2}i^{6}A synthesis is designated miaA (119).

Transcription of the trp operon in E. coli is regulated primarily at two sites. At the operator site, initiation of transcription is controlled by interaction with an aporepressor·L-tryptophan complex (120). The operon is additionally regulated by a mechanism that functions in series with the repressor-operator system (121). This system controls gene expression by terminating transcription at a specific attenuator site. This attenuator has been located in the leader sequence, between the operator and the first structural gene of the tryptophan operon. The termination process is clearly dependent on the proper functioning of tRNA^{Trp} and tryptophanyl-tRNA synthetase and is apparently dependent on the level of charged tRNA^{Trp} (122, 123).

The tRNA^{Trp} isolated from the miaA mutant has an unmodified A at position 37. Interestingly, the absence of the modification produces inefficient translation at the tandem tryptophan codons in the leader sequence of the corresponding mRNA. The inability to read those trp codons efficiently is thought to lead to an increased expression of the trp operon (123, 124).

An analogous mechanism is discussed for the expression of aminoacid operons in the his-T mutants of Salmonella typhimurium (125, 126), and E. coli (127). Strains carrying the his-T mutation produce tRNAs lacking two pseudouridines in the anticodon region. Many tRNAs normally have this modification; consequently, the syntheses of those amino acids that are cognate for several of these tRNAs become deregulated (128, 129). The regulatory region of the his operon contains seven consecutive his codons that seem to be involved in regulation (130, 131).

2. tRNA AND tRNA^{Ser} OF YEAST

A lack of i^{6}A or ms^{2}i^{6}A in suppressor tRNA^{Tyr} sup3$^+$ in yeast leads to a functional impairment during translation (113). Mutants of Saccharomyces cerevisiae that contain almost exclusively A in place of

i^6A at 37 in $tRNA^{Tyr}$ and $tRNA^{Ser}$ have been isolated (*132*). The cells of those mutants apparently grow as well as nonmutant cells under conditions where suppression is not required for growth. It was therefore suggested that the i^6A-deficient tRNAs do not limit the rate of protein synthesis in the mutant. Unfractionated tRNA from the mutant contains as much $tRNA^{Tyr}$ and $tRNA^{Ser}$ as nonmutant tRNA. The i^6A deficiency therefore does not affect the relative rates of tRNA synthesis and degradation. In addition, the mutant cells do not differ from nonmutant cells in the accumulation of tRNA precursors.

The modified base i^6A is also missing from tRNAs of the i^6A family isolated from the antisuppressor strain *sin1* of *Schizosaccharomyces pombe*. Sequence analysis of $tRNA^{Tyr}$ from this mutant shows an A instead of i^6A. $tRNA^{Ser}$ and $tRNA^{Trp}$ from the *sin1* mutant show a similar shift in RPC-5 chromatography elution profiles. From this it was concluded that these two tRNAs were also deficient in i^6A. The presence of the antisuppressor mutant *sin1* leads to an inactivation of the nonsense suppressor (*sup3-i*) by a mutated $tRNA^{Ser}$. These results suggest that the loss of i^6A in the suppressor $tRNA^{Ser}$ is responsible for the inactivation of suppression in *S. pombe* (*sup3-i*). The *sin1* mutation probably affects the structural gene of an enzyme in the isopentenyl biosynthetic pathway (*133*).

There is only a slight reduction in growth rate in *S. pombe*, which is caused by the i^6A deficiency in tRNAs, thus indicating that these tRNAs have not been made rate-limiting for growth by the loss of i^6A. It is therefore assumed that the interaction between the tRNAs, lacking i^6A, mRNA, and the ribosome must still be strong enough to permit elongation of the polypeptide chain (*134, 135*). The deficiency of i^6A in yeast has been suggested to affect sporulation or germination.

VI. Central Role of N^6-(2-Isopentenyl)adenosine and Derivatives in Growth and Development

A. Sporulation in *Bacillus subtilis*

Bacillus subtilis was the first organism used to study tRNA modifications, especially i^6A and ms^2i^6A during growth (glucose salt medium) and during development to spores. During sporulation of *B. subtilis*, several changes in transcriptional and translational processes occur (*136, 137*). Among these are alterations in the relative amounts and kinds of isoaccepting tRNA species (*138–141*). It was suggested that these are caused by the expression of new tRNA genes at various growth stages, and/or differences in the expression of the genes that code for tRNA-modifying enzymes. The primary structure of isoacceptors that occur in *B. subtilis* during growth and the stationary phase have been elucidated for tRNA^{Phe} (*142*), for tRNA^{Tyr} (*143*), and for tRNA^{Lys} (*144*).

The two species of tRNA^{Phe} of *B. subtilis* differ only in two posttranscriptional modifications. The minor tRNA$^{Phe(1)}$ (first peak eluting from an RPC-5 column) lacks 2'-O-methylguanosine at position 34 and has i^{6}A at position 37, whereas tRNA$^{Phe(2)}$ has a methylated G in the first position of the anticodon and ms^{2}i^{6}A next to the 3' end of the anticodon.

Also, two species of tRNA^{Tyr} appear during growth of *B. subtilis* in glucose salt medium. They have different chromatographic properties, and their relative proportions change during growth and sporulation. tRNA$^{Tyr(1)}$ is more important in vegetative cells, whereas tRNA$^{Tyr(2)}$ is predominant in stationary cells and in spores (*140, 141, 145, 146*). The nucleotide compositions differ only at position 37: i^{6}A is present in tRNA$^{Tyr(1)}$ and ms^{2}i^{6}A in tRNA$^{Tyr(2)}$ (*145, 147*).

In addition to tRNA^{Phe} and tRNA^{Tyr}, three tRNA^{Lys} isoacceptors change during development. tRNA^{Lys} isoacceptors 1, 2, and 3 share the same primary sequence, but differ in the degree of posttranscriptional modification in the anticodon loop. The first has an unmodified C in position 32; furthermore, the isoacceptors differ with respect to A-37 modification, having *N*-[(9-β-D-ribofuranosylpurin-6-yl)carbamoyl]-L-threonine (t^{6}A) or the hypermodified ms^{2}t^{6}A in position 37 (*144, 148*).

Although neither species shows any preference in responding to either of the lysine codons, there is an overall preference for tRNA$^{Lys(3)}$ in lysine incorporation during translation of rabbit globin mRNA in rabbit reticulocyte lysates (*149*).

These three examples show that alterations in the relative amounts and kinds of isoaccepting species for tRNA^{Phe}, tRNA^{Tyr}, and tRNA^{Lys} in *B. subtilis* can be caused by differences in the expression of the genes that code for specific tRNA-modifying enzymes.

B. Relation to Iron and Amino Acid Metabolism

In *Escherichia coli*, grown in iron-deficient media, tRNA^{Phe}, tRNA^{Tyr}, and tRNA^{Trp} accumulate with i^{6}A in place of ms^{2}i^{6}A (*150–152*). The same alterations of tRNA modification have been observed in enteropathogenic *E. coli* upon treatment with the iron-binding protein lactoferrin, obtained from human milk or bovine colostrum. Adding sufficient iron to saturate the iron-binding capacity of the lactoferrin reversed these chages. It was therefore concluded that iron is involved in methylthiolation of the isopentenyladenosine residue in tRNA (153). In *E. coli* cells, grown in the presence of the lactoferrin, tRNA^{Phe} and tRNA^{Trp} (i^{6}A) were found to relieve transcription termination of the Trp and Phe attenuators, so that increased operon expression apparently occurs under iron-restricted conditions (154). The increased expression of the Trp operon in *E. coli* was less than that produced when tRNA^{Trp} lacks both ms^{2}i^{6}A and i^{6}A. Interestingly, the undermodified tRNAs for Phe, Tyr, and Trp, accumulating under iron restriction, can also function as positive regulatory elements of the aromatic amino-acid transport system in *E. coli* (155).

These results suggest that the relaxation of attenuation in the Trp and Phe operons in *E. coli* might be related physiologically to an iron-dependent change in tRNA modification, and that this is important for the adaptation of pathogenic strains of *E. coli* for growth in mammalian body fluids containing iron-binding proteins (156, 157). Interestingly, the bacteriostatic action of human milk depends on the combined action of lactoferrin and an antibody. Lactoferrin itself cannot reverse the pathogenic effect of *E. coli*, because the bacteria secrete a high-affinity iron-chelator "enterochelin," which removes iron from lactoferrin and transports it to the bacterial cell. The antibody blocks this essential process, apparently by interfering with synthesis of enterochelin (158–161).

Enterochelin is synthesized from chorismic acid by way of a branch in the biosynthetic pathway of the aromatic amino acids. Any derepression of the enterochelin system during iron restriction is suggested to require adjustments to the whole aromatic pathway to ensure production of sufficient chorismic acid, and continued production of the aromatic amino acid and other aromatic compounds (162).

C. The Link between A-37 Modification and Major Routes of Metabolism

In mutant strains of *E. coli* K12 (PA1 relA$^+$ and PA2 relA$^-$), an auxotroph for arginine, the relative amounts of tRNA$^{Phe(1)}$ (i^{6}A) and tRNA$^{Phe(2)}$ (ms^{2}i^{6}A) are characteristically changed, depending on the growth rate and the availability of arginine. When the cells are grown at optimal concentrations of arginine, almost all of the tRNA^{Phe} is present as tRNA$^{Phe(1)}$ (i^{6}A); when the cells are grown at limiting concentrations of arginine, tRNA$^{Phe(2)}$ (ms^{2}i^{6}A) appears in the stringent strain, whereas the relaxed strain contains both counterparts (162a) (Fig. 8).

Surprisingly, the addition of arginine to *E. coli* MRE 600 (grown in glucose/salt medium) also causes an incomplete methylthiolation of

i^6A in the i^6A family of tRNAs (*162a*) reversible by the addition of Fe^{3+} (Kersten, unpublished). This result poses the question whether arginine might interfere with the iron-transport systems of *E. coli*.

Bacteria can take up iron from the medium by independent siderophore systems: hydroxamate-mediated (ferrichromes, mycobactin, hydroxamate-citrate complexes, ferrioxamines, albomycin, and aerobactin); phenolate-mediated (enterochelin); and citrate-mediated (iron-citrate complexes) (for reviews, see *162b,c*). The complex structures of some iron chelators are shown in Fig. 9a–e. Ferrichromes are specific for fungi and are not produced by *E. coli*, although *E. coli* synthesize a ferrichrome receptor, the TonA protein. Ferrichrome shares this receptor with albomycin, with phages, T1, T5, and $\phi 80$, and with colicin A. The enterochelin-mediated transport involves a protein of the outer membrane having a molecular weight of 81,000. This receptor is common for iron and colicin B; it has been referred to as ColB or feuB. The postulated mechanism of iron transport for *E. coli* involves several complex iron-transport systems with decreasing affinity for iron: enterochelin>citrate>ferrichrome/albomycin>aerobactin, and a low-affinity system of iron transport, as yet unidentified (*162c*).

The iron-chelating hydroxamate siderophores are derived from hydroxamic acids, which are common to most microorganisms. Although the structural diversity of these compounds is enormous, they all have a common precursor, namely ornithine (*162d*). In hydroxamic acids, the amino and carboxyl groups of the ornithine residues are linked by amide bonds in a cyclic hexapeptide, and the hydroxylamine groups are acetylated or acylated to yield the hydroxamic acids. The acyl groups of three hydroxamic acids of the ferrichrome type are close derivatives of mevalonic acid (*162d*). Interestingly, the isopentenyl group of mevalonic acid also serves as a precursor of i^6A in tRNA.

Arginine can interfere with the synthesis of hydroxamic acids (see Fig. 10). These steps involve glutamate, acetylglutamate, acetylornithine, and ornithine (*162e,f*). Arginine, at high levels, inhibits the first enzyme in this pathway; in addition, the conversion of acetylglutamate to acetylornithine is repressed by arginine (*161f*).

The conversion of tRNA (i^6A) to tRNA(ms^2i^6A) requires iron. Assuming that the iron level is lowered by reduced synthesis of a hydroxamic acid, it follows that the methylthiolation of tRNA (i^6A), becomes incomplete. This in turn causes a derepression of the synthesis of chorismic acid and aromatic amino acids by relaxation of attenuation in the "aro regulon" (*160*). Aromatic amino acids are then synthesized and the aromatic ring system can be oxidized by mono-

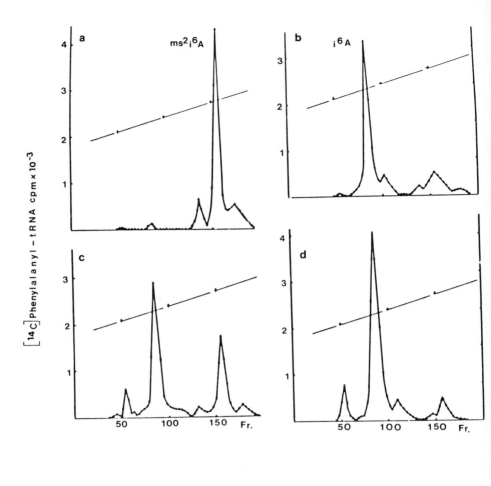
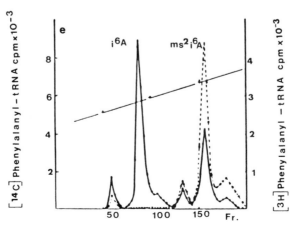

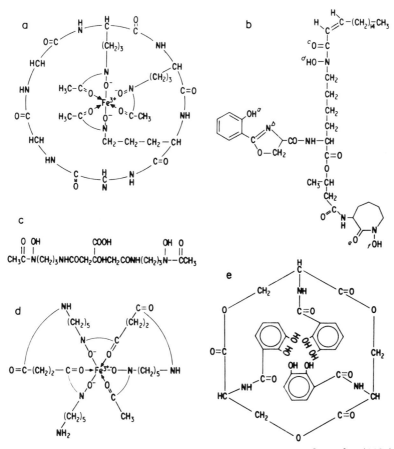

FIG. 9. Structure of siderophores according to Messenger and Barclay (162c): (a) ferrichrome, (b) mycobactin, (c) hydroxamate-citrate complex, (d) ferrioxamines, (e) enterochelin.

FIG. 8. Influence of arginine on the methylthiolation of i⁶A in tRNA^Phe of the arginine auxotrophic strain K12, PA1, relA⁺ (panels a, b) and PA2 relA⁻ (panels c, d) grown with 20 μg (panels a, c) and 60 μg of per milliliter arginine (panels b, d). tRNAs were purified, aminoacylated with [¹⁴C]Phe, and separated on RPC-5 columns. Panel e: The prototrophic strain of E. coli MRE 600 was used in an analogous experiment. tRNAs were purified, aminoacylated with [³H]Phe or [¹⁴C]Phe, respectively, and cochromatographed. ---, tRNAs from cells grown with 20 μg of arginine; ——, tRNAs from cells grown with 60 μg of arginine (162a and dissertation of H. Schellberger, Erlangen, 1975).

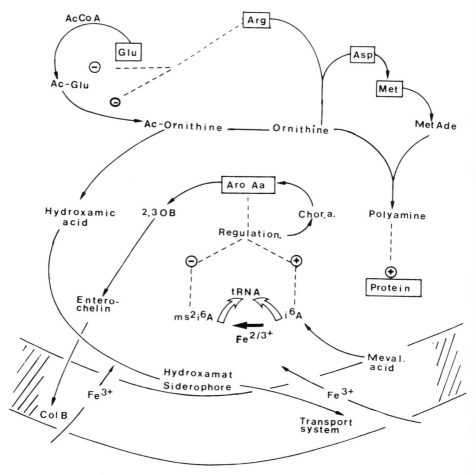

FIG. 10. Schematic representation of the relationship between arginine and ornithine metabolism and iron-transport mechanisms involving the tRNA (i^6A)-methylthiotransferase as a control element in the synthesis of the iron-transport systems.

oxygenases to provide the 2,3-dihydroxybenzoic acid, a moiety of the high-affinity iron-chelator, enterochelin (*162g*). Enterochelin will pass the membrane and accumulate iron from a surrounding that is low in Fe^{3+}, and will usually overcome iron starvation.

The possible role of alterations of the modification of i^6A to ms^2i^6A in corresponding tRNAs for the adaptation of an organism to an iron-restricted environment is summarized in the scheme of Fig. 10. This outline shows the possible linkage between iron-mediated modification of tRNA and the metabolism of arginine and ornithine, which plays a central role in the regulation of growth.

D. Cytokinin

1. Cytokinins and Growth

N^6-(2-Isopentenyl)adenines, called cytokinins, are found as nucleosides in tRNAs in all classes of cells, especially in plants (163). The free N^6-isopentenyladenines released by digestion of tRNA have been administered exogenously to plants in various test systems, and do stimulate growth, promote cell division, and stimulate germination and budding. Until recently, hydroxylation of the isopentenyl side chain of i^6A to form io^6A in the tRNA was thought to be restricted to higher plants. However, hydroxylated i^6A derivatives occur in the tRNAs of plant-associated bacteria, e.g., in tRNA of *Pseudomonas aeruginosa* (164) and of *Agrobacterium tumefaciens* (165), in *Rhizobium leguminosarum*, in *A. tumefaciens*, in *Corynebacterium fascians*, in *Erwinia amylovora* (166), and in *Salmonella typhimurium* (166a). The mechanism by which cytokinins regulate growth and development is not precisely known. Possible functions are discussed in the following sections.

2. Cytokinins and DNA Replication

Mevalonic acid serves as a precursor of the isopentenyl group of ms^2i^6A in tRNA of *Lactobacillus acidophilus* (167). Mevalonic acid also seems to be involved in the biosynthesis of i^6A in plant tRNAs as well as in the synthesis of free cytokinins (168). More recently, an interesting relation between cytokinin activity and mevalonic acid metabolism has been reported (169), which is based on the observation that mevalonate synthesis is essential for DNA replication (170). The enzyme 3-hydroxy-3-methylglutaryl-CoA reductase (HMG-CoA reductase) (EC 1.1.1.34 or 1.1.1.88), an enzyme that catalyzes mevalonate synthesis, approximately triples its activity in synchronized BHK cells during the S-phase of the cell cycle. Second, if this acitivity is blocked with the competitive inhibitor compactin, DNA synthesis is reduced to basal levels specifically during the S-phase of the cell cycle. Finally, the addition of mevalonic acid rapidly and entirely restores DNA replication in compactin-blocked cells. These results suggested that mevalonate serves as a precursor for an isoprenoid compound that specifically regulates S-phase DNA replication.

Further, isopentenyladenine, as well as its hydroxylated analog zeatin, not only can substitute for mevalonate in reversing the compactin-induced inhibition of S-phase DNA replication, but is at least

100 times more active than mevalonate in restoring DNA synthesis in this system. The specificity of isopentenyladenine and its 4'-hydroxy analog is indicated by the fact that the nucleoside isopentenyladenosine is totally unable to reverse the compactin-induced inhibition of DNA replication. Taken together, these observation suggest that isopentenyladenine, or a closely related isoprene purine derivative, may mediate the effect of mevalonate in controlling S-phase DNA replication (169).

3. CYTOKININS AND SPORE GERMINATION

Recently, an N^6-isopentenyladenine derivative with cytokinin activity in plants was isolated from the slime mold *Dictyostelium discoideum*. Discadenine is 3-(3-amino-3-carboxypropyl)-N^6-(2-isopentenyl)adenine (171) (Fig. 11). Because of the unique structure and interesting physiological activity of discadenine, attemps have been made to elucidate the mechanism of its biosynthesis and the mechanism of its action. Discadenine can be synthesized from N^6-(2-isopentenyl)adenine by transfer of the 3-amino-3-carboxypropyl group from S-adenosylmethionine (AdoMet) (172). The enzyme responsible for this pathway was named discadenine synthase. Fruiting bodies of *D. discoideum* contain a large amount of free i^{6}Ade, a precursor of discadenine (173). It is, however, not known whether i^{6}Ade has any biological function in *D. discoideum* or is merely a precursor of discadenine. 5'-AMP is the direct precursor of the cytokinin in *D. discoideum*, being the actual acceptor molecule for the isopentenyl group from 2-isopentenyl pyrophosphate; the i^{6}AMP synthesized is then converted to i^{6}Ade (174).

The spore germination inhibitor and derivatives lacking either the NH_2 or COOH group, or the isopentenyl residue, have been chemically synthesized (175). Only the synthetic product that includes the same functional groups as the natural compound shows biological activity.

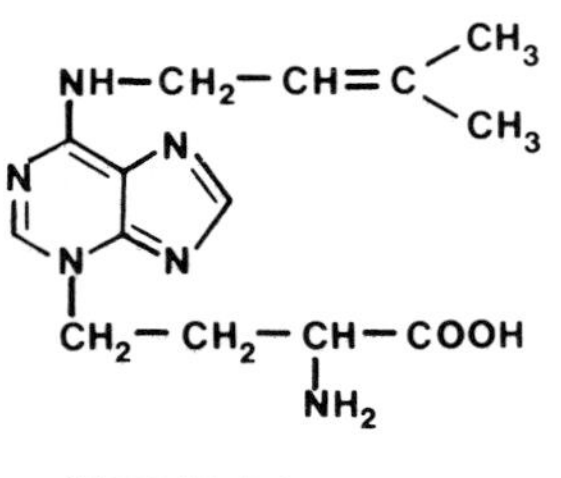

Discadenine

FIG. 11. The structure of discadenine, 3-(3-amino-3-carboxypropyl)-N^6-(2-isopentenyl)adenine.

Discadenine inhibits the expression of certain proteins that are formed in the early phase of spore germination (176). This inhibitory effect can be reversed by cyclic AMP, which also abolishes the inhibitory effect of discadenine on spore germination.

Spores can be activated to germinate in water by the addition of dimethyl sulfoxide (Me_2SO). The spores will germinate under these conditions even in the presence of discadenine. The mechanisms by which Me_2SO activates spore germination is not yet known. However, it has been observed in *E. coli* that the inducible enzyme of the *gal* operon, galactokinase, in synthesized constitutively in the presence of Me_2SO (177). This has been explained as an effect on the catabolite activator protein (CAP) in that Me_2SO relieves the necessity for cAMP. Catabolite repression has been postulated to occur in *Dictyostelium* as in prokaryotes (178). If this is true, it is possible that discadenine competes with cAMP for an activator protein analogous to prokaryotes.

VII. On the Function of Queuine in tRNA

The analysis of the primary sequence and modification of specific tRNAs has shown characteristic variations in differentiating aging or neoplastic transformed cells. The mechanisms by which alterations of tRNA modifications are related to metabolic control has yet to be clarified (for review, see 4, 10, 52, 84, 89, 179–181). New biological model systems may help to elucidate the predicted importance of tRNA modification in cellular regulation.

The modified nucleoside queuosine (Q) is most frequently discussed as being involved in cell differentiation (182–185).

Queuosine and glycosylated derivatives occur at position 34 in the anticodon of $tRNA^{Tyr}$, $tRNA^{His}$, $tRNA^{Asn}$, and $tRNA^{Asp}$ of eubacteria and of eukaryotes, except yeast. In eukaryotic $tRNA^{Asp}$ and $tRNA^{Tyr}$, glycosylated derivatives of Q (Q*) occur. In eukaryotes, the base of Q (queuine) is inserted into the corresponding tRNAs by exchange with the guanine at position 34 by tRNA-guanine ribosyltransferase (EC 2.4.2.29) (186). Eubacterial tRNAs are almost completely modified with respect to Q, whereas tRNAs of eukaryotes exhibit variable degrees of Q content (187–189). The Q content in tRNA changes during development of *Drosophila* (190, 191) and during erythroid differentiation of leukemia cells (183, 184). tRNAs with G in place of Q accumulate in fetal and regenerating rat liver and in a variety of animal tumors (52, 188). Germfree mice, fed on a defined diet, do not synthesize queuine, indicating that the modification is derived from nutrition or from the intestinal flora (191, 192).

The structure, biosynthesis, and alterations of Q-modification in tRNA during differentiation and neoplastic transformation have been described in detail in this series by Nishimura (193). I therefore confine the following sections to biological model systems that have turned out to be useful for investigations of the function of Q.

A. Queuine in *Dictyostelium discoideum*

The slime mold *Dictyostelium discoideum* is the simplest multicellular eukaryote. Wild-type strains grow naturally on bacteria, and axenic strains grow vegetatively as single cells on a partially or fully defined medium (for review, see 92). Developmental transition from unicellular to multicellular stages is induced by starvation, and proceeds when a certain amount of cells is placed on agar or filter supports in phosphate buffer. After an early developmental preaggregation stage, during which the cells become responsive to pulses of cAMP and form cAMP-binding sites on the cell surface (194, 195), the cells aggregate into mounds containing variable amounts of cells (average, 10^5). The mounds form slugs that subsequently differentiate into mature fruiting bodies, comprising spores at the top of a vacuolized stalk (for reviews, see 92) (Fig. 12). Early developmental changes also occur when the cells are starved in shaking suspension (195).

The Q content in specific tRNAs changes in *D. discoideum* during development and depends on the environmental conditions (10, 196, 197). Wild-type strains of the slime mold are supplied with queuine by bacteria and the Q family of tRNA is fully modified with respect to Q. When the axenic strain AX-2 is grown on a partially defined medium containing yeast extract and peptone about 60% of the Q family of tRNA is Q deficient. If peptone is omitted and replaced by amino acids, the tRNAs totally lack Q (196).

The addition of the modified base (at 10^{-7} M) to queuine-free medium does not change the growth rate of the amoebae, but causes an almost complete modification of the corresponding tRNAs, and an increase of tRNA isoacceptors containing Q* that are specific for Asp and Tyr.

Labeling of cells with [^{32}P]orthophosphate and quantitation of purified tRNA^{Asp} show definitely that the amount increases about 2-fold. Inhibition of RNA synthesis in logarithmically grown cells by actinomycin C, and analysis of the specific tRNAs, indicate that tRNA^{Asp} (Q*-34) is remarkably stable compared to tRNA^{Asp} (G-34) and to tRNA^{Asn} (Q-34). This suggests that further modification of Q to Q*

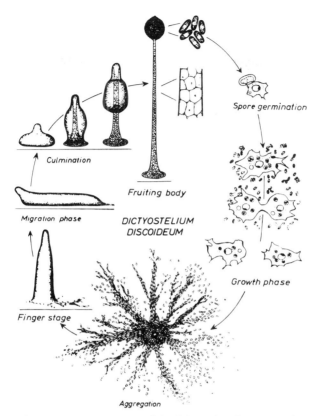

Fig. 12. Schematic representation of the life cycle of *Dictyostelium discoideum*. After growth has stopped, owing to the lack of nutrients, the cells start to aggregate and form a pseudo-plasmodium that contains up to 10^5 cells. The aggregate is subsequently transformed into a fruiting body that is 1–2 mm high and produces about 7×10^4 spores. The spores can germinate and give rise to single ameboid cells (*195*).

decreases the turnover rate of the respective tRNAs (*197a*). Interestingly, the kinetic analysis of aminoacylation of mammalian tRNA^{Asp} (Q*-34) reveals a higher V_{max} than for tRNA^{Asp} (G-34) (*197b*). The higher rate of the aminoacylation of the modified species might play a role in the reduced turnover of tRNA^{Asp} (Q*-34).

Development of amoebae, partly or totally lacking Q in their tRNAs, is favored by the addition of at most 10^{-7} M queuine. Morphological changes show an enlargement of aggregates and an increase in the size of fruiting bodies. The restoration of the Q content in tRNA also accelerates the germination of spores (*197, 198*).

During the preaggregation stage, about 8–10 hours after the onset of development, cAMP receptors are produced (199, 200). Restoration of the Q content in tRNA increases the level of cAMP binding to its surface receptors. In addition, the time course of the synthesis of some early regulated proteins is accelerated, but a few other remarkable changes are found in the overall pattern of protein synthesis (201).

When cells lacking Q in their tRNAs are kept for 24 hours in the preaggregation stage in suspension culture, they lose their capacity to aggregate, whereas queuine-treated cells become aggregation-competent and continue morphogenesis. These observations show that queuine is of survival advantage for D. discoideum. Similar results are described for E. coli: In a mutant strain, lacking the tRNA-guanine ribosyltransferase, and thus Q in corresponding tRNAs, the viability is remarkably reduced when the cells are kept under unfavorable conditions during the stationary phase (202).

B. A Possible Role of Queuine and Queuine-Containing tRNAs in Differentiation and Neoplastic Transformation

The synthesis and metabolism of queuine shares common metabolic pathways with pteridines (203). Some pteridines inhibit the insertion of queuine into tRNA by the tRNA-guanine ribosyltransferase in a competitive manner. Pteridines play a role in electron-transfer reactions involving mixed-function oxygenases during the hydroxylation of aromatic amino acids (204). A possible linkage between these metabolic routes and Q modification in tRNA is therefore postulated.

Q deficiency in E. coli tRNA can be caused by the addition of biopterin to resting, aerobically kept cells, conditions under which tRNAs are synthesized. Under anaerobic conditions, Q-lacking tRNAs are not formed (205). These results suggest that alterations in the Q content of tRNA can be caused either by an increase in the amount of biopterin (or related pteridines) and/or changes in their redox state.

The extent of Q modification in tRNAs fluctuates in rat tissues in an ordered manner during aging, being essentially complete at 9 months (206). Exactly at this time, the activities of lactate dehydrogenase isoenzymes reach maximum values (207). These observations, and the fact that yeast totally lacks Q in tRNA and does not form lactate in anaerobic conditions pose the question whether the observed metabolic changes in response to Q modification in tRNA might be involved in the regulation of the metabolism of lactate.

NAD-dependent lactate dehydrogenases (nLDH) are key enzymes

in energy metabolism in bacteria and eukaryotes. Two types specific either for D(−)-lactate (EC 1.1.1.28) or L(+)-lactate (EC 1.1.1.27) have long been known. D(−)-nLDHs have been detected in bacteria (208), fungi (209), in the slime mold *Polysphondylium* (210). L(+)-nLDHs occur in animals and microorganisms. L(+)-nLDHs from vertebrates are divided into five tetrameric isoenzymes. The subunits A (muscle type, anaerobic) and B (heart type, aerobic) are encoded by two different structural genes. Additionally, an LDH C gene has been discovered in mammals, birds, and fishes (for summarizing references see 211). Each tissue or organ has its characteristic L(+)-nLDH isoenzyme pattern, which is changed during embryogenesis, development, and aging (207, 212, 213).

NAD-independent LDHs (iLDH) are found in a variety of bacteria (for summarizing references, see 208), yeast (214), and also in ascites cells (215). iLDHs are found in complexes together with cytochromes of type b and convert lactate to pyruvate; the acceptor of electrons is cytochrome c (EC 1.1.2.3 and 1.1.2.4).

Until recently little was known about the occurrence and function of LDH enzyme in *D. discoideum*. It has been reported that the activity of NAD-dependent lactate dehydrogenase is rather low (216, 217) and that the enzyme is not developmentally regulated (218).

1. LACTATE AND LACTATE DEHYDROGENASES IN
   VEGETATIVE AMOEBAE

In vegetative cells of strain AX-2, grown with sufficient supply of queuine, a high level of D(−)-lactate (millimolar amounts) is present and a highly reproducible pattern of nine different forms of D(−)-nLDHs is observed (219). As described for other organisms (220), reactions catalyzed by D-LDHs may provide energy also in *D. discoideum*. In queuine-lacking cells, the level of D(−)-lactate is lower than in queuine-treated cells (see Table II, Vegetative cells); the pattern of D(−)-nLDH enzymes is not influenced by queuine.

In vegetative amoebae grown in the presence of queuine, small, but significant, amounts of L(+)-lactate are present. Correspondingly, a low overall activity of L(+)-nLDH is detectable. At least three multiple forms of L(+)-nLDH occur; they might reflect the different isoenzymes observed in other eukaryotes. If this assumption is correct, at least two different genes encoding subunits are present in *D. discoideum*.

Queuine-lacking, vegetative amoebae, compared to those grown with queuine have lower levels of L(+)-lactate (see Table II, Vegetative cells). The multiple forms in the enzyme pattern of L(+)-LDHs are much less pronounced.

It is therefore possible that queuine or Q-tRNAs influence the expression of genes encoding the different subunits or the expression of modifying enzymes that affect subunit association. For example, tyrosine residues of LDH can be modified by phosphorylation (221).

2. Lactate and Lactate Dehydrogenases during Development

Characteristic changes are found in the pattern of nLDH isoenzymes during the life cycle of *D. discoideum* AX-2 (222). Likewise, Q modification of tRNA is decreased during development (223–225).

D(−)-Lactate decreases after the onset of development in starved cells and predominantly when queuine is lacking (Table II, Starvation). Characteristic changes are also observed in the pattern of D(−)-nLDHs. The different D(−)-LDH enzymes may be part of redox systems in membranes that provide energy, similar to *E. coli* (226). (See addendum to this article at the end of the volume.)

The intracellular level of L(+)-lactate also decreases during starvation and is more pronounced when queuine becomes limiting (Table II). As in other organisms undergoing differentiation, the pattern of L(+)-nLDH isoenzymes is altered during development in *D. discoideum*.

TABLE II
Queuine-Dependent Changes of D(−)- and L(+)-Lactate: Levels of D(−)- and L(+)-Lactate in Cells during Growth and Preaggregation in Suspension Culture (24 Hours)[a]

	Vegetative cells		Starvation 4 hours				Starvation 24 hours			
	Q^-	Q^+	Q^{--}	Q^{+-}	Q^{-+}	Q^{++}	Q^{--}	Q^{+-}	Q^{-+}	Q^{++}
D(−)-lactate	15 $(1.35)^b$	26 $(2.25)^b$	13	19	20	22	8	9	15	19
L(+)-lactate	1.7 $(0.15)^b$	2.4 $(0.21)^b$	0.3	—	1.0 ± 0.2	—	0.2	—	1.0 ± 0.2	—

[a] The same results (expressed in nmol/mg protein) were obtained in two independent repeated experiments. Cells were grown in a defined synthetic medium, supplemented with yeast extract without (−) and with (+) 10^{-7} M queuine. (Q^-) or (Q^+) cells were starved in the absence—(Q^{--}) and $Q^{(+-)}$ cells—or in the presence—(Q^{-+}) and (Q^{++}) cells, respectively— 10^{-7}M queuine.

[b] Values in parentheses represent the approximate intracellular concentration of lactate, calculated from nanomoles per milligram of protein (1 mg of protein $\triangleq$ 1 × 10^7 vegetative cells), and a cell volume of 1.17 × 10^3 μm^3 (222).

Cytochromes b are components of iLDHs. In *D. discoideum*, cytochromes b apparently occur in microsomal fractions as well as in mitochondria (227). In axenically grown strains, partially lacking Q in their tRNAs, cytochrome b_{559} accumulates during transition from logarithmic growth to the stationary phase (Fig. 13A–D). The most pronounced differences appeared in the cytochrome spectra of queuine-lacking or of queuine-containing cells upon developmental transition (Fig. 13E–H). Cytochrome b_{559} exhibits an absorption maximum almost identical to that of cytochrome b_2, a component of the NAD-independent LDHs (iLDHs). Those enzymes, in *Saccharomyces cerevisiae* transfer electrons from D(−)- or L(+)-lactate directly to cytochrome c. The transport of the two electrons is coupled to a site of phosphorylation and 1 mol of ATP is formed per mole of lactate (214). Assuming that cytochrome b_{559} is a component of an iLDH in *Dictyostelium*, lactate might be oxidized in queuine-lacking cells during the whole starvation period by a similar pathway.

It is tentatively assumed that queuine or Q-tRNA normally repress the oxidation of lactate to pyruvate via this shunt and thereby are involved in the control of aerobiosis and lactate metabolism.

The functional differences of Q-containing and Q-lacking tRNAs have recently been summarized (3, 6). The most important finding in this respect is that a Q-lacking species of tRNA[Tyr] reads through a terminator codon of TMV RNA, whereas the Q-containing counterpart stops (228). An alteration in the Q content of the corresponding tRNAs might therefore be involved in the regulation of gene expression of proteins necessary for the above discussed pathways of lactate metabolism and redox reactions. Since cyclic AMP appears to be involved in the regulation of lactate dehydrogenases (229), and queuine or Q tRNAs influence the binding of cAMP to its receptors (201), the modified nucleoside might affect transcription or translation of LDHs in concert with cyclic AMP.

Morphogenesis is regulated in *Dictyostelium discoideum* by specific morphogens. The activity of the low-molecular-weight hydrophobic factor DIF, responsible for stalk cell differentiation is antagonized by ammonia (230). It is assumed that the choice of differentiation is mediated by the intracellular pH. Therefore LDHs are suggested to play an important role in establishing intracellular pH or proton gradients in membranes.

C. Queuine and the Melanophore System of Xiphophorine Fish

In one approach to ascertain the role of alterations in the Q content of tRNAs specific for Asn, Asp, His, and Tyr in relation to differentia-

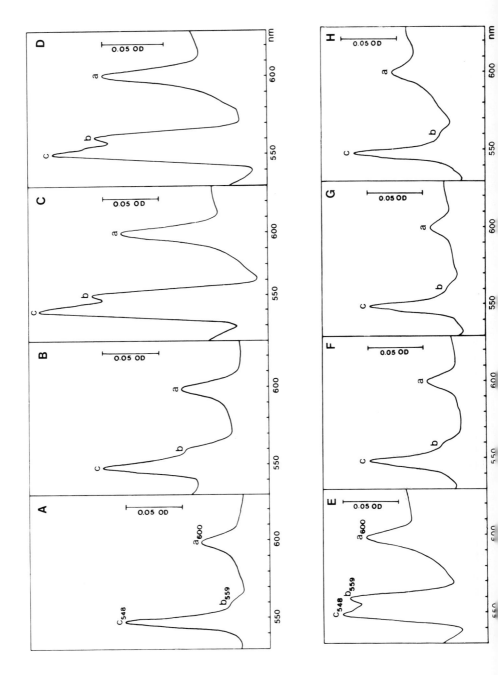

tion, neoplastic transformation, and pteridine metabolism, the melanophore system of *Xiphophorus* serves as a model. *Xiphophorus* is a genus of small viviparous freshwater fish. Crossing conditioned heritable melanoma can occur in the progeny of *X. helleri* (swordtail) and *X. maculatus* (platy) caused by the expression of the oncogene "Tu", which is normally repressed by regulatory "R" genes, e.g., the non-linked R_{Diff}. The oncogene Tu is probably identical with the *c-src* gene. Malignant melanoma develop in Tu-carrying hybrids, lacking the R_{Diff}-containing chromosomes (*231*).

Malignant melanomas contain variable, but always higher, amounts of Q-lacking tRNAs than "benign melanoma," in which the transformed melanoblasts finally differentiate to melanophores. In the normal skin of *Xiphophorus*, tRNAs with G in place of Q accumulate; this is more pronounced in *X. helleri* ($R_{Diff}-$) than in *X. maculatus* ($R_{Diff}+$). Thus the product of R_{Diff} might be somehow related to the insertion of Q into tRNA (*182*).

Pteridines [which inhibit in a competitive manner the transglycosylase reaction and thus the insertion of queuine into tRNA (*203*)] occur in variable amounts and patterns in the skin of different genotypes of *Xiphophorus*. tRNA isolated from two genotypes showed that high levels of Q^--tRNA are correlated with a high pteridine content of the skin, possibly favorable conditions for tumor development.

VIII. Summary and Outlook

In this chapter, the role of T-54, i^6A-37, and Q-34 and their derivatives in tRNA function is discussed. A total lack of any one of these modifications in an organism that usually contains the normal nucleoside in specific tRNAs decreases its ability to survive under unfavor-

---

Fig. 13. Changes of difference spectra of cytochromes in *D. discoideum* during growth and development and in response to queuine. Low-temperature spectra were taken from whole cells. About 1.3×10^9 cells in 2 ml of phosphate buffer (17 mM) were resuspended in the same volume of glycerol. One portion was reduced with $Na_2S_2O_4$, and one was oxidized with $K_3Fe(CN)_6$, each at 5 mM final concentration. The difference spectra were measured between 530 and 630 nm, using an Aminco spectrophotometer with low-temperature equipment. Spectra of axenically grown cells from (A) the early, (B) the middle, (C) the late log phase, and (D) the stationary phase.

(E–H) Spectra of preaggregating cells. Cells were grown in a defined queuine-free medium supplemented with yeast extract with and without added queuine and starved in phosphate buffer in a shaking suspension for 24 hours in the presence or the absence of queuine. Shown are growth and preaggregation without queuine (E); growth with, preaggregation without, queuine (F); growth without, preaggregation with, queuine (G); growth and preaggregation with queuine (H).

able environmental conditions. From the available information, answers are emerging to questions that were raised in the past: Are tRNAs so elaborate to achieve the remarkable accuracy of translation (232, 233)? Is nucleoside modification a procedure by which the adaptability of tRNA to new functions can be increased without introducing mutations into strongly selected tRNA genes (234)? Does translational ambiguity increase during cell differentiation, and does this involve alterations in tRNA modification (235)?

The modification T in elongator tRNAs of eubacteria and of eukaryotes increases the V_{max} of ribosomal A-site interaction of a cognate (EF-Tu)·GTP·(AA-tRNA) complex. These changes are probably important to regulate the overall rate of protein synthesis.

In cell-free systems of homopeptide synthesis, the absenc eof T in tRNA increases significantly the misincorporation of an amino acid by an error-prone tRNA. Several models have been proposed to explain the ribosomal accuracy (for review, see 236–239). Proofreading mechanisms are discussed, based on the finding that the binding of a noncognate (EF-Tu)·GTP·(AA-tRNA) complex to the programmed ribosome is accompanied by the cleavage of many more GTP molecules than that of the cognate complex. An alteration of methylation of U-54 in tRNA might therefore cause subtle changes in the energy requirement for the selection process.

Initiator tRNA of eubacteria, unlike eukaryotic initiator tRNA, contains T at position 54. A new tRNA$^{fMet(2)}$ partially lacking T-54 has been discovered in amino acid auxotrophs of E. coli. In contrast to the modified tRNA$^{fMet(1)}$ (T-54) that stops at AUG initiation codons when unformylated, the new form initiates, during MS-2 phage RNA translation in vitro, the coat protein and the replicase cistrons without prior formylation. Unformylated Met-tRNA$^{fMet(2)}$ accumulates in amino-acid auxotrophic strains of E. coli under H$_4$folate limitation. Under these conditions, RNA-polymerase retains its specificity for rRNA and for r-protein genes. In amino acid auxotrophs, H$_4$folate limitation gives rise to the formation of ppGpp that might partly counteract the expression of those genes and adapt their synthesis to the availability of H$_4$folate. Likely candidates for a metabolic control, involving tRNA$^{fMet(2)}$, are the initiation factors. The control is different from that in gram-positive bacilli that synthesise T in tRNA with H$_4$folate as coenzyme.

The modification of A-37 in specific tRNAs of microorganism also exhibits an all-or-none effect in codon recognition. The replacement of i^{6}A by A, or ms^{2}i^{6}A by i^{6}A reverses the suppression of particular codons by specific suppressor tRNAs. An incomplete modification of i^{6}A to ms^{2}i^{6}A in respective tRNAs of E. coli is caused in iron-restricted growth, because the enzymes involved in methylthiolation are dependent on iron. The available information supports the view that the tRNAs (i^6A), contrary to their counterparts with ms^2i^6A, will stop at corresponding codons for Phe and Trp in regulatory regions of mRNA, and that this relieves attenuation at the Phe and Trp operons. Intermediates of this pathway provide the aromatic ring system for the synthesis of the high-affinity iron-chelator enterochelin. This chelator overcomes iron starvation by accumulating iron from an environment already low in iron. Furthermore, the methylthiolation of i^6A in corresponding tRNAs is inhibited by repression of ornithine synthesis. Since ornithine is the precursor of hydroxamate siderophores—components of low-affinity iron-transport systems—a repression of ornithine synthesis could decrease the intracellular level of iron. Alterations of methylthiolation of i^6A in specific tRNAs are apparently involved in the control of expression of proteins involved in iron transport in microorganisms.

In eukaryotes, Q modification of tRNA has been suggested to play a role in processes of differentiation and aging and neoplastic transformation. Alterations of the Q content in the Q family of tRNAs can be caused intracellularly by limitation or alterations in the redox state of pteridines influencing the transglycosylase that inserts queuine into the tRNAs.

Dictyostelium discoideum proved to be a valuable model system to elucidate the function of Q tRNAs in growth and differentiation. Axenic strains grow in a defined medium as well in the absence as in the presence of queuine. If Q is lacking in tRNA, the cells are less tolerant of starvation during their developmental cycle. In Q-deficient, starving cells a specific cytochrome b_{559} accumulates; concomitantly, the levels of D(−)- and D(+)-lactate decrease. Cytochrome b_{559} is suggested to be a component of an NAD-independent lactate oxidoreductase, as in yeast, that transfers electrons from lactate to cytochrome c. Morphogenesis in *D. discoideum* is induced by specific morphogens; their activities probably depend on intercellular pH gradients. Queuine and Q tRNAs might exert their effect on differentiation by controlling the expression of enzymes of lactate metabolism and of membrane-bound oxidoreductases—important to establish intracellular pH gradients—and for the control of energy metabolism.

In general, alterations of specific modifications in corresponding tRNAs are linked to amino-acid metabolism, membrane transport, and energy metabolism (*130, 131, 240, 222*) and apparently involve alterations of the redox state or levels of coenzymes or cofactors. These in turn modulate the activity of tRNA-modifying enzymes; subsequently, the tRNA with an altered modification changes the expression of specific genes. The present state of knowledge supports the view that tRNA modification evolved primarily for regulatory functions and that the protein-synthesizing systems concomitantly developed with increasing flexibility for adapting cells to environmental changes, but likewise maintaining a high fidelity of translation.

At the molecular level, the presence or absence of a modified nucleoside in tRNA might change anticodon usage by respective codons in protein genes (241), but more likely codon/anticodon recognition is changed in regulatory regions of mRNA as shown in eubacteria for attenuator regions of amino-acid biosynthetic operons.

Extended models of attenuation have been presented for eukaryotes (242, 243). But what is the link between the control mechanisms in the nucleus and the cytosol? I emphasize that tRNA modifications could serve as a link between control of transcription and control of translation that has to be considered in addition to other regulatory mechanisms of gene expression in eukaryotes, reviewed recently (244).

Almost all of the modifying enzymes occur in the nucleus. Their activities are modulated by coenzymes or cofactors that can pass through the nuclear membrane. It is conceivable that the rate of transcription of a tRNA gene comes to a transient halt when the tRNA-modifying enzyme is blocked and that subsequently a signal (positive or negative) can be formed that permits or terminates transcription of a specific family of genes. Alternatively, the modifying enzyme in complex with the metabolite forms a "repressor-like" molecule. The undermodified tRNA is transported to the cytosol, where it stops or "reads through" a specific codon in the regulatory region of the corresponding mRNA. The modified base queuine is an exception. It is an essential nutrient, like folate, and becomes inserted into the corresponding tRNAs by a ribosyltransferase within the cytosol. In its free form, queuine is sensitive to oxidation in the cyclopentenediol ring and therefore is sensitive to alterations in the redox state of the cell. Free queuine might control the transcription of specific genes, encoding proteins involved in redox reactions, whereas alterations in the Q content of tRNA cause changes in the expression of corresponding genes at the translational level (for results that accord with this model, see Addendum).

This hypothesis is consistent with experimental results and theories of the control of gene expression. I hope that it stimulates further research on the role of modified nucleosides in tRNA and the molecular mechanisms that are involved in their regulatory functions.

References

1. E. Borek, *Control Processes in Neoplasia* 4, 147 (1974).
2. F. Nau, *Biochimie* 58, 629 (1976).
3. P. F. Agris and D. Söll, in "Nucleic Acid-Protein Recognition" (H. J. Vogel, ed.), p. 321. Academic Press, New York, 1977.
4. S. Nishimura, in "Transfer RNA" (S. Altman, ed.), p. 168. MIT Press, Cambridge, Massachusetts, 1978.
5. P. F. Agris, "A Bibliography of Biochemical and Biophysical Studies from 1970 to 1979." Liss, New York, 1980.
6. G. Dirheimer, G. Keith, A. P. Sibler, and R. P. Martin, in "Transfer RNA: Structure, Properties and Recognition" (P. R. Schimmel, D. Söll, and J. N. Abelson, eds.), p. 19. CSH Lab, CSH, New York, 1979.
7. S. Nishimura, in "Transfer RNA: Structure, Properties and Recognition" (P. R. Schimmel, D. Söll, and J. N. Abelson, eds.), p. 547. CSH Lab, CSH, New York, 1979.
8. R. P. Singhal and P. A. M. Fallis, This Series 23, 227 (1979).
8a. R. P. Singhal, E. F. Roberts, and V. N. Vakkaria, This Series 28, 211 (1983).
8b. G. R. Björk, in "Processing of RNA" (D. Apirion, ed.), p. 291. Boca Raton, Florida, CRC, 1984.
8c. J.-H. Alix and D. Hayes, *Biol. Cell.* 47, 139 (1983).
9. G. Dirheimer, in "Modified Nucleosides and Cancer" (G. Nass, ed.), p. 15. Springer-Verlag, Berlin and New York, 1983.
10. H. Kersten, in "Biochemistry of S-Adenosylmethionine and Related Compounds" (E. Usdin, R. T. Borchardt, and C. R. Creveling, eds.), p. 357. Macmillan, London, 1982.
11. J. Ofengand, in "Protein Biosynthesis in Eukaryotes" (R. Pérez-Bercoff, ed.), p. 1. Plenum, New York, 1982.
12. M. Sprinzl and D. H. Gauss, *NARes* 10 (#2), rl (1982).
13. L. R. Mandel and E. Borek, *Bchem* 2, 555 (1963).
14. E. Borek and P. R. Srinivasan, *ARB* 35, 275 (1966).
15. R. Cortese, R. Landsberg, R. A. von der Haar, H. E. Umbarger, and B. N. Ames, *PNAS* 71, 1857 (1974).
16. H. Hori and S. Osawa, *PNAS* 76, 381 (1976).
17. W. J. Sharrock, W. M. Gold, and J. C. Rabinowitz, *JMB* 135, 627 (1979).
18. H. Kersten, in "Transmethylation" (E. Usdin, R. T. Borchardt, and C. R. Creveling, eds.), p. 419. Elsevier/North-Holland, Amsterdam, 1979.
19. B. A. Roe, E. Y. Chen, and H. Y. Tsen, *BBRC* 68, 1339 (1976).
20. S. J. Kerr and E. Borek, in "The Enzymes" (P. D. Boyer, ed.), 3rd ed., p. 167. Vol. 9 Academic Press, New York, 1973.
21. H. H. Arnold, W. Schmidt, and H. Kersten *FEBS Lett.* 53, 258 (1975).
22. H. Kersten, L. Sandig, and H. H. Arnold, *FEBS Lett.* 55, 57 (1975).
23. A. S. Delk and J. C. Rabinowitz, *PNAS* 72, 528 (1975).
24. W. Schmidt, H. H. Arnold, and H. Kersten, *J. Bact.* 129, 15 (1977).
25. A. S. Delk, D. P. Nagle, Jr., and J. C. Rabinowitz, *BBRC* 86, 244 (1979).
26. Y. Kuchino, M. Ihara, Y. Yabusaki, and S. Nishimura, *Nature* 298, 684 (1982).
27. H. Pang, M. Ihara, Y. Kuchino, S. Nishimura, R. Gupta, C. R. Woese and J. A. McCloskey, *JBC* 257, 3589 (1982).
28. A. Schousboe, L. Herz, and B. Svenneby, *Neurochem. Res.* 2, 217 (1977).
29. K. P. Schäfer and D. Söll, *Biochemie* 56, 795 (1974).

30. L. K. Kwong, V. G. Moore, and I. I. Kaiser, *JBC* **252**, 6310 (1977).
31. G. R. Björk and F. C. Neidhardt, *J. Bact.* **124**, 99 (1975).
32. G. R. Björk and F. C. Neidhardt, *Cancer Res.* **31**, 706 (1971).
33. A. Zamir, R. A. Holley, and M. Marquise, *JBC* **240**, 1267 (1965).
34. J. Ofengand and C. Henes, *JBC* **244**, 6241 (1969).
35. N. Shimizu, H. Hayashi, and K.-I. Miura, *J. Biochem. (Tokyo)* **67**, 373 (1970).
36. J. Ofengand, *in* "Ribosomes, Structure, Function and Genetics" (G. Chambliss, G. R. Craven, J. Davies, K. Davis, L. Kahan, and M. Nomura, eds.), p. 497. Univ. Park Press, Baltimore, Maryland, 1980.
37. B. N. White and B. T. Bayley, *BBA* **272**, 583 (1972).
38. C. E. Samuel and J. C. Rabinowitz, *JBC* **249**, 1198 (1974).
39. A. S. Delk and J. C. Rabinowitz, *Nature (London)* **252**, 106 (1974).
40. H. H. Arnold, *BBA* **476**, 76 (1977).
41. P. Davanloo, M. Sprinzl, K. Watanabe, A. Albani, and H. Kersten, *NARes* **6**, 1571 (1979).
42. A. Hoburg, H. J. Aschhoff, H. Kersten, U. Manderschied, and H. G. Gassen, *J. Bact.* **140**, 408 (1979).
43. H. U. Petersen, A. Danchin, and M. Grunberg-Manago, *Bchem* **15**, 1357 (1976).
44. H. U. Petersen, A. Danchin, and M. Grunberg-Manago, *Bchem* **15**, 1362 (1976).
45. H. J. Rheinberger and K. H. Nierhaus, *Bchem. Int.* **1**, 297 (1980).
46. H. J. Rheinberger, H. Sternbach, and K. H. Nierhaus, *PNAS* **78**, 5310 (1981).
47. S. Watanabe, *JMB* **67**, 443 (1972).
48. P. Wurmbach and K. H. Nierhaus, *PNAS* **76**, 2143 (1979).
49. H. Kersten, M. Albani, E. Männlein, R. Praisler, P. Wurmbach, and K. H. Nierhaus, *EJB* **114**, 451 (1981).
50. J. L. Lessard and S. Pestka, *JBC* **247**, 6909 (1972).
51. P. C. Jelenc and C. G. Kurland, *PNAS* **76**, 3174 (1979).
52. H. Kersten, *in* "Modified Nucleosides and Cancer" (G. Nass, ed.), p. 255. Springer-Verlag, Berlin and New York, 1983.
53. R. Praisler, E. Männlein, H. Richter, G. Ott, and H. Kersten, *in* "Regard sur la Biochimie," p. 55. Société de Chimie Biologique, Strasbourg, France, 1981.
54. G. Ott, R. Praisler, E. Männlein, and H. Kersten, *ZpChem* **363**, 884 (1982).
55. T. Ruusala, M. Ehrenberg, and C. G. Kurland, *EMBO J.* **6**, 741 (1982).
55a. F. Bouadloun, D. Donner, and C. G. Kurland, *EMBO J.* **2**, 1351 (1983).
56. J. Parker, J. W. Pollard, J. D. Friesen, and C. P. Stanners, *PNAS* **75**, 1091 (1978).
57. P. Z. O'Farrell, M. H. Goodman, and P. H. O'Farrell, *Cell* **12**, 1133 (1977).
58. J. Gallant and D. Foley, *in* "Ribosomes, Structure, Function and Genetics" (G. Chambliss, G. R. Craven, J. Davies, K. Davis, L. Kahan, and M. Nomura, eds.), p. 615. Univ. Park Press, Baltimore, Maryland, 1980.
59. M. Nomura, E. A. Morgan, and S. R. Jaskunase, *ARGen* **11**, 297, 1977.
60. W. A. Haseltine and R. Block, *in* "Ribosomes" (M. Nomura, A. Tissiéres, and P. Lengyel, eds.), p. 747. CSHLab, CSH, New York, 1974.
61. F. Lipmann and J. Sy, This Series **17**, 1 (1976).
62. D. Richter and K. Isono, *Curr. Top. Microbiol. Immunol.* **76**, 83 (1977).
63. J. Sy, *Bchem* **15**, 606 (1976).
64. J. Sy and H. Akers, *Bchem* **15**, 4399 (1976).
65. S. Fehr, F. Godt, K. Isono, and D. Richter, *FEBS Lett.* **97**, 91 (1979).
65a. D. Richter, S. Fehr, and R. Harder, *EJB* **99**, 57 (1979).
66. O. Pongs, R. Bald, and E. Reinwald, *EJB* **32**, 117 (1973).

67. J. Ofengand and C. Henes, *JBC* **244**, 6241 (1969).
68. D. Richter, V. Erdmann, and M. Sprinzl, *Nature (London)* **246**, 132 (1973).
69. V. Erdmann, M. Sprinzl, and O. Pongs, *BBRC* **54**, 942 (1973).
70. D. Richter, V. Erdmann, and M. Sprinzl, *PNAS* **71**, 3226 (1974).
71. T. Ny, J. Tomalle, K. Halverson, G. Nass, and G. R. Björk, *BBA* **607**, 277 (1980).
72. G. R. Björk, A. S. Byström, T. G. Hagervoll, K. J. Hjalmarrson, K. Kjellin-Straby, and P. H. R. Lindström, in "Biochemistry of S-Adenosylmethionine and Related Compounds" (E. Usdin, R. T. Borchardt, and C. R. Creveling, eds.), p. 371. Macmillan, London, 1982.
72a. B. R. Baumstark, L. L. Spremulli, U. L. RajBhandary, and G. M. Brown, *J. Bact.* **129**, 457 (1977).
73. M. Grunberg-Manago, in "Ribosomes, Structure, Function and Genetics" (G. Chambliss, G. R. Craven, J. Davies, K. Davis, L. Kahan, and M. Nomura, eds.), p. 445. Univ. Park Press, Baltimore, 1980.
74. H. U. Petersen, E. Joseph, A. Ullmann, and A. Danchin, *J. Bact.* **135**, 453 (1978).
75. E. Männlein, R. Praisler, E. Stockhausen, and H. Kersten, *Z. Chem* **364**, 1176 (1983).
75a. P. G. Debenham, O. G. Pongs, and A. A. Travers, *PNAS* **77**, 870 (1980).
75b. R. E. Kingston, W. C. Nierman, and M. J. Chamberlin, *JBC* **256**, 2787 (1981).
75c. G. A. Kassavetis and M. J. Chamberlin, *JBC* **256**, 2777 (1981).
75d. M. Kajitani and A. Ishihama, *JBC* **259**, 1951 (1984).
76. H. H. Arnold and A. Ogilvie, *BBRC* **74**, 343 (1977).
77. S. R. Kahn and H. Yamazaki, *BBRC* **48**, 169 (1972).
77a. P. O. Olins and M. Nomura, *NARes* **9**, 1757 (1981).
77b. M. Nomura, S. Jinks-Robertson, and A. Miura, in "Interaction of Translational and Transcriptional Controls in the Regulation of Gene Expression" (M. Grunberg-Manago and B. Safer, eds.), p. 91. Elsevier Biomedical, New York, 1982.
77c. L. Lindahl, R. H. Archer, and J. M. Zengel, in "Interaction of Translational and Transcriptional Controls in the Regulation of Gene Expression" (M. Grunberg-Manago and B. Safer, eds.), p. 105. Elsevier, Amsterdam, 1982.
77d. St. Pedersen, J. Skouv, T. Christensen, M. Johnsen, and N. Fill, in "Interaction of Translational and Transcriptional Controls in the Regulation of Gene Expression" (M. Grunberg-Manago and B. Safer, eds.), p. 119. Elsevier, Amsterdam, 1982.
77e. J. W. Hershey, J. G. Howe, J. A. Plumbridge, M. Springer, and M. Grunberg-Manago in "Interaction of Translational and Transcriptional Controls in the Regulation of Gene Expression" (M. Grunberg-Manago and B. Safer, eds.), p. 91. Elsevier, Amsterdam, 1982.
78. B. A. Roe and H. J. Tsen, *PNAS* **74**, 3696 (1977).
79. R. Reszelbach, R. Greenberg, R. Pirtle, R. Prasad, K. Marcu, and B. Dudock, *BBA* **475**, 383 (1977).
80. K. B. Marcu and B. S. Dudock, *Nature (London)* **261**, 159 (1976).
81. K. Marcu, R. Mignery, R. Reszelbach, B. Roe, M. Sirover, and B. Dudock, *BBRC* **55**, 477 (1973).
82. D. P. Ma, W. Merrick and B. A. Roe, in "Biochemistry of S-Adenosylmethionine and Related Compounds" (E. Usdin, R. T. Borchardt, and C. R. Creveling, eds.), p. 385. Macmillan, London, 1982.
83. N. Sueoka and T. Kano-Sueoka, *This Series*, **10**, 23 (1970).
84. E. Borek and S. J. Kerr, *Adv. Cancer Res.* **15**, 163 (1972).
85. S. Nishimura, in "Transfer RNA" (S. Altman, ed.), p. 168. MIT Press, Cambridge, Massachusetts, 1978.

86. E. Randerath, A. S. Gopalakrishnan, and K. Randerath, in "Morris Hepatomas, Mechanism of Regulation" (H. P. Morris and W. E. Criss, eds.), p. 517. Plenum, New York, 1978.
87. L. A. Osterman, *Biochimie* **61**, 323 (1979).
88. R. Cortese, in "Biological Regulation and Development" (R. F. Goldberger, ed.), p. 401. Plenum, New York, 1979.
89. E. Borek, *Trends Biochem. Sci. (Pers. Ed.)* **2**, 3 (1977).
90. W. Schmidt, H. Kersten, and H. G. Schweiger, in "Progress in Acetabularia Research (C. L. F. Woodcock, ed.), p. 39. Academic Press, New York, 1977.
91. Dingermann, W. Schmidt, and H. Kersten, *FEBS Lett.* **80**, 205 (1977).
92. W. F. Loomis, "*Dictyostelium discoideum:* A Developmental System." Academic Press, New York, 1975.
93. T. Dingermann, F. Pistel, and H. Kersten, *EJB* **104**, 33 (1980).
94. T. Dingermann, M. Mach, and H. Kersten, *J. Gen. Microbiol.* **115**, 223 (1979).
95. D. Pillinger and E. Borek, *PNAS* **62**, 1145 (1969).
96. O. K. Sharma and E. Borek, *J. Bact.* **101**, 705 (1970).
97. B. Dudock, R. Greenberg, L. E. Fields, M. Brenner, and C. M. Palatnik, *BBA* **608**, 295 (1980).
98. E. Randerath, C.-T. Yu, and K. Randerath, *Anal. Biochem.* **48**, 172 (1972).
99. F. Pistel and H. Kersten, in "Regard sur la Biochimie," p. 65. Société de Chimie Biologique, Strasbourg, France, 1981.
100. D. D. Brown, *Science* **211**, 667 (1981).
101. B. D. Hall, St. G. Clarkson, and G. Tocchini-Valentini, *Cell* **29**, 3 (1982).
102. S. Sharp, D. DeFranco, T. Dingermann, P. Farrell, and D. Söll, *PNAS* **78**, 6657 (1981).
103. G. Ciliberto, L. Castagnoli, D. A. Melton, and R. Cortese, *PNAS* **79**, 1195 (1982).
104. H. Hofstetter, A. Kressmann, and M. L. Birnstiel, *Cell* **24**, 573 (1981).
105. G. Galli, H. Hofstetter, and M. L. Birnstiel, *Nature (London)* **294**, 626 (1981).
106. G. Ciliberto, C. Traboni, and R. Cortese, *PNAS* **79**, 1921 (1982).
107. W. R. Folk, H. Hofstetter, and M. L. Birnstiel, *NARes* **10**, 7153 (1982).
108. W. Gruissem, M. Prescott, B. M. Greenberg, and R. B. Itallick, *Cell* **30**, 81 (1982).
109. D. A. Melton, E. M. DeRobertis, and R. Cortese, *Nature (London)* **284**, 143 (1980).
110. R. C. Ogden, G. Knapp, C. L. Peebles, J. Johnson, and J. Abelson, *Trends Biochem. Sci. (Pers. Ed.)* **6**, 154 (1981).
111. J. J. Janzer, J. P. Raney, and B. D. McLennan, *NARes* **10**, 5663 (1982).
112. M. Buck, J. A. McCloskey, B. Basile, and B. N. Ames, *NARes* **10**, 5649 (1982).
112a. P. F. Agris, D. J. Armstrong, K. P. Schäfer, and D. Söll, *NARes* **3**, 691 (1975).
112b. J. Weissenbach and H. Grosjean, *EJB* **116**, 207 (1981).
112c. R. G. Pergolizzi, D. L. Engelhardt, and D. Grunberger, *NARes* **6**, 2209 (1979).
112d. R. G. Pergolizzi, D. L. Engelhardt, and D. Grunberger, *JBC* **253**, 6341 (1978).
113. M. Gefter and R. L. Russell, *JMB* **39**, 145 (1969).
114. H. Grosjean, D. G. Söll, and D. M. Crothers, *JMB* **103**, 499 (1976).
115. H. H. Arnold, R. Raettig, and G. Keith, *FEBS Lett.* **73**, 210 (1977).
116. H. Grosjean and H. Chantrenne, in "Codon/Anticodon Interactions on Biological Recognition" (C. Chapeville and A. Haenni, eds.), p. 347. Springer-Verlag, Berlin and New York, 1979.
117. A. Garen, *Science* **160**, 149 (1968).
118. L. Gorini, *ARGen* **4**, 107 (1970).
119. S. P. Eisenberg, M. Yarus, and L. Soll, *JMB* **135**, 111 (1979).
120. C. Yanofsky, in "Molecular Mechanisms in the Control of Gene Expression"

(D. P. Nierlich, W. J. Rutter, and C. F. Fox, eds.), p. 75. Academic Press, New York, 1976.
121. K. Bertrand, L. Korn, F. Lee, T. Platt, C. L. Squires, C. Squires, and C. Yanofsky, *Science* **189**, 22 (1976).
122. D. E. Morse and A. N. C. Morse, *JMB* **103**, 209 (1976).
123. T. Platt, *Cell* **24**, 10 (1981).
124. R. H. Buckingham and C. G. Kurland, *PNAS* **74**, 5496 (1977).
125. G. W. Chang, J. R. Roth, and B. N. Ames, *J. Bact.* **108**, 410 (1971).
126. C. E. Singer, G. R. Smith, R. Cortese, and B. N. Ames, *Nature NB* **238**, 72 (1972).
127. C. B. Bruni, V. Colantuoni, L. Sbordone, R. Cortese, and R. Blasi, *J. Bact.* **130**, 4 (1975).
128. J. E. Brenchley and L. S. Williams, *Annu. Rev. Microbiol.* **29**, 251 (1975).
129. L. Bossi and J. R. Roth, *Nature (London)* **286**, 123 (1980).
130. P. P. Di Nocera, F. Blasi, R. Di Lauro, R. Frunzio, and C. B. Bruni *PNAS* **75**, 4276 (1978).
131. W. Barnes, *PNAS* **75**, 4281 (1978).
132. H. Laten, J. Gorman, and R. M. Bock, *NARes* **5**, 4329 (1978).
133. F. Janner, G. Vögeli, and R. Fluri, *JMB* **139**, 207 (1980).
134. S. I. Feinstein and S. Altman, *Genetics* **88**, 201 (1978).
135. D. S. Colby, P. Schedl, and C. Guthrie, *Cell* **9**, 449 (1976).
136. A. L. Sonenshein and K. M. Campbell, *in* "Spores VII," p. 179. American Society for Microbiology, Washington, D.C., 1978.
137. J. Trowsdale, M. Shiflett, and J. Hoch, *Nature (London)* **272**, 179 (1978).
138. R. A. Lazzarini, *PNAS* **56**, 185 (1966).
139. R. H. Doi and I. Kaneko, *CSHSQB* **31**, 581 (1966).
140. J. L. Arceneaux and N. Sueoka, *JBC* **244**, 5959 (1969).
141. B. S. Vold, *J. Bact.* **113**, 825 (1973).
142. H. H. Arnold and G. Keith, *NARes* **4**, 2821 (1977).
143. B. Menichi, H. H. Arnold, T. Heyman, G. Dirheimer, and G. Keith, *BBRC* **95**, 461 (1980).
144. B. S. Vold, D. E. Keith, Jr., M. Buck, J. A. McCloskey, and H. Pang, *NARes* **10**, 3125 (1982).
145. B. S. Vold, *J. Bact.* **114**, 178 (1973).
146. R. A. McMillan and J. L. Arceneaux, *J. Bact.* **122**, 526 (1975).
147. G. Keith, H. Rogg, G. Dirheimer, B. Menichi, and T. Heyman, *FEBS Lett.* **61**, 120 (1976).
148. Y. Yamada and H. Ishikura, *NARes* **4**, 4291 (1977).
149. D. W. E. Smith, A. L. McNamara, and B. S. Vold, *NARes* **10**, 3117 (1982).
150. F. O. Wettstein and G. S. Stent, *JMB* **38**, 25 (1968).
151. A. H. Rosenberg and M. L. Gefter, *JMB* **46**, 581 (1969).
152. H. Juarez, A. C. Skjold, and C. Hedgcoth, *J. Bact.* **121**, 44 (1975).
153. E. Griffiths and J. Humphreys, *EJB* **82**, 503 (1978).
154. M. Buck and E. Griffiths, *NARes* **10**, 2609 (1982).
155. M. Buck and E. Griffiths, *NARes* **9**, 401 (1981).
156. H. Camakaris, J. Camakaris, and J. Pittard, *J. Bact.* **143**, 613 (1980).
157. J. M. Dolby, P. Honour, and H. B. Valman, *J. Hyg.* **78**, 85 (1977).
158. J. J. Bullen, H. J. Rogers, and L. Leigh, *Br. Med. J.* **1**, 69 (1972).
159. J. J. Bullen, H. J. Rogers, and E. Griffiths, *in* "Microbial Iron Metabolism" (J. B. Neilands, ed.), p. 517. Academic Press, New York, 1974.
160. H. J. Rogers, *Infect. Immun.* **7**, 445 (1973).

161. H. R. Rogers, C. Synge, B. Kimber, and P. M. Bayley, *BBA* **497**, 548 (1977).
162. H. Rosenberg and I. G. Young, in "Microbial Iron Metabolism" (J. B. Neilands, ed.), p. 67. Academic Press, New York, 1974.
162a. H. Schellberger and H. Kersten, *Abstr. Commun. Fed. Eur. Biochem. Soc. 9th Meet.*, p. 451 (1974).
162b. J. M. Di Rienzo, K. Nakamura, and M. Inouye, *ARB* **47**, 481 (1978).
162c. A. M. Messenger and R. Barclay, *Biochem. Educ.* **11**, 54 (1983).
162d. T. Emery, *Bchem* **4**, 1410 (1965).
162e. A. M. Albrecht and H. J. Vogel *JBC* **239**, 1872 (1964).
162f. E. Jones, *ARB* **34**, 381 (1965).
162g. C. Walsh, *ARB* **47**, 881 (1978).
163. R. H. Hall, This Series **10**, 57 (1970).
164. B. Thimmappaya and J. D. Cherayil, *BBRC* **60**, 665 (1974).
165. R. W. Chapman, R. O. Morris, and J. B. Zaerr, *Nature (London)* **262**, 153 (1976).
166. J. D. Cherayil and M. N. Lipsett, *J. Bact.* **131**, 741 (1977).
166a. M. Buck, J. A. McCloskey, B. Basile, and B. N. Ames, *NARes* **10**, 5649 (1982).
167. M. D. Litwack and A. Peterkofsky, *Bchem* **10**, 994 (1971).
168. M. Hellbach, M. Lineweaver, and T. Klambt, *Physiol. Plant.* **44**, 313 (1978).
169. V. Quesney-Huneeus, M. H. Wiley, and M. D. Siperstein, *PNAS* **77**, 5842 (1980).
170. V. Quesney-Huneeus, M. H. Wiley, and M. D. Siperstein, *PNAS* **76**, 5056 (1979).
171. H. Abe, M. Uchiyama, Y. Tanaka, and H. Saito, *Tetrahedron Lett.* **42**, 3807 (1976).
172. Y. Taya, Y. Tanaka, and S. Nishimura, *FEBS Lett.* **89**, 326 (1978).
173. Y. Tanaka, H. Abe, M. Uchiyama, Y. Taya, and S. Nishimura, *Phytochemistry* **17**, 543 (1978).
174. Y. Taya, Y. Tanaka, and S. Nishimura, *Nature (London)* **271**, 545 (1978).
175. F. Seela, D. Hasselmann, W. Bussmann, H. Kersten, and E. Schachner, *in* "Improved and New Synthetic Procedures, Methods and Techniques in Nucleic Acid Chemistry" (P. Townsend, ed.), in press.
176. E. Schachner, G. Ott, A. Ogilvie, H. Kersten, and S. Seela, *in* "Regard sur la Biochimie," p. 65. Société de Chimie Biologique, Strasbourg, France, 1981.
177. S. Nakanishi, S. Adhya, M. Gottesman, and J. Pastan, *Cell* **3**, 39 (1974).
178. H. V. Rickenberg, H. J. Rahmsdorf, A. Campbell, M. J. North, J. Kwasniak, and J. M. Ashworth, *J. Bact.* **124**, 212 (1975).
179. S. Nishimura and Y. Kuchino, *GANN Monogr. Cancer Res.* **24**, 245 (1979).
180. S. Nishimura, N. Shindo-Okada, H. Kasi, Y. Kuchino, S. Noguchi, M. Iigo, and A. Hoshi, *in* "Modified Nucleosides and Cancer" (G. Nass, ed.), p. 401. Springer-Verlag, Berlin and New York, 1983.
181. Y. Kuchino, H. Kasai, Z. Yamaizumi, S. Nishimura, and E. Borek, *BBA* **565**, 215 (1979).
182. H. Kersten, *in* "Biochemistry of Differentiation and Morphogenesis" (L. Jaenicke, ed.), p. 116. Springer-Verlag, Berlin and New York, 1982.
183. V. K. Lin, W. R. Farkas, and P. F. Agris, *NARes* **8**, 3467 (1980).
184. N. Shindo-Okada, M. Terada, and S. Nishimura, *EJB* **115**, 423 (1981).
185. E. Schachner, S. Nishimura, and H. Kersten, *ZpChem* **363**, 887 (1982).
186. N. K. Howes and W. R. Farkas, *JBC* **253**, 9082 (1978).
187. J. R. Katze, *BBA* **383**, 131 (1975).
188. N. Okada, N. Shindo-Okada, S. Sato, Y. H. Itoh, K. J. Oda, and S. Nishimura, *PNAS* **75**, 4247 (1978).
189. R. P. Singhal, R. A. Kopper, S. Nishimura, and N. Shindo-Okada *BBRC* **99**, 120 (1981).

190. B. White and G. M. Tener, *JMB* **74**, 635 (1973).
191. W. R. Farkas, *JBC* **255**, 6832 (1980).
192. J. P. Reyniers, J. R. Pleasants, B. S. Wostmann, J. R. Katze, and W. R. Farkas, *JBC* **256**, 11591 (1981).
193. S. Nishimura, This Series 28. 50 (1983). See also Singhal, *ibid.*, 75. [Eds.].
194. G. Gerisch, *Annu. Rev. Physiol.* **44**, 535 (1982).
195. G. Gerisch, *Curr. Top. Dev. Biol.* **3**, 159 (1968).
196. T. Dingermann, A. Ogilvie, F. Pistel, W. Mühlhofer, and H. Kersten, *ZpChem* **362**, 763 (1981).
197. G. Ott, H. Kersten, and S. Nishimura, *FEBS Lett.* **146**, 311 (1982).
197a. H. G. Ott, Dissertation, Univ. Erlangen-Nürnberg (1984).
197b. R. P. Singhal and V. N. Vakharia, *NARes* **11**, 4257 (1983).
198. E. Schachner, S. Nishimura, and H. Kersten, *ZpChem* **363**, 887 (1982).
199. A. A. Green and P. C. Newell, *Cell* **6**, 129 (1975).
200. T. H. Alton and H. F. Lodish, *Dev. Biol.* **60**, 180 (1977).
201. E. Schachner and H. Kersten, *J. Gen. Microbiol.* **130**, 135 (1984).
202. S. Noguchi, Y. Nishimura, Y. Hirota, and S. Nishimura, *JBC* **257**, 6544 (1982).
203. K. B. Jacobson, W. R. Farkas, and J. R. Katze, *NARes* **9**, 2351 (1981).
204. S. Kaufman and D. B. Fischer, in "Molecular Mechanisms of Oxygen Activation" (O. Hayaishi, ed.), p. 285. Academic Press, New York, 1974.
205. H. Kersten, E. Schachner, and G. Dess, in "Biochemical and Chemical Aspects of Pteridines" (H. Wachter, H. C. Curties, and W. Pfleiderer, eds.), p. 367. de Gruyter, Berlin, 1983.
206. R. P. Singhal, R. A. Kopper, S. Nishimura, and N. Sindo-Okada, *BBRC* **99**, 120 (1981).
207. M. S. Kanungo and S. N. Singh, *BBRC* **21**, 454 (1965).
208. E. Garvie, *Microbiol. Rev.* **44**, 106 (1980).
209. F. H. Gleason, R. A. Nolan, A. C. Wilson, and R. Emerson, *Science* **152**, 1272 (1966).
210. R. C. Garland and N. O. Kaplan, *BBRC* **26**, 679 (1967).
211. C. L. Markert, J. B. Shaklee, and G. Whitt, *Science* **189**, 102 (1975).
212. C. Arizmendi, J. M. Cuezva, and J. M. Medina, *Enzyme* **29**, 66 (1983).
213. S. N. Alahiotis, A. Onoufriou, M. Fotaki, and M. Pelecanos, *Biochem. Genet.* **21**, 199 (1983).
214. P. Pajot and M. L. Claisse, *EJB* **49**, 275 (1974).
215. S. R. Sarkar, *Acta Biol. Med. Ger.* **41**, 1085 (1982).
216. B. E. Wright and M. L. Anderson, *BBA* **31**, 310 (1959).
217. S. V. Cleland and E. L. Coe, *BBA* **156**, 44 (1968).
218. R. A. Firtel and R. W. Brackenbury, *Dev. Biol.* **27**, 307 (1972).
219. E. Schachner, H. J. Aschhoff, and H. Kersten, *ZpChem* **364**, 1204 (1983).
220. D. Dennis and N. O. Kaplan, *JBC* **235**, 810 (1960).
221. J. A. Cooper, N. A. Reiss, R. J. Schwartz, and T. Hunter, *Nature (London)* **302**, 218 (1983).
222. E. Schachner, H. J. Aschhoff, and H. Kersten, *EJB* **139**, 481 (1984).
223. H. Döbeli and G. A. Schoenenberger, *Experientia* **39**, 281 (1983).
224. H. Kersten, *Recent Res. Cancer Res.* **84**, 255 (1983).
225. H. Kersten, in "tRNA Methylation" (E. Usdin, R. T. Borchardt, and C. R. Creveling, eds.), p. 357. Pitman, London, 1982.
226. H. R. Kaback, *Science* **186**, 882 (1974).

227. C. Whoffendin, S. W. Edwards, and A. J. Griffiths, *Comp. Biochem. Physiol. B* **75B**, 53 (1983).
228. H. A. Hosbach and E. Kubli, *Mech. Ageing Dev.* **10**, 141 (1979).
229. R. Jungmann, D. C. Kelley, M. F. Miles, and D. M. Milkowski, *JBC* **258**, 5312 (1983).
230. J. D. Gross, J. Bradbury, R. R. Kay, and H. J. Peacey, *Nature (London)* **303**, 244 (1983).
231. F. Anders, in "Modern Trends in Human Leukemia V." Haematology and Blood Transfusion" (R. Neth, R. C. Gallo, M. F. Graeves, M. A. S. Moore, and M. Winkler, eds.), Vol. 28, p. 186. Springer-Verlag, Berlin and New York, 1983.
232. H. Chantrenne, *Mol. Cell. Biochem.* **21**, 3 (1978).
233. D. H. Grosjean, S. De Henau, and D. M. Crothers, *PNAS* **75**, 610 (1978).
234. E. Kubli, *TIPS* **90** (1980).
235. M. Picard-Bennoun, *FEBS Lett.* **149**, 167 (1982).
236. C. G. Kurland, in "Ribosomes, Structure, Function and Genetics" (G. Chambliss, G. R. Craven, J. Davies, K. Davis, L. Kahan, and M. Nomura, eds.), p. 597. Univ. Park Press, Baltimore, Maryland, 1980. See also Kurland and Ehrenberg in this volume. [Eds.]
236a. C. G. Kurland and M. Ehrenberg, *Prog. Mol. Biol. Nucleic Acids Res.* (1984, in press).
237. J. J. Hopfield and T. Yamane, in "Ribosomes, Structure, Function, and Genetics" (G. Chambliss, G. R. Craven, J. Davies, K. Davis, L. Kahan, and M. Nomura, eds.), p. 585. Univ. Park Press, Baltimore, Maryland, 1980.
238. J. Ninio, *Biochimie* **57**, 587 (1975).
239. R. C. Thompson and P. J. Stone, *PNAS* **74**, 198 (1977).
240. G. R. Björk, *JMB* **140**, 391 (1970).
241. T. Ikemura, *JMB* **146**, 1 (1981).
242. J. A. Remington, *FEBS Lett.* **100**, 225 (1979).
243. C. Yanofsky, *Nature (London)* **302**, 751 (1983).
244. M. Grunberg-Manago and B. Safer, eds., "Interaction of Translational and Transcriptional Controls in the Regulation of Gene Expression" Elsevier, Amsterdam, 1982.

The Organization and Transcription of Eukaryotic Ribosomal RNA Genes[1]

RADHA K. MANDAL

Department of Biochemistry
Bose Institute
Calcutta, India

I. Ribosomal RNA Genes 117
   A. Redundancy of rRNA Genes................................ 117
   B. Chromosomal Location 118
   C. Amplification and Extrachromosomal Location 118
II. Organization of rRNA Genes 120
   A. Arrangement of Coding and Noncoding Regions............... 120
   B. Transcription-Initiation Region 124
   C. Transcription-Termination Region 127
   D. Coding Regions 130
   E. Transcribed Spacers 132
   F. Nontranscribed Spacers 133
   G. Ribosomal-Insertion Sequences............................. 136
III. Transcription of rRNA Genes.................................... 138
   A. *In Vivo* Transcription 139
   B. *In Vitro* Transcription..................................... 142
   C. Processing of Pre-rRNA 145
   D. Regulation of Transcription 147
IV. Concluding Remarks... 150
   References 151

Life is characterized by the complexity of its architecture which is manifested at all levels of organization, from the molecules to the assemblages of specialized organelles and cells that make up higher organisms.[2] In view of their role in protein synthesis, ribosomes have been interesting organelles for the study of structure–function rela-

---

[1] Research by the author reported in this article was supported in part by the Council of Scientific and Industrial Research, the Department of Atomic Energy, and the Department of Science and Technology, Government of India and a Visiting Scientist Program of the United States National Institutes of Health.

[2] Abbreviations: bp, base-pair: kb, kilobases or kilobase-pairs; rRNA, ribosomal RNA; pre-rRNA, precursor of ribosomal RNA; rDNA, ribosomal DNA (gene for rRNA); ETS, external transcribed spacer; ITS, internal transcribed spacer; NTS, nontranscribed spacer.

tionships. The structure and biogenesis of the prokaryotic ribosome have been studied in great detail in *Escherichia coli* (1, 2), but our knowledge of eukaryotic ribosomes is much less complete (3, 4). A typical eukaryotic ribosome is composed of four RNA species (28 S, 18 S, 5.8 S, and 5 S) and more than 70 proteins. Despite the fact that about two-thirds of the mass of the ribosome is composed of rRNA, its role in the topography and function of the ribosomes is yet to be fully defined. The structure and function of the genes for rRNA are also of special interest, as they are transcribed by a distinct class of RNA polymerase (EC 2.7.7.6) (polymerase I)[3] (5, 6) and are under specific control during development (7–9) as well as in hormone action (10, 11).

In most eukaryotes, the genes for 18-S, 5.8-S, and 28-S rRNAs are cotranscribed from a unit of continuous genes together with internal and external transcribed spacers (ITS and ETS). Transcribed units are separated by nontranscribed spacers (NTS). The genes coding 5-S rRNA are not usually linked to the major rRNA genes. The rRNA gene repeating units (rDNA), each composed of a gene region and a spacer region, occur in most organisms in tandem head-to-tail arrays. A comprehensive review on the repetition and organization of rRNA (12) and three monographs (13) on different aspects of rDNA organization and transcription have appeared; 5-S RNA has been reviewed in this series (14). Earlier reviews cover different aspects of rRNA genes, such as redundancy and amplification (15, 16), developmental regulation (17), and transcription and processing (18–20).

In recent years, rRNA genes from different organisms have been studied at the nucleotide sequence level by using recombinant DNA techniques. This has been done with the hope that this family of genes may provide some insight into the biological implications of the diversity of sequence duplication and rearrangements to which the nuclear DNA of higher organisms is subject. The purpose of the present review is to summarize the current status of our knowledge of the organization and transcription of the rRNA genes of eukaryotes with emphasis on the nucleotide sequences in relation to function. The attempt is to be informative rather than comprehensive. Thus many otherwise important references may not be found in this article.

---

[3] Polymerases I, II, and III are differentiated and defined in the introduction to Section III.

I. Ribosomal RNA Genes

A. Redundancy of rRNA Genes

The genes coding for rRNA species are among the most extensively studied eukaryotic genes. Owing to the clustering of repeated rRNA genes and their high G+C content in some species, they can be easily isolated by buoyant density gradient centrifugation (21). Saturation hybridization of DNA immobilized on nitrocellulose filters with labeled rRNA (22) permits the determination of rRNA gene redundancy. The ribosomal cistrons are reiterated within the genome of all organisms examined except mycoplasma (23). In general, the increase in the number of rRNA genes is roughly correlated with the evolutionary increase in DNA content per haploid genome. Bacteria contain 5–10 copies of rDNA, and lower eukaryotic and animal cells have a few to several hundred copies per haploid genome, whereas in higher plants, the reiteration frequency may be as much as 10 times higher, thus reaching the highest rDNA amounts measured so far in somatic cells (23, 24). However, large variations in rRNA gene numbers exist not only between closely related species, but also between different strains of the same species and, in some cases, between individuals of the same species (12). This may arise from mutational loss of gene copies or selective over- or underreplication of specific genes compared to the total DNA.

Ribosomal RNA gene deletion mutants of *Xenopus* and *Drosophila* have been studied in great detail. In *Xenopus laevis*, the homozygous nucleolus-less mutants are devoid of any nucleolus and die at the swimming tadpole stage before feeding. These mutant animals fail to synthesize any rRNA. Heterozygous mutants possess only one nucleolus compared to the two present in normal wild-type animals, but they are capable of synthesizing rRNA at the same rate as the latter. RNA·DNA hybridization experiments show that nucleolus-less animals do not contain DNA sequences coding for rRNA, whereas heterozygous animals contain half the dose of wild-type rDNA (21, 25). A similar situation exists in the "bobbed" locus (bb) mutants (16, 26) of *Drosophila melanogaster*. Furthermore, when the dose of rDNA in the "bobbed" locus of an X chromosome is below a critical level, a differential replication of rRNA genes and a compensation of their number takes place in somatic cells during ontogenesis. This increased redundancy of rRNA genes due to "compensation" is not heritable. On the other hand, a reversion of the "bobbed" phenotype can occur with the concomitant heritable increase in the number of rRNA genes. This phenomenon is called "gene magnification" (27). Some of the rDNA is underreplicated in polytene cells of *Drosophila*. Thus, the number of rRNA genes per haploid genome is only a fraction of that in diploid cells (28, 29).

The level of rRNA genes along with some other DNA is increased severalfold in rat hepatoma cells (30). Indeed, apart from the massive amplification of rDNA in animal oocytes as discussed in Section I,C, limited amplification of some genes in developing or differentiating cells under demanding or selective conditions may be of wider occurrence than observed so far.

B. Chromosomal Location

In most eukaryotes, rRNA gene clusters are located at specific sites on one or several chromosomes (12). During mitosis, these clusters are cytologically observed as "secondary constrictions." These secondary constrictions are identified as the "nucleolus organizer" in interphase cells. The simple silver-staining technique (31) supplemented by the *in situ* RNA·DNA hybridization method (32) made possible the chromosomal localization of rRNA genes. In the *Saccharomyces* group of yeasts, most of the 100–120 rRNA genes per haploid genome are clustered along a single chromosome (33, 34). In maize also, the several thousand rRNA genes are clustered in one chromosome, namely, chromosome 6 (35). Compared to this, *Drosophila melanogaster* (26) has nucleolar organizers on two sex chromosomes; in man, 50–200 rRNA genes are distributed on five chromosomes (36). In closely related species like chimpanzee and man, the genes are found on homologous chromosomes (37). But great variations in the location and number of chromosomal sites of the rRNA genes occur throughout evolution (38, 39). Although all true nucleoli in the interphase nucleus contain rDNA, in some cases it can be detected also at chromosome loci outside the nucleolar organizer (40).

C. Amplification and Extrachromosomal Location

The first discovered, and still one of the best understood examples of differential replication is the amplification of rRNA genes during oogenesis (41, 42), a phenomenon best studied in insects and amphibians. Apparently, the differential replication takes place in response to high demands for ribosome production in oocytes for use during early embryonic development. The extra copies of rRNA genes generally remain extrachromosomal (see, however, bobbed mutants of *Drosophila*, Section I,A). The presence of large amounts of extrachromosomal rDNA in the oocytes of many species is now well documented, and different aspects of the process have been reviewed (*15, 16*). The species include amphibians (*41*), bony fishes (*43, 44*), and several insects with panoistic ovaries where oocytes lack the nurse cells, like the water beetle *Dytiscus marginalis* (*45*) and house cricket *Acheta domesticus* (*46*). In contrast to high levels of rDNA amplification in amphibians, bony fishes, and insects (as cited above), a low level (less than 10-fold) of rDNA amplification has been observed in oocytes of see urchin (*41*), brine shrimp *Artemia salina* (*47*), and man (*48*).

Some unicellular primitive eukaryotes present interesting examples of extrachromosomal rRNA genes. In *Tetrahymena pyriformis*, only a single copy of an rRNA gene is integrated in the chromosome of the micronucleus or the "genetic" nucleus, whereas the macronucleus or the "metabolic" nucleus contains many copies (about 200 per haploid genome) of extrachromosomal rDNA (*49*). Similarly, large numbers of extrachromosomal rDNA genes occur in the macronucleus of *Physarum polycephalum* (*50, 51*), *Dictyostelium discoideum* (*52*), *Paramecium tetraurelia* (*53*), and *Stylonychia* (*54*). The haploid number of extrachromosomal rRNA genes in the *Tetrahymena* macronucleus varies somewhat according to the growth phase (*55*).

Amplification of rDNA in the germ cells of amphibians and insects is a two-step process. A low level of rDNA amplification can be detected both in oogonia and spermatogonia (*42, 56, 57*). During the pachytene stage of meiosis in oocytes (but not in spermatocytes), a second round of massive amplification takes place resulting in a large amount of extrachromosomal rDNA. Thus, the amount of rDNA in *Xenopus laevis* oocyte is about 1000 times the haploid rDNA content (*23*). The amplified rDNA occurs mostly as circles containing many repeating units (*58*). The linear molecules observed probably arise by breakage of the circles. Individual oocytes in a single frog amplify a limited subset of repeats, and different oocytes amplify different subsets (*59*). But the repeating units along one rDNA molecule are homogeneous in length, indicating that each amplified rDNA molecule originated from a single integrated repeat (*59*). The question that still remains unanswered is whether the first extrachromosomal rDNA molecule is formed by selective excision or by selective replication of an integrated gene. Different replicative forms, including tailed circles, have been observed, suggesting a rolling-circle mechanism (*15, 60–62*). In insects also, circles of amplified rDNA have been demonstrated (*45, 46*). Gene-sized circular DNA molecules have been observed in the ciliate *Stylonychia* macronucleus (*63*). rDNA amplification in *Xenopus* is dosage-regulated. Thus, heterozygous females contaning only one nucleolus accumulate the same amount of rDNA in their oocytes as the wild-type frogs containing two nucleoli (*41*). Although there is a general correlation between rDNA amplification and the occurrence of multiple nucleoli in oocytes, there are exceptions to this rule. Amplification of rDNA has been demonstrated in the eggs of the surf clam *Spisula solidissima* and the echiurid worm *Urechis caupo*, although only a single large nucleolus in present in the oocytes of both species (*41, 64*).

II. Organization of rRNA Genes

A. Arrangement of Coding and Noncoding Regions

The recent techniques of gene cloning, mapping, and nucleotide sequencing combined with RNA·DNA hybridization and electron-microscopic visualization permit the study of the general organization and structure of rRNA genes from an ever-increasing number of eukaryotes. As new information is constantly arising, it is practically impossible to make a complete up-to-date review. Therefore, only the basic features of the structural organization of some typical and well-studied rRNA genes are presented here.

The rRNA genes (rDNAs) are organized in repeating pre-rRNA transcription units separated by nontranscribed spacers (NTS). The pre-rRNAs transcription unit consists of the coding sequences corresponding to the mature 18-S, 28-S, and 5.8-S rRNA or their equivalents, and both external and internal transcribed spacer (ETS and ITS) sequences, which are transcribed as parts of the pre-rRNA molecule, but are lost during maturation. One transcription unit and the adjoining NTS constitute one rDNA repeating unit. The tandem head-to-tail arrangement of the repeated units is the basic feature in the organization of rRNA genes in most eukaryotes. Exceptions to this are the palindromic tail-to-tail configuration of two repeating units in the *Tetrahymena* macronucleus (*49, 65*) and the alternate head-to-head and tail-to-tail arrangement in *Acetabularia* (*66*). The polarity of transcription is always (from 5' to 3') ETS-(18-S RNA)-ITS-(28-S-RNA), the 5.8-S RNA conventionally being taken as part of the ITS. This polarity is universal (*67–72*), including bacterial (*73*) and organelle rRNA genes (*74, 75*).

Although the general pattern in the topology of pre-rRNA transcription unit has remained stable during evolution, there are large variations in its size. Thus, the transcription unit has a size of about 6 kb in *Acetabularia* (76), about 8 kb in yeast (77), *Drosophila* (78), and *Xenopus* (79), 10.5 kb in birds (80), and 13 kb in mammals (81). The spacer lengths also vary considerably in different organisms. As a result, the length of the rDNA repeat unit (transcription unit plus spacer) varies from about 10 kb in yeast, *Drosophila*, and frog to about 40 kb in mammals.

Figure 1 presents the organization of rDNA repeat units of four well-studied organisms. These results have been obtained by a combination of techniques including cloning, restriction endonuclease mapping, heteroduplex and "R-loop" mapping under the electron microscope, and transcription mapping with S1 nuclease. Figure 1a shows the repeating unit structure in the yeast *Saccharomyces cerevisiae* (33, 82–84). This differs from the general pattern in higher eukaryotes in that the repeating unit also includes the coding sequence for 5-S rRNA. However, the 5-S RNA is transcribed from the opposite strand of DNA, representing a transcription unit separate from that for

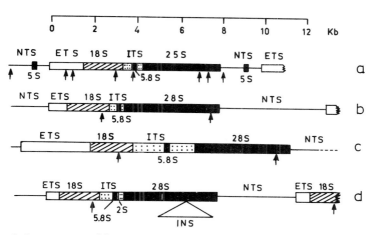

FIG. 1. Organization of the rDNA repeating units. Coding regions are indicated by a hatched box for small rRNA, a filled box for large rRNA, an empty box for external transcribed spacer (ETS), a dotted box for internal transcribed spacer (ITS), and a line for nontranscribed spacer (NTS) (a) rDNA of *Saccharomyces cerevisiae* (33, 82–84), which also contains the 5-S RNA gene; (b) rDNA of *Xenopus laevis* (86, 87); (c) rDNA of mouse (94, 95), showing only part of the NTS, which is much longer than the others shown; (d) rDNA of *Drosophila melanogaster* (98–101). In *Drosophila*, the ITS has also sequences coding for a 2-S RNA, which is absent in the other three. The small gap found in mature 28-S rRNA and the position of the ribosomal insertion sequence (INS) found in some of the 28-S RNA genes are indicated. Arrows indicate *Eco*RI sites.

pre-rRNA (68). A similar organization of rDNA has been found in *Saccharomyces carlsbergensis* (77, 85).

The rDNA repeating unit of *Xenopus laevis* (Fig. 1b) has been studied extensively with respect to its organization, structure, and regulation of transcription and has provided a typical model of rDNA in higher eukaryotes. The transcription unit (about 7.9 kb) is identical or nearly identical within cells and individuals of the species (86, 87). But there is considerable length heterogeneity in the nontranscribed spacer in the repeat units, even in those originating from a single repeat cluster (86). This length heterogeneity is seen also in the amplified rRNA of oocytes (59). The position of the 5.8-S RNA has been definitely established in the internal transcribed spacer region between 18-S and 28-S RNA (88, 89, 201).

Repeating rDNA units of the *Xenopus* type occur in most eukaryotes, including *Leishmania* (90), plants (91), insects (92, 93), birds (80), mouse (94, 95), cattle (96), and man (97). Although the pre-rRNA transcription unit of the mouse rDNA repeat (Fig. 1c) is only slightly larger than that of *Xenopus* (Fig. 1b), the total rDNA repeat length is much larger (40 kb), owing to the increased length of the nontranscribed spacer segment (94, 95).

The same general pattern of the coding and noncoding sequences is also present in *Drosophila melanogaster* rDNA (Fig. 1d) but it has some special features. Of the approximately 250 rRNA genes clustered on X and Y chromosomes (16), about half of the repeating units are "normal" or continuous with a size distribution of 10.5 to 12.5 kb (69, 98–101), the limited length heterogeneity being due to the variable size of the nontranscribed spacer (102). The rest of the rRNA genes are interrupted by "intervening sequences" (IVS) or "insertions" (called "ribosomal insertions"). The structural features of the ribosomal insertions are discussed in Section II,G. The ribosomal insertions in *Drosophila melanogaster* are predominantly of three different lengths, resulting in rDNA repeating units that are 0.5, 1.0, and 5.0 kb longer than the continuous units (99–102). In the mature 28-S rRNA in *Drosophila*, there is a hidden gap in the polynucleotide chain, giving rise to two pieces of large ribosomal subunit RNA (28-S α and 28-S β) held together by hydrogen bonds. The position of this break in the rDNA map (Fig. 1d) is indicated by the open bar in 28-S RNA. Another feature of *Drosophila* rRNA transcription unit is that the internal transcribed spacer also contains the coding sequence of a ribosomal 2-S RNA in addition to 5.8-S RNA. The occurrence of "pseudo rRNA genes" with about 4-kb transcription units has been demonstrated (103). Further mapping and determination of nucleotide sequences of standard and pseudo-rRNA genes are necessary to resolve the questions regarding their origin and function.

Interrupted rRNA genes of the *Drosophila* type have been found in some, but not all, dipteran insects. Thus, other *Drosophila* species (*104, 105*), *Caliphora* (*106*), *Sciara* (*107*), and *Sarcophaga* (*108*) have interrupted rRNA genes. On the other hand, no interrupted rRNA genes have been found in the dipteran *Chironomus tentans* (*92*) or the lepidopteran insect *Bombyx mori* (*93*).

The extrachromosomal rRNA genes in *Tetrahymena* and some other lower eukaryotes represent a separate type for their unique organization. In *Tetrahymena*, the macronucleus contains many copies of extrachromosomal rDNA in the form of palindromic molecules, each containing two transcription units (Fig. 2). The size of all palindromes is the same (about 19 kb). Most of these exist as linear molecules, and the remaining ones exist as circles (*49, 65, 109*). Electron microscopy combined with hybridization and "R-loop" mapping showed that the two transcription units have their 17-S rRNA sequences proximal to the center. Divergent transcription of opposite coding strands takes place starting from the centrally located promoters (*65, 110, 111*). Starting from the 5' end, the polarity of transcription is ETS, 17-S rRNA, and ITS, which also contains 5.8-S rRNA (not shown in Fig. 2), and the 26-S rRNA, common to other eukaryotes (*70–72*). The NTS sequences have an unusual feature in that at about 1 kb from the 3' end of the transcription unit, a tandemly repeated sequence of 5'-d(CCCCAA)$_n$-3' is found in the coding strand, where n is between 20 and 70 (*112*). Another special feature of rDNA structure found in some, but not all, strains of *Tetrahymena* is the presence of a 400-bp ribosomal insertion that interrupts the coding sequence of 26-S rRNA, at a distance of about 1 kb from the 3' end (*110, 113*). Barring the existence of the intervening sequence, all *Tetrahymena* species and strains have the same general organization of rDNA. Strains that have ribosomal insertions in their rDNA therefore have an RNA splicing mechanism (*110, 114*).

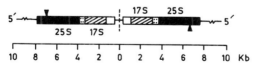

FIG. 2. Organization of the palindromic extrachromosomal rDNA of *Tetrahymena* (*65, 110, 111*). The linear molecule is composed of two rDNA repeating units showing divergent transcription from opposite strands. The repeated-sequence regions in the NTS are shown by small wiggles. The arrowheads indicate the positions of the ribosomal insertion found in some *Tetrahymena* species.

Large palindromic rDNA molecules occur in several lower eukaryotes. In *Physarum polycephalum*, palindromes about 60 kb long have the rRNA coding regions close to the ends, with about 15-kb NTS located at the center (*50, 51, 115, 116*). In this organism, all copies of rRNA genes are interrupted by two insertions (*51, 116*). In *Dictyostelium discoideum*, palindromic rDNA molecules are about 88 kb long, with one 5-S RNA gene in each rRNA gene (*52, 117*).

B. Transcription-Initiation Region

Xenopus laevis rRNA gene was the first whose transcription-initiation site is precisely determined at the nucleotide level. The two *Eco*RI sites in the rDNA repeating unit facilitated the first cloning of this gene (*87, 118*). Electron microscopic and restriction analysis studies (*70, 86, 87, 119*) indicated that the 5' end of pre-rRNA is located about 2.5 kb upstream from the *Eco*RI site in the 18-S coding region of the gene (Fig. 1b). The precursor begins with about 900 nucleotides of transcribed spacer sequence, which is discarded during processing (*79*). However, the precise localization of the 5' end had to wait until the isolation of unprocessed primary pre-rRNA transcripts and nucleotide sequencing were achieved. For various technical reasons (for discussion, see *20*), initial attempts to isolate intact pre-rRNA led to heterogeneous 5' ends (*20, 120*). It is now known that transcription starts with ATP or GTP (*121*) and the 5'-terminal nucleoside triphosphate can be "capped" by RNA guanylyltransferase (EC 2.7.7.50) (*122*). Thus, identification of 5'-terminal triphosphate (pppN) in pre-rRNA and its ability to be capped would indicate the absence of any processing at the 5' end. Direct demonstration of 5'-terminal triphosphates, and in some cases capping as well, was soon achieved with respect to the pre-rRNAs of several eukaryotes, such as *Xenopus* (*123*), yeast (*77, 124*), *Tetrahymena* (*125*), *Drosophila* (*126*), *Dictyostelium* (*127*), mouse (*128*), and rat (*129*).

Once the primary pre-rRNA transcript was isolated, the initiation site could be localized by the techniques of S1-nuclease protection (*130*) and reverse-transcriptase elongation mapping (*131*). In the first method, a small DNA fragment containing the presumptive initiation site is labeled at the 5' terminus and the coding strand is isolated, or the labeled end in the opposite strand is removed by a restriction enzyme. The coding strand is hybridized with the pre-rRNA, and the hybrid is digested with S1 nuclease (Fig. 3a) The length of the protected DNA fragment and the position of the 5'-terminal nucleotide can be determined (± 1 nucleotide) by electrophoresis of the denatured fragment on a sequencing gel together with the original end-

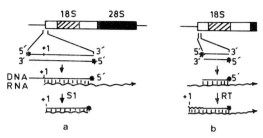

FIG. 3. Mapping of the pre-rRNA transcription initiation site by (a) S1 protection and (b) reverse transcriptase (RT) elongation. In the first method, a small DNA fragment containing the initiation site from the coding strand and labeled at its 5′ end is hybridized with pre-rRNA. The hybrid is digested with S1, and the size of the protected fragment is determined. In the second method, the coding strand of a small 5′-terminally labeled DNA fragment several nucleotides downstream from the initiation site is hybridized with pre-rRNA, and the DNA strand in the hybrid is extended by reverse transcriptase.

labeled coding fragment cleaved by base-specific chemical reactions (132). In the second method, the coding strand of a small 5′-terminally labeled DNA fragment several nucleotides downstream from the initiation site is hybridized with pre-rRNA, and the DNA strand in the hybrid is extended by reverse transcriptase (Fig. 3b). The extended DNA is then "sized" on a sequencing gel. The first few 5′-terminal nucleotides can be determined also by direct enzymatic (133) or chemical (134) sequencing of the pre-RNA.

Using these techniques, the transcription-initiation site of *Xenopus laevis* rRNA has been located about 2250 bp upstream from the *Eco*RI site in the 18-S rRNA coding sequence (135), in agreement with previous measurements by electron microscopy (67, 119). However, a region surrounding the initiation site (−127 to +4, with +1 being the nucleotide corresponding to the 5′-terminal nucleotide of pre-rRNA) is duplicated with about 10% mismatch at about 1 kb upstream (−1147 to −1017). Since the sequences immediately surrounding the transcription initiation site constitute the presumed promoter, these results indicate a promoter duplication, as already suggested (136) (see Section II,F).

Subsequently, the rRNA transcription-initiation sites were mapped and sequenced in a large number of eukaryotes, including *Xenopus* species (137), yeast (138–140), *Tetrahymena* (141, 142), *Dictyostelium* (143), *Drosophila* (144), mouse (128, 145), rat (129, 146, 147), and man (148, 149). It was expected that a comparison of nucleotide sequences at initiation sites and flanking regions of different rRNA transcription units might reveal the similarity of regulatory sequences for RNA polymerase I.[3] Although the number of rRNA genes studied in this respect is limited, there seems to be no homology in the initiation regions of different organisms. This is in contrast to the genes transcribed by RNA polymerase II,[3] where a more or less conserved sequence at a constant distance from the cap site (presumed initiation site) occur and seem to be essential for proper transcription (*150, 151*).

However, the organisms studied comprise evolutionarily widely distant species, so that the possibility that more closely related species share some sequence homologies at the 5' end of the ribosomal RNA transcription unit cannot be eliminated. In fact, in three species of *Xenopus* [*X. laevis*, *X. borealis*, and *X. clivii* (*137*)], 13 nucleotides at the NTS–ETS boundary are identical, pointing to a highly conserved and important function. More revealing is the fact that the transcription-initiation site in rat (*129, 146, 147*) is flanked by homologous sequences, also found in mouse (*128, 145*) and man (*148, 149*), the homology being more perfect between rat and mouse (Fig. 4). However, the sequence in mouse rDNA was reportedly not at the initiation site, but about 450 (*128*) or 650 (*145*) nucleotides upstream. Since S1-protection experiments, using pre-rRNA·DNA hybrids, cannot distinguish downstream initiation sites from processing sites, it cannot be said that rDNA transcription really starts from different promoter sites. *In vitro* transcription is being used currently to resolve such issues (see Section IV). In a subsequent report using *in vitro* transcription (*152*), the initiation point in mouse rDNA has been located about 650 nucleotides upstream from the site originally reported (*145*), exactly at the point where rat rRNA transcription starts. This unique conservation flanking the initiation site must have some significance regarding function of the sequence in binding RNA polymerase or some initiation factors.

Some general features of the initiation sites of rDNAs of different organisms may be pointed out. All the eukaryotic rRNA genes mentioned in the preceding paragraph, except human and rat, start with

FIG. 4. Nucleotide homologies surrounding the transcription-initiation sites of rat (*129, 146, 147*), mouse (*128, 145, 152*), and human (*148, 149*) rDNAs. +1 indicates the transcription-initiation site. Bars connect homologous nucleotides.

adenine. Human and rat rRNAs start with guanine. In all organisms studied, the putative promoter site for the rRNA gene is duplicated in the nontranscribed spacer preceding the initiation point (see Section II,F), as in *E. coli* (*153*). In *Drosophila*, *Xenopus*, mouse, rat, and human, the sequence d(CTTT) at around the −30 position appears to be common to all five rRNA genes. In mouse (*128, 145*), rat (*129, 146, 147*), and *Drosophila* (*144*), the several-hundred-nucleotide sequence around the initiation site is identical in independent rDNA clones of the same species, except for a few nucleotide changes. The two highly (A+T)-rich stretches (40–45 T's out of 45–50 nucleotides) in rat at +352 and +501 nucleotides downstream are strikingly similar to those found in mouse (*128, 145*), and are reminiscent of the (A+T)-rich nature of *Tetrahymena* (*141*) and yeast (*140*) rDNA initiation regions. These findings suggest that the sequences may be involved in initiation or processing of transcripts. In *Drosophila* (*144*), since the entire 250-nucleotide region sequenced downstream of the initiation point is (A+T)-rich (except for a small stretch from +120 to +150), the significance of the base composition in the region cannot be assessed.

C. Transcription-Termination Region

As described for the transcription-initiation region (Section II,B), the S1-protection experiment has been used to map the location of the 3' terminus of pre-rRNA and mature 28-S rRNA on the rDNA transcription unit. These studies combined with electron microscopic transcription mapping indicate that the 3' site of the *Xenopus laevis* pre-rRNA transcript is located at the *Hind*III site just before an oligo-T cluster (*134, 154*). Subsequently, the transcription termination sites of the rDNAs of *Saccharomyces carlsbergensis* (*155*), *Drosophila melanogaster* (*156*), *Physarum polycephalum* (*157*), *Neurospora* (*158*), *Tetrahymena thermophila* (*159*), *T. pyriformis* (*160*), and mouse (*161*) have been determined. It was generally assumed that the termination site is at or close to a cluster of thymidine residues preceded by a hyphenated diad symmetry, in analogy with the termination sites of prokaryotic genes (*162, 163*) and the eukaryotic genes transcribed by polymerase III[3] (*164*).

A comparison of the termination sites of all the above-mentioned eukaryotic rDNAs reveals no obvious homology or consensus sequence that can be termed a terminator signal. But a cluster of three or more thymidylates is at or near the termination site in most organisms. In *Drosophila* (*154*), *Neurospora* (*158*), mouse (*161*), and *Physarum* (*157*), the termination site does not include, or is not followed by, a stretch of T's. An (A+T)-rich region is thought to facilitate the

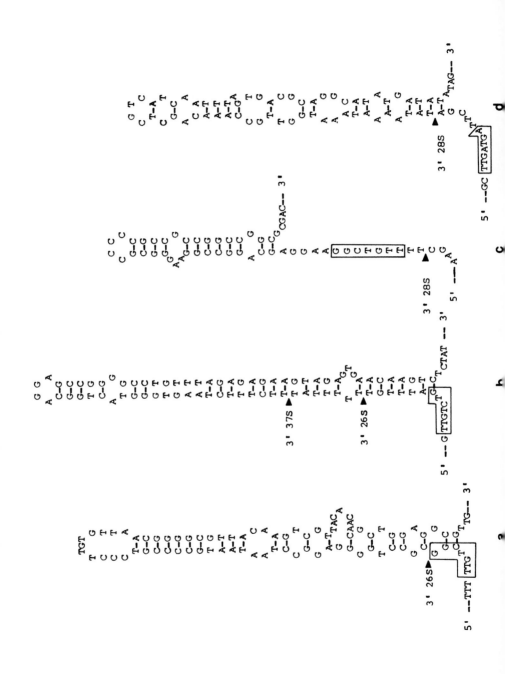

release of the transcript by localized melting of the transcript-template duplex (*165*). It should be pointed out here that a run of T's by itself is not sufficient for transcription termination. Some termination factor(s) and/or overall secondary and tertiary structures of DNA around the termination region may be more important. In *Physarum polycephalum* (*157*), the 26-S rRNA reads through a stretch of seven T's and terminates at six nucleotides beyond this. In rat rDNA, the external transcribed spacer includes runs of many T's (45 T's out of 50 nucleotides) at positions +352 and +501 nucleotides downstream from initiation site (*129*).

The small hyphenated diad symmetry (consisting of four to six bp) can form a "hairpin" structure just before the termination site. This type of symmetry is present in many organisms, including *Xenopus* (*135*), *Drosophila* (*156*), and mouse (*161*). An extended diad symmetry can be found downstream from the termination site in the form of a somewhat mismatched hairpin loop in *Physarum* (*157*), *Drosophila* (*156*), and *Xenopus* (*135*), as was originally proposed for yeast (*155*). (Fig. 5). Also, there is a common sequence of TTGTCNG in these organisms, except for *Drosophila* (where the sequence is TTGATGA), upstream from the 3′ terminus. The structural similarity suggests that sequences beyond the site of termination may be involved in the termination process. Comparison of the 3′-terminal 100-nucleotide sequence of the large subunit rRNA or pre-rRNA shows considerable homology between these sequences in yeast, *Xenopus*, and *Drosophila* (see Fig. 7 in *155* and Fig. 6 in *156*). In the region downstream from the termination site, no homology can be detected.

There is some difference in the processing of the 3′ ends of pre-rRNAs of different organisms. The 3′-terminal processing of pre-rRNA in mouse and rat has been reported (*166*). Subsequent S1-protection experiments also show that the pre-rRNA in mouse terminates 30 nucleotides downstream from the 3′ end of mature 28-S RNA (*161*). Also, 3′-terminal processing for two other organisms has been reported. In yeast (*155*) and *Tetrahymena pyriformis* (*160*), the pre-rRNA terminates about 7 and 15 nucleotides, respectively, beyond the 3′ end of mature 25-S rRNA. In contrast, *Xenopus* (*135*) and *Drosophila* (*156*) do not show any 3′-end processing of pre-rRNA. Unlike the 5′ end, where the purine nucleoside di- or triphosphate is diagnostic for an unprocessed pre-rRNA terminus, it is difficult to

---

FIG. 5. Extended diad symmetry in the noncoding strand beyond the rRNA transcription termination sites of (a) *Physarum polycephalum* (*157*), (b) *Saccharomyces cerevisiae* (*155*), (c) *Xenopus laevis* (*135*), and (d) *Drosophila melanogaster* (*156*). The common heptanucleotide sequences are boxed.

exclude the possibility of transcription termination at a site farther downstream, followed by rapid processing of the transcript. In any case, the general conclusion is that, unlike the 5' end, the 3' end of an rRNA transcription unit requires no extensive processing to become mature rRNA.

D. Coding Regions

As the initiation and termination regions of rRNA are of interest with respect to the regulation of transcription, the coding regions are of interest because of their participation in the organization of the functional centers of ribosomes involved in the interactions with mRNA, tRNA, and factors of protein synthesis. As a result, different aspects of organization, function, and evolution of rRNA have frequently been reviewed (1, 3, 167–169). The determination of the complete nucleotide sequence of coding regions for *E. coli* 16-S (172, 173) and 23-S RNA (174) has permitted the construction of secondary structure models with sites for interaction with 5-S and 5.8-S rRNAs, whose sequences are also known (175). As ribosomes perform the same function in both prokaryotes and eukaryotes, the location of conserved sequences within rRNA is one method of identifying the regions important for function. To date, the complete nucleotide sequences of both subunit of large rRNAs or their corresponding rDNAs are known only for one eukaryote, namely yeast (176–178). However, enough partial sequences of the rRNA coding regions in a number of eukaryotes are available for meaningful comparisons.

Heterologous hybridization between rRNA and DNA from various eukaryotic species is possible (179, 180). Studies carried out with the rRNA of the small ribosomal units from several organisms reveal a highly conserved region at the 3' end (172, 173, 176, 181–187). In many eukaryotes, the *Eco*RI site near the 3' end of 18-S RNA (Fig. 1) is conserved. Roles proposed for this 3'-end region in both prokaryotic and eukaryotic rRNA (16-S or 18-S rRNA) include interaction with mRNA in protein synthesis (188, 189), and, with 5-S, rRNA, as a mechanism of ribosomal subparticle binding (190).

When the 3' end of available small rRNA (or rDNA) sequences are compared, large parts of the genes in different eukaryotes are seen to be identical. Figure 6 illustrates one such comparison between the 3' end of the small rRNA in *Saccharomyces cerevisiae* (176), *Neurospora crassa* (191), and *Xenopus laevis* (184). With more available data, there is greater than 90% homology in the first 230 nucleotides, starting from the 3' end of the 18-S RNA and extending to the conserved *Eco*RI site in the rDNAs of *X. laevis* (184), *S. cerevisiae* (176), *Bombyx*

110	110	90	80	70	60
CCGGACUGGG	GCAGCACCAC	CAGGCGGAA	AGCUAUC	CAA ACUCGGUCAU	UUAGAGGAAG
GAAGGGGGCA	ACUCCAUCUC	AGAGCGGAGA	AUUUGGA	CAA ACUUGGUCAU	UUGGAGGAAC
GUCGGCCACG	GCCCUGGCGG	AGCGCCGAGA	AGACGAU	CAA ACUUGACUAU	CUAGAGGAAG

50	40	30	20	10	
UAAAAGUCGU	AACAAGGUAU	CCGUUGGUGA	ACCAGCGGAU	GGGAUCAUUA$_{OH}$	
UAAAAGUCGU	AACAAGGUUU	CCGUAGGUGA	ACCUGCGGAA	GGAUCAUUA$_{OH}$	
UAAAAGUCGU	AACAAGGUUU	CCGUAGGUGA	ACCUGCGGAA	GGAUCAUUA$_{OH}$	

FIG. 6. The nucleotide sequence at the 3' end of 18-S rRNA from *Neurospora crassa* (NC) (*191*), *Saccharomyces cerevisiae* (SC) (*176*), and *Xenopus laevis* (XL) (*184*). Numbering is from the 3'-terminal nucleotide. Homologous sequences are boxed.

mori (*182*), and rat (*192, 192a*). In the region between nucleotide 60 and 130 (numbering from the 3' end of the 18-S RNA), the sequences are highly divergent (Fig. 4 in *192*). The homology of the 3' ends of small rRNAs extends to organisms of greater evolutionary distances, like *E. coli* and *Xenopus*, as found by molecular hybridization using restriction fragments of cloned rDNA (*193*). This small 3'-end region of 18-S RNA can form a highly conserved hairpin structure. Under appropriate conditions, this region can form a stable complex with part of the 5-S RNA sequence (*190, 191*).

Like the 3' end, the 5' end of the 18-S RNA is homologous in yeast (*194*), *Xenopus* (*195*), and rat (*196*). This homology is absent in both the external and internal transcribed spacer regions flanking 18-S RNA. In eukaryotes, the coding region for 18-S RNA also has long stretches of internal homologous sequences interspersed with smaller stretches of divergent sequences (*193*). More extended nucleotide sequences at the 5' (400 bp) and 3' (300 bp) ends in rabbit 18-S RNA (*197*) reveal quite extensive homology at both the ends between yeast (*176*), *Xenopus* (*184*), and rabbit (*197*), interspersed with small tracts with little homology. Recently, using the experimentally established secondary structures of *E. coli* 16-S rRNA as a basis, structures have been derived for yeast and *Xenopus* 18-S rRNA (*198*). These models support the concept that the overall secondary structures of rRNA have been highly conserved during evolution, in spite of considerable changes in their primary structures.

The somewhat conserved nature of the 3'-terminal region of the 28-S rRNA is discussed in Section II,C. The homology of the 3'-terminal 110 nucleotides of 28-S rRNA in several eukaryotes is greater than 60% compared to about 50% homology with the corresponding region of *E. coli* 23-S RNA (*155, 156, 199*). Secondary structure models with four helices and three hairpin loops of this part of the 28-S RNA have been proposed (*157, 199*). The eukaryotic 3' end of 28-S RNA has some "5.8-S RNA-like" sequences. Based on this, a model for the interaction between 5.8-S RNA and the 3' end of large rRNA in eukaryotes has been proposed (*158*).

At the 5' end of 28-S rRNA, the sequence of the first 70 nucleotides of rat (*192*) is highly homologous with yeast (*200*) and *Xenopus* (*201*) sequences. Just like the 3' end of 25-S RNA proposed for *Neurospora* (*158*), the 5' end of yeast 26-S rRNA has been proposed to interact with the 3' end of yeast 5.8-S rRNA (*200*). A similar structure can be constructed for the 5' end of rat 28-S rRNA and the 3' end of 5.8-S rRNA (*192*). Moreover, two conserved nucleotide stretches involved in 5.8-S rRNA binding have been found in the 5'-terminal region of mouse 28-S rRNA (*202*). These two alternative proposals, based on the sequence data, imply that interactions between different regions of the mature rRNA species in ribosomes needed for subunit interaction and protein synthesis may differ from those in pre-rRNA during processing. An extraordinary evolutionary conservation in large subunit rRNA in the area where ribosomal RNA insertions are found in some organisms has been observed (*203*). This region has been proposed to be involved in interaction with tRNA^{fMet} (*203*).

The complete nucleotide sequence of the 5.8-S and 26-S rRNA genes and the ITS of *Physarum polycephalum* has been determined (*203a*). Comparison of the sequences with those of the *E. coli* 23-S and yeast 26-S rDNAs shows 16 highly homologous regions. This comparison also shows that the eukaryotic 5.8-S rRNA is the counterpart of the 5'-terminal region of *E. coli* 23-S rRNA, indicating that the eukaryote-specific 5.8-S rRNA gene is derived from the 5'-terminal region of the prokaryotic large rRNA gene (*203a*).

E. Transcribed Spacers

In contrast to the coding regions, there is very little, if any, homology between the ETSs of different organisms (*193*), except for small blocks of nucleotides near the transcription-initiation region (see Section II,B). Enough sequence data are not available to make any comparison among different eukaryotes. In vertebrates, it is highly (G+C)-rich (*204, 205*) except for blocks of (A+T)-rich sequences (*128, 129*). In *Drosophila*, however, the ETS proximal to the initiation site is (A+T)-rich (*144*). In *Tetrahymena*, there is a block of 16 consecutive

A's in the ETS near the initiation site (*141*). In rat, there is a duplication of 25-bp sequences at positions +408 to +432 and +546 to +570 [numbered from the initiation site (*129*)]. Such blocks of (A+T)-rich sequences and duplicated sequences may be important for transcription regulation or processing of pre-RNA.

The ITSs containing the 5.8-S rRNA genes of cloned yeast (*200*), *Xenopus* (*201*), rat rDNAs (*192*), and silk worm (*182*) rDNAs have been sequenced. In general, there is no homology among the ITSs of these organisms. It was already pointed out that even in two closely related *Xenopus* species, there is very little homology in the spacer (*206*). However, a comparison of rat sequences with those of *Xenopus* shows the occurrence of stretches 10–25 nucleotides long with more than 75% homology between the two species. Apart from this, the ITS of each species has its own characteristics (*192*). For example, in rat there are few A's in the spacer and there are "unique" stretches of 11 T's and 11 G-A dinucleotides, and a large number of oligo(G)'s. *Xenopus* ITS is richer in C stretches, and yeast ITS is U-rich. In both rat and *Xenopus*, the ITS is more (G+C)-rich (75–80%) than the coding regions (60%). The ITS sequences can be folded into a series of hairpin structures, leaving the gene-spacer boundaries single-stranded in the pre-rRNA molecule. This type of structure in ITS RNA, presumed to be involved in pre-rRNA processing, contrasts with the situation in *E. coli*, in which the spacers on the sides of both 16-S and 23-S rRNA can form a long base-paired stem, which contains the sites for the processing enzyme (*207*).

F. Nontranscribed Spacers

The most characteristic feature of the NTS of many eukaryotes is the variation in its absolute length among rDNA repeating units, both within and between individuals. Also, the comparison of rDNA sequences between species shows that the sequence divergence of NTS is far greater than that found for the 18-S and 28-S rRNA coding regions (for review see *12, 208, 209*). The basis of length heterogeneity is due to the number of short repeated sequences in the NTS (*59, 86*). Restriction analysis has helped to identify these repeated elements. For example, *X laevis* NTS has many repeating sequences recognized by *Hae*III and *Hpa*II (*210*). Similarly, the repeated elements are recognized by *Alu*I and *Mnl*I in *Drosophila melanogaster* (*211, 212*), *Taq*I in *Physarum polycephalum* (*157*), and *Hae*III in *Chironomus* (*213, 214*).

The best-studied ribosomal NTSs are those of *Xenopus laevis* and *Drosophila melanogaster*. In *Xenopus*, nucleotide sequence analysis indicates three repetitive regions, separated by nonrepetitive areas containing *Bam*HI sites and termed "Bam islands" (*136, 204*). The repeats are homologous, the sizes differing by approximately 10 nucleotides (60, 80, 90, and 100 bp). The Bam islands contain a 145-bp sequence also found immediately upstream from the pre-rRNA initiation site. Initiation of transcription at these sites in the NTS has been reported (see Section III,A). In some clones of rDNA, the repeat region 3 (nearest to the 5' end of pre-rRNA) differs from repeat region 2 in that every 21-bp unit contains an *Sma*I site. In different clones, either one unit of all analogous repetitive regions or all units of one repetitive region contains the *Sma*I mutation. None of the proposed evolutionary mechanisms—"independent mutation with unequal crossing-over" (*215*) and "reduplication and insertion" (*216*)—can satisfactorily explain all the data on *Xenopus* NTS.

Compared to *Xenopus*, the organization of NTS in *Drosophila* is more regular and simpler. Digestion of several cloned rDNA fragments of *Drosophila melanogaster* with *Alu*I revealed three types of fragments (*211*). The left-hand (proximal to the 3' end of pre-rRNA) is of varied length between 1150 and 2650 bp; the right-hand fragment from an *Alu*I site to *Hinf*I in the ETS (see Fig. 7) is constantly 650 kb, and the middle region of varied length gives 5–12 repeats of equal size (later found to be 235 bp). As a matter of convenience, the small *Alu*I repeats are called "240-bp repeats." Three kinds of periodicity have been noted in the entire NTS (*211*). In the small region (a in Fig. 7), 16 bp downstream from the 3' end of 28-S RNA, a 90-bp periodicity

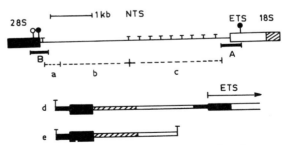

FIG. 7. Map of the external transcribed spacer (ETS) and nontranscribed spacer (NTS) of *Drosophila* rDNA clone Y22 (*78*). Restriction enzyme symbols are ♀, *Hind*III; ●, *Hinf*I; ⊤, *Alu*I. Regions A (*144*), B (*156*), and some of the *Alu*I repeats in C (*212, 218, 219*) have been sequenced. Regions a, b, and c, respectively, have 90-, 340-, and 240-bp periodicity (*212*). The lower part shows, in expanded scale, the homology of one *Alu*I repeat (e) with *Alu*I/*Hinf*I NTS/ETS fragment (d). Extensive homology is indicated by filled boxes, and less extensive homology by crosshatched boxes. The arrow indicates the initiation site and direction of transcription.

recognized by *Mnl*I can be discerned by computer analysis of the known sequence (156) in this region. Following this is the region (b) with a 340-bp periodicity recognized by *Mnl*I and *Dde*I, but not by *Alu*I. This is followed by the major repeated region (c) recognized by *Alu*I, *Mnl*I, and *Dde*I. Previous transcription studies *in vitro* have shown that transcription can initiate somewhere within a 240-bp repeat (217).

A number of the 240-bp repeats have been subcloned and sequenced (212, 218–219a). On alignment of the sequences, perfect homology of the 240 repeats is noted, except for one or two base changes. Strikingly, the *Alu*I repeats contain essentially perfect copies of the sequence from nucleotide −30 to +30 (+1 being the initiation site of ETS). The homologies are shown in Fig. 7 (d, e). With only slight mismatch, this homology of the repeats with the region upstream from the ETS extends to nucleotide −140. The presence of multiple "promoter-like" sequences in the NTS of genera as distant as *Xenopus* and *Drosophila* suggests that the arrangement is of some functional significance. Repetition of short sequences of 21 and 33 bp is found in the NTS preceding the transcription-initiation site of *Tetrahymena* (141).

A 90-bp periodicity has been found in the NTS of *D. mauritiana* and *D. simulans*, which are sibling species of *D. melanogaster* (220). The 240-bp periodicity is also found in all sibling species of the *D. melanogaster* subgroup (220), in *D. virilis* (221), and in *D. hydei* (222). Despite the similar periodicity, *D. melanogaster* NTS has little or no homology with *D. virilis* spacer. This contrasts with the rRNA-coding sequences, which have been highly conserved during the divergence of *Drosophila* species. Faster evolutionary changes in NTS compared to the rRNA-coding sequences is also found in *Xenopus* (137).

In the mouse, a variably sized region of repeated DNA (VrDNA) lies only 210 bp upstream from the origin of transcription (223). These VrDNA regions are composed of variable number of tandemly repeated copies of an approximately 135-bp subrepeats. The same features are present in both the inbred Balb/c mouse and a distantly related *Mus pahari* species (224). Sequence analysis reveals regious of inverted repeats and large (11–27 dT's) poly (dT) tracts. The VrDNA appears to be genetically labile and unstable during propagation in *E. coli* (223). These elements are very similar to *Cla*I elements in *Chironomus* NTS (213). Lying 5' to the VrDNA in mouse is an additional 6-kb NTS region, sequences of which are found interspersed throughout the genome in addition to rDNA repeats (225). Sequences homologous to regions of NTS in rat are also found elsewhere in the genome (L. I. Rothblum, personal communication). This property of NTS is not shared by the NTSs of other organisms (see 225 for discussion). It should be pointed out here that, unlike *Xenopus* (*136, 204*) and *Drosophila* (*212, 218, 219*), mouse (*224*) and human (*149*) rDNAs show no extensive duplication of the sequences at their respective initiation sites. Thus, the significance of "promoter" duplication in the region 5' upstream of many rRNA genes cannot be explained by the general model of sequestering of polymerase I molecules in preparation for rRNA transcription (*136*).

G. Ribosomal-Insertion Sequences

The discovery of insertion sequences ("introns" or "intervening sequences") was in the gene coding for rRNA in *Drosophila* (*98–101*). Since, then, ribosomal insertions have been identified in other insects and some lower eukaryotes (see Section II,A) and in the mitochondrial rDNA genes of yeast (*226, 227*) and *Neurospora* (*228, 229*). Recently, some of *Trypanosma* species have been found to contain rRNA genes with a mobile insertion element (*230*).

In *Drosophila melanogaster*, these insertions form a complex group of sequences of several size classes ranging from 0.5 to about 6 kb (*12, 98*). These insertions are of two types, based on sequence homology. Type-1 insertions of 0.5, 1, and 5 kb and other sizes share homologous regions and occur exclusively on the X chromosomes. About one-third of all rDNA repeats in *D. melanogaster* contain the major 5-kb type-1 insertion. Type-2 insertions are also complex, share no homology with type-1 insertions and occur in 15% of the rDNA units of both X and Y chromosomes (*101, 231, 232*). A large proportion of type-1 sequences occurs in chromosomal loci outside the nucleolus organizer (*233–235*). As far as is known, the type-2 insertions are found only in rRNA genes repeats. Similar types of ribosomal insertions have been found in other *Drosophila* species. Here, the more recent data available on nucleotide sequences are discussed.

The nucleotide sequences in the region of 28-S rRNA, where ribosomal insertions occur are conserved (*203*). The nucleotide sequences at the boundaries between gene and insertion regions of *Drosophila* have been determined by three groups (*236–238*). The two types of insertions occur at different sites of 28-S-β rRNA gene that are separated by 51–78 bp in different rDNA clones. In *Drosophila melanogaster*, there is a small deletion of the 28-S rRNA coding sequence at the site of the 5-kb type-1 insertion (*236, 238*), whereas there are duplications of 11 and 14 nucleotides at the target sites of 1-kb and 0.5-kb type-1 insertions (*237*). On the other hand, there is no such deletion or duplication of rDNA sequences flanking the type-2 insertions (237, 238). The 5-kb type-1 insertion of *D. virilis* is inserted at the same region as for type-1 insertion in *D. melanogaster*, but there is a 14-bp duplication (239). Interestingly, at the left-hand gene-insertion junction, the insertion sequences show no homology within the type, whereas at the right-hand junctions, the insertion sequences are highly homologous. As many as 50% of the type-1 insertion sequences are not inserted in rDNA, but are located within the chromocentric heterochromatin in tandem arrays (235, 240). These are also flanked by very short segments of the 28-S rRNA coding sequence (238). These data suggest that *Drosophila* rDNA insertions arose as transposable elements. The chromocentric ribosomal insertion-like elements are frequently interrupted by other sequences characterized as a new class of transposable elements in *Drosophila* (241, 242).

In *Drosophila*, the insertion-containing rRNA genes do not contribute significantly to the production of rRNA (243–247) (however, see Section III,B). In contrast, in some strains of *Tetrahymena* and *Physarum*, all copies of rDNA have "introns" that are transcribed as part of the pre-rRNA and subsequently removed by a splicing mechanism (51, 110, 112, 113). These types of introns are of particular interest with respect to the splice junctions. In both *Tetrahymena pigmentosa* (248) and *T. thermophila* (249), the insertion occurs exactly at the same site of conserved 28-S rRNA, which is 5 and 79 nucleotides away from *Drosophila* type-2 and type-1 insertions, respectively. In the case of polypeptide-coding genes, the intron sequences bordering the coding sequences are highly conserved and may be involved in recognition of splice junctions (250, 251). Comparison of the sequences at the rRNA splice junctions of *Tetrahymena* (248, 249), *Physarum polycephalum* (252), yeast (253) mitochondria, and *Chlamydomonas reinhardii* chloroplast (254) reveal no such "consensus" sequence. Sequences closely related to the sequence YTCAGAGACTA (Y = a pyrimidine nucleotide) are common to ribosomal introns of both mitochondrial and nuclear origin (255, 256) and are thought to be important for RNA splicing.

The function of ribosomal insertions is still a matter of conjecture. Although insertion sequences are not a universal feature of large rRNA genes, sequences and structures conserved between *Tetrahymena* nuclear rDNA introns and fungal mitochondrial introns suggest that they are of ancient origin. One obvious possibility is that the introns had their origin in transposable elements that were able to enter both nuclear and mitochondrial genes of a cell. In *Drosophila*, these are in fact members of families of mobile elements as already discussed. There are large open reading frames corresponding to some mitochondrial ribosomal proteins in the introns of mitochondrial rRNAs of *N. crassa* (256) and *Aspergillus nidulans* (257). The longest reading frame in the ribosomal introns of *Tetrahymena* and *Drosophila* can code for polypeptides of 26 and 90 amino acids, respectively. There is no evidence that these are actually expressed as proteins (Igor Dawid, personal communication).

III. Transcription of rRNA Genes

The process of transcription is mediated by the key enzyme, DNA-dependent RNA polymerase (EC 2.7.7.6), which catalyzes the synthesis of RNA by forming $3'$—$5'$ phosphodiester bonds using ribonucleoside triphosphates as substrates and DNA as template, in the presence of divalent cation (Mg^{2+} or Mn^{2+}). In prokaryotes, a single RNA polymerase catalyzes the synthesis of all types of RNA including rRNA. But in eukaryotes, multiple forms of polymerases catalyzing the synthesis of different types of RNA have been identified. Of the three major types in eukaryotic RNA polymerases, form I (or A) is responsible for the transcription of rRNA, form II (or B) for heterogeneous nuclear RNA (hnRNA) which includes mRNA, and form III (or C) for "low-molecular-weight" RNAs like 5-S RNA and tRNA. Whereas all the DNA-dependent RNA polymerases are inhibited by low concentrations of actinomycin D, form II is selectively inhibited by low concentrations (10^{-9} to 10^{-8} M) of α-amanitin, form III by higher concentrations (10^{-5} to 10^{-4} M) of the same drug, while form I is fully resistant to it. This selective inhibition has been used to characterize eukaryotic RNA polymerases (5, 6). The specialization of eukaryotic RNA polymerases in the transcription of different types of genes is a remarkable evolutionary acquisition. Although all three forms of polymerases have similar types of subunit structures, the molecular recognition mechanisms involved in selective transcription still remain unclear. Many reviews on both prokaryotic (258, 259) and eukaryotic (5, 6, 260, 261) RNA polymerases and their regulatory factors have appeared.

Since, in eukaryotes, DNA in the form of chromatin complex rather than naked DNA is the template for RNA synthesis, the sequence of events involved in the process of transcription by RNA polymerase I can be summarized as follows: (i) the activation of rDNA chromatin; (ii) the binding of RNA polymerase I at specific site of rDNA template ("promoter" or initiation site) and formation of the initiation complex; (iii) elongation of the pre-rRNA chain in the $5'$ to $3'$ direction; (iv) termination and release of the pre-rRNA; (v) processing of pre-rRNA (endonucleolytic cleavage, exonucleolytic trimming, and also splicing in the case of intron-containing genes). Several enzymes and protein factors may be involved in these steps, details of which have not been worked out yet. Modification of conserved rRNA sequences by methylation is simultaneous with step iii. Steps iv and v are not always sequential, but may be simultaneous in some cases; e.g., splicing may precede maturation cleavage of pre-rRNA. The exhaustive literature on these aspects have been reviewed (*13, 18–20*). Here I summarize some of the more recent information.

A. *In Vivo* Transcription

1. Ultrastructure of Active Transcription Units

One approach to understanding eukaryotic gene regulation involves studying the structure, sequence and expression of various genes *in vivo*. This approach is often limited in its ability to define the precise molecular features required for accurate, regulated transcription. In spite of the limitation, this approach has been particularly useful for studying rDNA structure and ultrastructure of transcriptionally active chromatin.

The chromatin spreading technique (*9, 58, 262*) allows the direct visualization of transcriptionally active rRNA genes under the electron microscope. Since the original studies with amphibian oocytes, extensive studies have been carried out with numerous cells and organisms (for review, see *13, 20, 262*). The active transcription units or matrix units are arranged in tandem on the rDNA axis and contain closely packed laterals fibrils of growing RNA chains of increasing length presenting a "Christmas-tree" pattern. Between the matrix units lie fibril-free zones of the nontranscribed spacers of variable length that are the basis of the size variation of the rDNA repeating units.

The lateral fibrils of transcription units are growing pre-rRNA chains that, in *Xenopus*, have a large size of 0.4 μm (*263, 264*). This length is severalfold shorter than that expected for a full grown pre-rRNA transcript. This observation, combined with enzymatic and specific staining (*264*) studies, led to the conclusion that growing pre-rRNA chains are already coated with proteins and have secondary structures. By using antibodies against specific ribosomal proteins, ribosomal proteins have also been localized on the pre-rRNA fibrils (*265*). The termini of fibrils always show large knobs (up to 30 nm in diameter) indicating a tighter RNA-protein packing at the 5' end.

Along the rDNA axis of the matrix units, the growing RNA fibrils are attached to particles 12–15 nm across, which are identified as RNA polymerase particles. Full-length RNA fibrils with the terminal knobs (presumably unprocessed pre-rRNA released from RNA polymerase) are sometimes observed close to the vicinity of the 3' end of the transcription unit (266). In a broad variety of organisms, a packing of 40–50 RNA polymerase molecules per micrometer of rDNA axis in the matrix unit indicates a high efficiency of rDNA transcription.

The chromatin spreading technique has been used to study the transcription initiation and termination regions of both endogenous rDNA and cloned rDNA in plasmids injected into oocytes (154, 266–268). These studies have shown that each individual rDNA unit has its own promoter site contained within a relatively shorts NTS segment upstream of the initiation site (267, 268). At the termination site, transcript release and dissociation of the RNA polymerase from the rDNA is not necessarily coupled. In the oocyte-injection system, active promoter elements have been located within the rDNA fragment beginning 145 bp upstream and ending 16 bp downstream (154, 269–271). Small transcription units or "prelude" sequences found in oocytes of some frogs initiate at the two promoter-like "Bam island" regions in the NTS (262, 266). The more frequent type of NTS transcription unit is about 0.86 μm (or 3.2 kb) long and starts at about 0.4 μm (1.5 kb of DNA) downstream from the 3'-termination site of the preceding pre-rRNA transcription unit. The smaller type of NTS transcription unit starts at the promoter sites downstream from the first one. The RNA fibrils in the NTS transcription units are characterized by the absence of dense terminal knobs identified at the termini of true pre-rRNA fibrils (268). This observation indicates that NTS transcripts are not the products of read-through transcription from pre-rRNA.

2. STRUCTURE OF ACTIVE CHROMATIN

In view of the present-day nucleosomal or "beads-on-a-string" structure for chromatin (272, 273), there is a renewed interest in the structure of active and inactive chromatin of all amenable genes including rDNA. Both electron microscopic (262) and enzymatic (274) techniques have been used to study the mechanism of chromatin "activation" for transcription, which is still little understood. The packing ratio of DNA (the ratio of the length of DNA in the B form to the same DNA segment in chromatin form) in inactive ribosomal chromatin is 2.1–2.4 (262, 267, 275, 276). The data indicate that the nucleosome structure of inactive ribosomal chromatin is the same as the bulk chromatin, i.e., the same general arrangement of nucleosomes containing about 200 bp of rDNA (140 bp rDNA plus two molecules each of histones H2A, H2B, H3, and H4 in the core nucleosome and 20–30 bp of linker rRNA plus histone H1). Measurement of the length of ribosomal transcription units in active chromatin yields an average rDNA packing ratio of 1.1 to 1.4 (262, 275, 276). This lower packing ratio indicates an unfolding of chromatin for transcription. Digestion of chromatin active in rDNA transcription with micrococcal nuclease reveals the same pattern of repeating nucleosome units (277–279). All the evidence indicates that rDNA in the form of chromatin, rather than naked DNA, is the template for rRNA transcription.

RNA polymerase and other protein factors present in the nonhistone chromosomal proteins may be involved in activation of chromatin (280–282). Unfolding of ribosomal chromatin is not simultaneous with transcription, but actually may precede or follow it (262, 275). The initiation site of an active chromatin is more sensitive to DNase (274). Thus, the DNase-I-hypersensitive sites at the promoters are present only in the expressed rDNA of *X. laevis* and *X. borealis* hybrids (283). The first promoter-like sequence in the NTS downstream from the rRNA transcription-termination site is also hypersensitive, which is corroborated by the finding of transcription units at these NTS sites (266). Differential nucleosome spacing and accessibility to micrococcal nuclease in transcribed and nontranscribed regions of *Tetrahymena* ribosomal gene chromotin have been found (T. R. Cech, personal communication). It may be mentioned that in the well-studied SV40 minichromosome model system, the active chromatin has nucleosomal structure except for about 19% of the DNA, which is found as nonnucleosomal DNA (284, 285).

3. CHAIN-ELONGATION RATE

The rate of rRNA synthesis is geared to the demands of the cell (17, 19). In optimally growing cells, there is little turnover of rRNA, and the rate of synthesis approaches the possible maximum. The absolute synthesis rate in rRNA can be measured on the basis of incorporation of labeled precursor and the specific activity of the precursor nucleotide pool in the cell. By knowing the number of active rRNA genes and the number of RNA polymerase molecules transcribing the genes (by the spreading technique), the "step time," i.e., the transcription rate per polymerase molecule, can be calculated. By this method, this rate in *Xenopus* oocytes under *in vitro* culture conditions was found to be about 18 nucleotides per second per polymerase molecule (286). Using the same calculations, the rate is about 25–40 nucleotides per second per polymerase molecule in *Tetrahymena* (287, 288). Considering the uncertainties involved in such calculations, the rates are in good agreement with estimates for transcription rates in other eukaryotic systems (289, 290).

B. *In Vitro* Transcription

1. TRANSCRIPTION IN ISOLATED NUCLEI AND NUCLEOLI

A successful *in vitro* transcription system would be characterized by high fidelity (i.e., initiation and termination at proper sites) and if possible, a transcription rate comparable to that found *in vivo*. Initial attempts to develop such a system using purified RNA polymerase I and DNA or chromatin resulted in aberrant transcription (291–293). Therefore, a number of studies have been made on rDNA transcription in isolated nuclei or nucleoli using the endogenous RNA polymerase I and template DNA. In such systems, transcription of rDNA still continues (5, 6, 294–300). However, the limitation of such systems is that the observed rRNA synthesis is primarily due to elongation of rRNA chains initiated *in vivo*. The conclusion is supported by the kinetics of rRNA synthesis, which plateaus after 15–30 minutes. The effects of initiation inhibitors such as heparin (294–296, 298) and Sarkosyl (301) also indicate little or no reinitiation.

In spite of the limitation, isolated nuclei can give information on transcription or RNA polymerase loading on specific genes. Thus, nuclei and nucleoli isolated from *Tetrahymena* have been used to study rRNA transcription and splicing (302, 302a, 303). The chain-elongation rate is about one order of magnitude less (4 nucleotides per second) compared to the *in vivo* rate of 25–40 nucleotides per second. It seems likely that some RNA polymerase and/or other protein factors are lost during isolation of nuclei or nucleoli. In contrast to initiation, accurate termination of transcription occurs in isolated nuclei or nucleoli. Termination occurs at or near the *in vivo* termination site (303). Moreover, the *in vitro* transcripts do not hybridize to terminal NTS regions of rDNA (303), indicating the absence of any read-through transcription. Some protein factor(s) that can be solubilized from nucleoli at high ionic strength is necessary for correct termination. This study clearly indicates that the presence of some nucleotide sequence specifying termination is not enough for correct termination of the transcript.

We used transcription in nuclei isolated from cultured *Drosophila melanogaster* cells to answer a more intriguing question. As already mentioned (Section II,G), rDNAs containing insertions are inactive in the production of rRNA (243–247). Yet, these rDNAs have the same nucleotide sequence at their initiation (144) and termination (156) regions as the active uninterrupted genes. In *in vitro* transcription experiments using truncated cloned *Drosophila* rDNA, the interrupted genes correctly initiate transcription (217). Why then are these not contributing to rRNA production? By hybridizing labeled nuclear transcripts to restriction fragments of rDNA upstream or downstream from the point of insertion, we obtained an indication that interrupted genes are loaded with nascent RNA chains up to the point of insertion (242, 305).

S1-nuclease mapping of labeled nascent RNA suggested that many rRNA molecules have 3' ends at or near the point of insertion in the rDNA sequence. These experiments support a model in which interrupted rRNA genes initiate transcription just like uninterrupted repeats, but terminate at or close to the point of insertion. The resulting incomplete precursor of rRNA must be rapidly degraded probably owing to lack of the stabilization imparted to normal pre-rRNA transcript by secondary structure and/or by binding with proteins including ribosomal proteins, which bind to the highly conserved region of 28-S RNA (203). Very little transcript corresponding to the type-1 ribosomal insertion was found.

The results are supported by the distribution of histones along transcribed and nontranscribed regions of rDNA chromatin in *Drosophila* (306). The method consists of hybridization of DNA, previously cross-linked to histones in nuclei, with various cloned DNA probes. The histone:DNA ratio was found to be one for sequences upstream from insertion, three for insertion and two for sequences downstream from insertion, indicating that about half of the rRNA genes are not transcribed in the downstream region (306). We also found transcripts corresponding to NTS sequences that were not read-through transcripts beyond normal termination points (305). The initiation points of these NTS transcripts have been characterized (212, 218, 219).

2. Transcription in Solubilized System

The RNA synthetic capacity of *Xenopus* oocytes was used to develop a true *in vitro* system for studying the accurate transcription of genes transcribed by RNA polymerases of class II and III (307, 308). The system consisted of an extract from manually isolated oocyte nuclei. The homogenate of oocyte nuclei initiated and transcribed endogenous rRNA genes and continued to do so for at least 8 hours (309). An RNA chain-elongation rate of 2–5 nucleotides per second is obtained, which is comparable to that in isolated nuclei. However, this extract did not use cloned rDNA added to the system. Soon after the successful development of an *in vitro* system for transcription of genes catalyzed by RNA polymerases II and III (*310–312*), cloned rDNA transcription was demonstrated in a 100,000 g supernatant fraction (S-100) of the extract made from actively growing mouse cells (*313*). The use of such an *in vitro* system correctly localized the initiation site of mouse rDNA (*152, 314, 315*), thus ending the anomaly regarding the initiation site indicated by earlier S1-mapping experiments (*128, 145*).

Using mutant rDNA clones carrying BAL-31 deletions of definite sequences upstream from the initiation site as templates, the nucleotide position −39 to −34 is indicated as the essential promoter region (*316*) analogous to the Goldberg–Hogness (TATA) box in genes transcribed by polymerase II (*150, 151*). The control region may extend up to nucleotide position −12. In addition, sequences located farther upstream (positions −45 to −169) may also exert some influence in efficient transcription, a situation analogous to the CCAAT box for polymerase II (*150, 151, 317*). Some protein factors form a stable initiation complex by binding with rDNA and remain bound to the template during several rounds of transcription. These factors have not been characterized (*318, 318a*). Further, the expression of SV40 T antigen under the control of mouse rDNA promoter (sequence containing nucleotides −169 to +56) linked to the T-antigen coding sequence has been obtained after injection of the chimeric plasmid into mouse cell nuclei (S. Fleischer and I. Grummt, personal communication). This shows that a gene transcribed by RNA polymerase II can be expressed under the control of the polymerase I promoter.

In parallel studies, *Drosophila* cell-free extracts have been used to locate the rDNA promoter site (*217*). Nucleotides between positions −34 bp and about +30 bp of the initiation site are sufficient for low levels of accurate transcription *in vitro* (*218*). Using BAL 31-deleted subclones of rDNA as template, the promoter site has been delimited between nucleotides −43 to −27. Moreover, the region between −18 and +20 bp sustains a low level of accurate transcription (*304*). Thus, the *in vitro* studies with both mouse and *Drosophila* systems locate the promoter site in the vicinity of −30 bp upstream, which is the same general area of the TATA box for polymerase II (*317*), though the recognized nucleotide sequences are different. Another major component of the polymerase I promoter in *Drosophila* lies within the first four nucleotides of the external transcribed spacer (B. D. Kohorn and P. M. M. Rae, personal communication).

3. Species Specificity in rDNA Transcription

Accurate transcription of rRNA genes may involve both common as well as species-specific regulatory sequences recognized by protein factors. The mouse rDNA is transcribed only by mouse ascites cell extract, not by human KB cell extract (313). *Drosophila melanogaster* KC cell extract similarly transcribes *D. melanogaster* rDNA, but not *D. virilis* rDNA (217). However, this strict species specificity is not observed in *Xenopus*, as is evidenced by the transcription of *X. laevis* rDNA in *X. borealis* oocytes (269). The expression of human rDNA in human–mouse heterokaryons is normally repressed (318b). In some hybrid lines, the inactive rDNA can be reactivated (319). The apparent discrepancy between *in vivo* and *in vitro* systems cannot be resolved at present. However, it may be pointed out that free rDNA is used as template *in vitro*, while rDNA in the form of chromatin is the template *in vivo*. *In vitro* transcription of cloned yeast rDNA in the presence of a stimulatory factor (320) and several factors stimulating eukaryotic RNA polymerases have been reported (6). An initiation factor in plant nuclei, which stimulates transcription of both plant and animal RNA polymerases (321, 322), has been isolated.

C. Processing of Pre-rRNA

The primary pre-rRNA transcript undergoes a series of modification and processing reactions ultimately giving rise to the 28-S, 18-S, and 5.8-S rRNA species. Since the nascent rRNA chains are already complexed with proteins (265, 268), the processing involves important RNA–protein interactions in addition to the enzymatic reactions.

1. Modifications of Pre-rRNA

Ribosomal RNA contains 2'-O-methylribose and pseudouridine. These modifications take place during or shortly after pre-rRNA transcription, and the methyl groups found in the pre-rRNA are conserved in the mature rRNA. The distribution of methyl groups is nonrandom. Comparative studies with vertebrates show that the oligonucleotides containing methylated nucleotides are more conserved than the rest of the polynucleotide chain (323, 324). The occurrence of pseudouridine is similar. A striking observation is that the numbers of pseudouridine and methylated nucleotides in rRNA are almost equal, almost 100 each. The highly specific pattern of these two modifications indicate some important function of methylation and pseudouridylation in the processing, structure and function of rRNA. The subject has been reviewed (325, 326).

2. Maturation of Pre-rRNA

The organization of pre-rRNA transcription units (Section II,A) immediately suggests that a basically similar mechanism in pre-rRNA maturation is shared by all eukaryotes. Kinetic studies on the labeling of pre-rRNA with radioactive precursors followed by characterization of the RNA species by ultracentrifugation and electrophoretic separation allow the identification of discrete intermediates in pre-rRNA processing. The subject has been extensively reviewed (18, 19). Further refinement in the method of identification of RNA intermediates, such as chromatography, Southern hybridization, and sequencing, have helped in identifying uncertain intermediates and alternative pathways of maturative processing (327, 328). Normally, the folded structure of pre-rRNA (or pre-rRNP) may specify the marker sites at the gene-spacer boundaries for selective cleavage that is the basis of the general scheme illustrated in Fig. 8 (80, 126, 127, 200, 327–334). The sequence of the cleavages at different sites of pre-rRNA is not the same in all organisms. Even in a single species, simultaneous operation of alternate pathways may be operating. For a detailed discussion on the processing of pre-rRNA and preribosomes in eukaryotes, see Hadjiolov (20).

3. Splicing of Pre-rRNA

Excision of the intervening sequence (IVS) located in the 26-S rRNA near the end of the transcription unit is the first discernible step in the processing of the primary rRNA transcript in *Tetrahymena* (110, 114, 335). The subsequent endonucleolytic processing occurs as in other eukaryotes. In *Physarum*, the picture is not yet as clear. The available information suggests that the two introns in the 26-S rRNA are transcribed, but are then spliced out in a random order even before rRNA synthesis has reached the 3′ end of the primary transcript. In any case, splicing precedes the subsequent processing steps.

Pulse-chase labeling experiments in isolated nuclei from *Tetrahymena* show that the IVS is excised as a linear molecule and subse-

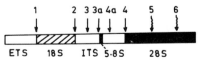

Fig. 8. The most general model of eukaryotic pre-rRNA processing sites is shown. The order of processing is sites 1, (2 + 3), 4. 3a, and 4a are additional processing sites leading to the formation of mature 5.8-S RNA, but uncertainty lies in the order and location of these sites. Site 5 is special to dipteran insects (e.g., *Drosophila*), and indicates the gap site in 28-S rRNA; site 6 is the splicing site in *Tetrahymena*.

quently cyclized (336). The splicing is inhibited at low salt concentrations (114) or in the absence of Mg^{2+} (337), permitting the isolation of the unspliced pre-rRNA. On incubation of the isolated unspliced RNA *in vitro* with guanosine or guanosine phosphates, Mg^{2+}, and a monovalent cation, the IVS is excised (114). During excision, IVS RNA acquires a 5'-terminal guanosine residue not encoded by the DNA (338). It appears that the IVS RNA itself has the intrinsic property of autoexcision and autocyclization without the intervention of any enzymes (339).

D. Regulation of Transcription

The synthesis of rRNA (and the formation of ribosomes) is regulated according to the requirements of protein synthesis of the cells under different conditions. Stimulation of the growth of cultured cells results in a proportionately greater increase in the synthesis of rRNA compared to other classes of RNA (18, 341–343). Since efficient regulation of ribosome formation involves the coordinate regulation of all the components of ribosome (rRNAs and r-proteins), the mechanism is very complex, involving the interaction of various proteins—enzymatic, structural, and regulatory—at different stages of rRNA transcription, processing, and assembly. Several reviews deal with different aspects of this subject (3, 4, 18–20).

1. Number of Active Genes

Transcriptional control may be exerted primarily through the number of active rRNA genes, the level of RNA polymerase I, and the rate of transcription (including initiation, chain elongation, termination, and release/processing of pre-rRNA). Excepting the special cases of rDNA amplification (Section I,C), the number of rRNA genes is more or less fixed for the cells of a particular organism. Several lines of evidence suggest that the number of rRNA molecules are in excess of cell requirements, and that only a part of the total number of genes are active. For instance, strains of *Drosophila* and *Xenopus* with varying rRNA gene redundancy produce the same amount of rRNA (Section I,A). The gene-spreading technique used to study the number of active rRNA genes indicates that, in lower eukaryotes in the growing stage (76) and in the growing oocytes (262, 263, 344), switching on of all or most of the rRNA genes takes place. The number of active genes are greatly reduced as the oocyte matures or the animal hibernates (17, 286, 345, 346). The mechanism of activation of eukaryotic genes is as yet poorly understood. It has been suggested that structural modification of the genes in the chromosomal matrix (chromatin) could regulate the accessibility of specific genes and hence their ability to be transcribed (see Section III,A,2).

2. METHYLATION AND TRANSCRIPTION OF rDNA

In recent years, the undermethylation of cytosine in d(C-G) sequences has been correlated with transcriptional activity. For example, the genes for chicken ovalbumin and conalbumin (347, 348), chicken β-globin (347), and rabbit β-globin (349) are undermethylated in those tissues where they are expressed. In contrast, the genes in sperm are fully methylated. The methylation of rDNA has been studied in a number of animals by means of restriction endonucleases that distinguish d(m^{5}C-G) from d(C-G). The rDNA in vertebrate somatic cells is mostly methylated; the undermethylated fraction is in the transcribed chromatin. In contrast, the rDNA in nucleated erythrocytes is fully methylated (350, 351). In Xenopus, the rDNA is heavily methylated except in two regions on the NTS, and loss of rDNA methylation accompanies the onset of rRNA gene activity in early development (352). Amplified rDNA in oocytes is unmethylated (353). Methylation of the rDNA in the rat is lowest in growing cells and highest in totally repressed cells like spermatozoa (354). The most heavily methylated region of flax rDNA is close to the 5' end of the transcription unit (355). All these studies indicate, but do not establish, that fully methylated rDNA is not transcribed, and that transcription is somehow correlated with the "removal" of methyl groups from the sites of initiation, or close to it, in the NTS region.

3. LEVEL OF RNA POLYMERASE I

Normally, the level of RNA polymerase I is not limiting for rRNA transcription. This is exemplified by the transcription of excess genes injected into oocytes (266–268) and by relatively minor changes in the level of RNA polymerase I from fertilization through the gastrula stage of amphibian (356, 356a) and fish (357) embryos, a period when the rRNA transcription rate changes considerably. However, there are cases where an increased rate of rRNA synthesis is accompanied by a proportionate increase in the level of polymerase I, such as in developing sea-urchin embryo (358), in rooster liver during estradiol action (359), and in mitogen-stimulated lymphocytes (360, 361).

4. TRANSCRIPTION FACTORS

In vitro transcription studies with extracts of mouse cells indicate that growing cells contain factors required for the faithful transcription of rRNA genes. Extracts from nongrowing or slowly growing cells contain very little of this activity (313). These factors appear to be species-specific (see Section III,B,3). It is possible that such factors control the formation of the initiation complex (318). Subsequently, species-specific factors for the accurate transcription of mammalian rDNA have been partially characterized (362). Similarly, factor(s) may be involved in the correct termination and release of the transcript. There is evidence for some protein factors in *Tetrahymena* nucleoli needed for correct termination (303).

The differential activation and transcription of rRNA genes during early embryonic development have been subjects of intensive study for a long time (17). It is generally believed that in amphibian embryos no significant rRNA transcription takes place before gastrulation, the time when formation of the nucleolus starts (7–9). This notion has been questioned on the basis of calculation of the actual rates of rRNA synthesis corrected for the number of cells in the embryo at the blastula and gastrula stages of development (17). To obviate this criticism, dissociated cells from the embryos were used to show differential synthesis of rRNA (363); this experiment was not reproducible (364, 365).

Later studies implicated various nonprotein and protein factors as regulators of rRNA synthesis (366–369). There is a specific stimulatory factor for rRNA synthesis that is present at high level in active oocytes and gastrula cells, but is low in early embryonic cells, which make little rRNA (370). Also, the loss of rDNA methylation on the onset of transcriptional activity in early embryonic development (352) and progressive changes in template-bound RNA polymerase I (371) clearly point out that rRNA genes are in fact under transcriptional control during early embryonic development. We have found a similar differential synthesis of rRNA in the early development of catfish (*Heteropneustes fossilis*) embryos (357). Cell extracts prepared from gastrula-stage embryos stimulate rRNA synthesis in isolated blastula cell nuclei (372, 373).

Apart from the regulation by transcription factors at the initiation and termination steps, rRNA transcription may also be influenced by the chain-elongation rate and by the stability and processing of pre-rRNA involving many proteins, including ribosomal proteins. When growing cells of *Tetrahymena* are shifted to a nongrowth or starvation medium, the rates of transcription and processing are decreased by factors of 10 to 60 (287, 288). The overall reduction in the processing rate is not limited by the splicing rate under starvation conditions, and the pathway of processing may be altered (287). The inhibition of protein synthesis in growing cells does not alter the transcription rate of pre-rRNA synthesis immediately, but processing of pre-rRNA is blocked (*19, 341, 374–379*). A block in the release/processing of pre-rRNA will ultimately alter the rate of transcription by hindering the movement of RNA polymerase I molecules (*380*).

The proteins involved in such posttranscriptional control have not been characterized. Considering the coupling between transcription, translation, and degradation of mRNA in prokaryotes as proposed in various models (*381–383*), ribosomal proteins are likely candidates to play such role. In both prokaryotes and eukaryotes, the synthesis of rRNA and ribosomal proteins are regulated together, though not in a strictly coordinate way (*384–386*). In the absence of protein synthesis, the transcription of pre-rRNA is reduced and its processing is slowed down or abolished. In the absence of rRNA synthesis, ribosomal proteins are synthesized and degraded (*386*). In nongrowing *E. coli* (*384*) and lymphocytes (*387*), there is "wastage" or degradation of newly made rRNA.

The rate of rRNA synthesis as well as ribosomal protein synthesis is programmed during embryogenesis in *Xenopus* (*388*). It is also known that some ribosomal proteins bind to growing pre-rRNA chains even before their transcription is complete (*265*). This protein binding may be necessary for the stability of pre-rRNA before processing. There is a dramatic increase in ribosomal protein mRNAs during *Xenopus* development (*389*). Thus, the availability of ribosomal proteins may directly or indirectly control the differential transcription of rRNA in early development. We have found a stimulation of fish RNA polymerase I activity *in vitro* upon addition of homologous ribosomal proteins (*372*; R. K. Mandal and A. Chaudhuri, unpublished observation). Whether the stimulations of rRNA synthesis by the gastrula cell extracts and ribosomal proteins are parts of the same control system or of independent ones can be decided only by further studies.

IV. Concluding Remarks

In the last few years, the new techniques of gene cloning and sequencing have led to information about the organization and transcription of ribosomal RNA genes from an ever-increasing number of eukaryotic organisms. The general arrangement of the coding and noncoding sequences in the rDNA repeating units has been determined, down to the nucleotide level in some cases. Determination of the nucleotide sequences at the transcription initiation and termination sites and at the gene-spacer boundaries has revealed some general features regarding sequence conservation and divergence in rDNAs. In general, the structural and functional domains of rRNA have been conserved in evolution in spite of wide divergence in the transcribed and nontranscribed spacers. Some characteristic properties of the repeating units in the nontranscribed spacer sequences have come to light, but their functional significance is not understood at present.

The development of crude cell-free systems for accurate *in vitro* transcription has been instrumental in determining the nucleotide signals acting as promoters. Further dissection and reconstitution of the transcription system will reveal many factors responsible for regulation of rRNA transcription. The ultimate aim of such studies is to understand the mechanisms by which the ribosomal genes in the chromatin complex are regulated *in vivo* according to the requirements of cell growth, development, and differentiation.

Acknowledgments

I am grateful to Drs. Igor Dawid, S. Ghosh, and B. B. Biswas for going through the manuscript, to Drs. Thomas Cech, Gabriel Dover, Ingrid Grummt, Bruce Kohorn, Masami Muramatsu, Ronald Reeder, Lawrence Rothblum, and Michael Trendelenburg for making reprints and unpublished work available, and to Kalidas Saha for help in preparing the manuscript.

References

1. M. Nomura, A. Tissières, and P. Lengyel, "Ribosomes," CSHLab, CSH, New York, 1974.
2. M. Nomura, E. A. Morgan, and S. R. Jaskunas, *ARGen* 11, 297 (1977).
3. B. E. H. Maden, *Prog. Biophys. Mol. Biol.* 23, 129 (1971).
4. I. G. Wool, *ARB* 48, 719 (1979).
5. P. Chambon, *ARB* 44, 613 (1975).
6. B. B. Biswas, A. Ganguly, and A. Das, This Series 15, 145 (1975).
7. D. D. Brown and E. Littna, *JMB* 20, 81 (1966).
8. J. B. Gurdon and H. R. Woodland, *Proc. R. Soc. London Ser. B* 173, 99 (1969).
9. S. L. McKnight and O. L. Miller, Jr., *Cell* 8, 305 (1976).
10. E. E. Baulieu, *MCBchem* 7, 157 (1975).
11. J. R. Tata, *in* "Handbook of Physiology" (R. O. Greep and E. B. Astwood, eds.), Vol. 3, p. 469, Am. Physiol. Soc., Washington, D. C., 1974.
12. E. O. Long and I. B. Dawid, *ARB* 49, 727 (1980).
13. H. Busch and L. Rothblum, eds., "The Cell Nucleus," Vols. 10 and 11 (1982) and Vol. 12 (1982) Academic Press, New York.
14. R. P. Singhal and J. K. Shaw, This Series 28, 177 (1983).
15. H. Tobler, *in* "Biochemistry of Animal Development" R. Weber, ed.), Vol. 3, p. 91. Academic Press, New York, 1975.
16. K. D. Tartof, *ARGen* 9, 355 (1975).
17. E. Davidson, "Gene Activity in Early Development," 2nd ed. Academic Press, New York, 1976.

18. R. P. Perry, *ARB* **45**, 605 (1976).
19. A. A. Hadjiolov and N. Nikolaev, *Prog. Biophys. Mol. Biol.* **31**, 95 (1976).
20. A. A. Hadjiolov, *Subcell. Biochem.* **7**, 1 (1980).
21. H. Wallace and M. L. Birnstiel, *BBA* **114**, 296 (1966).
22. D. Gillespie and S. Spiegelman, *JMB* **12**, 829 (1965).
23. M. L. Birnstiel, M. Chipchase, and J. Speirs, This Series **11**, 351 (1971).
24. J. Ingles, J. N. Timmis, and J. Sinclair, *Plant Physiol.* **55**, 496 (1975).
25. D. D. Brown and J. B. Gurdon, *PNAS* **51**, 139 (1964).
26. F. Ritossa, *in* "The Genetics and Biology of Drosophila (M. Ashburner and E. Novitski, eds.), Vol. 1A, p. 801, Academic Press, New York, 1976.
27. F. Ritossa, *PNAS* **60**, 509 (1968).
28. B. B. Spear and J. G. Gall, *PNAS* **70**, 1359 (1973).
29. S. A. Endo and D. M. Glover, *Cell* **17**, 597 (1979).
30. O. J. Miller, R. Tantravahi, D. A. Miller, L.-C. Wu, P. Szabo, and W. Prensky, *Chromosoma* **71**, 183 (1979).
31. C. Goodpasture and S. E. Bloom, *Chromosoma* **53**, 37 (1975).
32. J. G. Gall and M. L. Pardue, *PNAS* **63**, 378 (1969).
33. P. Philippsen, M. Thomas, R. A. Kramer, and R. W. Davis, *JMB* **123**, 387 (1978).
34. T. D. Petes, *PNAS* **76**, 410 (1979).
35. B. McClintock, *Z. Zellforsch. Mikroskop. Anat.* **21**, 294 (1934).
36. A. S. Henderson, D. Warburton, and K. C. Atwood, *PNAS* **69**, 3394 (1972).
37. R. Tantravahi, D. A. Miller, V. G. Dev, and O. J. Miller, *Chromosoma* **56**, 15 (1976).
38. H. Busch and K. Smetana, "The Nucleolus," Academic Press, New York, 1970.
39. A. Lima-da-Faria, *Hereditas* **83**, 1 (1976).
40. R. Batistoni, F. Andronico, I. Nardi, and G. Barsacchi-Pilone, *Chromosoma* **65**, 231 (1978).
41. D. D. Brown and I. B. Dawid, *Science (Washington, D.C.)* **160**, 272 (1968).
42. J. G. Gall, *Genetics* **61** (Suppl.), 121 (1969).
43. W. S. Vincent, H. O. Halvorson, H. R. Chen, and D. Shin, *Exp. Cell Res.* **57**, 240 (1969).
44. A. Chaudhuri and R. K. Mandal, *Nucleus (Calcutta)* **23**, 78 (1980).
45. J. G. Gall and J. D. Rochaix, *PNAS* **71**, 1819 (1974).
46. M. F. Trendelenburg, U. Scheer, H. Zentgraf, and W. W. Franke, *JMB* **108**, 453 (1976).
47. L. Lison and N. Fautrez-Firlefyn, *Nature* (London) **166**, 610 (1950).
48. D. J. Wolgemuth, G. M. Jagiello, and A. S. Henderson, *Exp. Cell Res.* **118**, 181 (1979).
49. K. Karrer and J. G. Gall, *JMB* **104**, 421 (1976).
50. V. M. Vogt and R. Braun, *JMB* **106**, 567 (1976).
51. G. R. Campbell, V. C. Littau, P. W. Melera, V. G. Allfrey, and E. M. Johnson, *NARes* **6**, 1433 (1979).
52. A. F. Cockburn, W. C. Taylor, and R. A. Firtel, *Chromosoma* **70**, 19 (1978).
53. R. C. Findlay and J. G. Gall, *PNAS* **75**, 3312 (1978).
54. H. J. Lipps and G. Steinbrück, *Chromosoma* **69**, 21 (1978).
55. J. Engberg and R. E. Pearlman, *EJB* **26**, 393 (1972).
56. M. L. Pardue and J. G. Gall, *PNAS* **64**, 600 (1969).
57. M. Buongiorno-Nardelli and F. Amaldi, *Nature (London)* **225**, 946 (1970).
58. O. L. Miller, Jr. and B. R. Beatty, *Genetics* **61** (Suppl.), 133 (1969).
59. P. K. Wellauer, R. H. Reeder, I. B. Dawid, and D. D. Brown, *JMB* **105**, 487 (1976).

60. D. Hourcade, D. Dressler, and J. Wolfson, *PNAS* **70**, 2926 (1973).
61. J. D. Rochaix, A. Bird, and A. Bakken, *JMB* **87**, 473 (1974).
62. M. Buongiorno-Nardelli, F. Amaldi, and P. A. Lava-Sanchex, *Exp. Cell Res.* **98**, 95 (1976).
63. G. Steinbrück, I. Haas, K.-H., Hellmer and D. Ammermann, *Chromosoma* **83**, 199 (1981).
64. I. B. Dawid and D. D. Brown, *Dev. Biol.* **22**, 1 (1970).
65. J. Engberg, P. Andersson, V. Leick, and J. Collins, *JMB* **104**, 455 (1976).
66. S. Berger, D. M. Zellmer, K. Kloppstech, G. Richter, W. L. Dillard, and H. G. Schweiger, *Cell Biol. Int. Rep.* **2**, 41 (1978).
67. I. B. Dawid and P. K. Wellauer, *Cell* **8**, 443 (1976).
68. R. A. Kramer, P. Philippsen, and R. W. Davis, *JMB* **123**, 405 (1978).
69. D. M. Glover and D. S. Hogness, *Cell* **10**, 167 (1977).
70. P. K. Wellauer and I. B. Dawid, *Brookhaven Symp. Biol.* **26**, 214 (1975).
71. B. Lewin, "Gene Expression" (B. Lewin, ed.), 2nd ed., Vol. 2, p. 570. Wiley, New York, 1980.
72. R. H. Reeder, T. Higashinakagawa, and O. J. Miller, Jr, *Cell* **8**, 449 (1976).
73. M. Nomura, E. A. Morgan and S. R. Jaskunas, *ARGen.* **11**, 297 (1977).
74. I. B. Dawid, C. K. Klukas, S. Ohi, J. L. Ramirez, and W. B. Upholt, in "The Genetic Function of Mitochondrial DNA" (C. Saccone and A. M. Kroon, eds.), p. 3. Elsevier/North-Holland, Amsterdam, 1976.
75. P. R. Whitfeld and W. Bottomley, *Annu. Rev. Plant Physiol.* **34**, 279 (1983).
76. H. Spring, G. Krohne, W. W. Franke, U. Scheer and M. F. Trendelenburg, *J. Microsc. Biol. Cell.* **25**, 107 (1976).
77. J. Klootwijk, P. de Jonge, and R. J. Planta, *NARes* **6**, 27 (1979).
78. I. B. Dawid, P. K. Wellauer, and E. O. Long, *JMB* **126**, 749 (1978).
79. P. K. Wellauer and I. B. Dawid, *JMB* **89**, 379 (1974).
80. U. Schibler, T. Wyler, and O. Hagenbuchle, *JMB* **94**, 505 (1975).
81. P. K. Wellauer and I. B. Dawid, *PNAS* **70**, 2827 (1973).
82. K. Nath and A. P. Bollon, *JBC* **252**, 6562 (1977).
83. G. I. Bell, L. J. DeGennaro, D. H. Gelfand, R. J. Bishop, P. Valenzuela, and W. J. Rutter, *JBC* **252**, 8118 (1977).
84. J. H. Cramer, F. M. Farelly, J. T. Barnitz, and R. H. Rownd, *MGG* **151**, 229 (1977).
85. J. H. Meyerink and J. Retel, *NARes* **3**, 2697 (1977).
86. P. K. Wellaner, R. H. Reeder, D. Carrol, D. D. Brown, A. Deutch, T. Higashinakagawa, and I. B. Dawid, *PNAS* **71**, 2823 (1974).
87. P. K. Wellauer, I. B. Dawid, D. D. Brown, and R. H. Reeder, *JMB* **105**, 461 (1976).
88. T. A. Walker and N. R. Pace, *NARes* **4**, 595 (1977).
89. P. G. Bosley, A. Tuyns, and M. L. Birnstiel, *NARes* **5**, 1121 (1978).
90. W. Leon, D. L. Fouts, and J. Manning, *NARes* **5**, 491 (1978).
91. W. L. Gerlach and J. R. Bedbrook, *NARes* **7**, 1869 (1979).
92. A. Degelmann, H. D. Royer, and C. P. Hollenberg, *Chromosoma* **71**, 263 (1979).
93. R. F. Manning, D. R. Samols, and L. P. Gage, *Gene* **4**, 153 (1978).
94. S. Cory and J. M. Adams, *Cell* **11**, 795 (1977).
95. N. Arnheim and E. M. Southern, *Cell* **11**, 363 (1977).
96. N. Blin, E. C. Stephenson, and D. W. Stafford, *Chromosoma* **58**, 41 (1976).
97. P. K. Wellauer and I. B. Dawid, *JMB* **128**, 289 (1979).
98. P. K. Wellauer and I. B. Dawid, *Cell* **10**, 193 (1977).
99. R. L. White and D. S. Hogness, *Cell* **10**, 177 (1977).
100. M. Pellegrini, J. Manning, and N. Davidson, *Cell* **10**, 213 (1977).

101. I. B. Dawid, P. K. Wellauer, and E. O. Long, *JMB* **126**, 749 (1978).
102. P. K. Wellauer and I. B. Dawid, *JMB* **126**, 769 (1978).
103. W. Y. Chooi and K. R. Leiby, *MGG* **182**, 245 (1981).
104. T. Barnett and P. M. M. Rae, *Cell* **16**, 763 (1979).
105. W. Kunz and K. H. Glatzer, *ZpChem* **360**, 313 (1979).
106. K. Beckingham and R. White, *JMB* **137**, 349 (1979).
107. R. Renkawitz, S. A. Gerbi, and K. H. Glatzer, *MGG* **173**, 1 (1979).
108. C. K. French, D. L. Fouts, and J. E. Manning, *JCB* **83**, 193 (1979).
109. J. G. Gall, *PNAS* **71**, 3078 (1974).
110. T. R. Cech and D. C. Rio, *PNAS* **76**, 5051 (1979).
111. J. Engberg, N. Din, W. A. Eckert, W. Kaffenberger, and R. E. Pearlman, *JMB* **142**, 289 (1980).
112. E. H. Blackburn and J. G. Gall, *JMB* **120**, 33 (1978).
113. M. A. Wild and J. G. Gall, *Cell* **16**, 565 (1979).
114. T. R. Cech, A. J. Zaug, and P. J. Grabowski, *Cell* **27**, 487 (1981).
115. W. M. Steer, H. V. Molgaard, E. M. Bradbury, and H. R. Mathews, *EJB* **88**, 599 (1978).
116. U. Gubler, T. Wyler, T. Seeback, and R. Braun, *NARes* **8**, 2647 (1980).
117. N. Maizels, *Cell* **9**, 431 (1976).
118. J. Morrow, S. Cohen, A. Chang, H. Boyer, H. Goodman, and R. Hellings, *PNAS* **71**, 1743 (1974).
119. R. H. Reeder, T. Higashinakagawa, and O. J. Miller, Jr, *Cell* **8**, 449 (1976).
120. D. Rungger and M. Crippa, *Prog. Biophys. Mol. Biol.* **31**, 247 (1977).
121. U. Maitra and J. Hurwitz, *PNAS* **54**, 815 (1965).
122. B. Moss, *BBRC* **74**, 374 (1977).
123. R. Reeder, B. Sollner-Webb, and H. Wahn, *PNAS* **74**, 5402 (1977).
124. N. Nikolaev, O. I. Georgiev, P. V. Venkov, and A. A. Hadjiolov, *JMB* **127**, 297 (1979).
125. E. G. Niles, *Bchem* **16**, 3215 (1978).
126. R. Levis and S. Penman, *JMB* **121**, 219 (1978).
127. B. Batts-Young and H. Lodish, *PNAS* **75**, 740 (1978).
128. R. Bach, I. Grummt, and B. Allet, *NARes* **9**, 1559 (1981).
129. L. R. Rothblum, R. Reddy, and B. Cassidy, *NARes* **10**, 7345 (1982).
130. A. J. Berk and P. A. Sharp, *Cell* **12**, 721 (1977).
131. M. Bina-stein, M. Thoren, N. Salzman, and J. Thompson, *PNAS* **76**, 73 (1979).
132. A. Maxam and W. Gilbert, *Methods Enzymol.* **65**, 499 (1980).
133. H. Donis-Keller, A. Maxam, and W. Gilbert, *NARes* **4**, 2527 (1977).
134. D. Peattie, *PNAS* **76**, 1760 (1979).
135. B. Sollner-Webb and R.-H. Reeder, *Cell* **18**, 485 (1979).
136. P. Bosley, T. Moss, M. Mächler, R. Portmann, and M. Birnstiel, *Cell* **17**, 19 (1979).
137. R. Bach, B. Allet, and M. Crippa, *NARes* **9**, 5311 (1981).
138. P. Valenzuela, G. I. Bell, A. Venegas, E. T. Sewell, F. R. Masiarz, L. J. DeGennaro, F. Weinberg, and W. J. Rutter, *JBC* **252**, 8126 (1977).
139. R. Klemenz and E. P. Geiduschek, *NARes* **8**, 2679 (1980).
140. A. A. Bayev, E. I. Georgiev, A. A. Hadjiolov, M. B. Kermekchiev, N. Nikolaev, K. G. Skyrabin, and V. M. Zakharyev, *NARes* **8**, 4919 (1980).
141. E. G. Niles, J. Sutiphong, and S. Haque, *JBC* **256**, 12849 (1981).
142. H. Saiga, K. Mizumoto, T. Matsui, and T. Higashinakagawa, *NARes* **10**, 4223 (1982).
143. Y. Hoshikawa, Y. Iida, and M. Iwabuchi, *NARes* **11**, 1725 (1983).

144. E. O. Long, M. L. Rebbert, and I. B. Dawid, *PNAS* **78**, 1513 (1981).
145. Y. Urano, R. Kominami, Y. Mishima, and M. Muramatsu, *NARes* **8**, 6043 (1980).
146. C. A. Harrington and D. M. Chikaraishi, *NARes* **11**, 3317 (1983).
147. I. Financsek, K. Mizumoto, and M. Muramatsu, *Gene* **18**, 115 (1982).
148. I. Financsek, K. Mizumoto, Y. Mishima, and M. Muramatsu, *PNAS* **79**, 3092 (1982).
149. R. Miesfeld and N. Arnheim, *NARes* **10**, 3933 (1982).
150. J. Corden, B. Wasylyk, A. Buchwalder, P. Sassone-Corsi, C. Kedinger, and P. Chambon, *Science* **209**, 1406 (1980).
151. R. Breathnach and P. Chambon, *ARB* **50**, 349 (1981).
152. Y. Mishima, O. Yamamoto, R. Kominami, and M. Muramatsu, *NARes* **9**, 6773 (1981).
153. E. Lund and J. E. Dahlberg, *PNAS* **76**, 5480 (1979).
154. A. H. Bakken, G. Morgan, B. Sollner-Webb, J. Roan, S. Busby, and R. H. Reeder, *PNAS* **79**, 56 (1982).
155. G. M. Veldman, J. Klootwijk, P. de Jonge, R. J. Leer and R. J. Planta, *NARes* **8**, 5179 (1980).
156. R. K. Mandal and I. B. Dawid, *NARes* **9**, 1801 (1981).
157. T. Kukita, Y. Sakaki, H. Nomiyama, T. Otsuka, S. Kuhara, and Y. Takagi, *Gene* **16**, 309 (1981).
158. J. M. Kelly and R. A. Cox, *NARes* **9**, 1111 (1981).
159. N. Din, J. Engberg, and J. G. Gall, *NARes* **10**, 1503 (1982).
160. E. G. Niles, K. Cunningham, and R. Jain, *JBC* **256**, 12357 (1981).
161. R. Kominami, Y. Mishima, Y. Urano, M. Sakai, and M. Muramatsu, *NARes* **10**, 1963 (1982).
162. S. Adhya and M. Gottesman, *ARB* **47**, 967 (1978).
163. M. Rosenberg and D. Court, *ARGen* **13**, 319 (1979).
164. L. J. Korn and D. D. Brown, *Cell* **15**, 1145 (1978).
165. F. H. Martin and I. Tinoco, *NARes* **8**, 2295 (1980).
166. H. Hamada, R. Kominami, and M. Muramatsu, *NARes* **8**, 889 (1980).
167. A. A. Bogadanov, A. M. Kopylov, and I. N. Shatsky, *Subcell. Biochem.* **7**, 81 (1980).
168. R. Brimacombe, G. Stofler, and H. G. Wittmann, *ARB* **47**, 217 (1978).
169. R. Brimacombe, P. Maly, and C. Zweib, This Series **28**, 1 (1983).
172. P. Carbon, C. Ehresmann, and J. P. Ebel, *FEBS Lett.* **94**, 152 (1978).
173. J. Brosius, M. L. Palmer, P. J. Kennedy, and H. F. Noller, *PNAS* **75**, 4801 (1978).
174. J. Brosius, T. Dull, and H. F. Noller, *PNAS* **77**, 201 (1980).
175. V. A. Erdmann, E. Huysmans, A. Vandenberghe, and R. De Wachter, *NARes* **11**, r105 (1983).
176. P. M. Rubtsov, M. M. Musakhanov, V. M. Zakharyev, A. S. Krayev, K. G. Skyrabin and A. A. Bayev, *NARes* **8**, 5779 (1980).
177. O. I. Georgiev, N. Nikolaev, A. A. Hadjiolov, K. G. Skyrabin, V. M. Zakharyev, and A. A. Bayev, *NARes* **9**, 6953 (1981).
178. G. M. Veldman, J. Klootwijk, V. C. H. F. de Regt, R. J. Planta, C. Branlant, A. Krol, and J. P. Ebel, *NARes* **9**, 6935 (1981).
179. A. J. Bendich and B. J. McCarthy, *PNAS* **65**, 340 (1970).
180. J. H. Sinclair and D. D. Brown, *Bchem* **10**, 2761 (1971).
181. R. A. Young and J. A. Steitz, *PNAS* **75**, 2593 (1978).
182. D. R. Samols, O. Hagenbuchle, and L. P. Cage, *NARes* **7**, 1109 (1979).
183. A. A. Azad and N. J. Deacon, *NARes* **8**, 4365 (1980).
184. M. Salim and B. E. H. Maden, *Nature (London)* **291**, 205 (1981).
185. B. R. Jordan, M. Latil-Damotte, and R. Jourdan, *FEBS Lett.* **117**, 227 (1980).

186. M. N. Schnare and M. W. Gray, *FEBS Lett.* **128**, 298 (1981).
187. R. Van Charldorp and P. H. Van Knippenberg, *NARes* **10**, 1149 (1982).
188. J. Shine and L. Dalgarno, *PNAS* **71**, 1342 (1974).
189. H. F. Noller and C. R. Woese, *Science* **212**, 403 (1981).
190. A. A. Azad, *NARes* **7**, 1913 (1979).
191. J. M. Kelley and R. A. Cox, *NARes* **10**, 6733 (1982).
192. C. S. Subrahamanyam, B. Cassidy, H. Busch, and L. I. Rothblum, *NARes* **10**, 3667 (1982).
192a. A. T. Torczynski, A. P. Bollon, and M. Fuke, *NARes* **11**, 4879 (1983).
193. R. L. Gourse and S. A. Gerbi, *JMB* **140**, 321 (1980).
194. A. A. Bayev, A. S. Drayev, P. M. Rubtsov, K. G. Skyrabin, and V. M. Zakharyev, *Dokl. Akad. Nauk SSSR* **247**, 1275 (1979).
195. M. Salim and B. E. H. Maden, *NARes* **8**, 2871 (1980).
196. B. G. Cassidy, C. S. Subrahmanyam, and L. I. Rothblum, *BBRC* **107**, 1571 (1982).
197. R. E. Lockard, J. F. Connaughton, and A. Kumar, *NARes* **10**, 3445 (1982).
198. C. Zwieb, C. Glotz, and R. Brimacombe, *NARes* **9**, 3621 (1981).
199. M. A. Machatt, J. P. Ebel, and C. Branlant, *NARes* **9**, 1533 (1981).
200. G. M. Veldman, J. Klootwijk, H. V. Heerikhuizen and R. J. Planta, *NARes* **9**, 4847 (1981).
201. L. M. C. Hall and B. E. H. Maden, *NARes* **8**, 5993 (1980).
202. B. Michot, J. P. Bachellerie, and F. Raynal, *NARes* **10**, 5273 (1982).
203. E. L. Gourse and S. A. Gerbi, *NARes* **8**, 3623 (1980).
203a. T. Otsuka, H. Nomiyama, H. Yoshida, T. Kukita, S. Kuhara, and Y. Sakaki, *PNAS* **80**, 3163 (1983).
204. T. Moss, P. G. Bosley, and M. L. Birnstiel, *NARes* **8**, 467 (1980).
205. B. E. H. Maden, M. Moss, and M. Salim, *NARes* **10**, 2387 (1982).
206. D. D. Brown, P. C. Wensink, and E. Jordan, *JMB* **63**, 57 (1972).
207. R. J. Bram, R. A. Young, and J. A. Steitz, *Cell* **19**, 393 (1980).
208. N. V. Federoff, *Cell* **16**, 697 (1979).
209. D. Treco, E. Brownell, and N. Arnheim, *in* "The Cell Nucleus" (H. Busch and L. Rothblum, eds.), Vol. 12, p. 101. Academic Press, New York, 1982.
210. P. Botchan, R. H. Reeder, and I. B. Dawid, *Cell* **11**, 599 (1977).
211. E. O. Long and I. B. Dawid, *NARes* **7**, 205 (1979).
212. E. S. Coen and G. Dover, *NARes* **10**, 7017 (1982).
213. E. R. Schmidt, E. A. Godwin, H. G. Keyl, and N. Israelewski, *Chromosoma* **87**, 389 (1982).
214. N. Israelewski and E. K. Schmidt, *NARes* **10**, 7689 (1982).
215. G. P. Smith, *CSHSQB* **38**, 507 (1973).
216. H. G. Keyl, *Chromosoma* **17**, 139 (1965).
217. B. D. Kohorn and P. M. M. Rae, *PNAS* **79**, 1501 (1982).
218. B. D. Kohorn and P. M. M. Rae, *NARes* **10**, 6879 (1982).
219. J. R. Miller, D. C. Hayward and D. M. Glover, *NARes* **11**, 11 (1983).
219a. A. Simeone, A. de Falco, G. Macino, and E. Boncinelli, *NARes* **10**, 8263 (1982).
220. E. Coen, T. Strachan, and G. A. Dover, *JMB* **158**, 17 (1982).
221. P. M. M. Rae, T. Burnett, and V. L. Murtit, *Chromosoma* **82**, 637 (1981).
222. R. Renkawitz-Pohl, K. H. Glätzer, and W. Kunz, *NARes* **8**, 4593 (1980).
223. N. Arnheim and M. Kuehn, *JMB* **134**, 743 (1979).
224. M. Kuehn and N. Arnheim, *NARes* **11**, 211 (1983).
225. N. Arnheim, P. Seperack, J. Banerji, R. B. Lang, R. Miesfeld, and K. B. Marcu, *Cell* **22**, 179 (1980).

226. J. L. Boss, C. Heyting, P. Borst, A. C. Arnberg, and E. F. J. Van Bruggen, *Nature (London)* **275**, 336 (1978).
227. G. Faye, N. Denebuoy, C. Kujawa, and C. Jacq, *MGG* **168**, 101 (1979).
228. J. E. Heckman and U. L. RajBhandary, *Cell* **17**, 583 (1979).
229. U. Hehn, C. M. Lazarus, H. Lünsdorf, and H. Kuntzel, *Cell* **17**, 191 (1979).
230. G. Hasan, M. J. Turner, and J. S. Cordingley, *NARes* **10**, 6747 (1982).
231. H. Roiha and D. M. Glover, *JMB* **140**, 341 (1980).
232. E. O. Long, M. Rebbert and I. B. Dawid, *CSHSOB* **45**, 667 (1981).
233. I. B. Dawid and P. Botchan, *PNAS* **74**, 4233 (1977).
234. I. B. Dawid and P. K. Wellauer, *CSHSQB* **42**, 1185 (1978).
235. S. J. Kidd and D. M. Glover, *Cell* **19**, 103 (1980).
236. P. M. M. Rae, *NARes* **9**, 4997 (1981).
237. I. B. Dawid and M. L. Rebbert, *NARes* **9**, 5011 (1981).
238. H. Roiha, J. R. Miller, L. C. Woods, and D. M. Glover, *Nature (London)* **290**, 749 (1981).
239. P. M. M. Rae, B. D. Kohorn, and R. P. Wade, *NARes* **8**, 3491 (1980).
240. I. B. Dawid, M. Lauth, and P. K. Wellauer, *Miami Winter Symp.* **17**, 217 (1980).
241. I. B. Dawid, E. O. Long, P. P. DiNocera, and M. L. Pardue, *Cell* **25**, 399 (1981).
242. I. B. Dawid, P. P. DiNocera, and R. K. Mandal, *Proc. Int. Congr. Genet. 15th, Symp. VI*, New Delhi, 1983 (in press).
243. E. O. Long and I. B. Dawid, *Cell* **18**, 1185 (1979).
244. D. J. Jolly and C. A. Thomas, Jr, *NARes* **8**, 67 (1979).
245. W. Y. Chooi, *Chromosoma* **74**, 57 (1979).
246. K. H. Glätzer, *Chromosoma* **75**, 161 (1979).
247. E. O. Long, M. Collins, B. I. Kiefer, and I. B. Dawid, *MGG* **182**, 377 (1981).
248. M. A. Wild and R. Sommer, *Nature (London)* **283**, 693 (1980).
249. N. C. Kan and J. G. Gall, *NARes* **10**, 2809 (1982).
250. R. Breathnach, C. Benoist, K. O'Hara, F. Gannon, and P. Chambon, *PNAS* **75**, 4853 (1978).
251. I. Seif, G. Khoury, and R. Dhar, *NARes* **6**, 3387 (1979).
252. H. Nomiyama, Y. Sakaki, and Y. Takagi, *PNAS* **78**, 1376 (1981).
253. J. L. Bos, K. A. Osinga, G. Van der Horst, N. B. Hecht, H. F. Tabak, G.-J. B. Van Ommen, and P. Borst, *Cell* **20**, 207 (1980).
254. B. Allet and J.-D. Rochaix, *Cell* **18**, 55 (1979).
255. T. R. Cech, N. K. Tanner, I. Tinoco, Jr., B. R. Weir, M. Zuker, and P. S. Perlman, *PNAS* **80**, 3903 (1983).
256. J. M. Burke and U. L. RajBhandary, *Cell* **31**, 509 (1982).
257. R. Netzkar, H. G. Köchel, N. Basak, and H. Küntzel, *NARes* **10**, 4783 (1982).
258. M. Chamberlin, *ARB* **43**, 721 (1974).
259. M. Chamberlin and R. Losick, eds., "RNA Polymerase" CSHLab, CSH, New York, 1976.
260. S. T. Jacob, This Series **13**, 93 (1973).
261. B. B. Biswas, R. K. Mandal, A. Stevens, and W. E. Cohn, eds., "Control of Transcription" Plenum, New York, 1974.
262. W. W. Franke, U. Scheer, H. Spring, M. F. Trendelenburg, and H. Zentgraf, in "The Cell Nucleus" (H. Busch, ed.), Vol. 7, p. 49. Academic Press, New York, 1979.
263. U. Scheer, M. F. Trendelenburg, G. Krohne, and W. W. Franke, *Chromosoma* **60**, 147 (1977).
264. O. L. Miller, Jr. and B. A. Hamkalo, *Int. Rev. Cytol.* **33**, 1 (1972).

265. W. Y. Chooi and K. R. Leiby, *PNAS* **78**, 4823 (1981).
266. M. F. Trendelenburg, *Chromosoma* **86**, 703 (1982).
267. M. F. Trendelenburg and J. B. Gurdon, *Nature (London)* **276**, 292 (1978).
268. M. F. Trendelenburg, *Biol. Cell.* **42**, 1 (1981).
269. B. Sollner-Webb and S. L. McKnight, *NARes* **10**, 3391 (1982).
270. T. Moss, *Cell* **30**, 835 (1982).
271. T. Moss, *Nature (London)* **302**, 223 (1983).
272. R. D. Kornberg, *ARB* **46**, 931 (1977).
273. J. D. McGhee and G. Felsenfeld, *ARB* **49**, 1115 (1980).
274. S. C. R. Elgin, *Cell* **27**, 413 (1981).
275. V. E. Foe, *CSHSQB* **42**, 723 (1978).
276. S. L. McKnight, M. Bustin, and O. L. Miller, Jr., *CSHSQB* **42**, 741 (1978).
277. E. M. Johnson, H. R. Mathews, V. C. Littau, L. Lothstein, E. M. Bradburg, and V. G. Allfrey, *ARB* **191**, 537 (1978).
278. R. Reeves, *Bchem* **17**, 4908 (1978).
279. D. J. Mathys and M. L. Gorovsky, *CSHSQB* **42**, 773 (1978).
280. R. K. Mandal, A. Ganguly, and H. K. Mazumder, *J. Sci. Ind. Res.* **33**, 533 (1974).
281. T. Y. Wang and N. C. Kostraba, in "The Cell Nucleus" (H. Busch, ed.), Vol. 4, p. 289. Academic Press, New York, 1978.
282. S. Weisbrod, M. Groudine, and H. Weintraub, *Cell* **19**, 289 (1980).
283. A. La Volpe, M. H. Taggart, M. McStoy, and A. P. Bird, *NARes* **11**, 5361 (1983).
284. S. Saragasti, G. Meyne, and M. Yaniv, *Cell* **20**, 65 (1980).
285. G. C. Das and S. K. Niyogi, This Series **25**, 187 (1981).
286. D. M. Anderson and L. D. Smith, *Dev. Biol.* **67**, 274 (1978).
287. C. A. Sutton, P. Sylvan, and R. L. Hallberg, *J. Cell. Physiol.* **101**, 503 (1979).
288. W. A. Eckert and W. Kaffenberger, *Eur. J. Cell. Biol.* **21**, 53 (1980).
289. H. Greenberg and S. Penman, *JMB* **21**, 527 (1966).
290. F. Kafatos, *Curr. Top. Dev. Biol.* **7**, 125 (1972).
291. R. G. Roeder, R. H. Reeder, and D. D. Brown, *CSHSQB* **35**, 727 (1970).
292. T. Honjo and R. H. Reeder, *Bchem* **13**, 1896 (1974).
293. N. R. Ballal, Y. C. Choi, R. Mouche, and H. Busch, *PNAS* **74**, 2446 (1977).
294. A. A. Hadjiolov and G. I. Milchev, *BJ* **142**, 263 (1974).
295. A. Udvardy and K. H. Seifart, *EJB* **62**, 353 (1976).
296. B. E. H. Coupar, J. A. Davies, and C. J. Chesterton, *EJB* **84**, 611 (1978).
297. F. Scalenghe, M. Buscaglia, C. Steinheil, and M. Crippa, *Chromosoma* **66**, 299 (1978).
298. R. H. Reeder, H. L. Wahn, P. Botchan, R. Hipskind, and B. Sollner-Webb, *CSHSQB* **42**, 1167 (1978).
299. T. Onishi and M. Muramatsu, *Methods Cell Biol.* **19**, 301 (1978).
300. I. Grummt, S. H. Hall, and R. J. Crouch, *EJB* **94**, 437 (1979).
301. M. H. Green, J. Buss, and P. Gariglio, *EJB* **53**, 217 (1975).
302. A. J. Zaug and T. R. Cech, *Cell* **19**, 331 (1980).
302a. E. Gocke, J. C. Leer, O. F. Nielsen, and O. Westergaard, *NARes* **5**, 3993 (1977).
303. J. C. Leer, D. Tiryaki, and O. Westergaard, *PNAS* **76**, 5563 (1979).
304. B. D. Kohorn and P. M. M. Rae, *PNAS* **80**, 3265 (1983).
305. R. K. Mandal and I. B. Dawid, *Proc. Indo-Soviet Symp. on Biological Macromolecules*, Madurai, 1983 (in press).
306. A. D. Mirzabekov, *Proc. Indo-Soviet Symp. on Biological Macromolecules*, Madurai, 1983 (in press).
307. E. H. Birkenmeier, D. D. Brown, and E. Jordan, *Cell* **15**, 1077 (1978).

308. O. Schmidt, J. I. Mao, S. Silverman, B. Hovemann, and D. Söll, *PNAS* **75**, 4819 (1978).
309. R. A. Hipskind and R. H. Reeder, *JBC* **255**, 7896 (1980).
310. P. A. Weil, D. S. Luse, J. Segall, and R. G. Roeder, *Cell* **18**, 469 (1979).
311. J. L. Manley, A. Fire, A. Cano, P. A. Sharp, and M. L. Gefter, *PNAS* **77**, 3855 (1980).
312. S. Sakonju, D. F. Bogenhagen, and D. D. Brown, *Cell* **19**, 13 (1980).
313. I. Grummt, *PNAS* **78**, 727 (1981).
314. K. G. Miller and B. Sollner-Webb, *Cell* **27**, 165 (1981).
315. I. Grummt, *NARes* **9**, 6093 (1981).
316. I. Grummt, *PNAS* **79**, 6908 (1982).
317. T. Shenk, *Curr. Top. Microbiol. Immunol.* **93**, 25 (1981).
318. C. Wandelt and I. Grummt, *NARes* **11**, 3795 (1983)
318a. A. Cizewski and B. Sollner-Webb, *NARes* **11**, 7403 (1983).
318b. T. Onishi, C. Bergland, and R. H. Reeder, *PNAS* **81**, 484 (1984).
319. K. J. Soprano and R. Baserga, *PNAS* **77**, 1566 (1980).
320. M. Sawadogo, A. Sentenac, and P. Fromageot, *BBRC* **101**, 250 (1981).
321. H. Mondal, A. Ganguly, A. Das, R. K. Mandal, and B. B. Biswas, *EJB* **28**, 143 (1972).
322. R. K. Mandal, H. K. Mazumder, and B. B. Biswas, *in* Biswas *et al.* **261**, p. 295.
323. B. E. H. Maden and M. Salim, *JMB* **88**, 133 (1974).
324. R. C. Brand and S. A. Gerbi, *NARes* **7**, 1497 (1979).
325. B. E. H. Maden, M. S. N. Khan, D. G. Hughes, and J. P. Goddard, *Biochem. Soc. Symp.* **42**, 165 (1977).
326. M. S. N. Khan, M. Salim, and B. E. H. Maden, *BJ* **169**, 531 (1978).
327. E. O. Long and I. B. Dawid, *JMB* **138**, 873 (1980).
328. R. Reddy, L. I. Rothblum, C. S. Subrahmanyam, M.-H. Liu, D. Henning, B. Cassidy, and H. Busch, *JBC* **258**, 584 (1983).
329. P. K. Wellauer, I. B. Dawid, D. E. Kelley, and R. P. Perry, *JMB* **89**, 397 (1974).
330. M. D. Dabeva, K. P. Dudov, A. A. Hadjiolov, I. Emanuilov, and B. N. Todorov, *BJ* **160**, 495 (1976).
331. K. P. Dudov, M. D. Dabeva, A. A. Hadjiolov, and B. N. Todorov, *BJ* **171**, 375 (1978).
332. C. R. Pousada, L. Marcaud, M. M. Portier, and D. D. Hayes, *EJB* **56**, 117 (1975).
333. J. Trapman, P. DeJonge, and R. J. Planta, *FEBS Lett.* **57**, 26 (1975).
334. T. J. Hall and M. R. Cummings, *Insect Biochem.* **7**, 347 (1977).
335. N. Din, J. Engberg, W. Kaffenberger, and W. E. Eckert, *Cell* **18**, 525 (1979).
336. P. J. Grabowski, A. J. Zaug, and T. R. Cech, *Cell* **23**, 467 (1981).
337. M. Carin, B. F. Jensen, K. D. Jentsch, J. C. Leer, O. F. Nielsen, and O. Westergaard, *NARes* **8**, 5551 (1980).
338. A. J. Zaug and T. R. Cech, *NARes* **10**, 2823 (1982).
339. K. Kruger, P. J. Grabowski, A. J. Zaug, J. Sands, D. E. Gotschling, and T. R. Cech, *Cell* **31**, 147 (1982).
341. R. P. Perry, *Biochem. Soc. Symp.* **37**, 114 (1973).
342. H. L. Cooper, *Nature (London)* **227**, 1105 (1970).
343. P. S. Rudland, S. Weil, and A. R. Hunter, *JMB* **96**, 745 (1975).
344. U. Scheer, M. F. Trendelenburg, and W. W. Franke, *JCB* **69**, 465 (1976).
345. H. C. McGregor, *Biol. Rev.* **47**, 177 (1972).
346. M. F. Trendelenburg and R. G. McKinnell, *Differentiation* **15**, 73 (1979).
347. J. L. Mandel and P. Chambon, *NARes* **7**, 2081 (1979).

348. M. T. Kuo, J. L. Mandel, and P. Chambon, *NARes* **7**, 2105 (1979).
349. C. J. Shen and T. Maniatis, *PNAS* **77**, 6634 (1980).
350. A. P. Bird and E. M. Southern, *JMB* **118**, 27 (1978).
351. A. P. Bird, M. H. Taggart, and C. A. Gehring, *JMB* **152**, 1 (1981).
352. A. Bird, M. Taggart, and D. Macleod, *Cell* **26**, 381 (1981).
353. I. B. Dawid, D. D. Brown, and R. H. Reeder, *JMB* **51**, 341 (1970).
354. L. Kunnath and J. Locker, *NARes* **10**, 3877 (1982).
355. T. H. N. Ellis, P. B. Goldsborough, and J. A. Castleton, *NARes* **11**, 3047 (1983).
356. R. G. Roeder, *JBC* **249**, (1974).
356a. T. G. Hollinger and L. D. Smith, *Dev. Biol.* **51**, 86 (1976).
357. A. Chaudhuri, P. K. Sarkar, and R. K. Mandal, *Indian J. Exp. Biol.* **17**, 1016 (1979).
358. R. G. Roeder and W. J. Rutter, *Bchem* **9**, 2543 (1970).
359. J. A. Van-den-Berg, T. Kooistra, A. B. Geert, and M. Gruber, *BBRC* **61**, 367 (1974).
360. H. K. Mazumder and R. K. Mandal, *Indian J. Biochem. & Biophys.* **13**, 132 (1976).
361. J. A. Jaehning, C. C. Stewart, and R. G. Roeder, *Cell* **4**, 51 (1975).
362. Y. Mishima, I. Financsek, R. Kominami, and M. Muramatsu, *NARes* **10**, 6659 (1982).
363. K. Yamana and K. Shiokawa, *Exp. Cell Res.* **44**, 283 (1966).
364. R. Landesman and P. R. Gross, *Dev. Biol.* **18**, 571 (1968).
365. R. N. Hill and E. H. McConkey, *J. Cell. Physiol.* **79**, 15 (1972).
366. K. Shiokawa, A. Kawahara, Y. Misumi, Y. Yasuda, and K. Yamana, *Dev. Biol.* **57**, 210 (1977).
367. R. A. Laskey, J. C. Gerhart, and J. S. Knowland, *Dev. Biol.* **33**, 241 (1973).
368. A. Hildebrandt and H. Sauer, *BBRC* **74**, 466 (1977).
369. M. Crippa, *Nature (London)* **227**, 1138 (1970).
370. J. M. Crampton and H. R. Woodland, *Dev. Biol.* **70**, 467 (1979).
371. C. Thomas, V. Heilporn-pohl, F. Hanocq, E. Pays, and M. Boloukhere, *Exp. Cell Res.* **127**, 63 (1980).
372. A. Chaudhuri, Control of Transcription in Developing Fish Gametes and Embryos, Ph.D. thesis, Calcutta Univ., 1981.
373. R. K. Mandal, S. Dasgupta, and A. Chaudhuri, *Int. Congr. Genet. 15th, Abstr.* C,II,A (1983).
374. H. L. Ennis, *Mol. Pharmacol.* **2**, 543 (1966).
375. R. Soeiro, M. H. Vaughan, and J. E. Darnell, *J. Cell Biol.* **36**, 91 (1968).
376. R. K. Mandal, *BBA* **182**, 375 (1969).
377. J. L. Farber and R. Farmar, *BBRC* **51**, 626 (1973).
378. T. Onishi, T. Matsui, and M. Muramatsu, *J. Biochem. (Tokyo)* **82**, 1109 (1977).
279. B. B. Stoyanova and A. A. Hadjiolov, *EJB* **96**, 349 (1979).
380. V. E. Foe, *CSHSQB* **42**, 723 (1978).
381. G. S. Stent, *Proc. R. Soc. London Ser. B* **164**, 181 (1966).
382. J. M. Zimmerman and R. Simha, *J. Theor. Biol.* **13**, 106 (1966).
383. U. N. Singh, *J. Theor. Biol.* **40**, 553 (1973).
384. K. Gausing, *JMB* **115**, 335 (1977).
385. B. E. H. Maden, M. H. Vaughan, J. R. Warner, and J. E. Darnell, *JMB* **45**, 265 (1969).
386. J. R. Warner, *JMB* **115**, 315 (1977).
387. H. L. Cooper, *Nature (London)* **227**, 1105 (1970).
388. R. L. Hallberg and D. D. Brown, *JMB* **46**, 393 (1969).
389. Y. C. Weiss, C. A. Vaslet, and M. Rosbash, *Dev. Biol.* **87**, 330 (1981).

Structure, Function and Evolution of 5-S Ribosomal RNAs

NICHOLAS DELIHAS
AND JANET ANDERSEN

*Department of Microbiology
School of Medicine
SUNY at Stony Brook
Stony Brook, New York*

RAM P. SINGHAL

*Chemistry Department
Wichita State University
Wichita, Kansas*

I.	Generalized Structures of the 5-S Ribosomal RNAs	161
II.	"Hot Spots" of Insertions and Deletions in 5-S RNAs	169
III.	Conformation of 5-S RNA Derived from Enzymatic, Chemical, and Physical Studies	174
IV.	5-S RNA Interactions with Protein	177
V.	Phylogeny and Endosymbioisis	181
VI.	Summary	187
	References	188

I. Generalized Structures of the 5-S Ribosomal RNAs

With the recent publication of nucleotide sequences[1] of the ribosomal 5-S RNAs from over 100 different sources (see *1, 2*) and the correction of several previously reported sequences (*3, 4*), it is now evident that 5-S RNA has been highly conserved during the course of evolution. All known 5-S RNAs derived from eubacteria (prokaryotes), archaebacteria, eukaryotes, and organelles conform to a universal structure (Fig. 1). When purines and pyrimidines[2] are considered as

[1] The authors plan to update the 5-S RNA sequence list very soon. Researchers are requested to send new 5-S (and 5.8-S) RNA sequences and changes in known structures to RPS.

[2] Abbreviations: R, purine (either G or A); Y, pyrimidine (either C or U). Nucleases mentioned in the text: RNase T_1, ribonuclease T_1 from *Aspergillus oryzae* (EC 3.1.27.3); RNase T_2, ribonuclease T_2 from *Aspergillus oryzae* (EC 3.1.27.1); RNase A, ribonuclease A from bovine pancreas (EC 3.1.27.5); nuclease S_1 from *Aspergillus oryzae*; and RNase V_1, ribonuclease V_1 from *Naja naja oxiana* cobra venom (venom exonuclease, EC 3.1.15.1).

invariant residues, approximately 35% of the residues appear to be universally conserved.

The secondary-structure model shown in Fig. 1 is the consensus of structures[3] proposed by several investigators (5–10) and is based on comparisons of sequences. Generalized structures for the eubacterial and eukaryotic 5-S RNAs showing conserved positions (2, 6–8, 11) are shown in Fig. 2. Included in the models shown here are the extended base-pairing schemes and A∘G bonding between helices IV and V that were first formulated by Stahl et al. (5). The consensus secondary structure represents essentially a restatement of structures originally proposed by Nishikawa and Takemura in 1974 (12) and Fox and Woese in 1975 (13), but it includes extended base-pairing. These models were derived by comparisons of 5-S RNA sequences available during that period. Although no crystallographic data confirm the proposed structure (14), its strength stems from a nearly perfect fit of all known 5-S RNA sequences to it (8). As discussed below, experimental data from partial nuclease digestions provide support for the generalized secondary structure shown in Fig. 1.

1. Non-Watson–Crick Base-Pairs in Helices IV and V

There are phylogenetic reasons to include A∘G base-pairs in the stem of 5-S RNA that contains helices IV and V. Gram-positive and most other eubacterial 5-S RNAs have the conserved pair $A^{104}\circ G^{72}$. In *Streptococcus cremoris*, helix IV appears to be "rearranged" with an insertion of one base-pair but the A∘G pairing is maintained at position 103 and 73 (11). The significance of this transposition lies in the apparent conservation of the A∘G pair as opposed to the conservation of nucleotides A and G at positions 104 and 72, respectively. As pointed out by Stahl et al. (5), helix V can be extended in most 5-S RNAs by the inclusion of conserved A∘G pairing. The exceptions are the cyanobacterial and chloroplast 5-S RNAs, but these RNAs have the following Watson–Crick or non-Watson–Crick pairing at the same positions: A·U (*Anacystis nidulans*), A∘C (*Synechococcus lividus*), and A∘A (chloroplast).

---

[3] Use of the terms "helix" and "stem:" Regions of the 5-S RNA presumed to be double-helical are referred to as helix I, etc., the customary usage. The term "helix" does not necessarily denote a complete double helix. The term "stem" denotes the double-helical and adjacent internal loop regions. The terminal stem is helix I, the internal stem closest to the 5' end includes helices II and III, and the internal stem closest to the 3' end includes helices IV and V.

5-S RIBOSOMAL RNAs

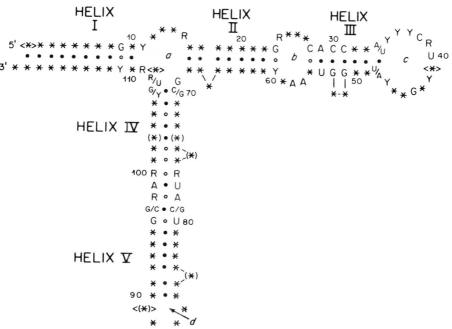

FIG. 1. Universal 5-S RNA structure.[3] The numbering system is that of the *Escherichia coli* 5-S RNA. (*) represents insertions in the eukaryotic 5-S RNA; ⟨*⟩ represents eubacterial 5-S RNA positions not present in most eukaryotic 5-S RNAs. Helix I refers to the double-helical region of the 5′, 3′ end terminal stem. Helices II and III are in the stem closest to the 5′ end, and helices IV and V are in the stem closest to the 3′ end. R^{100}∘R^{76} is arbitrarily chosen as the dividing region between helices previously referred to as IV and V. For simplicity, insertions and deletions in archaebacterial and organelle 5-S RNAs have not been included. The conserved positions were derived from a comparison of 5-S RNA sequences compiled by Singal and Shaw (2) and Erdmann *et al.* (1). The following criteria were used to assign the universally conserved positions: (i) The position is found in all classes of 5-S RNAs, i.e., from eubacteria, archaebacteria, eukaryotes, and organelles. (ii) The position is found in 90% or more of the sequences compared. If a conserved position in eukaryotic 5-S RNAs differed in three or more unrelated species (i.e., yeast, algae, protozoa), it was not included as a universally conserved position.

Experimental evidence for the presence of A∘G pairs in nucleic acids comes from cobra venom ribonuclease V_1 cleavages at sites having A∘G pairs in 16-S RNA (15), and the detection of A∘G base pairs in oligodeoxynucleotides by proton magnetic resonance spectroscopy (16). In eukaryotic 5-S RNAs, ribonuclease V_1 cleaves in the region between helices IV and V, where there is putative A∘G pairing (17).

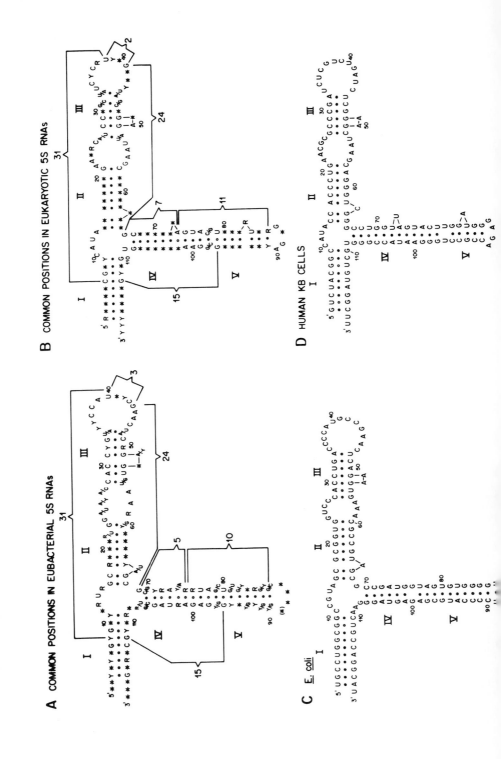

2. Helix II

Several regions of the 5-S RNA molecule are sometimes drawn in a form different from the generalized model when the form is more favorable energetically than the generalized scheme. In most cases, however, the sequence can be shown to conform to the consensus model. One example is helix II of the 5-S RNA of the red alga *Gracilaria compressa* (*18*) (Fig. 3). Structure A has been calculated to be energetically more stable than structure B. Nevertheless, the RNA sequence can be shown to conform to the generalized structure of helix II in both conserved positions and secondary structure with the exception of mispairing between residues 20 and 58 (see Fig. 3, B and C). Thus, helix II of *G. compressa* 5-S RNA may have evolved of necessity into another more stable conformation, but the "skeleton" of the generalized structure is still evident.

3. Helix III

Alternate secondary structures have been proposed for helix III by De Wachter *et al.* (*9*). These forms are considered to be in equilibrium with the generalized form and are energetically as stable as the proposed generalized structure. An alternate form for helix III of the *Escherichia coli* 5-S RNA is shown in Fig. 4. Alternate schemes are attractive in that they allow for the occurrence of more than one conformation and a dynamic 5-S RNA structure. The alternate forms are not as universal as the generalized structure for helix III. The inability to form an alternate structure to which all 5-S RNAs will conform is primarily due to nucleotide insertions that are present in helix III of cyanobacterial, organelle, and certain fungal 5-S RNAs (*19–22*). However, these RNAs may have their own particular conformation of helix III.

4. Helix V

Two secondary structures have been considered for helix V of metazoan 5-S RNAs (Fig. 5) (*9, 10, 23*). These forms have been calculated to have approximately the same energetic stabilities (*8*). Walker

---

FIG. 2. Generalized models showing common positions in eubacterial and eukaryotic 5-S RNAs. (A) Eubacterial 5-S RNA after Böhm *et al.* (*6*), Singhal and Shaw (*2*), and Neimark *et al.* (*11*) showing conserved positions. (B) Eukaryotic 5-S RNA after Böhm *et al.* (*7*), Singhal and Shaw (*2*), and Delihas and Andersen (*8*), showing conserved positions. Numbers outside of brackets represent constant chain lengths between universal nucleotide positions. (C) Structure of *E. coli* 5-S RNA. (D) Structure of human KB cell 5-S RNA.

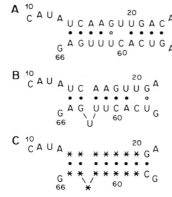

FIG. 3. Helix II structure of *Gracilaria compressa* 5-S RNA. (A) Secondary structure of helix II according to Lim *et al.* (*18*). (B) Sequence of *G. compressa* 5-S RNA drawn according to the generalized model for helix II. (C) Generalized structure of helix II. The numbering system used is that of the generalized eukaryotic model (see Fig. 2B).

and Doolittle (*24*) analyzed the structure of helix V in several newly determined 5-S RNA sequences. The 5-S RNA sequences of the starfish *Asterias vulgaris* and the marine worm *Lineus geniculatus* (*25*) do not fit model B well, but do fit model A. Model A (Fig. 5) appears to be a structure to which all known 5-S RNAs conform. In addition, Walker and Doolittle (*24*) argue for an ancient origin of a U-U pairing in helix V. This pairing has been observed in 5-S RNAs of certain eubacteria, of several basidiomycetes, and of the metazoans.

5. Alternate 5-S RNA Structures

An alternate base-pairing scheme for the 5-S RNA involving the pairing of nucleotides between the regions encompassed by positions 30–50 and 75–95 has been proposed (*26*). This scheme is of interest in that again it shows that alternate base-pairing is possible. However, there is a considerable amount of variability in the location and number of base-pairs that can be formed with different 5-S RNAs, and thus this alternate base-pairing cannot be considered universal. In addi-

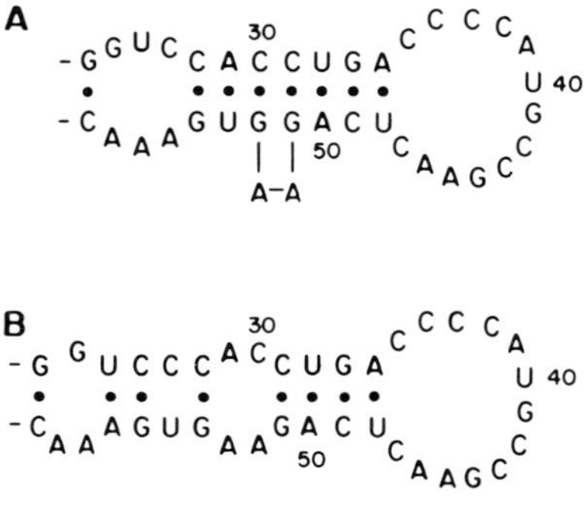

FIG. 4. Two possible structures for helix III in *Escherichia coli* 5-S RNA. (A) According to the generalized model; (B) alternate form according to De Wachter *et al.* (*9*).

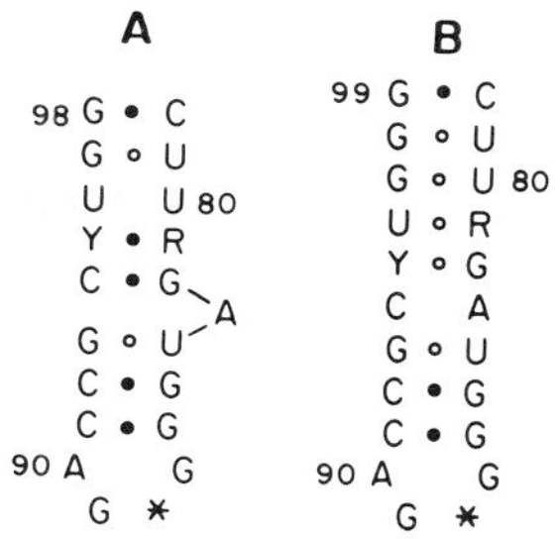

5-S RIBOSOMAL RNAs

```
                              A                    B
                        98 G • C            99 G • C
                           G ○ U               G ○ U
                           U   U 80            G ○ U 80
                           Y • R               U ○ R
                           C • G               Y ○ G
                                  \A           C   A
                           G ○ U               G ○ U
                           C • G               C • G
                           C • G               C • G
                        90 A   G            90 A   G
                           G *                 G *
```

FIG. 5. Two possible structures for helix V of metazoan 5-S RNAs.

tion, there is little experimental evidence for the existence of this alternate form. No alternate models thus far compare with the consensus generalized structure (Fig. 1) in the terms of the preciseness of the fit of 5-S RNAs analyzed from a vast range of organisms and organelles.

Two structural forms of the *E. coli* 5-S RNA have been detected by nuclear magnetic resonance (27). Interconversion of the two forms is dependent upon Mg^{2+}, and both forms appear to be of the "native" 5-S RNA type as opposed to the denatured B form. It has been suggested that these two forms correspond to the high- and low-temperature states of the *E. coli* 5-S RNA reported previously (28).

6. Tertiary Interactions

An interesting three-dimensional model of the *E. coli* 5-S RNA is one in which the arm of helix III folds to form base-pair contact points with a region of the arm encompassing helices IV and V (positions 41–44 and 74–77) (29). These positions partially encompass conserved positions. For many (but not all) eubacterial 5-S RNAs, at least three base-pairs from these two regions can be formed with Watson–Crick and G∘U pairing. It should be pointed out that universal tertiary interactions can be formulated by using the adjacent invariant residues $R^{39}∘U^{79}$ and $U^{40}∘R^{78}$. However, some of these positions form part of the S1 nuclease cleavage sites (29).

In another study (30), a cross-link between G^{41} and G^{72} of the *E. coli* 5-S RNA was obtained with the bifunctional phenyldiglyoxal reaction. A tertiary interaction between residues 37–40 and 73–76 has been proposed.

7. Conserved 5-S RNA Structure

The universal 5-S RNA model (Fig. 1) shows 42 out of approximately 120 nucleotide residues conserved in RNAs of eubacteria, archaebacteria, eukaryotes, and organelles. This is an updated esti-

TABLE I
GRAM-POSITIVE AND GRAM-NEGATIVE BACTERIAL 5-S RNA SIGNATURES[a]

Nucleotide position	Gram-positive bacteria	Gram-negative bacteria
27	A	C
34	U	A
48	A	U
56	U	G
59	G	A

[a] Exceptions to these signatures include 5-S RNAs from *Micrococcus lysodeikticus*, *Bacillus acidocaldarius*, and *Thermus aquaticus*.

mate from that previously reported (8) and is based on new and corrected sequences. The greatest degree of conservation in sequence appears to be among the eubacterial 5-S RNAs (Fig. 2A). The gram-positive and gram-negative generalized 5-S RNA signatures of bacteria differ by only five base substitutions (11) (Table I). A comparison of the conserved positions in the eubacterial and eukaryotic 5-S RNAs

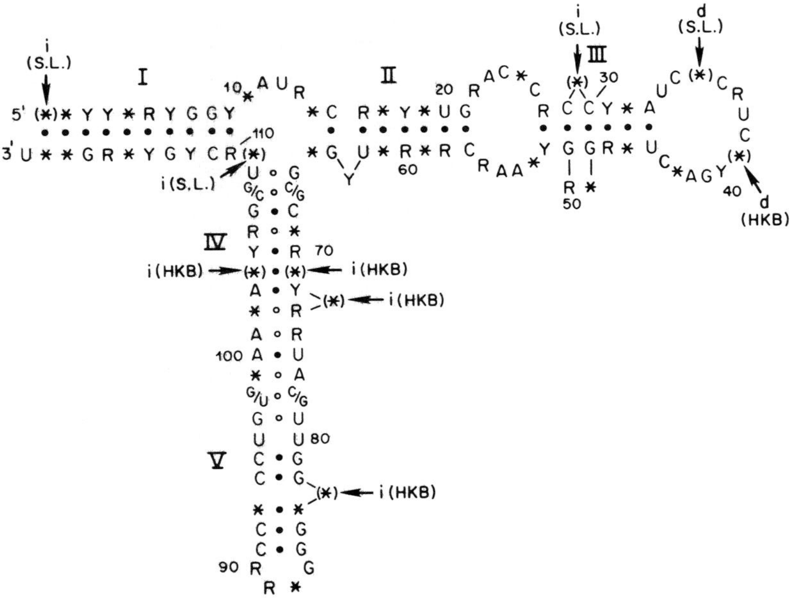

FIG. 6. Structural homology between *Synechococcus lividus* III and human KB-cell 5-S RNAs. The two structures are superimposed, and the homology in sequence is shown. Arrows indicate insertions and deletions. The numbering system used is that of human KB-cell 5-S RNA. i = insertion, d = deletion, S.L. = *S. lividus* 5-S RNA, and HKB = human KB-cell 5-S RNA.

reveals that more positions are conserved in the double-helical regions of eubacterial 5-S RNAs relative to the eukaryotic 5-S RNAs. A high concentration of conserved G∘Y pairs in double-helical regions prevails in both eubacterial and eukaryotic 5-S RNAs. The prevalence of G·C and G∘U pairs in these double-helical regions may be related to a requirement for either stability or flexibility of structure in the double-helical regions of the 5-S RNAs.

An example of the extent of conservation found in the 5-S RNAs from two widely different sources—the thermophilic cyanobacterium *Synechococcus lividus* and human KB cells—is shown in Fig. 6. The primary differences between these two RNAs are single nucleotide insertions or deletions found in various regions of the RNAs.

The graphic representations of the 5-S RNA used by different authors vary considerably (*12, 29, 31*). We propose a standard form, as shown in Fig. 1. This form was originally used by Fox and Woese (*13*) and is the one most commonly used. In addition, it constitutes a highly readable representation of the 5-S RNA (see Figs. 1 and 2).

II. "Hot Spots" of Insertions and Deletions in 5-S RNAs

Conserved chain-lengths between universal positions of the eubacterial and eukaryotic 5-S RNAs are included in Fig. 2. One of the major distinguishing features between eubacterial and eukaryotic 5-S RNAs are the chain-lengths between universal residues U^{40} and G^{44}; G^{69} and R^{76}; and U^{80} and G^{96} (*8*). Archaebacterial 5-S RNAs appear to have a combination of eubacterial and eukaryotic 5-S RNA chain-length signatures (*8, 32*).

Although there are conserved chain-lengths between universally conserved residues of most 5-S RNAs studied, insertions and deletions are found in RNAs from a variety of sources; these are depicted in Fig. 7. Most of these chain-length deviations represent single nucleotide changes scattered throughout the molecule, but there are three major regions where large insertions or deletions are found.

1. Major Deletions in Helix V

Block deletions of 6–10 nucleotides that appear to be base-pair losses are seen in helix V in all known mycoplasma 5-S RNA sequences (*33, 34*) and in the wheat mitochondrial 5-S RNA (*19*). This block of deletions results in a truncated structure of helix V. Both the mycoplasma and mitochondrial 5-S RNAs have small genome sizes and represent examples of efficient uses of genetic information. The

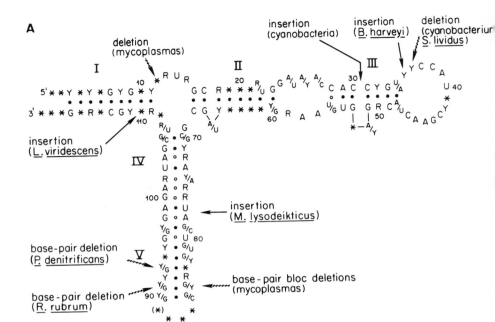
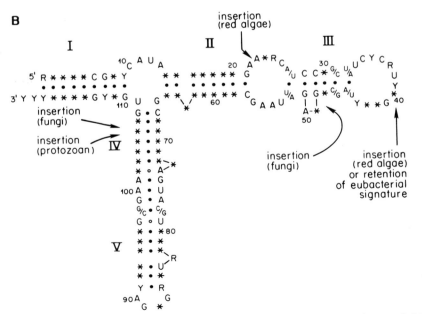

FIG. 7. Examples of major insertions and deletions found in (A) eubacterial 5-S RNA; (B) eukaryotic 5-S RNA. Wavy arrows denote deletions; straight arrows denote insertions.

mycoplasma 5-S RNAs, which are approximately 107 nucleotides in chain-length, are the smallest 5-S RNAs known and may constitute a minimal 5-S RNA structure that can function. Helix V appears to be the only segment of the 5-S RNA partially missing in known functional 5-S RNAs of cells and organelles. There are no examples of gross deletions in any other parts of the molecule.

2. Major Insertions in Helix IV

Another segment of the 5-S RNA where changes are apparently tolerated is the 3' side of helix IV, where large insertions are found. The most striking example is in the 5-S RNA of *Halococcus morrhuae*. Luehrsen *et al.* (35) described an insertion of a 108-nucleotide segment between positions 105 and 106 (using the generalized eukaryotic 5-S RNA numbering system, (Fig. 2B) in the 5-S RNA of the archaebacterium *H. morrhuae*). Other examples are an insertion of three nucleotides between positions 105 and 106 of the 5-S RNA of the dinoflagellate *Crypthecodinium cohnii* (36), which is at the same location as the insertion in the 5-S RNA of *H. morrhuae*, and a single nucleotide insertion between positions 106 and 107 in the 5-S RNAs of several species of the basidiomycetes (37).

3. Changes in Helix III

A third region where several internal changes in polynucleotide length have been observed is the 3' side of helix III. The wheat mitochondrial 5-S RNA has a six-nucleotide insertion between positions 51 and 54, and the 5-S RNAs from several fungi have a single-nucleotide insertion in approximately the same region (see Table II).

4. Insertions and Deletions in General

A schematic summary of insertions and deletions found in various 5-S RNA is shown in Fig. 7. In some cases, the locations of internal chain-length deviations appear to be characteristic of related organisms. For example, an insertion between positions 30 and 31 appears only in cyanobacterial and green-plant chloroplast 5-S RNAs. There is a deletion in the 5-S RNAs of chloroplasts and the cyanobacterium *Synechococcus lividus* between positions 35 and 39 (20). The red algae *Porphyra yezoensis*, *Porphyra tenure*, and *Gracilaria compressa* 5-S RNAs have an insertion between positions 21 and 22 (18). These red algae are also most unusual in that they are the only eukaryotic organisms known whose 5-S RNAs have the eubacterial signature of three positions between the universal position U^{40} and G^{44}. In this instance, they parallel the 5-S RNAs of the archaebacteria in having

TABLE II
EXAMPLES OF INSERTIONS AND DELETIONS FOUND IN 5-S RNAS

Source of 5-S RNA	Deletion location	Insertion location
Cyanobacteria and chloroplasts		
Two species of cyanobacteria and Prochloron	—	One nucleotide between residues 30 and 31
Four species of plant chloroplasts	One nucleotide between residues 34 and 39	One nucleotide between residues 30 and 31
Synechococcus lividus (cyanobacterium)	One nucleotide between residues 34 and 39	One nucleotide between residues 30 and 31
Mitochondria and aerobic eubacteria		
Paracoccus denitrificans	One base-pair in helix V	—
Rhodospirillum rubrum	One base-pair in helix V between $G^{96}U^{80}$ and loop d	—
Wheat mitochondria	Three bases-pairs in helix V between $G^{96}U^{80}$ and loop d	Two nucleotides between residues 11 and 14
	One nucleotide at looped-out position 66 of helix II	Two nucleotides between residues 23 and 25
	—	Six nucleotides between residues 51 and 54
	—	One nucleotide between residues 56 and 57
	—	Four nucleotides between residues 106 and 108
Other bacteria		
Three species of mycoplasmas	Three to five base-pairs in helix V between $G^{96}U^{80}$ and loop d	—
Mycoplasmas capricolum	One nucleotide at position 12	—
Beneckea harveyi	—	One to two nucleotides between residues 34 and 35

Micrococcus lysodeikticus	—	One nucleotide between residues 77 and 78
Lactobacillus viridescens	—	One nucleotide between residues 109 and 110
Fungi		
Torulopsis utilis, Saccharomyces cerevisiae (yeasts)	—	One nucleotide between residues 47 and 48
Aspergillus nidulans	—	One nucleotide between residues 48 and 49
Neurospora crassa	—	One nucleotide between residues 48 and 49
Four species of basidiomycetes	—	One nucleotide between residues 106 and 107
Protozoans		
Crypthocodinium cohnii (dinoflagellate)	—	One nucleotide between residues 105 and 106
Red algae		
Three species	—	One nucleotide between residues 21 and 22
		One nucleotide between residues 38 and 41 (eubacterial like)
Archaebacteria		
Halococcus morrhuae	—	Insertion of 108 nucleotides between residues 105 and 106
Sulfolobus acidocaldarius	One nucleotide at looped-out position 63 of helix II	
Halobacterium cutirubrum	—	One nucleotide between residues 108 and 110
		One nucleotide between residues 66 and 74

both prokaryotic and eukaryotic chain-length signatures between these universal positions.

A summary of examples of insertions and deletions in the 5-S RNA structure is given in Table II. It appears that eubacterial 5-S RNAs, although they have maintained the largest number of conserved nucleotides (Fig. 2), also have accumulated more types of internal chain-length changes than the eukaryotic 5-S RNAs. There are no deletions in any known eukaryotic 5-S RNA sequences (see also Table II). There are no insertions in metazoan 5-S RNAs, but, as stated above, insertions are found in the 5-S RNAs of several fungi, the red algae, and in one protozoan (*C. cohnii*). In the archaebacterial 5-S RNAs, there are a number of examples of both insertions and deletions.

There is minor chain-length heterogeneity at both the 5′ and 3′ ends that is evident when comparing 5-S RNA structures. This probably relates to differences in processing mechanisms in the eubacteria. It is of interest that chain-length differences between gram-positive and gram-negative bacteria (38) primarily arise from differences at the 5′ and 3′ ends (11).

III. Conformation of 5-S RNA Derived from Enzymatic, Chemical, and Physical Studies

The reaction of a chemical or an enzyme on 5-S RNA depends on several factors: (a) reactivity of the probing agent, (b) reaction conditions, and (c) accessibility of the residues to the probing agent. While certain sites (residues, sequences) may be "exposed," hence readily accessible to the agent, other sites may be inaccessible owing to tertiary interactions or masking by a protein or the enormous size of the nuclease itself. The results of various studies involving different means (chemicals, oligonucleotide binding, and others) dealing with the conformation of 5-S RNA have been discussed recently (2). In this section, we discuss recent studies on 5-S RNA conformation.

1. EUBACTERIAL 5-S RNA STRUCTURE

Chemical reactivities presumably dictate the multiplicity of structures (conformations). The denatured conformation of 5-S RNA may not be important, since denatured RNA fails to bind to ribosomal proteins under reconstitution conditions (39).

Temperature-jump experiments indicate that a transition exists between two conformations. "Low"- and "high"-temperature forms of *E. coli* 5-S RNA occur between 0°C and 40°C and are characterized by cation uptake, proton release, and overall compaction (28). The low form (conformation) can be converted to a high form by addition of cations, including hydrogen ion (28). Recently, Rabin et al. studied 5-S RNA (E. coli) in the low and high forms with RNase T1 and measured the relative nucleolytic rates at various sites in the molecule (40). Residue G^{13} became more exposed in the high form, whereas residues G^{54}, G^{56}, G^{72}, and G^{83} through G^{86} became less exposed in a low- to high-temperature transition accompanied by increases in Mg^{2+} and salt concentrations. No change in the rate of cleavage at other sites was observed. They suggested that this transition could involve movement of the prokaryotic and stem helices away from each other, revealing residue G^{13} more fully in the high form. They proposed that this kind of movement could easily provide protection from nuclease (observed in this study), result in compaction of the molecule as suggested by laser light scattering studies (28), and explain the absorbance increase and NMR spectral changes (27).

The downfield (9–15 ppm) proton NMR spectrum of a RNAse-A-resistant fragment of E. coli 5-S RNA has been studied by nuclear-Overhauser methods (NOE) (41). The fragment comprises residues 1 to 11 and 69 to 120 of the E. coli 5-S RNA. The results characterize two double-helical segments in the fragment having the sequences: (G·C)₃(A·U)(G·U)₃ and (G·C)₂(A·U)(G∘U). These sequences correspond to the central portion of the terminal stem (helix I) at residues 2 to 8 base-pairing with residues 112 to 118 and to part of helix V at residues 81 to 84 base-pairing with residues 92 to 95, respectively. The results of a series of NOE experiments with this fragment of E. coli 5-S RNA establish tentative assignments for a large number of resonances in the fragment spectrum and suggest the existence of two helical segments in the molecule that match in sequence to the a portion of terminal stem and of helix V proposed for intact 5-S RNA by Fox and Woese (13) and others. The authors claim evidence for the existence of these structures in 5-S RNA.

2. Eukaryotic 5-S RNA Structure

Andersen et al. studied 5-S RNA from Xenopus laevis oocytes with single-strand-specific nucleases (RNAse A, T_1, and T_2) and a double-strand-specific RNase V_1 from cobra venom (17). The results support the generalized 5-S RNA secondary structure model derived from comparative sequence analysis (Fig. 1); however, three putative single-stranded regions of the molecule showed unexpected cleavages by RNase V_1 at residues C^{36}, U^{73}, U^{76}, and U^{102}. The authors believe that these cleavages may indicate additional conformation features of the 5-S RNA including A∘G base pairs. (See Fig. 8 for details of cleav-

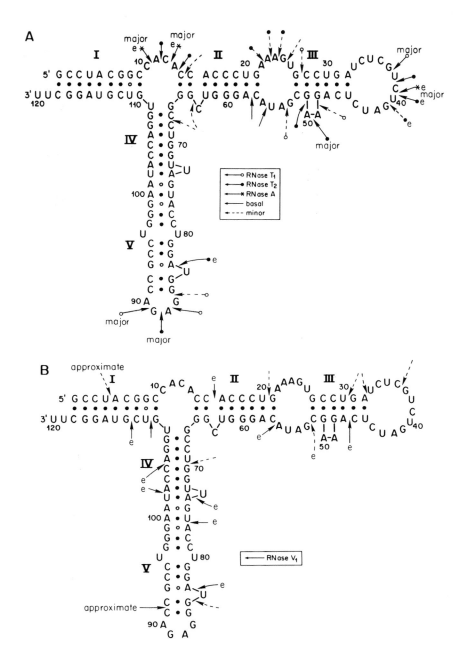

FIG. 8. *Xenopus* oocyte 5-S RNA drawn in the generalized secondary structural model according to Andersen et al. (17). (A) Arrows mark the cleavages that occur during partial digestions using single-strand-specific ribonucleases. The insert explains the symbols used to distinguish between RNases T1, T2, and A. "Basal" cleavages are those that occur during renaturation prior to ribonuclease treatment. (B) Arrows are drawn to indicate cleavages that occur during partial digestions using double-strand-specific RNase V1.

age sites by single- and double-strand-specific nucleases in 5-S RNAs.)

Partial nuclease digestion of *E. coli* and other eubacterial 5-S RNAs are also consistent with the generalized eubacterial 5-S RNA model (29, 42, 43). However, additional data are needed to confirm the structure of helix IV.

Ohta *et al.* studied melting of local ordered structures in yeast 5-S RNA in aqueous salt solutions (44). Equilibrium and kinetics of thermal melting of the RNA were studied with or without Mg^{2+} by differential thermal-melting and temperature-jump methods. Two peaks and a shoulder were observed in each of the melting curves. The local structures were stabilized considerably by Mg ions. The authors concluded from these observations that native yeast 5-S RNA has at least three local structures that melt between 10°C and 95°C at physiological ionic strength and are stabilized by Mg^{2+}. However, the data fail to indicate whether each structural stabilization is due to binding of Mg^{2+} to specific sites.

IV. 5-S RNA Interactions with Protein

Eukaryotic and eubacterial 5-S ribosomal RNAs interact with a number of different types of proteins. The small, highly conserved RNA is an integral part of the ribosome in all organisms. *Escherichia coli* 5-S RNA associates *in vitro* with three ribosomal proteins designated L5, L18, and L25 (20, 13, and 11 kDa, respectively) (44a). The binding of L5 is cooperatively stimulated by L18 while the binding of L25 is about 10-fold less than that of L18 (45). Ribosomal proteins that bind 5-S RNA occur in other eubacteria, and some of these are equivalent to the *E. coli* ribosomal proteins (46, 47). In eukaryotic ribosomes, 5-S RNA associates with a large protein (38–42 kDa) designated L3 in *Xenopus* and in *Saccharomyces* and L5 in rat liver (48–51), although 5-S RNA interactions with other ribosomal proteins have been demonstrated by affinity chromatography (52, 53). The 5-S RNA·L3 complex can be extracted from the larger ribosomal subunit in the presence of EDTA, and so is the most characterized of the eukaryotic 5-S RNA complexes (54–56).

Proteins other than ribosomal proteins also recognize and associate with 5-S RNA. Eubacterial 5-S RNA processing reactions occur within ribonucleoprotein particles (57–59). Eukaryotic 5-S RNAs do not undergo extensive posttranscriptional modification. The 5-S RNA genes form an independent transcriptional unit and are transcribed by RNA polymerase III (59a). Yeast 5-S RNA transcripts do undergo precise trimming at their 3' ends by an exonuclease (60). Eukaryotic 5-S RNAs appear to associate with nonribosomal proteins before being transported out of the nucleus and before being complexed with ribosomal proteins. In *Xenopus laevis*, the oocyte 5-S RNA is found complexed with the transcription factor protein A (TF-IIIA) in the 7-S particle and is also found complexed with two nonribosomal protein and tRNA in the 42-S storage particle (48, 61).

1. INTERACTIONS OF EUBACTERIAL 5-S RNA WITH PROTEIN

The complexes of *E. coli* and *B. stearothermophilus* 5-S RNAs with *E. coli* ribosomal proteins L18 and L25 have been probed with ribonucleases in order to ascertain the protection sites of these proteins on the RNA (62). Few differences in the protection sites on *E. coli* 5-S RNA and the heterologous *B. stearothermophilus* 5-S RNA were observed. The protection sites of ribosomal protein L18 are not contiguous along the 5-S RNA molecule and include (using the *E. coli* numbering system) C^{63} through U^{65} (next to the looped-out position A^{66}) on the 3' side of helix II, C^{38} through C^{43} found in the 13-membered loop region bounded by helix III, and C^{113} on the 3' side of helix I. Minor protection by L18 was found for residues in positions U^{14} and A^{52} (one of the looped-out positions of helix III). Protection sites of ribosomal protein L25 include only two positions, G^{81} and C^{94}, both in helix V.

Kethoxal modification data support the protection sites of helix II by L18 (63). The looped-out position of helix II, A^{66}, has been proposed to be a primary contact site for ribosomal protein L18 owing to its protection and the protection of surrounding nucleotide positions from RNase-A cleavage when the RNA is complexed with L18. Also, chemical modification of A^{66} inhibits the free 5-S RNA from complexing with L18 (64). The 13-membered loop bounded by helix III (around position 40) has been suggested to be important in the binding of L18 (43).

Earlier ribonuclease probe data of the L25·5-S-RNA complex have implicated positions 79 to 97 as the protein protection sites (65). Ribosomal protein L25 complexes with a 5-S RNA fragment that includes positions G^{69} to U^{87} and C^{90} to C^{110} (66), although the protein has a lower affinity for the fragment than it does for the whole RNA. This fragment encompasses the protection domain for L25 mentioned above.

A probe of the complexes of *E. coli* 5-S RNA with ribosomal protein with the endonuclease α-sarcin (67) partially confirms the postulated protection sites of L18 and L25 (62) and indicates that the attachment site of ribosomal protein L5 includes most of helix I (67).

The binding of *E. coli* ribosomal proteins L18 and L25 to two eubacterial 5-S RNAs not only causes protection of certain nucleotide positions from ribonuclease activity, but also stimulates certain cleavages and produces new cleavages at other nucleotide positions (62). The stimulated cleavages occur within and outside of proposed protein-binding domains. The stimulated cleavages and the appearance of new cleavages have been interpreted as indications of protein-induced conformational changes in the 5-S RNA.

Near-ultraviolet circular dichroisim spectra and melting profile data indicate conformational changes of *E. coli* 5-S RNA complexed with all three *E. coli* ribosomal proteins (68). The configuration of *E. coli* 5-S RNA complexed with each of the three ribosomal proteins separately was investigated using circular dichroisim (45). The data show that a protein-induced shift in secondary structure occurs in the 5-S RNA when it is complexed with L18 but not with L5 or L25 (45). The association constants (K_a) for the binding of the three *E. coli* ribosomal proteins to *E. coli* 5-S RNA are 2.3×10^8 M^{-1} for L18; 1.5×10^7 M^{-1} for L25, and 2.3×10^6 M^{-1} for L5 (45). Optimal 5-S RNA binding conditions have been defined for each protein, and the stoichiometry of each complex has been confirmed as 1:1 at optimal conditions (69).

Bacillus subtilis ribosomal protein BL16 (equivalent to *E. coli* L18) is the 5-S-RNA-binding β-subunit of maturation RNase M5, a 5-S RNA processing enzyme. (47). The subunit binds to the 5-S RNA and is necessary for the catalytic α component to finalize the processing of the RNA (47). However, Me$_2$SO can substitute for the β-subunit, suggesting that the β-subunit may affect the conformation of the precursor 5-S RNA.

Bacillus licheniformis 5-S RNA inhibits maturation of *Bacillus* precursor 5-S RNA by interacting with the processing enzyme RNase M5 (70). *Escherichia coli* 5-S RNA and eukaryotic *Saccharomyces carlsbergensis* 5-S RNA can also inhibit the maturation of *Bacillus* precursor 5-S RNA, but to varying degrees. Chimeric 5-S RNAs constructed to reduce base-pairing at the terminus of helix I did not inhibit maturation of *Bacillus* precursor 5-S RNA.

RNase M5 recognizes not only overall conformation of the potential substrate, but also the sequence UAGG found twice in the *B. subtilis* precursor 5-S RNA (59). The distance between the UAGG sequence (positions 101 to 104 in *B. subtilis* 5-S RNA) and the internal portion of helix I is correlated with the efficiency of inhibition of precursor maturation by the mature 5-S RNAs from E. coli and S. carlsbergensis (70).

2. Interactions of Eukaryotic 5-S RNA with Protein

The ribosomal protein L3 from *Saccharomyces cerevisiae* that binds 5-S RNA has been examined (49). The protein is acidic except for a basic region near its N terminus. This region of the protein shows amino-acid sequence homology to the N terminus of the eubacterial 5-S RNA binding ribosomal proteins L18 from *E. coli* and L13 from *Halobacterium cutirubrum*. An internal fragment of the protein L3 from yeast shows amino-acid sequence homology to the 5-S-RNA-binding ribosomal protein L5 from *E. coli*. Nazar et al. (49) suggested L3 (42 kDa) to be a fusion of the three known eubacterial 5-S-RNA-binding ribosomal proteins whose molecular weights add up to about 44,000.

Purified L3 from yeast will not reassociate with free 5-S RNA *in vitro*. However *S. cerevisiae* 5-S RNA and a variety of heterologous eukaryotic 5-S RNAs have been found to exchange into the L3·5-S complex. A chemical modification exclusion assay has shown that three helical domains of eukaryotic 5-S RNA, but not nucleotide sequence, are essentially involved in 5-S RNA exchange into the ribonucleoprotein complex (71). Specific modifications of bases in helices I, II, or IV cause the 5-S RNA to be excluded from exchange into the complex. It has previously been determined by limited RNase-A digestions that regions in yeast 5-S RNA protected by L3 include nucleotides 1 to 12 and 84 to 121 (72). These positions encompass helices I, IV, and V. Nucleotides in helix II important in the exchange of the 5-S RNA into the L3 ribonucleoprotein complex evidently do not remain bound to the protein after partial ribonuclease treatment of the complex.

The effects of rat liver ribosomal protein L5 on the conformation of rat liver 5-S RNA by lasar Raman spectroscopy have been studied (73). The data suggest that pyrimidine residues in single-stranded regions and the N-7 of guanine residues in the RNA are involved in the interaction with L5 protein. The data also show small protein-induced conformational changes in the 5-S RNA; these changes do not appear to involve the base-paired regions. Eukaryotic 5-S RNA has been found complexed in nonribosomal protein particles in the oocytes of amphibians and teleosts (48). The ribonucleoprotein particle containing 5-S RNA and a 37-kDa protein (74) is the 7-S storage particle. Recent data suggest that 7-S particles exist in HeLa cells (75). TF-IIIA protein in the *X. laevis* 7-S particle is not ribosomal protein L3 (48).

TF-IIIA protein binds to an intragenomic transcriptional control region of the *X. laevis* oocyte 5-S RNA gene during transcription of the 5-S RNA as well as to the RNA (61). The two proteins of the 42-S particle are different from both ribosomal L3 and TF-IIIA (61).

Labeled 5-S RNA complexed in 7-S particles from *X. laevis* has been probed by ribonucleases (76). The RNA is not extensively protected from ribonuclease cleavage and appears to be in a different conformation from renatured *X. laevis* oocyte 5-S RNA. The stem of the RNA encompassing helices IV and V is partially denatured. Also the looped-out position C^{63} in helix II is highly exposed to ribonuclease cleavage when the RNA is complexed with TF-IIIA but is protected in renatured RNA. *Xenopus laevis* 5-S RNA and heterologous 5-S RNAs can be exchanged into the 7-S particle *in vitro* (96). Thus, TF-IIIA can recognize a generalized 5-S RNA structure.

All 5-S RNAs in previtellogenic oocytes of *X. laevis* not complexed with ribosomal proteins are complexed in either 42-S or 7-S particles; the 5-S RNA is not free of protein (48).

3. Significance of Interactions of 5-S RNA with Protein

The interaction of eubacterial and eukaryotic 5-S RNA with protein suggest that the proteins recognize domains of secondary and tertiary structures. Primary structure may play a role in protein recognition of the RNA at looped-out positions and at highly conserved single-stranded residues. Changes in the conformation of 5-S RNA occur when it associates with certain 5-S-RNA binding proteins. Such changes may prepare the RNA for further protein interactions, such as those that occur during processing or those that occur during ribosome assembly. The ability of the RNA to function may depend on the protein-induced changes in conformation. It is noteworthy that the 5-S RNA appears not to be free of protein in either eubacterial or eukaryotic systems.

V. Phylogeny and Endosymbiosis

The 5-S ribosomal RNA provides information on phylogeny and endosymbiosis (77). Because of its high degree of stability during the course of evolution, the 5-S gene product provides a suitable means of interrelating organisms of ancient origin phylogenetically. Studies of the phylogeny of the 5-S RNAs have been valuable in the reclassification of several lower eukaryotes and in identifying possible prokaryotic descendants of the organelle precursors. In addition, detailed phylogenetic trees that provide an overview of the interrelationships of a broad range of organisms have been constructed. On the other hand, 5-S RNA has a small number of signatures, and this limits its usefulness in the determination of phylogeny within closely related groups of organisms. This is especially true with metazoans, where 5-S RNA reveals few signatures that can be used as phylogenetic markers.

Two methods have been used to provide information on phylogeny and endosymbiosis. One is a comparison of overall sequence homology and the formation of similarity matrices; another is the use of group-specific signatures, which includes insertion and deletion signatures as phylogenetic markers. Quantitative estimates of phylogenetic relationships have been obtained by comparisons of overall sequence homologies, but signature markers have been used so far to provide qualitative information. Since the overall sequence comparisons take into account each nucleotide position equally and include the highly variable positions of the molecule, the convergence of sequences is more likely to complicate a phylogenetic analysis by overall sequence homology than the more specific type of comparison of signatures. To obtain a numerical 5-S RNA phylogeny of greater accuracy than a simple comparison of overall nucleotide sequences, the application of a statistical formulation such as pattern recognition or correspondence analysis (78–80) to the 5-S RNA would be useful, but this type of analysis has not yet been presented.

1. 5-S RNA Phylogenetic Trees

Hori (81) was one of the first to use aligned 5-S RNA sequences to construct phylogenetic trees. Others since then have also constructed phylogenetic relationships from 5-S RNA sequences. Using the criterion of minimal mutations, such trees were constructed for several eubacterial and chloroplast 5-S RNAs (82). These trees provide definitive groupings of the gram-positive, gram-negative, and chloroplast-cyanobacterial 5-S RNAs. In another tree construction, the major taxonomic grouping of eukaryotes, eubacteria, archaebacteria, and organelles have been mapped by nucleotide differences within and between these groups (10). In yet another 5-S RNA evolutionary tree (83), a cluster analysis, using the method of weighted-pair grouping and arithmetic averaging (84), was utilized to form the tree. One important conclusion made from this analysis is that the fungi and protozoa were each derived from several different ancestral types.

These phylogenetic trees are useful in interrelating the broad categories of extant organisms and determining the approximate branching points in time. Limitations in tree analyses reside in problems of determining interrelatedness quantitatively within a genus or between genera because of too few differences in the sequences. Also, estimates of times of divergence between organisms are subject to errors arising from differences in rates of mutations. For example, there is a surprisingly high and variable rate of mutation in insect 5-S RNAs relative to 5-S RNAs of other eukaryotes (23, 85).

2. TAXONOMIC RECLASSIFICATIONS SUGGESTED BY 5-S RNA COMPARISONS

Several reclassifications of organisms into new taxonomic groupings on the basis of 5-S RNA sequence homologies have been proposed. Sequences of 5-S RNAs from two classes of basidiomycetes, Homobasidiomycetae and Heterobasidiomycetae contain several inconsistencies in sequence homology within each class (37). A matrix of 5-S RNA sequence homology clearly shows a sharp distinction between the two clusters of fungi, and a redivision of these organisms into the two clusters is proposed (37). This division is consistent with a division of these organisms based on the presence or absence of cell-wall septal pores. The 5-S RNAs of fungi from the two clusters also differ markedly in a number of group-specific signatures (see below). Additional studies on the 5-S RNAs of the basidiomycetes (83) are compatible with a grouping of these fungi into three major taxonomic classes as proposed earlier (86). A further assignment of the basidiomycetes species into a particular order can only partially be made because the 5-S RNA sequences from some of these organisms are not sufficiently dissimilar (24, 83, 87).

From a comparison of 5-S RNA sequences from several lower fungi, it was concluded that organisms of the class Zygomycetes are derived from the chytrid water molds (order Chytridiales) (37). This conclusion was based on the high degree of similarity in sequence between *Phycomyces blakesleeanus* (88) and the water molds *Phlyctochytrium irregulare* and *Blastocladiella simplex* (37).

A comparison of 5-S RNA sequences from echinoderms and vertebrates does not support the close phylogenetic correlations made by zoologists in the past (89). On the other hand, the determination of accurate phylogenetic relationships among the metazoans can be difficult since metazoan 5-S RNAs have not diverged sufficiently and, as mentioned above, can have variable rates of substitutions (24, 85).

The 5-S RNA sequences from several archaebacteria have been determined (5, 32). Comparisons of these 5-S RNAs support the assignment of the archaebacterium *Caldariella acidophilia* to the species of *Sulfolobus* (4).

Physarum polycephalum has in the past been classified in the phylum Myxomycota of the fungi. The comparisons of overall sequence homologies of 5-S RNAs from *P. polycephalum* and other organisms suggest a phylogenetic relationship closer to the protozoans (23). Comparisons of the 5-S RNA sequences of the two slime molds *Physarum polycephalum* and *Dictyostelium discoideum* also suggest that these two organisms diverged very early during evolution (23).

3. USE OF 5-S RNA SIGNATURE ANALYSIS IN
   ASSESSING EVOLUTIONARY RELATIONSHIPS

The 5-S RNAs of closely related organisms exhibit homologies in given regions of the molecule; i.e., they have group-specific signatures. Sequences in specific regions of the RNAs, structures of helices (lengths, positions, and nature of looped-out nucleotides), insertions, and deletions are examples of signatures that are shared by related organisms (Fig. 7). Some examples of group-specific signatures follow. (i) Green plant 5-S RNAs (sequences from mosses and flowering plants) share the sequence CAGAAC[44], an unpaired U residue at position 63, and an A∘C pair in helix IV. In helix V, they have an unpaired U at position 84 and U at position 93 (residue 93 is a purine in most eukaryotes). (ii) The red algal 5-S RNAs have an insertion between position 21 and 22 and the prokaryotic signature of three positions between U^{38} and G^{41}. (iii) The cyanobacterial and chloroplast 5-S RNAs have an insertion between C^{30} and C^{31}. (iv) The adoliporous basidiomycetes 5-S RNA sequences show an insertion between positions 106 and 107. (v) The purple nonsulfur bacteria have a base-pair deletion in helix V. (vi) The 5-S RNAs of organisms from three kingdoms—eubacteria, archaebacteria, and eukaryotes—exhibit chain-length signatures between universal positions that are specific to the RNAs from each kingdom.

A comparison of the 5-S RNAs of the green algae (division, Chlorophyta) and the green plants reveals a number of interesting similarities in the signatures mentioned above. The 5-S RNAs from both phyla share the sequence CAGAAC[44], an unpaired U in position 63 in helix II, and an A∘C pair in helix IV. Helix V presents some curious and intriguing partial homologies. It has been proposed that unicellular green algae, multicellular green algae, and green plants have a common ancestor and have diverged relatively recently (90).

Neither the brown nor the red algal 5-S RNAs share signatures specific to the RNAs of the green algae. For example, the sequence

CGCUC[43] of the brown algal 5-S RNAs is more like certain fungal 5-S RNAs (e.g., *Neurospora*, water molds) in this region.

The 5-S RNA of the photosynthetic protist *Euglena gracilis* has signatures homologous to those of the protozoans, specifically to those signatures of the 5-S RNA of the trypanosomatid protozoan *Crithidia fasciculata* (*91, 92*). A close phylogenetic relationship between these two organisms has been suggested (*91, 92*).

Of special significance is the prokaryotic-type chain-length between U^{38} and G^{41} (i.e., three positions) in the red algal 5-S RNAs. This prokaryotic signature has not been seen in any other eukaryotic 5-S RNA and may be a consequence of sequence convergence, or, viewed in a more provocative manner, a relic of a prokaryotic signature maintained during a possible evolutionary pathway from prokaryotes to eukaryotes. The archaebacterial 5-S RNAs also reveal signatures that raise the same questions. These 5-S RNAs vary in structure, but in general appear eukaryotic-like, with the exception of having the prokaryotic chain length between U^{38} and G^{41}. Here again the same question can be asked as to whether this prokaryotic signature represents a retention of a prokaryotic 5-S RNA trait by the archaebacteria during an evolutionary transition of prokaryotes to eukaryotes.

All mycoplasma 5-S RNAs have a truncated helix V (*33, 34*). In other parts of the molecule, the mycoplasma have the gram-positive signatures described in Table I. This signature homology is consistent with the proposal that these organisms originated from the gram-positive bacteria (*93–95*).

4. 5-S RNA Signature Analysis and Endosymbiosis

Strong evidence for the cyanobacterial origin of the green plant chloroplasts comes from a comparison of 5-S RNAs from these sources. The 5-S RNAs from the cyanobacteria and chloroplasts share unique insertion and deletion signatures. An insertion between C^{30} and C^{31} is found in cyanobacterial and chloroplast 5-S RNAs. In addition, the cyanobacterial species *Synechococcus lividus* III shares with the chloroplast 5-S RNAs a unique deletion between A^{34} and A^{39} and the base-substitution signatures of helix III (*20*). The 5-S RNAs of the green-plant chloroplasts are exceedingly stable and differ by only a few base substitutions (*96*). The chloroplast 5-S RNA from the ancient ferns whose fossil records date from $\sim 4 \times 10^8$ years shares this stability. The uniqueness of the shared deletion and insertion signatures together with the extreme stability of the chloroplast 5-S RNAs argue against the idea that these shared signatures arose from convergence.

Synechococcus lividus III is a thermophilic cyanobacterium found in Yellowstone National Park (97). Like other cyanobacteria and the green-plant chloroplasts, its photosynthetic machinery includes photosystems I and II. The cyanobacteria contain biliproteins that are accessory pigments and aid in the absorption of light. Biliproteins with these functions are not known to be present in green-plant chloroplasts, but it is not known whether proteins of analogous structures are present in the green-plant chloroplasts. *Synechococcus lividus* and other cyanobacteria do not have chlorophyll *b*, which the green-plant chloroplasts have, but this difference involves only the side-chains of the tetrapyrrole nucleus of chlorophyll and is of minor significance in the total scheme of chlorophyll synthesis. The chlorophylls found in different organisms probably evolved in response to wavelength requirements for efficient light-gathering processes. On the basis of general similarities in photosynthetic properties and the sharing of unique 5-S RNA signatures, *S. lividus* III can be considered a candidate for the "missing link" in the origin of the green-plant chloroplasts. *Synechococcus lividus* and the green-plant chloroplasts also share an amino-acid deletion signature near the amino-terminal end of the ferredoxin protein (77). This homology further supports the hypothesis that *S. lividus* and the green-plant chloroplasts have a common ancestor. Prochloron, which has been considered to be a direct descendant of the green-plant chloroplast endosymbiont, shares the cyanobacterial insertion signature between C^{30} and C^{31} (98), but does not have the deletion or helix III signatures specific to the green-plant chloroplast and *S. lividus* 5-S RNAs. The 5-S RNA sequences from chloroplasts of the protozoa and the red and brown algae will undoubtedly yield a great deal of information on possible polyphyletic origins of the chloroplast.

The wheat mitochondrial 5-S RNA, more than any other known 5-S RNA, differs from the generalized structure in terms of the variety of insertions and deletions present (19). This is probably the result of a high rate of mutation in the mitochondrial DNA and presents problems in finding homologous 5-S RNAs among the prokaryotes where presumably the rates of mutation of chromosomal genes in most species are much less than that of the mitochondrial genome. Nevertheless, some hints may be obtained from 5-S RNA comparisons concerning the identification of possible prokaryotic precursors to the mitochondrion, particularly if two or more of the mitochondria-specific insertion and deletion signatures are found in a prokaryotic 5-S RNA.

Members of the family Rhodospirillaceae (the purple nonsulfur photosynthetic bacteria), which also includes the nonphotosynthetic *Paracoccus denitrificans*, have been considered as possible descendants of the bacterial ancestors of the mitochondrion (*99*). There are homologies in cytochrome *c* structures between members of this group of bacteria and the cytochromes *c* of eukaryotes (*100, 101*). The 5-S RNA sequences of *Paracoccus denitrificans* and *Rhodospirillum rubrum* have been determined (*98, 102*). It is significant that both of these RNAs are the only prokaryotic 5-S RNAs (other than the mycoplasma) known to have a base-pair deletion in helix V relative to the generalized eubacterial 5-S RNA structure. The wheat mitochondrial 5-S RNA also bears a closer similarity in overall sequence homology to *Paracoccus denitrificans* than to 10 other bacterial 5-S RNAs, although this similarity may not be significant by itself (*98*). Taken together, the tendency to sustain base-pair deletions in helix V and the overall sequence homology of their 5-S RNAs adds support to a proposal that there was a common ancestor to the mitochondria and Rhodospirillaceae. The 5-S RNAs of other members of the Rhodospirillaceae family should be sequenced to search for the possible presence of more signatures specific to the wheat mitochondrial 5-S RNAs.

VI. Summary

The recent culmination of new and corrected nucleotide sequences of 5-S RNAs derived from the eubacteria, archaebacteria, eukaryotes, and the organelles reveals that the 5-S ribosomal RNA has been extremely conserved in nucleotide sequence and chain-length during the course of evolution. Eubacterial and eukaryotic 5-S RNAs differ primarily by the presence of a small number of insertions or deletions that are scattered throughout the molecule. Certain regions of the molecule tend to sustain chain-length changes more readily than other regions, e.g., helix V.

The nucleotide sequences from a wide range of organisms conform to a generalized secondary structural model. Probes of the 5-S RNA structure by partial nuclease digestions largely confirm this model. In *E. coli* 5-S RNA, two double-helical regions (helices I and V) have been detected by NMR spectroscopy. There is both phylogenetic and experimental evidence for the presence of non-Watson–Crick A∘G base-pairing in the stem encompassing helices IV and V of the 5-S RNA. There is phylogenetic evidence also for the presence of A∘A and A∘C pairing but as yet no experimental data reveal the presence of these base-pairs in the 5-S RNA.

The 5-S RNA appears to undergo conformational changes when it binds to proteins. Domains of interaction involving secondary and tertiary structure appear to be prevalent in binding.

The 5-S RNA appears to be a good molecule to ascertain the relatedness of ancient organisms. Progress has been made in refining the taxonomic classifications of several fungi, protozoans, and algae by 5-S RNA comparisons. The 5-S RNA group-specific signatures have proved to be particularly useful in revealing possible origins of the chloroplasts. Although the mitochondrial 5-S RNA genes have accumulated a large number of mutations, a hint of possible origins of wheat mitochondria comes from both signature analysis and overall homology comparisons.

Acknowledgments

This work was supported by Grant NSF 83-02127. We thank Karen Hendrickson for elaborate drawings.

References

1. V. A. Erdmann, E. Huysmans, A. Vandenberghe, and R. De Wachter, *NARes* **11**, r105 (1983).
2. R. P. Singhal and J. K. Shaw, This Series **28**, 177 (1983).
3. K. R. Luehrsen and G. E. Fox, *PNAS* **78**, 2150 (1981).
4. E. Dams, A. Vandenberghe, and R. De Wachter, *NARes* **11**, 1245 (1983).
5. D. A. Stahl, K. R. Luehrsen, C. R. Woese, and N. R. Pace, *NARes* **9**, 6129 (1981).
6. S. Böhm, H. Fabian, and H. Welfle, *Acta Biol. Med. Ger.* **40**, k19 (1981).
7. S. Böhm, H. Fabian, and H. Welfle, *Acta Biol. Med. Ger.* **41**, k1 (1982).
8. N. Delihas and J. Andersen, *NARes* **10**, 7323 (1982).
9. R. De Wachter, M. Chen, and A. Vandenberghe, *Biochimie* **64**, 311 (1982).
10. H. Kuntzel, B. Piechulla, and U. Hahn, *NARes* **11**, 893 (1983).
11. H. Neimark, J. Andersen, and N. Delihas, *NARes* **11**, 7569 (1983).
12. K. Nishikawa and S. Takemura, *J. Biochem. (Tokyo)* **76**, 935 (1974).
13. G. E. Fox and C. R. Woese, *Nature (London)* **256**, 505 (1975).
14. K. Morikawa, M. Kawakami, and S. Takemura, *FEBS Lett.* **145** (1983).
15. S. Douthwaite, A. Christensen, and R. A. Garrett, *JMB* **169**, 249 (1983).
16. L.-S. Kan, S. Chandrasegaran, S. M. Pulford, P. S. Miller, *PNAS* **80**, 4263 (1983).
17. J. Andersen, N. Delihas, J. S. Hanas, and C.-W. Wu, *Bchem*, in press.
18. B. L. Lim, H. Hori, and S. Osawa, *NARes* **11**, 5185 (1983).
19. D. F. Spencer, L. Bonen, and M. W. Gray, *Bchem* **20**, 4022 (1981).
20. N. Delihas, W. Andresini, J. Andersen, and D. Berns, *JMB* **162**, 721 (1982).
21. H. Kuntzel, M. Heidrich, and B. Piechulla, *NARes* **9**, 1451 (1981).
22. E. U. Selker, C. Yanofsky, K. Driftmier, R. L. Metzenberg, B. Alzner-DeWeerd, and U.-L. RajBhandary, *Cell* **24**, 819 (1981).
23. H. Komiya, N. Shimizu, M. Kawakami, and S. Takemura, *J. Biochem. (Tokyo)* **88**, 1449 (1980).
24. W. F. Walker and W. F. Doolittle, *NARes* **11**, 5159 (1983).
25. T. Kumazaki, H. Hori, and S. Osawa, *NARes* **11**, 3347 (1983).

26. E. N. Trifonov and G. Bolshoi, *J. Mol. Biol.* **169**, 1 (1983).
27. M. J. Kime and P. B. Moore, *NARes* **10**, 4973 (1982).
28. T. H. Kao and D. M. Crothers, *PNAS* **77**, 3360 (1980).
29. T. Pieler and V. A. Erdmann, *PNAS* **79**, 4599 (1982).
30. J. Hancock and R. Wagner, *NARes* **10**, 1257 (1982).
31. A. Troutt, T. J. Savin, W. C. Curtiss, J. Celentano, and J. N. Vournakis, *NARes* **10**, 653 (1982).
32. G. E. Fox, K. R. Luehrsen, and C. R. Woese, *Zentralbl. Bakteriol. Hyg., Abt. I, Orig. C* **3**, 330 (1982).
33. H. Hori, M. Sawada, S. Osawa, K. Murao, and H. Ishikura, *NARes* **9**, 5407 (1981).
34. R. T. Walker, E. T. J. Chelton, M. W. Kilpatrick, M. J. Rogers, and J. Simmons, *NARes* **10**, 6363 (1982).
35. K. R. Luehrsen, D. E. Nicholson, D. C. Eubanks, and G. E. Fox, *Nature (London)* **293**, 755 (1981).
36. A. G. Hinnebusch, L. C. Klotz, R. L. Blanken, and A. R. Loeblich III, *J. Mol. Evol.* **17**, 334 (1981).
37. W. F. Walker and W. F. Doolittle, *Nature (London)* **299**, 723 (1982).
38. H. Hori and S. Osawa, *PNAS* **76**, 381 (1979).
39. M. Aubert, G. Bellemare, and R. Monier, *Biochimie* **55**, 135 (1973).
40. D. Rabin, T. H. Kao, and D. M. Crothers, *JBC* **258**, 10813 (1983).
41. M. J. Kime and P. B. Moore, *Bchem* **22**, 2615 (1983).
42. S. Douthwaite and R. A. Garrett, *Bchem* **20**, 7301 (1981).
43. M. Speek and A. Lind, *NARes* **10**, 947 (1982).
44. S. Ohta, S. Marnyuma, K. Nitte, and S. Suga, *NARes* **11**, 3363 (1983).
44a. C. E. Kurland, *ARB* **41**, 377 (1972). Also see Kurland and Ehrenberg, this volume.
45. P. Spierer, A. A. Bogdanov, and R. A. Zimmermann, *Bchem* **17**, 5394 (1978).
46. T. Pieler, I. Kumagai, and V. A. Erdmann, *Zentralbl. Baktariol. Hyg., Abt. I, Orig. C* **3**, 69 (1982).
47. D. A. Stahl, B. Pace, T. Marsh, and N. R. Pace, *JBC*, in press.
47a. B. Pace, D. A. Stahl, and N. R. Pace, *JBC*, in press (1984).
48. B. Picard and M. Wegnez, *PNAS* **76**, 241 (1979).
49. R. N. Nazar, M. Yaguchi, G. E. Willick, C. W. Rollin, and C. Roy, *EJB* **102**, 573 (1979).
50. K. Terao, Y. Takahashi, and K. Ogata, *BBA* **402**, 230 (1975).
51. E. H. McConkey, H. Bielka, J. Gordon, S. M. Lastick, A. Lin, K. Ogata, J.-P. Reboud, J. A. Traugh, R. R. Traut, J. R. Warner, H. Welfle, and I. G. Wool, *MGG* **169**, 1 (1979).
52. N. Ulbrich and I. G. Wool, *JBC* **253**, 9049 (1978).
53. A. Metspalu, M. Saarma, R. Villems, M. Ustav, and A. Lind, *EJB* **91**, 73 (1978).
54. G. Blobel, *PNAS* **68**, 1881 (1971).
55. B. Lebleu, G. Marbaix, M. G. Huez, J. Temmerman, A. Burny, and H. Chantrenne, *EBJ* **19**, 264 (1971).
56. F. Grummt, I. Grummt, and V. A. Erdmann, *EJB* **43**, 343 (1974).
57. M. L. Sogin and N. R. Pace, *Nature (London)* **252**, 598 (1974).
58. M. L. Sogin, B. Pace, and N. R. Pace, *JBC* **252**, 1350 (1977).
59. B. Meyhack and N. R. Pace, *Bchem* **17**, 5804 (1978).
59a. E. H. Birkenmeier, D. D. Brown, and E. Jordan, *Cell* **15**, 1077 (1978).
60. P. A. Tekamp, R. L. Garcea, and W. J. Rutter, *JBC* **255**, 9501 (1980).
61. H. R. B. Pelham and D. D. Brown, *PNAS* **77**, 4170 (1980).
62. S. Douthwaite, A. Christensen, and R. A. Garrett, *Bchem* **21**, 2313 (1982).

63. R. A. Garrett and H. F. Noller, *JMB* **132**, 637 (1979).
64. D. A. Peattie, S. Douthwaite, R. A. Garrett, and H. F. Noller, *PNAS* **78**, 7331 (1981).
65. J. Zimmerman and V. A. Erdmann, *MGG* **160**, 247 (1978).
66. S. Douthwaite, R. A. Garrett, R. Wagner, and J. Feunteun, *NARes* **6**, 2453 (1979).
67. P. W. Huber and I. G. Wool, *PNAS* **81**, 322 (1984).
68. J. W. Fox and K.-P. Wong, *JBC* **253**, 18 (1978).
69. P. Spierer and R. A. Zimmermann, *Bchem* **17**, 2474 (1978).
70. W. J. Stiekema, H. A. Raue, M. M. C. Duin, and R. J. Planta, *NARes* **8**, 5411 (1980).
71. R. N. Nazar and A. G. Wildeman, *NARes* **11**, 3155 (1983).
72. R. N. Nazar, *JBC* **254**, 7724 (1979).
73. H. Fabian, S. Bohm, W. Carius, R. Misselwitz, and H. Welfle, *FEBS Lett.* **155**, 285 (1983).
74. D. R. Engelke, S.-Y. Ng, B. S. Shastry, and R. G. Roeder, *Cell* **19**, 717 (1980).
75. W. Gruissem and K. H. Seifart, *JBC* **257**, 1468 (1982).
76. J. Andersen, N. Delihas, J. S. Hanas, and C.-W. Wu, *Bchem*, in press (1984).
77. R. M. Schwartz and M. O. Dayhoff, *Science* **177**, 395 (1978).
78. M. O. Hill, *Appl. Statist.* **23**, 340 (1974).
79. J.-P. Benzecri, in "Methodologies of Pattern Recognition" (S. Watanabe, ed.), p. 35. Academic Press, New York, 1969.
80. J. Frank, A. Verschoor, and M. Boublik, *JMB* **161**, 107 (1982).
81. H. Hori, *J. Mol. Evol.* **7**, 75 (1975).
82. D. Sankoff, R. J. Cedergren, and W. McKay, *NARes* **10**, 421 (1982).
83. E. Huysmans, E. Dams, A. Vandenberghe, and R. De Wachter, *NARes* **11**, 2871 (1983).
84. R. R. Sokal and C. D. Michener, *Univ. Kansas Sci. Bull.* **38**, 1409 (1958).
85. G. Xian-Rong, K. Nicoghosian, and R. J. Cedergren, *NARes* **10**, 5711 (1982).
86. P. H. B. Talbot, *Taxon* **17**, 620 (1968).
87. A. Templeton, *Nature (London)* **303**, 731 (1983).
88. J. Andersen, W. Andresini, and N. Delihas, *JBC* **257**, 9114 (1982).
89. T. Ohama, T. Kumazaki, H. Hori, S. Osawa, and M. Takai, *NARes* **11**, 473 (1983).
90. B. L. Lim, H. Hori, and S. Osawa, *NARes* **11**, 1901 (1983).
91. N. Delihas, J. Andersen, W. Andresini, L. Kaufman, and H. Lyman, *NARes* **9**, 6627 (1981).
92. T. Kumazaki, H. Hori, and S. Osawa, *J. Mol. Evol.* **18**, 293 (1982).
93. H. Neimark, in "The Mycoplasmas" (M. F. Barile and S. Razin, eds.), Vol. II, p. 43. Academic Press, New York, 1979.
94. H. Neimark and J. London, *J. Bact.* **150**, 1259 (1982).
95. C. R. Woese, J. Maniloff, and L. B. Zablen, *PNAS* **77**, 494 (1980).
96. F. Takaiwa and M. Sugiura, *NARes* **10**, 5369 (1982).
97. D. L. Dyer and R. D. Gafford, *Science* **134**, 616 (1961).
98. R. M. Mackay, D. Salgado, L. Bonen, E. Stackebrand, and W. F. Doolittle, *NARes* **10**, 2963 (1982).
99. R. F. Whalley, in "Origins and Evolution of Eukaryotic Intracellular Organelles" (J. F. Fredrick, ed.), *Ann. N. Y. Acad. Sci.* 361, 330 (1982).
100. R. P. Ambler, T. E. Meyer, and M. Kamen, *PNAS* **73**, 72 (1976).
101. R. J. Almassy and R. E. Dickerson, *PNAS* **75**, 2674 (1978).
102. N. Newhouse, K. Nicoghosian, and R. J. Cedergren, *Can. J. Biochem.* **59**, 921 (1981).

Optimization of Translation Accuracy

C. G. Kurland and
Måns Ehrenberg

*Department of Molecular Biology
Biomedical Center
Uppsala, Sweden*

I.	Gene Expression by Ambiguous Translation...................	192
II.	Correlations of Speed and Accuracy in Translation	194
III.	Codon–Anticodon Selectivity on the Ribosome	197
IV.	Kinetic Proofreading: One More Time.......................	200
V.	The Problem Solved by Kinetic Proofreading: Rate or Accuracy?..	205
VI.	Translational Accuracy and Exponential Growth	208
VII.	Kinetic Options for Inexpensive Proofreading and Error Regulation..	212
VIII.	Error Feedback Increases Accuracy: End of Error Catastrophe	216
	References..	217

The first issue of this series contains an unusual article by Crick (*1*). What is remarkable about that article is that it really describes "work in progress." The tentative character of the conclusions as well as the frank discussion of weak arguments did much to demystify the work on the code that then preoccupied molecular biologists. There is another aspect of this paper that brings us closer to our current preoccupation.

An important part of the data reviewed by Crick (*1*) concerns the identification of amino acids in the products translated *in vitro* from synthetic mRNA species. What is so striking in retrospect is that the incorporation of an amino acid only several times background in those experiments could provide the basis for a codon assignment. Thus, one of the problems discussed by Crick (*1*) was the low signal-to-noise ratios of cell-free systems, and he cautions the reader by relating observations of what appeared to be missense incorporation of leucine for phenylalanine. Nevertheless, the ambiguities of translation were put aside by students of the code as at most a second-order perturbation.

The conceptual and technical circumstances have changed so radically in the intervening 20 years that it now seems more appropriate to set the ambiguities of translation at the center of our focus. Thus, it is at last meaningful to ask how the translation system functions so that its errors are held to marginal frequencies. Indeed, we can even begin to discuss some of the parameters that determine what levels of error are optimal. A general description of the problem as we see it now can be outlined as follows.

It is well known that bacteria can mutate so that their translational error frequencies are changed in characteristic ways. For example, mutations affecting the ribosomal protein S4 can raise the error frequencies, whereas mutations affecting the ribosomal protein S12 can lower these frequencies (2). Clearly, the existence of these different ribosome phenotypes defines a problem of mechanism: How does the structure of the ribosome influence the fidelity with which aminoacyl-tRNA species are matched with codons? This much is well appreciated; however, we wish to suggest that this mechanistic problem is but one aspect of a broader problem.

What seems most interesting about the S12 mutants is that they contradict a prejudice that is rarely examined, namely that, in general, mutants should be less accurate than wild-type forms (3). That wild-type ribosomes normally function at accuracies demonstrably below what is attainable by mutants leads to our central concern: What are the parameters that limit the optimal levels of translational fidelity? In particular, why have the streptomycin-resistant ribosomes not been selected over the wild type during the evolution of *Escherichia coli* (*3a*).

I. Gene Expression by Ambiguous Translation

One engaging view of this problem is that ambiguous translation of the code can be used to multiply the number of distinct proteins that can be encoded by a given nucleotide sequence. According to this view, not all of the ambiguities of translation represent errors, and, therefore, there is selective value for some ambiguities. Recent studies show such an interpretation to be realistic.

Chakrabarti and Gorini (4) found that almost all of their streptomycin-resistant derivatives of an HFr strain of *Escherichia coli* had become resistant to the phage MS2. Their analysis suggested that the mutants were "unable to be lysed" by the phage. Similar observations were made with the related RNA phage f2 by Cody and Conway (5), who noted the similarity between this phenomenon and the behavior of the lysis-defective mutant of f2, *op3*, which is unable to produce the so-called lysis protein (6, 7). Cody and Conway (5) could then show that the streptomycin-resistant bacteria actually accumulated infective f2, but could not release them. Accordingly, they suggested that the streptomycin-resistant bacteria are unable to synthesize the lysis protein coded by the RNA phage. They recognized (5) the significance of the fact that the nucleotide sequence coding the lysis protein partially overlaps that for coat protein, but is out of phase by one nucleotide. According to them, one problem for the ribosome would be to initiate the translation of the lysis protein for which the ribosome-binding site might be exposed only after partial translation of the coat protein. Another problem is that the translation of the lysis protein by the same ribosome would require a reading-frame shift of +1 from the frame of the coat-protein coding sequence. Thus, "the synthesis of the lysis peptide may depend on the natural occurrence of mistakes" (5). Furthermore, the less error-prone streptomycin-resistant bacteria might restrict the infection by RNA phage because they do not produce sufficient lysis protein to release the phage progeny.

This interpretation has recently received strong support from the experiments of van Duin and his colleagues (8), who studied the expression of MS2 copy DNA clones of various lengths and structures. Their results show that the translation of the lysis protein depends on the prior initiation and translation of coat-protein sequences. Furthermore, it seems that the transit from the coat-protein sequence to that of the lysis protein involves a reading-frame shift that brings the ribosomes into register with two termination codons as well as the subsequent initiating sequence for the lysis protein. Kastelein *et al.* (8) emphasize the virtues of this mechanism for regulating the expression of a protein in relatively low copy-numbers. In this particular case, the expression of the lysis protein is coupled to that of the coat protein, but the rate of its accumulation is limited by the dependence on the frameshift event.

There are a number of other examples of viruses both in prokaryotic and eukaryotic hosts that seem to depend on translational errors for their further propagation (9–14). Nevertheless, the phenomenon is not limited to viruses, and therefore it may have virtues beyond that of expanding the number of proteins codable by the limited sequences normally packaged in viruses.

Particularly provocative are the arguments that support the thesis that, among the lower eukaryotes, critical events of cell differentiation such as gametogenesis and sporulation require a minimum level of translational ambiguity (15). For example, there are mutants of the fungus *Podospora* that restrict translational errors, and these are found to be sterile, except in the presence of the error-enhancing antibiotic paromomycin. Also suggestive are the observations in a number of lower eukaryotes indicating that the ribosome population experiences more or less extensive changes at particular stages of cellular differentiation. It does not require the powers of a clairvoyant to forecast a growing interest in the relationship between translational fidelity and the expression of the cellular program. Nevertheless, we believe that there are more fundamental aspects of the translation mechanism that tend to reduce its accuracy; these are discussed in Section VI. It is, however, worth emphasizing that the recent work summarized in this section goes far to oppose the mistaken view that all ambiguities of gene expression are destructive, and that optimally functioning translation systems are necessarily minimally error-prone. In effect, we are arguing against the view developed by Orgel (3) in his important attempt to place the accuracy of gene expression in a broader biological setting; we return to this issue in Section VIII.

II. Correlations of Speed and Accuracy in Translation

A suitable way to motivate our concern with the problems of mechanism is to introduce the "ribosomal screen." This device was constructed by Gorini (2) to account for an important dichotomy. Thus, it is possible to distinguish at least two sorts of phenotypic suppression events at nonsense codons: one, in which an incorrectly matched wild-type tRNA translates a nonsense codon; the other, in which a mutant tRNA with a correctly matched anticodon translates the same codon. That mutations affecting ribosomes could influence the frequency of the first type of suppression event seemed not remarkable to Gorini (2). On the other hand, that the same mutant ribosomes could influence the frequency at which mutant tRNAs with matching anticodons could translate "stop" codons seemed to him to require a special explanation because it implies that codon–anticodon interactions alone do not steer the selection.

Two conclusions emerge from Gorini's data. One is that there must be sites on the tRNA besides the anticodon that influence the acceptability of a tRNA by the codon-programmed ribosome. Second, the ribosome must have a way of "antagonizing" correct codon–anticodon interactions when the rest of the tRNA is not canonically matched with the codon. The latter function could be provided by a "ribosomal screen" that distinguishes the different tRNAs, mutant and wild type, from each other (2).

Such a construction was viewed with barely veiled scorn by Ninio (16), who was moved to offer an alternative to these "mysterious properties of ribosomes." His alternative (16, 17) was to treat the ribosome as though it were a timing device. For example, any given tRNA species would be associated with a characteristic "sticking time" in its interaction with a particular codon-programmed ribosome, itself characterized by an idiosyncratic "transition time" for the formation of the peptide bond. Ninio (16) showed convincingly that with these two characteristic times as genetically determined variables, the interactions between particular mutant tRNAs and ribosomes can be rationalized in a consistent way.

From the vantage point of much subsequent work, it seems strange that these two models could be viewed as mutually exclusive alternatives. In our present view, they are wholly compatible and relevant, at least in modified forms. Thus, the one metaphor describes what is recognized by the codon-programmed ribosome, and the other describes how that recognition is developed in time.

It is relatively easy to imagine a physically reasonable mechanism to realize Gorini's ribosomal screen. For example, we can imagine that the aminoacyl-tRNA can occupy different conformational states, some of which are stabilized both by cognate codon–anticodon interactions and by a matching configuration of ribosomal binding sites that interact with sites apart from the anticodon of the tRNA (18). Such a model could satisfy the demands of the data for which the ribosomal screen was created. In a similar vein, Yarus (19) has summarized data for both naturally occurring and artificially constructed tRNA species that are consistent with this sort of complex tRNA recognition pattern. Thus, he shows that nucleotide sequences in the anticodon loop and stem are naturally correlated with the 3' nucleotide of the anticodon, and further, that alterations of these correlated sequences change the performance characteristics of the tRNA. In other words, sequences apart from the anticodon influence the acceptability of the tRNA to a codon-programmed ribosome, just as Gorini deduced.

Ninio's metaphor has fared less well, at least in the extreme interpretations of others. Nevertheless, it will become evident that his insistence on formulating questions about accuracy within a kinetic framework introduced an essential, new element into such discussions. The heuristic value of this contribution cannot be overestimated.

The kinetic model of ribosome function outlined by Ninio (16) would require some sort of destructive interaction between the rate process associated with tRNA selection and that associated with the promotion of peptide bond formation. The physical picture is that the sorting out of correctly and incorrectly matched tRNA species depends on the time allowed for the expression of their respective binding constants at the codon-programmed ribosome site. If this time is shortened by a more rapid forwarding of the tRNA to the next phase in the cycle, the error will increase as in a ribosomal ambiguity mutant (Ram). In contrast, if this selection time is lengthened, the accuracy of the tRNA selection could be increased, as in streptomycin-resistant mutants. In other words, if the rate-limiting step for translation is identified with the characteristic transition time of the ribosome, there should be a clear tendency for the speed and accuracy of translation to be inversely related in the relevant mutants (20).

Initial results testing this kinetic correlation with streptomycin-resistant mutants gave the expected results (20, 21). In contrast, a reappraisal in carefully constructed isogenic strains failed to reveal any difference in rate of β-galactosidase synthesis between some streptomycin-resistant and wild-type ribosomes *in vivo* (3a). The expectation for Ram mutants has also been explored *in vivo* as well as *in vitro* (3a, 23, 24). The data reveal no significant differences in the rates of elongation either *in vivo* or *in vitro* between Ram ribosomes and wild-type ribosomes even though their fidelities of function may differ by a factor of fifty.

The falsification of the extreme interpretations of Ninio's model is unfortunate, not the least reason being that they are kinetic models that are quite accessible to nonkineticists. Furthermore, this type of model is of great interest in the present context because it exemplifies models in which the sacrifice of accuracy yields some other advantage, in this case greater speed. Most generally, it illustrates in a simple way how a detail of mechanism might set the boundary conditions for the optimization of cell growth. We return to this issue in Section VI after introducing another set of ideas.

Finally, it should be abundantly clear that the replacement of a lyrical construction such as the "ribosome screen" with two Greek letters is by itself less than helpful. Thus, to the extent that such kinetic formalities obscure the important problems of mechanism identified by Gorini (2), they are destructive. On the other hand, the emphasis placed by Ninio (16) on the kinetic perspective is fruitful because it leads to consideration of such things as energy, concentration, and catalysis. Inevitably, we are led to connections between the accuracy problem and a central problem of molecular biology, which is how the absolute rates of catalysis are determined by the structures of proteins. In other words, we need to know about the relevant kinetics in order to understand both the physical limitations that define the problems as well as the strategies that provide the solutions to these problems for biological systems.

III. Codon–Anticodon Selectivity on the Ribosome

Concern about the intrinsic limitations on the sequence-specificity of codon–anticodon interactions has motivated much recent effort. A related stimulus is to be found in the tension between Gorini's ribosomal screen and the simplest interpretation of the adapter (25) as well as the messenger hypothesis (26). The question here is whether or not the accuracy of translation can be supported by unassisted codon–anticodon interactions, or whether something more complicated is involved. One would think that this issue could be settled by comparing the figures for the accuracy of translation with those for codon–anticodon interactions. The problem is that we do not have wholly reliable figures to compare.

The number usually quoted as the translational error frequency is 3×10^{-4}. This canonical figure is based on a single measurement of the replacement of isoleucine by valine at one position in ovalbumin (27). The study usually quoted as confirming this figure for *E. coli* is that of Edelmann and Gallant (28), who measured the missense incorporation of cysteine into flagellin, assumed that all of this incorporation represents substitutions for arginines, and obtained data consistent with a cysteine/arginine missense rate that is, in fact, anywhere between 2×10^{-5} and 6×10^{-4}. A figure remarkably close to the magic number comes up again as the "global estimate" of 2×10^{-4} obtained by Ellis and Gallant (29) in an electrophoretic study of three proteins of unknown sequence from *E. coli*.

In contrast, it has long been known that the misreading frequencies of nonsense codons vary by orders of magnitude up to values in the 10^{-2} range (23, 30–32). Not only does the phenotype of the ribosome influence these suppression frequencies, but even in the wild-type background, such error rates depend on which particular codons are involved, and what their neighboring nucleotides are. A parallel variability for the missense error frequencies is seen in studies of electrophoretic microheterogeneity of MS2 coat-protein coded either by the MS2 RNA (33) or by a copy-DNA (J. Parker, personal communication). Indeed, Parker has noted missense error frequencies in the 10^{-3} range, as have Bouadloun *et al.* (34) for the cysteine/arginine substitution in ribosomal protein L7/L12 and the cysteine/tryptophan substitution in ribosomal protein S6.

Two conclusions concerning the error frequencies seem to be in order. One is that *E. coli* turns out now to be more tolerant of translational errors than molecular biologists previously thought. The other is that there is no canonical error frequency. Instead, the error frequencies appear to be widely distributed. Indeed, there is every reason to believe that detailed studies of the error distributions will reveal some new biology.

The upper limits of the error frequencies, based on the nonsense suppression frequencies, could be as high as in the 10^{-2} range. In *E. coli*, the lower limit would seem to be set by the error rate of RNA polymerase. Thus, estimates of the RNA polymerase transcriptional error rate also reveal significant variability, and they range between 10^{-5} and 2.5×10^{-4} per nucleotide (*35, 36*). It would seem from these estimates that the amino-acid missense rate due to transcriptional error in *E. coli* would be in the range 5×10^{-5} to 5×10^{-4}. Indeed, when the codon-specific missense frequencies of a restrictive streptomycin-resistant mutant strain were measured *in vivo*, they were found to be close to 5×10^{-4} (*34*).

Unfortunately, a more profound uncertainty characterizes the available estimates of the maximum sequence-specificities of codon–anticodon interactions. Initially, the studies of trinucleotide interactions with complementary polymers or the appropriate anticodons of tRNA species seemed unambiguous: such triplet interactions are quite weak and the maximum stability difference between correctly matched and nearly correctly matched triplets is approximately 10-fold (*37–39*). Since the missense errors of translation are well below 0.1, something was obviously missing.

Eisinger and his colleagues (*37*) have questioned the relevance of such comparisons to the problem of translational accuracy. They argue that the sequence specificity of trinucleotide interactions should depend on their conformational state. In particular, they note that a trinucleotide free in solution will occupy more conformational states than one fixed in some optimized configuration on the ribosome. Therefore, they expect the formation of a complex with a complementary sequence in free solution to be less favored than the same interaction on the ribosome because the decrease in entropy will be greater in solution than on the ribosome. They further suggest that the anticodon of a tRNA molecule should present its triplet in a restricted and favored state. Therefore, they predicted that tRNA species with complementary anticodons should interact with greater stability than free trinucleotides. This prediction was verified (*37, 40*), and, in addition, it could be shown that in many cases the stability of matched pairs was at least 100 times greater than that of nearly matched pairs.

So far, we have considered only the conformational states of the anticodon and codon as relevant to the tRNA selection on the ribosome. However, the data discussed in connection with the ribosomal screen of Gorini (2) suggest that tRNA interacts with ribosomes at sites distinct from the anticodon. It is, therefore, questionable whether experiments done to determine the maximum sequence specificity of codon–anticodon interactions in the absence of ribosomes are relevant to protein synthesis. Thus, the interaction of codon and anticodon in the absence of the ribosome may yield a population of complexes of varying stabilities. Furthermore, measurements of the sequence specificities of these complexes in solution can yield only an average value for the population. In contrast, during translation, the interaction of non-anticodon sites on the tRNA with conjugate sites on the ribosome might trap the tRNA in a conformation that maximizes the sequence specificity of the codon–anticodon interaction (18). Such a ribosome-dependent selection mechanism could function as a "ribosome screen," and it could equally well serve as the basis for a kinetic model in which tRNA sticking times as well as ribosome transition times control the accuracy of translation.

As mentioned above, the most recent data concerning the functions of non-anticodon positions in tRNA, as summarized by Yarus (19), support this view of a tRNA state selection by the ribosome. Indeed, Yarus has suggested that the critical region of the tRNA involved in the selection is the "extended anticodon," meaning the anticodon loop and stem regions. It is, however, possible that more distant regions are involved.

A fascinating and relevant suppressor of the UGA codon was identified by Sambrook et al. (41) and sequenced by Hirsh (42). The mutant tRNA^{Trp} turns out to have a wild-type anticodon, and an A replacing a G in the D-stem. One interpretation of this remarkable mutant is that its D-stem alteration changes the conformational coupling between the anticodon and other sites on the tRNA. Here, it is suggested that tRNA normally enters a ribosome-stabilized conformation favored by correct matching of all three anticodon nucleotides, but in the mutant this conformation is favored when only two out of three nucleotides are properly matched (18).

This interpretation of the behavior of tRNA^{Trp} differs too much from the extended anticodon hypothesis of Yarus (19). He suggests that in reality the Hirsh (42) sequence determination is wrong in that the suppressor contains a modified nucleotide in the third position of the anticodon instead of a C. The modified nucleotide suggested by Yarus is one recently observed in a minor species of tRNA^{Ile} (44). However, the latter is thought to be specific for interactions with A's, whereas the suppressor tRNA^{Trp} functions with codons of the form UGR (43) as well as with UGU (45). Of course, it is possible to postulate the existence of yet another, so far unidentified, nucleotide in the suppressor, but there does not seem to be any need, nor is there any justification, for such a postulate at this time.

Finally, the data summarized in this section suggest that it is not yet possible to estimate the maximum sequence-specificity of codon–anticodon interactions in the absence of the ribosome. Accordingly, it is not possible to argue from model system experiments for or against the proposition that some sort of sophisticated mechanism is required to amplify the accuracy of codon–anticodon interactions during translation. Nevertheless, as we shall see, this ambiguity has not deterred theoreticians from definitively solving this problem of limited specificity, which may very well not exist.

IV. Kinetic Proofreading: One More Time

The deep prejudice of the mid-1970s was that codon–anticodon interactions by themselves are inadequate and that some sort of amplification is needed to support the still undefined accuracy of translation. This questionable challenge was taken up in an undeniably important way by Hopfield (46) and Ninio (47). Their kinetic model for a physically suitable way to amplify a modest selectivity is extremely important to the present discussion for two reasons. First, their theory provides a way of understanding the present problem, namely, why streptomycin resistance is less adaptive than sensitivity. Second, the demands of testing their kinetic model has had an enormous impact on the development of experimental systems for the analysis of translation.

The basic idea of substrate proofreading (46, 47) is that the selection is repeated one or more times. If this repetition can be done in such a way that the competitive advantage of one substrate over another is retained at each stage, the aggregate advantage can be compounded at each successive step so that the final discrimination of one substrate over another can be much greater than that obtained at any one step. A selection system working like this must be constructed in rather special ways.

One requirement is that at certain intermediate stages of the selection there must be branch points. Each branch point will provide an alternative flow for the substrate: either the substrate can proceed forward toward product formation, or it can be discarded from the system. The point of the proofreading scheme is to obtain substrate flows in which the incorrect substrate is preferentially discarded at the branch points while the correct substrate is preferentially forwarded to product. For example, the simplest proofreading scheme would consist of an initial selection step, in which correct and incorrect substrates compete for occupation of a binding site, from which the substrate is forwarded to a new intermediate state. If the second intermediate provides a proofreading branch, the accuracy of the selection can be improved over that obtained in the initial discrimination. In the limit of maximum discrimination in the initial selection as well as at the proofreading branch, we can expect up to a squaring of the initial accuracy of selection. More extensive amplification can be expected from more complicated schemes.

We expect there to be a cost associated with the enhanced accuracy of the substrate flows, because, in general, we do not expect to get something for nothing. This primitive sense of economic justice is applicable to the proofreading schemes because they are kinetic schemes, and that means that there must be an energy source to drive the flows. Both Hopfield (46) and Ninio (47) identify this driving force with the role of the nucleoside triphosphates in protein synthesis. In the Hopfield scheme, it is suggested that the nucleoside triphosphates are used at what we have called a branch point to form a "high-energy" intermediate, the decay of which drives the proofreading flow. This is an unfortunate, though colorful, metaphor, and in its place we will use a conventional chemical kinetic formulation (48).

We begin by recognizing the elementary fact that in order for there to be a net flow of substrate to product in an enzymatic reaction, the ratio of substrate concentration to product concentration must be greater than that at equilibrium for the reactants. Furthermore, if we have a succession of intermediates on the path to product, each must be displaced from equilibrium more than the next one on the path toward product in order for a net flow to exist. Now, if one of these intermediates is at a branch point in a proofreading scheme, the substrate will be displaced farther from equilibrium than the intermediate; therefore, in the absence of another driving force, substrate would tend to flow onto the enzyme at that branch point. Therefore, there is a requirement at the proofreading branch point for a second displacement from equilibrium, besides that for substrate and product, to drive the discard pathway so that there is a preferential net flow for noncognate substrate off the enzyme. It is in connection with this second driving force (or displacement from equilibrium) that the nucleoside triphosphates play their role in proofreading.

The aminoacyl-tRNA enters the selection scheme on the ribosome in a ternary complex with elongation factor (EF) Tu and GTP. During the selection, but prior to peptide bond formation, the GTP is hydrolyzed and the Tu along with the GDP is released from the ribosome (49). Since the GDP is not permitted to accumulate in the bacterium, but is used rapidly to regenerate GTP, the GTP : GDP ratios are displaced far from equilibrium (48). Therefore, by coupling the selection cycle for the tRNA to the hydrolysis of GTP, a second driving force, which is the displacement from equilibrium of the GTP : GDP ratio, is established, and this can in principle drive the discard reaction at a proofreading branch point.

For the sake of completeness, we should point out that according to this view of a proofreading scheme, the energy that drives the discard flow is the irreversible free-energy loss associated with the displacement from equilibrium of the guanine nucleotides. This energy term is not to be confused with the reversible standard-state free-energy change associated with the hydrolysis of GTP, which can be thought of as the hydrolysis of a "high-energy bond." [A fuller discussion of the thermodynamics of proofreading can be found in Blomberg et al. (50).]

If there is proofreading during the selection of tRNAs on the messenger-programmed ribosome, there should be some fraction of the ternary complexes that is dissipated on the ribosome without a concomitant formation of peptide bonds. The fraction of dissipated ternary complex should, of course, be higher for noncognate than for cognate tRNA species if the effect of the proofreading is to increase the accuracy. Accordingly, the lower ratio of peptide bonds formed per dissipated ternary complex for correct aminoacyl-tRNAs compared to that for incorrectly matched aminoacyl-tRNAs provides the experimental signal by which a proofreading function on the ribosome reveals itself.

The first claim to have identified such a signal on ribosomes was made by Thompson and Stone (51), and this was followed by a series of similar studies (52–55). All of these studies are based on an estimation of GTP hydrolysis mediated by EF-Tu in the presence of either cognate or noncognate aminoacyl-tRNA, but in the absence of EF-G. The relevant observation made in all of these studies is that the number of hydrolyzed GTPs normalized to the number of aminoacyl-tRNAs accepted by codon-programmed ribosomes is up to 25 times greater for noncognate than for cognate tRNA species. In other words, these data suggest that there might be a proofreading branch point coupled to GTP hydrolysis on ribosomes. Nevertheless, it must be said that the most significant aspect of these experiments is the tension created between their suggestiveness and their ambiguity.

Thus, the critical detail of a proofreading branch point is summarized in the ratio between the rates of discard and the rates of product formation for both cognate and noncognate species. The absence of the EF-G from the Thompson protocol means that there is no real product formation; i.e., the formation of the peptide bond is a kinetic dead end. Therefore, it is impossible to compare the critical ratios for cognate and noncognate flows in this sort of experiment; indeed, it is not possible even to obtain the error rates by this protocol. The problem can be stated in still another way: The sort of single factor experiment performed by Thompson *et al.* would certainly yield valuable data if there were a prior, independent demonstration that there is proofreading on ribosomes in a complete system. There are other aspects of the interpretations of these experiments to which we return in Section V.

An alternative approach to the experimental search for proofreading has been taken in this laboratory. The rationale of this approach rests on the realization that the accuracy of aminoacyl-tRNA selection on ribosomes is a kinetic phenomenon. Therefore, experiments designed to explore the mechanism of this selection must be carried out in systems with appropriate kinetic characteristics. The relevant characteristics are those found in bacteria and cells. For example, *E. coli* elongates polypeptides at a rate of 15–20 amino acids per second per ribosome, and with missense frequencies most likely in the range 10^{-4} to 10^{-3} (see Section III). In contrast, experimentalists have been content to elongate polypeptides *in vitro* at rates apparently less than 0.05 amino acid per second per ribosome and with missense frequencies anywhere between a few to 50% (56). It certainly seems a questionable undertaking to try to discover how the errors of selection can be kept in the 10^{-4} to 10^{-3} range *in vivo* with the aid of an *in vitro* system that has an error rate close to 10^{-1}.

Systematic biochemical studies of the ionic dependencies and the effects of varying the displacements from equilibrium of the substrates yielded a poly(U) translation system with leucine missense frequencies close to 4×10^{-4} (57–59). Under conditions that optimize the accuracy, a preliminary incubation of the poly(U)-programmed ribosomes with N-acetylphenylalanyl-tRNA (60) eliminates the rate-limiting step of polyphenylalanine initiation, and this reveals elongation rates of 8–12 peptide bonds per second per ribosome (61). Such an optimization is not specific for poly(U)-dependent polyphenylalanine synthesis. Under precisely the same *in vitro* conditions poly-(U-G)-programmed ribosomes elongate poly(Cys-Val) at a rate of 8–12 peptide bonds per second per ribosome with missense substitution of Cys by Trp at a frequency close to 10^{-4} and of Val by Met at a frequency close to 10^{-3} (62).

One additional biochemical study was required as preparation for a test of the proofreading functions; this concerned the response of the Tu cycle to Ts. Knowledge of the kinetics of the Tu·Ts interaction were clearly important, since the ternary complex containing Tu, GTP, and aminoacyl-tRNA is a likely candidate for the substrate species that initiates proofreading on the ribosome. Indeed, it has been clearly established in model system studies that Ts catalyzes the release of GDP from the binary complex with Tu (63, 64). What was less clear was the effect of Ts on the kinetics of polypeptide synthesis.

Ruusala et al. (65) studied the effects of Ts on poly(U) translation in the optimized system when Tu limits the rate of poly(Phe) elongation. The data show that the cycle time of Tu in polypeptide synthesis is stimulated by more than a factor of 1000 by Ts, and the rate-limiting step of GDP release from Tu is shortened from 90 seconds in the absence of Ts to 30 msec in its presence. Thus, it is easy to limit a polypeptide synthesizing system to the rate at which GDP dissociates from Tu simply by omitting Ts from the system. Under such steady-state conditions, nearly all of the Tu is found as binary complex with GDP and the maximum number of Tu cycles per unit time is fixed in such a way that it is possible to determine the stoichiometric efficiency with which Tu·GTP supports peptide bond formation with different aminoacyl-tRNA species.

If there is proofreading of noncognate tRNA species on the ribosome, the number of Tu·GTP cycles will be greater for noncognate amino-acid incorporation into peptide than for cognate amino acid. In particular, this difference in efficiency should be evident when the system is limited by the availability of Tu·GTP, i.e., in the absence of Ts. Hence, by varying the steady-state levels of cognate and noncognate aminoacyl-tRNA species in the system minus Ts, the variation of the efficiency of peptide bonds formed per Tu cycle should be revealed as the error ratio of the polypeptide product is manipulated (66). Indeed, it was observed that with bulk *E. coli* tRNA the incorporation by poly(U)-programmed ribosomes of a leucine requires about 50 times as many Tu cycles than are required for the incorporation of a phenylalanine during steady-state polypeptide synthesis. In other words, there seems to be a 50-fold proofreading of the Leu-tRNAs in this system (66).

Unfortunately, this result is not unambiguous. The problem is that only a fraction, usually between 10% and 25%, of the ribosomes used in these experiments is active in polypeptide elongation. Therefore, there is the possibility that nonelongating ribosomes are dissipating the ternary complexes and that elongating ribosomes do not really proofread ternary complexes (66). The question then is how to identify or eliminate the nonelongating "killer" ribosome?

This question has not been answered definitively. Instead, there are a large number of circumstantial observations that persuade us that the "killer" ribosome is only a remote possibility. For example, the proofreading factors for three different isoacceptor tRNA^{Leu} species from *E. coli* are significantly different during the translation of poly(U) (66). Similarly, the proofreading factors seem to be under strict genetic control: three different Ram ribosomes translate poly(U) with three characteristically lower proofreading factors (24), while some streptomycin-resistant ribosomes translate with characteristically higher proofreading factors compared to wild-type ribosomes (67). The effects of particular antibiotics further narrow the possibilities: Erythromycin inhibits polypeptide synthesis without affecting the proofreading factor (P. C. Jelenc, personal communication) kanamycin enhances the error of the initial selection step as well as the proofreading step, and streptomycin selectively disrupts the proofreading part of the tRNA selection (68a). All of these observations suggest that the excess dissipation of ternary complexes observed during polypeptide synthesis *in vitro* is indeed an expression of a proofreading function.

V. The Problem Solved by Kinetic Proofreading: Rate or Accuracy?

Although we must bear in mind the reservations we still have about the existence of proofreading on ribosomes, there are two aspects of the Hopfield–Ninio model that are genuinely compelling.

One is that participation in a proofreading function is the only straightforward suggestion that has been made for the function of GTP during polypeptide elongation. Thus, the formation of a peptide bond from aminoacyl-tRNA is thermodynamically a highly favored reaction, as far as we can tell now (48). In addition, we have not come across any rigorous arguments that would necessitate a GTP function in overcoming kinetic barriers (see below). In contrast, the requirement for a second substrate displacement to drive discard flows at a branch point is an unavoidable physical demand on a proofreading system that could be met by the guanine nucleotides.

Another attraction of proofreading is that it can be used to explain why the accuracy of translation is limited in wild-type bacteria. In principle, proofreading systems are capable of supporting substrate selections at arbitrarily high accuracy levels, but they do so at very high dissipative costs. Clearly, if there is an increased energetic cost associated with increased translational accuracy, one effective strategy in the design of the ribosome would be to optimize the trade-off between the accuracy and the cost of translation. We shall consider the energetic costs of proofreading in more detail below, but before that we must discuss what problems proofreading may be solving.

As mentioned above, we seriously doubt that data from model-system experiments in the absence of ribosomes can justify the conclusion that codon–anticodon interactions are inadequate to support the accuracy of tRNA selection outside of a proofreading system. On the contrary, we have favored a model in which the ribosome traps tRNAs in states that maximize the sequence-specificity of their interactions with codons (18). It was, therefore, of considerable interest when Thompson and Karim (55) offered a reevaluation of the functions of a ribosomal proofreading mechanism that accords with our earlier conjectures.

In effect, Thompson and Karim (55) have returned to the view held by Galas and Branscomb (20) that the accuracy of the translation process is limited by its speed. Their conclusion is based on two measurements (55). The first consists of a comparison of the stability of the complex formed between poly(U)-programmed ribosomes and ternary complexes made up with a virtually noncleavable GTP analog as well as with either Phe-tRNA^{Phe} or with Leu-tRNA^{Leu}. This comparison suggested that an error of less than 2.5×10^{-5} could be supported by the difference in binding stabilities between those two species; i.e., proofreading is not needed to amplify a limited sequence specificity of the codon–anticodon interactions on the poly(U)-programmed ribosome.

Second, such accuracy cannot be expressed in a single binding step because measurements of what is thought to be the next step, namely, Tu-dependent GTP hydrolysis, show it to be so fast, that the full expression of the differences in dissociation rates of different tRNA species is not possible. Therefore, an intermediate branch point, driven by the GTP, is required fully to express the sequence specificity of the codon–anticodon interactions. The alternative, that of slowing down the GTP hydrolysis step, while helping with the accuracy, would limit the rate of elongation. It is in this sense that there is a rate limitation of accuracy, according to Thompson and Karim (55).

This view may be correct, but neither the experimental evidence nor the theoretical grounds for this interpretation bear close scrutiny. In the first place, the difference in stability of cognate and noncognate interactions on the ribosome is measured by studying the decay of preformed complexes with the different acceptor species. No evidence is presented to support the assumption that the two different tRNAS are associated only with mutually accessible sites on the ribosome. Such evidence might have been obtained by doing competition experiments between the two tRNA species. Without such competition experiments, the stability comparisons reported (55) must be viewed as quite tentative. Likewise, the destructive kinetic coupling between the GTPase reaction and the dissociation of noncognate aminoacyl-tRNA could very well be an artifact of the sort seen before. For example, Pettersson and Kurland (58) have observed precisely such destructive kinetic effects when poly(U) translation is carried out in Tris-Mg^{2+} buffer of the sort favored by Thompson and Karim (55); however, these effects disappear in what we have referred to as optimized buffers.

The peculiar thing about Thompson and Karim's kinetic arguments, which in fact are based on two different notions, is that they do not seem to be terribly relevant to the conditions under which bacteria synthesize proteins. First, their calculations concern rather small rate effects when the ribosomes are functioning far from saturation by the ternary complexes. In contrast, all the available data suggest that the ribosomes are close to saturation by ternary complexes in growing bacteria (69). Therefore, the kinetic effects that preoccupy Thompson and Karim are, at most, relevant to a minute class of extremely minor isoacceptor tRNA species. Second, the notion that the forward rate constant of GTP hydrolysis on Tu has to be slowed down for the codon–anticodon selectivity to be more fully expressed appears doubtful, since a better alternative exists. There is no theoretical argument against an increase in the dissociation rates of ternary complexes from the ribosome instead of a decreased forward rate constant. Thus, the same accuracy could be obtained at an unchanged rate of GTP hydrolysis and therefore at an unchanged k_{cat} for protein synthesis.

Finally, it should be noted that if, as Thompson and Karim suggest, there is sufficient sequence specificity in the codon–anticodon interaction on ribosomes, a multistep selection process in which the accuracy of the selection is compounded at successive steps, could be driven in principle without the aid of a proofreading mechanism (18). In other words, if the specificity is there, why bother with the extra cost of hydrolyzing a GTP during the selection? For that matter, why incur the extra cost of investing so much of the bacteria's protein in the form of Tu.

We close this section by considering yet another discrepancy. The steady-state assays for proofreading described by Ruusala et al. (66) indicate proofreading factors greater than 100 for tRNA^{Leu2} competing with tRNA^{Phe} in a poly(U)-programmed system. In contrast, the largest proofreading factor seen for tRNA^{Leu2} in the single factor experiments is about 25 (54). It is conceivable that this lower number obtained in the absence of EF-G is not an artifact. Thus, it is possible that part of the accuracy of aminoacyl-tRNA selection is associated with EF-G function; this would account for the higher proofreading factor observed by Ruusala et al. (66). Indeed, there is a potentially decisive advantage to such an arrangement. If the maximum accuracy obtainable in a single step is small compared to the aggregate accuracy to be obtained ultimately in a proofreading selection, it can be shown (see Section VII) that increasing the number of branch points beyond the absolute minimum number will lower the dissipative losses associated with the selection (50).

VI. Translational Accuracy and Exponential Growth

The discussion of substrate proofreading models does not exhaust by any means the possibilities of an editing function for the ribosome. There is another class of editing models, introduced initially for DNA by Kornberg (70), which we call product editing. The particular forms of product editing that seem relevant to translational accuracy involve the rejection of the growing polypeptide. Therefore, they may appear at first glance to be rather extravagant ways of dealing with missense errors. Nevertheless, we shall argue that the cost of such an editing function under certain conditions is negligible. In order to make this argument, we must first introduce error-coupling strategies and then present the outline of a theory for the evaluation of editing costs.

A number of scattered observations in the earlier literature are consistent with the interpretation that a missense event on the ribosome raises the probability that the reading frame of the ribosome will be disrupted (71). This correlation can be explained with the intuitive argument that a mismatched tRNA will create a distorted codon–anticodon interaction. According to this view the geometry of the codon–anticodon interaction determines the accuracy of messenger RNA movement; therefore, the distortion of this geometry would be expected to perturb the reading frame of the ribosome. A recent direct test of this hypothesis has shown that frameshift errors can be increased when the frequency of missense errors is increased in the bacteria by physiological manipulations (72).

This error-coupling mechanism is relevant to the present discussion because there is a very substantial probability that, following a reading-frame error, the ribosome will encounter an out-of-phase termination codon. The resulting release of the polypeptide chain will then expose the aborted error-containing polypeptide to a battery of scavenger enzymes capable of degrading the proteins to the amino acids (73, 74). In other words, error coupling can promote editing of missense errors. Likewise, the error that we can refer to as polypeptidyl-tRNA drop-off may have a significantly higher probability of occurrence when missense events occur (75). If this is so, then the drop-off event will also function as an editing step. Indeed, measurements of incomplete β-galactosidase fragments accumulated by bacteria suggest that at least 30% of all the β-galactosidase chains that are initiated are eventually aborted one way or another, and that the frequency of abortive events is close to one per three thousand translated codons (76).

This frequency is suggestive because it is nested well within the range of missense error frequencies estimated above. We are therefore interested in comparing the advantages of these sorts of product-editing functions with those of substrate proofreading. Such a comparison is impossible without a suitable measure.

In our earlier studies of this problem, we focused on the free-energy change associated with the accuracy of biosynthesis. In addition to providing an unambiguous measure of the thermodynamic cost of a reaction, the free-energy change sets upper limits on the accuracy of the substrate selections (50, 77). On the other hand, it is by no means obvious that this thermodynamic limit is ever approached by biological systems. Therefore, we have searched for kinetic limitations on the accuracy of biosynthesis within the context of one particular set of boundary conditions, namely, those associated with the optimization of bacterial growth (78).

The starting point for this analysis is the perspective developed by Maaløe (79), who stresses the growth-media-dependent relationships between the various molecular compartments of exponentially growing bacteria. According to this perspective, the bacterial growth rates in different media are determined by the number of different enzymatic reactions required to convert medium into nucleic acid and protein; the larger the number of reactions, the slower the growth rate.

Implicit in this view is the usually tacit assumption that there are physical limits on the maximum catalytic rates attainable by enzymes. For example, the specific rate of consumption of O_2 by bacteria growing at different rates in different media is virtually invariant. This striking observation is consistent with the supposition that the rate of aerobic metabolism is limited, and, hence, that the generation rate of nucleoside triphosphate per unit mass is fixed. Hence, the ATP and GTP requirements for protein synthesis will be competitive with those for other reactions, such as nucleic acid biosynthesis. Accordingly, the cost of translational proofreading must be expressed in this competition with the other reactions supporting bacterial growth. Here, the extra cost of producing accurate gene products by proofreading must be weighed against the loss of nucleoside triphosphates to other reactions, a loss that in general will depress the growth rate (78). Therefore the optimum proofreading rate is defined as that supporting the maximum rate of bacterial growth. It is worth stressing here that for proofreading systems optimal accuracy is not maximum accuracy (see Section VII).

We have used the internal competition for nucleoside triphosphates to illustrate a notion that is completely general: all the components of the bacterium must be present in amounts that are optimal in the sense that they support the maximum growth rate. Furthermore, the size of the macromolecules should be optimized, since increased size corresponds to an increased cost in terms of the limited availability of resources, such as substrates, time on ribosomes. From this perspective the concentrations and kinetic characteristics of bacterial components should be limited by the physical extremes at which the different proteins can function. Indeed, that ribosomes are numerous, large, and slow suggests that they are carrying out rather complex physical processes just about as efficiently as it is possible to do.

Clearly then, a genetic lesion that strikes at the proofreading functions of the ribosome does not simply raise the error frequency of translation, it also lowers the metabolic cost of translation. This may explain why the Ram-type mutants are so easily obtained. The converse is that the more accurate streptomycin-resistant ribosome, is a luxury that wild-type bacteria cannot afford in the absence of antibiotic, because the metabolic cost associated with the extra proofreading is prohibitive.

The optimal macromolecular composition for the bacteria growing on different media will vary with the quality of the media (79). From this it follows that the metabolic cost of accurate gene expression will depend on the growth conditions.

To illustrate this conclusion, we can compare a bacterium growing in a medium containing the 20 amino acids with one growing in a minimal medium. The latter will be obliged to synthesize all its amino acids, from which it follows that the amino acids themselves are a major cost of polypeptide synthesis for this bacterium (79). Further, the opportunity to reutilize the amino acids after proteolysis reduces the cost of product editing to negligible levels. For this reason, the value of product editing of defective polypeptides will be much greater for this bacterium than for one growing on the rich medium. We therefore expect the proteolytic scavenger systems to be more active in bacteria growing on poorer media compared to those supported by rich media (78).

In contrast, in a rich medium the primary performance characteristics of the translation system are dominant. In particular, it will be necessary to maximize the rates of function for ribosomes, tRNA, and factors. One appropriate strategy involves the preferential use of a subset of degenerate codons to specify the sequences of those proteins that are produced at the highest concentrations in media supporting the fastest growth. This strategy will raise the effective concentration of individual codon-programmed ribosomes. Consequently, it is possible to approach the maximum rate of ribosome function at a lower total ternary-complex concentration with such a codon preference strategy.

According to this view, the codon preferences of *E. coli* are advantageous because they support a minimization of the total tRNA and Tu concentration, with the consequent optimization of growth rate. This interpretation is consistent with the otherwise surprising finding that the guanine content of mRNAs seems to have little or no effect on the performance characteristics of the ribosomes during polypeptide elongation (62). In other words, the data suggest that it is not the physical characteristics of the codon–anticodon interaction that are involved in the codon-preference strategies. Accordingly, the tendency of the highly preferred codons to have intermediate G + C compositions (80) is in our view a reflection of the composition of *E. coli* DNA rather than vice versa.

Discussions of translational accuracy are usually motivated by concern about the physical constraints on substrate selections. In contrast, we suggest here that there are many ways around these constraints and that a realistic assessment of the cost of these different strategies will depend on the particular growth conditions supporting the bacterium. In particular, we suggest that for exponentially growing bacteria the relevant cost of a given strategy can be assessed only through its impact on the growth rate. We present a more rigorous and detailed analysis of the these kinetic strategies elsewhere (78). We now conclude our discussion with an analysis of the kinetic factors that determine the magnitude of the substrate-proofreading flows.

VII. Kinetic Options for Inexpensive Proofreading and Error Regulation

Following Hopfield (46) and Ninio (47), we can describe the flows on the ribosome in a simple one-step proofreading scheme as Eq. (1).

$$C_1 \underset{k_{-a}}{\overset{k_a}{\rightleftharpoons}} (T_3 + R_A) \xrightarrow{k_1} C_2 \underset{k_{-b}}{\overset{k_b}{\rightleftharpoons}} (T_2 + AT + R_A) \underset{k_{-2}}{\overset{k_2}{\rightleftharpoons}} C_3 + T_2 \xrightarrow{k_3} \quad (1)$$

Here, T_3 is the ternary complex consisting of EF-Tu, GTP, and aminoacyl-tRNA (AT); R_A is the codon-programmed ribosome; T_2 is the EF-Tu binary complex with GDP; the ribosomal intermediate states are designated as C_i; and the rate constants describing the various transitions are designated as the corresponding k_i's. When this system is driven by very large displacements of the guanine nucleotide pool from equilibrium, the concentration of T_2 can be kept negligibly small so that the scheme can be effectively described as in Eq. (2).

$$C_1 \underset{k_{-a}}{\overset{k_a}{\rightleftharpoons}} (T_3 + R_A) \xrightarrow{k_1} C_2 \overset{k_{-b}}{\leftarrow} (T_2 + AT + R_A) \xrightarrow{k_2} C_3 + T_2 \xrightarrow{k_3} \quad (2)$$

For simplicity we assume that all the discrimination between cognate (c) and noncognate (w = wrong) ternary complexes depends on the corresponding dissociation constants k^c_{-a} and k^w_{-a}, as well as k^c_{-b} and k^w_{-b}, respectively; this assumption can be suspended without affecting in any way the conclusions that follow.

We define the intrinsic selectivity "d" as the ratio of the rate constants for the dissociation of noncognate and cognate species from the ribosome complexes ($k^w_{-a}/k^c_{-a} = d = k^w_{-b}/k^c_{-b}$). Hence, d sets the upper limit for the selection specificity at each step and d^2 is the maximum substrate specificity of the system. The question then is: How close to these limits can the system operate?

OPTIMIZATION OF TRANSLATIONAL ACCURACY 213

Inspection of Eq. (2) reveals that the critical parameter determining the extent to which this limit is approached is the ratio of the above dissociation constants to the corresponding forward rate constants (k_{-a}/k_1 and k_{-b}/k_2); we refer to these ratios as the a_i's. Clearly, the a_i's must approach infinity if the limiting discrimination levels are to be reached. On the other hand, the magnitude of a_2 also provides a measure of the dissipation of ternary complex associated with proofreading. Clearly, if a_2 goes to infinity, the system will be converted into a very efficient GTPase as the rate of polypeptide synthesis vanishes. Therefore, we can be certain that the value of a_2 will be contained at some reasonable level, and that the accuracy of the flows will be nowhere near the maximum levels. In order to get some insight into how the value of a_2 influences the system in detail, we will calculate some simple functions for Eq. (2).

The accuracy of the flows in Eq. (2) can be described as the ratio of the cognate flow to the noncognate flow when the corresponding ternary complexes are at equal concentrations. This ratio (A) can be described as in Eq. (3).

$$A = I \cdot F = \frac{1 + da_1}{1 + a_1} \cdot \frac{1 + da_2}{1 + a_2} \qquad (3)$$

Here, I is the initial discrimination and F is the proofreading factor. The number of ternary complexes dissipated per cognate peptide is ($1 + a_2$) and the corresponding figure for noncognate events is ($1 + da_2$). For reasonably efficient schemes, the value of a_2 should be less than 1, and when the system operates with values of d much greater than 1, the average number of ternary complexes dissipated per peptide bond will be dominated by the cognate term ($1 + a_2$). For example, if $d = 1000$, and $a_1 = a_2 = 0.1$, Eq. (2) will function at $A = 8500$ and with close to 1.1 ternary complexes dissipated per peptide bond. If more accuracy is demanded of the system, and d is not improvable, two options are available. One option is to increase a, which would raise the accuracy as well as substantially increase the dissipative losses. For example, if d remains at 1000, but a_1 and a_2 are raised to 0.5, the accuracy of the flows is now 110,000, but the stoichiometric ratio of ternary complexes dissipated per peptide bond is close to 1.5.

The second option is much cheaper in terms of dissipative losses (50, 77). This consists of simply adding a second proofreading branch. For example, with a_1, a_2, and a_3 reduced to 0.05 and with d at 1000, the accuracy can be raised to 120,000 at a stoichiometric ratio of only 1.1 ternary complexes per peptide bond.

At least one factor that should influence the relative merits of these two options is their relative cost in terms of their effects on growth rates. From this point of view the introduction of a second discard branch should be the best solution, particularly since inspection of Eq. (2) indicates how simple such an innovation could be.

We have described in Eq. (2) a discard branch in which the aminoacyl-tRNA (AT) is released from the intermediate C_2 together with Tu·GDP (T_2), as an alternative to the simple release of T_2 and the formation of the intermediate C_3 with a "naked" aminoacyl-tRNA still bound. If the forward processing of C_3 to form a peptide bond is slow enough, and if the free concentration of AT is kept sufficiently low, the second proofreading branch could function as in Eq. (4)

$$\begin{array}{cccccc} T_3+R_A & & T_2+AT+R_A & & AT+R_A & \\ k_a \Big\Uparrow k_{-a} & & \Big\Uparrow k_{-b} & & \Big\Uparrow k_{-c} & \\ C_1 & \xrightarrow{k_1} & C_2 & \xrightarrow{k_2} & C_3+T_2 & \xrightarrow{k_3} \end{array} \quad (4)$$

Indeed, we are currently trying to determine experimentally how many branches are used in the proofreading flows on the ribosome.

The experimental assessment of the stoichiometric ratio of ternary complexes dissipated per peptide bond is easy in principle (66). In practice, there is a serious snag: We do not have an unambiguous way to measure the absolute number of active Tu molecules in our assays. Therefore, our best estimate now places the stoichiometric ratio roughly between 1.1 and 1.3 Tu·GTP cycles per peptide bond with wild-type ribosomes translating poly(U) (T. Ruusala, unpublished data).

Another ambiguity in our experimental analysis comes from the uncertain relationship between the *in vitro* conditions we use and those maintained by *E. coli in vivo*. In particular, we have assumed up to now that the displacements from equilibrium are virtually infinite. However, since approximately one-tenth of the guanine nucleotide pool is GDP *in vivo,* and since the rate of Tu·GDP recycling is close to the limit set by the availability of Ts (65), it is possible that significant amounts of Tu·GDP are present in the steady state *in vivo*. In other words, there could be an influence of Tu·GDP on the performance characteristics of the system.

One way to search for such an effect of Tu·GDP is implicit in our discussion of the optimization of the accuracy. Thus, we have considered the advantages of reducing the dissipative cost of proofreading at the expense of a less than maximal expression of the discrimination capacity of the system. The point here is that, under certain kinetic conditions, raising the Tu·GDP concentrations should lead to an increased proofreading flow, which results in a higher accuracy of the selection.

This point is illustrated in Eq. (5).

$$C_1 \underset{}{\overset{k_1}{\longrightarrow}} C_2 \underset{k_{-2}}{\overset{k_2}{\rightleftarrows}} C_3 + T_2 \overset{k_3}{\longrightarrow} \quad (5)$$

with $T_3 + R_A$, $k_a \Updownarrow k_{-a}$ on C_1, and $T_2 + AT + R_A$, k_{-b} on C_2.

Here, we have assumed that k_b is too small to influence the flows in the concentration range of T_2 that concerns us here. In contrast, we assume that the back reaction governed by k_{-2} does take place in the relevant concentration range. It can be shown that when the concentration of T_2 is raised, Eq. (5) will function at a greater accuracy. The reason for this is simply that as the (T_2) term increases, the net flow of C_2 to C_3 decreases, with the consequence that there is more time for the discard reaction governed by k_{-b} to be expressed. In other words, in such a scheme, an increase in the concentration of Tu·GDP results in an effect equivalent to an increase of the ratio a_2.

Accordingly, we have looked for such an effect in our *in vitro* system, and, indeed, we have discovered a systematic variation of the proofreading factors that may reflect the putative influence of Tu·GDP (M. Ehrenberg, T. Ruusala, K. Bohman, D. I. Andersson, and C. G. Kurland, unpublished). This observation has two consequences: first, a realistic assessment of the proofreading flows for the different ribosome phenotypes will require more detailed information about the functioning of the EF-Tu cycle *in vivo*, something that is not going to be easy to get; second, there is now clearly the possibility that the proofreading functions can be regulated physiologically.

We have discussed in Section (VI) the advantages of regulating the different editing functions associated with translation. Although it is clear that such regulation could be effected by Tu·GDP, we have been curious about the possible regulatory function *in vivo* of the guanine nucleotide analog ppGpp, the "magic spot" (*80a*). Indeed, the *in vitro* data suggest that the Tu·ppGpp complex can increase the proofreading flows over the ribosome, and, thereby, increase the accuracy of the translation of poly(U) (A. M. Rojas, H. Pahverk, M. Ehrenberg, and

C. G. Kurland, unpublished). It therefore seems conceivable that this effect will explain the functions of Tu in the stringent control of translational accuracy (81).

Finally, the detailed effects of streptomycin on a particular resistant mutant strongly suggest that the sort of kinetic analysis we have discussed here is relevant to the physiological function of the bacteria. Zengel et al. (21) discovered a streptomycin-resistant mutant that behaved as though it were partially streptomycin-dependent: Its growth rate could be stimulated approximately 2-fold by the antibiotic. More recently, we have studied the effects of this mutational alteration on the ribosome performance in our kinetic assays. The data suggest that these mutant ribosomes in the absence of antibiotic have a stoichiometric ratio of Tu cycles per peptide bond between 1.5 and 2.0. When streptomycin is added, the stoichiometric ratio decreases to wild-type levels, i.e., close to 1.2 (81a). By reducing the proofreading flows, the antibiotic decreases the accuracy of the mutant ribosomes' function by a factor of 40. So, here we have a mutant, the growth of which is stimulated by an antibiotic that seems to increase the error frequency but lower the dissipative losses of translation. In other words, the mutant illustrates in a clear fashion the potential significance of the optimization principle we have presented here, an essential feature of which is a balance between accuracy and dissipative losses.

VII. Error Feedback Increases Accuracy: End of Error Catastrophe

The conjecture that, for example, a ribosome containing an error might tend to be more error-prone in its functions than a correctly constructed ribosome has provided the basis of much speculation concerning the stability of inherited structured (3, 82–88). Thus, it is conceivable that erroneously constructed ribosomes that are particularly error-prone might create a lethal error-cascade. It is also undeniable that biological systems must avoid such instabilities. What is questionable in this scenario is the supposition that, when an error in construction has any effect on accuracy of function, it must be a negative effect. Such a supposition is valid if, and only if, we assume that the accuracy of ribosome or polymerase function is normally at its maximum.

In fact, we know that the accuracy of ribosome function is not normally at its maximum, because there are mutants with more accurate ribosomes than those of wild type. Therefore, we suggest here that accuracy of ribosome function is optimized, so that an adequate degree of accuracy is obtained at an acceptable dissipative cost. Another aspect of this optimization is its long-term stability.

Clearly, any selection system such as the one depicted in Eq. (2) operating with a_2's at values less than 1 can operate more accurately than is normal, if an error of construction increases the value of one of the a_i's. In other words, it seems feasible for ribosomes to evolve, so that the kinetic effects of different errors of construction oppose one another (*3a, 89, 89a*). In effect, the population of ribosomes in a bacterium may be distributed with respect to individual accuracies, but the average accuracies may be damped with respect to error cascades, because a random error of construction that raises the accuracy is as likely or even more likely than one that decreases the accuracy. (For a more thorough discussion, see *78*).

It is worth stressing our earlier conclusion, that there is no compelling evidence for physical limits on the intrinsic discrimination of tRNA attainable in a single step by codon-programmed ribosomes (*18*). Furthermore, even if there were such a limitation, proofreading systems can provide virtually unlimited accuracy to such selections (*50*). Our point is that it is not the accuracy per se that presents the problem for biological systems, but rather it is such considerations as the dissipative costs, construction costs, and long-term stabilities of the systems that have shaped their evolution.

REFERENCES

1. F. H. C. Crick, *This Series* **1**, 163 (1963).
2. L. Gorini, *Nature NB* **234**, 261 (1971).
3. L. E. Orgel, *PNAS* **49**, 517 (1963).
3a. W. Piepersberg, V. Noseda, and A. Böck, *MGG* **171**, 23 (1979)
4. S. Chakrabarti and L. Gorini, *J. Bact.* **121**, 670 (1975).
5. J. D. M. Cody and T. W. Conway *J. Virol.* **37**, 813 (1981).
6. K. Horiuchi, in "RNA Phages" (W. D. Zinder ed.), p. 29. CSHLab, CSH, New York, 1975.
7. P. Model, R. E. Webster, and N. D. Zinder, *Cell* **18**, 235 (1979).
8. R. A. Kastelein, E. Remaut, W. Fiers, and J. van Duin, *Nature (London)* **295**, 35 (1982).
9. M. Bienz and E. Kubli, *Nature (London)* **294**, 188 (1981).
10. H. Engelberg-Kulka, L. Dekel, M. Israeli-Reches and M. Belfort, *MGG* **170**, 155 (1979).
11. H. R. B. Pelham, *Nature (London)* **272**, 469 (1978).
12. L. Philipson, P. Andersson, U. Olshevsky, R. Weinberg, D. Baltimore, and A. Gesteland *Cell* **13**, 189 (1978).
13. A. M. Weiner and K. Weber, *Nature NB* **234**, 206 (1971).
14. J. L. Yates, W. R. Gette, M. E. Furth, and M. Nomura, *PNAS* **74**, 689 (1977).
15. M. Picard-Benoun, *FEBS Lett.* **149**, 167 (1982).

16. J. Ninio, *JMB* **84**, 297 (1974).
17. J. Ninio, This Series **13**, 301 (1973).
18. C. G. Kurland, R. Rigler, M. Ehrenberg, and C. Blomberg, *PNAS* **72**, 4248 (1975).
19. M. Yarus, *Science* **218**, 646 (1982).
20. D. J. Galas and E. W. Branscomb, *Nature (London)* **262**, 617 (1976).
21. J. M. Zengel, R. Young, P. P. Dennis, and M. Nomura, *J. Bact.* **129**, 1320 (1977).
23. D. I. Andersson, K. Bohman, L. A. Isaksson and C. G. Kurland, *MGG* **187**, 467 (1982).
24. D. I. Andersson, and C. G. Kurland, *MGG* **191**, 378 (1983).
25. F. H. C. Crick, *Symp. Soc. Exp. Biol.* **12**, 138 (1958).
26. F. J. Jacob and J. Monod, *JMB* **3**, 318 (1961).
27. R. B. Loftfield and D. Vanderjagt, *BJ* **128**, 1353 (1972).
28. P. Edelmann and J. Gallant, *PNAS* **74**, 3396 (1977).
29. N. Ellis and J. Gallant, *MGG* **188**, 169 (1982).
30. M. M. Fluck, W. Salser, and R. H. Epstein *MGG* **151**, 137 (1977).
31. L. Bossi and J. R. Roth, *Nature (London)* **286**, 123 (1980).
32. H. Engelberg-Kulka, L. Dekel, and M. Israeli-Reches, *BBRC* **98**, 1008 (1981).
33. J. Parker, T. C. Johnston, and P. T. Boriga, *MGG* **180**, 275 (1980).
34. F. Bouadloun, D. Donner, and C. G. Kurland, *EMBO J.* **2**, 1351 (1983).
35. R. F. Rosenberger and G. Foskett, *MGG* **182**, 561 (1981).
36. R. F. Rosenberger and S. H. Hilton, *MGG* **191**, 207 (1983).
37. J. Eisinger, B. Feuer, and T. Yamane, *Nature NB* **231**, 126 (1971).
38. G. Högenauer, *EJB* **12**, 527 (1970).
39. O. C. Uhlenbeck, J. Baker, and P. Doty *Nature (London)* **225**, 508 (1970).
40. H. Grosjean, D. G. Söll, and D. M. Crothers, *JMB* **103**, 499 (1976).
41. J. F. Sambrook, D. P. Fan, and S. Brenner, *Nature (London)* **214**, 452 (1967).
42. D. Hirsh, *JMB* **58**, 439 (1971).
43. D. Hirsh, and L. Gold, *JMB* **58**, 459 (1971).
44. Y. Kuchino, M. Kato, H. Sugisaki, and S. Nishimura, *NARes* **6**, 3459 (1979).
45. R. H. Buckingham and C. G. Kurland, *PNAS* **74**, 5496 (1977).
46. J. J. Hopfield, *PNAS* **71**, 4135 (1974).
47. J. Ninio *Biochimie* **57**, 587 (1975).
48. C. G. Kurland, *Biophys. J.* **22**, 373 (1978).
49. J. Lucas-Lenard and F. Lipmann, *ARB* **40**, 409 (1971).
50. C. Blomberg, M. Ehrenberg, and C. G. Kurland, *Q. Rev. Biophys.* **13**, 231 (1980).
51. R. C. Thompson and P. J. Stone, *PNAS* **74**, 198 (1977).
52. J. L. Yates, *JBC* **254**, 11550 (1979).
53. R. C. Thompson, D. B. Dix, and J. F. Eccleston, *JBC* **255**, 11088 (1980).
54. R. C. Thompson, D. B Dix, R. B. Gerson, and A. M. Karim, *JBC* **256**, 81 (1981).
55. R. C. Thompson and A M. Karim, *PNAS* **79**, 4922 (1982).
56. C. G. Kurland, *Cell* **28**, 201 (1982).
57. P. C. Jelenc and C. G. Kurland, *PNAS* **76**, 3174 (1979).
58. I. Pettersson and C. G. Kurland, *PNAS* **77**, 4007 (1980).
59. E. G. H. Wagner and C. G. Kurland, *MGG* **180**, 139 (1980).
60. J. Lucas-Lenard and F. Lipmann, *PNAS* **57**, 1050 (1967).
61. E. G. H. Wagner, P. C. Jelenc, M. Ehrenberg, and C. G. Kurland, *EJB*, **122**, 193 (1982).
62. S. G. E. Andersson, R. H. Buckingham, and C. G. Kurland, *EMBO J.* **3**, 91 (1984).
63. D. L. Miller, and H. Weissbach, in "Molecular Mechanisms of Protein Biosynthesis" (H. Weissbach and S. Pestka, eds.), p. 323. Academic Press, New York, 1977.

64. V. Chau, G. Romero, and R. L. Biltonen, *JBC* **256**, 5591 (1981).
65. T. Ruusala, M. Ehrenberg, and C. G. Kurland, *EMBO J.* **1**, 75 (1982).
66. T. Ruusala, M. Ehrenberg, and C. G. Kurland, *EMBO J.* **1**, 741 (1982).
67. K. Bohman, T. Ruusala, P. C. Jelenc, and C. G. Kurland, *MGG*, submitted (1984).
68. P. C. Jelenc and C. G. Kurland, in press.
68a. T. Ruusala and C. G. Kurland, *MGG*, submitted (1984).
69. M. Gouy and R. Grantham, *FEBS Lett.* **115**, 151 (1980).
70. A. Kornberg, *Science* **163**, 1410 (1969).
71. C. G. Kurland, in "Nonsense Mutations and tRNA Suppressors" (J. E. Celis and J. D. Smith, eds.), p. 97 Academic Press, New York, 1979.
72. R. Weiss and J. Gallant *Nature (London)* **302**, 389 (1983).
73. A. L. Goldberg and J. F. Dice *ARB* **43**, 835 (1974).
74. A. I. Bukhari and D. Zipser *Nature NB* **243**, 238 (1973).
75. J. R. Menninger, *Mech. Ageing Dev.* **6**, 131 (1977).
76. J. L. Manley, *JMB* **125**, 407 (1978).
77. M. Ehrenberg and C. Blomberg, *Biophys. J.* **31**, 333 (1980).
78. M. Ehrenberg and C. G. Kurland, *Q. Rev. Biophys.* **17** (1974).
79. O. Maaløe, in "Biological Regulation and Development (R. F. Goldberger, ed.), p. 487. Plenum, New York, 1979.
80. H. Grosjean and W. Fiers, *Gene* **18**, 199 (1982).
80a. M. Cashel and J. Gallant, *Nature (London)* **221**, 838 (1969).
81. E. G. H. Wagner, M. Ehrenberg, and C. G. Kurland, *MGG* **185**, 269 (1982).
81a. T. Ruusala, D. I. Andersson, M. Ehrenberg, and C. G. Kurland, *EMBO J.*, submitted (1984).
82. L. E. Orgel, *PNAS* **67**, 1476 (1970).
83. G. W. Hoffman, *JMB* **86**, 349 (1974).
84. N. S. Goel and M. Ycas, *J. Theor. Biol.* **55**, 245 (1975).
85. N. S. Goel and S. Islam, *J. Theor. Biol.* **68**, 167 (1977).
86. T. B. L. Kirkwood and R. Holiday, *JMB* **97**, 257 (1975).
87. T. B. L. Kirkwood, *Nature (London)* **270**, 301 (1977).
88. J. A. Gallant and J. Prothero, *J. Theor. Biol.* **83**, 561 (1980).
89. J. Ninio, "Molecular Evolution," p. 116. Pitman, London, 1982.
89a. C. G. Kurland, D. I. Andersson, S. G. E. Andersson, K. Bohman, F. Bouadloun, M. Ehrenberg, P. C. Jelenc, and T. Ruusala, in "Gene Expression" (Alfred Benzon Symp. 19), p. 413. Munksgaard, Copenhagen, 1984.

Molecular Aspects of Development in the Brine Shrimp *Artemia*[1]

ALBERT J. WAHBA AND
CHARLES L. WOODLEY

Department of Biochemistry
University of Mississippi Medical Center
Jackson, Mississippi

I. Polypeptide-Chain Initiation 224
   A. eIF2 and the Initiation of Protein Synthesis 224
   B. Comparison of eIF2 from Dormant and from Developing Embryos 225
   C. Regulation of eIF2 Activity 227
   D. Comparison of *Artemia* and Rabbit Reticulocyte eIF2 237
II. Status of Message in *Artemia* Embryos 245
   A. mRNP Particles 246
   B. mRNA 248
III. Questions for the Future.. 255
   References 260

    The brine shrimp *Artemia* possesses an unusual reproductive cycle (*1*, *2*). After fertilization, cleavage of the zygote results in a blastula, which then undergoes gastrulation. The gastrula may continue development, giving rise eventually to an adult via formation of the prenaupliar and free-swimming naupliar stages. Alternatively, the gastrula may encyst, enter a dormant state, and be released into the environment, where it becomes desiccated (for review, see *3*). When cysts are hydrated under proper environmental conditions, the encysted embryos rapidly resume metabolic activity and development.
    The 8–18-hour period of development following hydration, but prior to emergence of the embryos from the cyst, is characterized by

---

[1] Abbreviations: eIF, eukaryotic polypeptide-chain-initiation-factor; eIF2(α-P), the form of eIF2 in which the smallest (α) subunit is phosphorylated; Co-eIF2(A), -(B), -(C), ancillary factors for eukaryotic polypeptide-chain-initiation-factor 2 (eIF2); EF1, eukaryotic protein-synthesis-elongation-factor 1, composed of three subunits; eEF-Tu, eukaryotic protein-synthesis-elongation-factor Tu, the high-M_r subunit of EF1; EF2, eukaryotic protein-synthesis-elongation-factor 2, composed of a single polypeptide; HCR, heme-controlled-repressor, which phosphorylates the α-subunit of eIF2 to yield eIF2(α-P); GEF, the guanine nucleotide exchange factor, also known as RF (rescue factor); ATA, aurintricarboxylic acid.

relatively synchronous development, the synthesis of polypeptides on newly formed polysomes, and the absence of DNA synthesis and cell division (1, 2, 4–6). Since embryonic development is divorced from cell division throughout preemergence development, *Artemia* offers a unique system for examining the relationship between protein synthesis, its regulation, and differentiation.

Oocytes from sea urchin (7–9), *Drosophila* (10–12), and silkmoth (13, 14), among many others, contain the elements required for translation, but remain inactive until fertilization. There is an increase in the proportion of ribosomes in polysomes and a sharp rise in protein synthesis when fertilization occurs. Initial development can proceed in the absence of RNA synthesis, but transcription generally occurs by early blastulation.

Highly differentiated cells such as rabbit reticulocytes produce essentially one type of mRNA and may, therefore, be inappropriate for the study of translational regulation. Similarly, oocytes, which have not yet started to differentiate, are primed protein-synthesizing systems and may not require specific factors to select different classes of mRNA. On the other hand, developing *Artemia* embryos contain large amounts of stored mRNA and must select and translate mRNA appropriate for each particular stage of development. Differentiating systems of this kind are ideal for the study of translational regulation of mRNA activity and the influence this may have on development.

In terms of technical manipulations, there are important differences between oocytes and dormant *Artemia* embryos. Prior to emergence, *Artemia* embryos are impermeable to RNA and protein precursors, whereas these are readily introduced during development in oocytes. *Artemia* cysts are available in much larger quantities than are oocytes, and this provides an advantage in their use.

All cellular processes stop when *Artemia* embryos enter dormancy in the maternal ovisac (3). At this stage, dormant embryos contain competent mRNA (15–20), large quantities of 80 S ribosomes (21, 22), elongation factors (23, 24), peptide-chain-termination factors (25), and aminoacyl-tRNA synthetases (26). Cell-free extracts of dormant embryos, encysted at gastrulation, are inactive in endogenous protein synthesis and do not translate exogenous natural mRNAs (27). Polysomes are observed within 15 minutes when hydrated cysts resume development (21, 22, 28, 29). It is during this period that embryos must overcome the factors that repress protein synthesis and select, from the store of mRNPs or newly synthesized mRNA, transcripts to be translated.

Changes in the *in vivo* level of numerous enzymes during development of *Artemia* embryos have been reported and are summarized in recent reviews (*3, 30–33*). These articles also describe the morphological and molecular changes associated with developing *Artemia* embryos.

The molecular aspects of development we have chosen to examine are (a) the amount and activity of polypeptide-chain-initiation-factors in embryos before and after resumption of development; (b) the nature of the stored message; (c) how transcripts are selected for translation; and (d) changes in mRNA content during the early stages of development.

Polypeptide-chain initiation is outlined in Fig. 1. The first two steps in the initiation process are recognition of the initiator tRNA by chain-initiation-factor 2 (eIF2) (*34–37*) and the selection of a message in the formation of the 40-S initiation complex. Several ancillary factors stimulate eIF2 activity in the formation and binding of the ternary complex (eIF2·GTP·Met-tRNA$_f$) to 40-S ribosomal subunits (for recent reviews, see *38, 39*). After formation of the 40-S preinitiation complex, the message with its cap-binding proteins binds to the ribosomal subunit to form the 40-S initiation complex (*41, 42*). Recruitment of a message by mRNA-binding proteins is influenced by the nature of the message, which may include capping, methylation, secondary structure, or proteins associated with the RNA (*43–50*).

Other factors may also play a role in positioning the message on the ribosomal subunit (*41, 51–53*). Hydrolysis of ATP is required for message binding, although the role of the hydrolysis, as well as the movement of the ribosome from the 5' cap structure to the initiation codon, are not as yet clearly understood (*42, 54–58*). In the presence of eIF5, the 60-S ribosomal subunit attaches to the 40-S initiation complex, the eIF2-bound GTP is hydrolyzed, and eIF2 is released (*59–62*). At this stage, the chain-initiation factor is presumably released as an eIF2·GDP complex (*63, 64*). A protein found in the cytosol facilitates the exchange of bound GDP for GTP (*64–68*). The presence of such a factor would allow eIF2 to enter another round of initiation. A pool of active 40-S ribosomal subunits, necessary for the initiation process, may be maintained by another cytosol factor, eIF6, which prevents reassociation of ribosomal subunits (*69, 70*). After formation of the 80-S initiation complex, protein synthesis continues through the elongation and termination steps.

We chose eIF2 as the starting point for our studies in *Artemia*, since this factor catalyzes the first step of the initiation process and may serve as a control point for polypeptide-chain initiation.

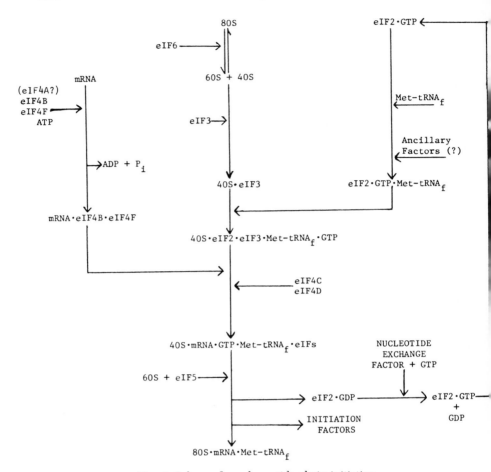

FIG. 1. Scheme for polypeptide-chain initiation.

I. Polypeptide-Chain Initiation

A. eIF2 and the Initiation of Protein Synthesis

The eukaryotic initiation-factor 2 (eIF2) could be the primary point of control for regulation of protein synthesis under a variety of environmental and developmental conditions (for recent reviews, see 65, 66, 71). There are several discrete steps in the process of polypeptide-chain initiation at which eIF2 activity could potentially be regulated. Several reports indicate that eIF2 activity, and thus the level of protein synthesis, changes under different physiological conditions. Sensitivity to Mg^{2+} (*39, 40*), the GTP/GDP ratio (*72–75*), the relative activities of protein kinases and phosphatases (*74, 76–79*), and interactions with ancillary factors (*39, 40, 68*) may all play roles in determining eIF2 activity. In addition, the ability of several of these ancillary factors to interact with eIF2 is dependent upon the phosphorylation state of eIF2. Phosphorylation of the α-subunit of eIF2 by one of several specific protein kinases may regulate eIF2 recycling by preventing the exchange of GDP for GTP in the eIF2·GDP binary complex (*64–67*). Therefore, polypeptide chain initiation would be inhibited by blocking the recycling of eIF2. The activation of these protein kinases can be brought about by double-stranded RNA, heme deficiency, oxidized glutathione, high pO_2, and interferon. The inhibition of polypeptide chain initiation can also be prevented or overcome by addition of various factors, including heme, eIF2, and GTP. Recent reviews on the regulation of eIF2 and its interaction with ancillary factors are available (*39, 40, 65, 66, 80, 81*).

In 1976, Filipowicz *et al.* (*53*) reported that dormant embryos of *Artemia* are deficient in eIF2 activity, but the level of eIF2 increases over 20-fold upon resumption of development. Therefore, a close examination of eIF2 in dormant and developing *Artemia* embryos seemed warranted, as this protein may play a major role in regulating protein synthesis activity in developing or differentiating systems.

In determining the role of eIF2 during development of *Artemia* embryos, a systematic examination of the isolation procedures and cellular distribution of the chain-initiation factor was undertaken.

B. Comparison of eIF2 from Dormant and from Developing Embryos

1. Activity of Cellular Fractions

The activity of eIF2 in the various cellular fractions is assayed by ternary complex formation. Significant eIF2 activity can be detected in extracts of both dormant and developing embryos in the postribosomal supernatant and ribosomal salt-wash fractions (*81, 83*). When the 35–60% ammonium sulfate fractions are assayed, 53% of the eIF2 activity in dormant embryos is found in the ribosomal high-salt wash and 47% in the postribosomal supernatant. Similarly, in developing embryos, 64% of the eIF2 activity is located in the ribosomal wash and 36% in the postribosomal supernatant. The total amounts of eIF2 activity, in contrast to earlier reports (*53*), are very similar in dormant and in developing embryos. Per 100 g net weight of embryos, 116 and

110 units of eIF2 are obtained from dormant and 11-hour developing embryos, respectively.

In a second series of experiments, the ribosomal salt wash of embryos allowed to hydrate, or to develop for different time intervals, was tested for eIF2 activity. Neither the time of hydration at 4°C nor the time of development up to emergence had any significant effect on the amount of eIF2 recovered in the 35–60% saturated ammonium sulfate fraction of the ribosomal salt wash. Thus, the overall level of eIF2 per se is not limiting during *Artemia* development.

Preparations of eIF2 from dormant and developing embryos were also tested for their ability to stimulate *in vitro* polypeptide synthesis in a cell-free system that contained *Artemia* 80-S ribosomes, globin mRNA, and a limiting level of reticulocyte ribosomal salt wash (Table I). Both preparations are similarly active in stimulating polypeptide synthesis at low levels of the factor, although eIF2 from dormant embryos is approximately 30% less active at the highest levels tested.

2. Structural Similarity of eIF2

On dodecyl sulfate/polyacrylamide gel electrophoresis, eIF2 preparations from dormant and developing embryos appear to be identical, each having three polypeptides with M_r's of 52,000, 43,000, and 42,000 (Fig. 2; 83).

The structural relationship between eIF2 isolated from dormant and from developing embryos was further examined by preparing antibodies against eIF2 from 11-hour developing embryos. Immunological identity between both eIF2 preparations is demonstrated in

TABLE I
Stimulation of Polypeptide Synthesis by eIF2

Source of eIF2[a]	eIF2 added (μg)	Leucine incorporated (pmol)	Stimulation (%)
None	0	70	0
Dormant embryos	0.8	89	27
	1.9	105	50
	3.8	121	72
Developing embryos	0.8	86	23
	1.9	116	66
	3.8	146	108

[a] Each reaction mixture contained 360 μg of reticulocyte unfractionated salt wash, 0.5 A_{260} unit of *Artemia* ribosomes, and 0.15 A_{260} unit of poly(A)-rich mRNA. From MacRae *et al.* (83).

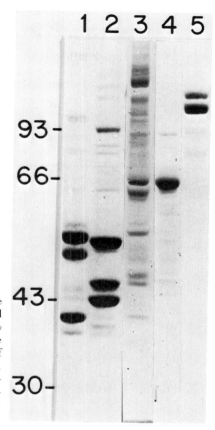

FIG. 2. Dodecyl sulfate/polyacrylamide gel electrophoresis of eIF2, Co-eIF2(A), and Co-eIF2(B). The electrophoresis was in 10% gels containing dodecyl sulfate (139). Lane 1, 6.6 of μg reticulocyte eIF2; lane 2, 10 μg of *Artemia* eIF2; lane 3, 11 μg of Co-eIF2(A); lane 4, 12 μg of Co-eIF2(A) after affinity chromatography; lane 5, 3.8 μg of Co-eIF2(B), Ultrogel ACA-34 fraction. From Woodley et al. (68).

Ouchterlony plates when these antibodies are tested for their ability to cross-react with eIF2 prepared from dormant or developing embryos. Furthermore, mouse antibodies against eIF2 decrease, to the same extent, the formation of the ternary complex (eIF2·GTP·Met-tRNA$_f$) and the 40-S initiation complex when eIF2 from either dormant or developing embryos is used (Table II). The observed inhibition in the presence of immune serum and normal serum is 90% and 7%, respectively (83).

C. Regulation of eIF2 Activity

1. Isolation of Factors That Stimulate eIF2

Several factors that enhance eIF2 activity have been isolated from extracts of rabbit reticulocytes and wheat germ (39, 40, 68, 84, 85). We

TABLE II
INHIBITION OF 40-S INITIATION COMPLEX BY ANTISERUM AGAINST eIF2

Source of eIF2[a]	Serum	Met-tRNA$_f$ Bound (pmol)	Inhibition (%)
Dormant embryos	None	0.24	0
	Normal	0.22	8
	Immune	0.03	87
Developing embryos	None	0.18	0
	Normal	0.17	7
	Immune	0.01	92

[a] Each reaction contained 6 µg of phosphocellulose-step eIF2. Where indicated, 20 µl of normal or immune serum from mouse was added. A blank value in the absence of eIF2 (0.03 pmol) was subtracted from each value. From MacRae et al. (83).

have characterized three distinct factors from the wash of *Artemia* ribosomes that stimulate ternary complex formation and Met-tRNA$_f$ binding to 40-S ribosomal subunits (68, 86). At this point it is not possible to correlate any of the *Artemia* factors with a specific reticulocyte or wheat germ counterpart. Some of the similarities in activities are discussed in this section. We designate the three *Artemia* factors Co-eIF2(A), Co-eIF2(B), and Co-eIF2(C).

Co-eIF2(A) and Co-eIF2(B) can be isolated from the 40–60% saturated ammonium sulfate fraction of the ribosomal salt wash. Neither factor can be assayed in the initial steps because of the presence of eIF2. Our preparations of Co-eIF2(A) (68) are at least 90% pure and have a single polypeptide of M_r 65,000 (Fig. 2, lanes 4 and 5). Analysis of the Co-eIF2(B) preparation in dodecyl sulfate/polyacrylamide gels shows two polypeptides of M_r 105,000 and 112,000 in a 1 : 1 ratio (Fig. 2, lane 6).

Co-eIF2(C), unlike Co-eIF2(A) and Co-eIF2(B), is isolated from the 0–40% saturated ammonium sulfate fraction of the ribosomal high-salt wash of 12-hour developing embryos (86). Fractionation of the phosphocellulose fraction of Co-eIF2(C) by sucrose-gradient centrifugation gives two peaks of activity sedimenting at 12 and 15 S. The 12-S fraction shows three polypeptides of M_r 115,000, 71,000, and 49,000 upon electrophoresis in dodecyl sulfate/polyacrylamide gels. In contrast, the 15-S fraction contains 9–11 polypeptides, 3 of which are identical to those of the 12-S fraction. In all assays the 15-S fraction behaves similarly to the 12-S fraction.

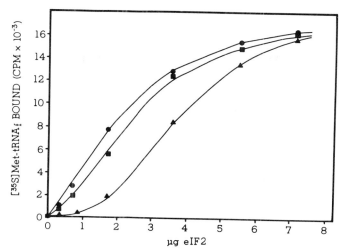

FIG. 3. Effect of eIF2 concentration on the stimulation of ternary complex formation by Co-eIF2(A) and Co-eIF2(B). [^{35}S]Met-tRNA$_f$ bound with eIF2 alone (▲——▲) or in the presence of 3.5 µg of Co-eIF2(A) (hydroxyapatite fraction) (■——■) or of 3.9 µg of Co-eIF2(B) (hydroxyapatite fraction) (●——●). [^{35}S]Methionine was 14,000 cpm/pmol. From Woodley et al. (68).

2. Characterization of eIF-2 Ancillary Proteins

a. Co-eIF2(A) and Co-eIF2(B)

i. Ternary complex formation. Stimulation of ternary complex formation is used as an assay for Co-eIF2(A) and Co-eIF2(B) (68). As shown in Fig. 3, binding of Met-tRNA$_f$ to eIF2 increases sigmoidally with increasing levels of eIF2. At levels of eIF2 below 2–3 µg per 75 µl of reaction mixture, the protein is most sensitive to stimulation by Co-eIF2(A) and Co-eIF2(B). Each factor stimulates eIF2 activity to a similar extent, and this effect is observed for both the rate and extent of ternary complex formation. Either protein produces maximum stimulation, and no additive effect is observed in the presence of saturation levels of either cofactor.

Artemia eIF2, like its reticulocyte counterpart, is strongly inhibited by low concentrations (4–10 µM) of aurintricarboxylic acid (ATA) (Table III; 39, 68, 72). With 0.74 µg of the initiation factor, inhibition of ternary complex formation by ATA is 85–90%. The addition of 3.9 µg of Co-eIF2(A) does not protect eIF2 from inhibition, whereas a comparable amount of Co-eIF2(B) provides 70–80% protection.

TABLE III
EFFECT OF AURINTRICARBOXYLIC ACID ON Co-eIF2(A) AND Co-eIF2(B) ACTIVITIES[a,b]

Additions	[35S]Met-tRNA$_f$ bound		
	−ATA	+ATA (4.5 μM)	+ATA (9.0 μM)
eIF2	625	95	75
eIF2 + Co-eIF2(A)	2765	550	215
eIF2 + Co-eIF2(B)	3050	2475	2215

[a] From Woodley et al. (68).
[b] Incubation mixtures for ternary complex formation contained 100 mM KCl, 1 mM Mg^{2+}, 0.16 mM GTP, 2 pmol [35S]Met-tRNA$_f$, 0.74 μg of eIF2, 3.9 μg of Co-eIF2(A) or 3.7 μg of Co-eIF2(B). After 5 minutes at 30°C, aurintricarboxylic acid was added; the reaction was continued for an additional 5 minutes and terminated by addition of cold buffer.

ii. *Formation of 40-S initiation complex.* Binding of the initiator tRNA (Met-tRNA$_f$) to 40-S ribosomal subunits is dependent upon the addition of eIF2 and GTP (68, 83). This reaction is stimulated 3- to 4-fold by the addition of the template AUG (Table IV). Maximum binding of Met-tRNA$_f$ to *Artemia* 40-S ribosomal subunits requires the addition of approximately 4 μg of Co-eIF2(A) or Co-eIF2(B) per 75 μl of reaction mixture (68). Addition of either protein results in approximately a 2-fold increase in the extent of formation of the 40-S initiation complex, and no additive effect is observed in the presence of saturating levels of either cofactor.

b. *Co-eIF2(C)*

i. *Formation of ternary complex.* Stimulation of ternary complex formation by the phosphocellulose fraction of Co-eIF2(C) was deter-

TABLE IV
AUG-DEPENDENT BINDING OF [35S]MET-TRNA$_f$ TO 40-S RIBOSOMES: STIMULATION BY Co-eIF2(A) AND Co-eIF2(B)[a,b]

Additions	[35S]Met-tRNA$_f$ bound (cpm)	Stimulation (fold)
eIF2	1000	1.0
eIF2 + Co-eIF2(A)	2030	1.8
eIF2 + Co-eIF2(B)	2725	2.3

[a] From Woodley et al. (68).
[b] The 40-S initiation complex assay contained 0.5 A$_{260}$ unit of 40-S ribosomal subunits, 1.1 μg of eIF2, and, where indicated, 3.9 μg of Co-eIF2(A) or 3.7 μg of Co-eIF2(B). A blank of 300 cpm for Met-tRNA$_f$ binding in the absence of AUG was subtracted from each value.

TABLE V
EFFECT OF Co-eIF2(C) ON TERNARY COMPLEX FORMATION WITH INCREASING AMOUNTS OF eIF2[a]

eIF2 (μg)	[^{35}S]Met-tRNA$_f$ bound		Stimulation (fold)
	$-$ Co-eIF2(C) (cpm)	$+$ Co-eIF2(C) (cpm)	
0.8	2,500	8,400	3.3
2.1	7,900	15,000	1.9
5.0	25,000	27,000	1.1
9.9	35,000	36,000	1.0

[a] Blanks in the absence of eIF2 (110 cpm) and with Co-eIF2(C) alone (330 cpm) were subtracted from the appropriate values. eIF2 activity was assayed by the nitrocellulose membrane filtration technique.

mined with different levels of eIF2 (Table V; 86). With 0.8 μg of eIF2, Co-eIF2(C) increases Met-tRNA$_f$ binding 3.3-fold. With increasing concentrations of the initiation factor, the degree of stimulation by Co-eIF2(C) gradually decreases, so that at high levels of eIF2 there is no stimulation of Met-tRNA$_f$ binding. This is similar to the effect of Co-eIF2(A) and Co-eIF2(B) (Fig. 3).

ii. Reversal of the inhibition by mRNA of ternary-complex formation. The binding of Met-tRNA$_f$ is decreased by 75% by the addition of 1 μg of mRNA to the ternary complex assay (Table VI). This inhibition is substantially relieved by Co-eIF2(C), and the addition of 1 mM Mg^{2+} aids in reversing this inhibition. Co-eIF2(C) is more effective

TABLE VI
EFFECT OF mRNA ON TERNARY COMPLEX FORMATION IN THE PRESENCE OF Co-eIF2(C)[a]

Factor	[^{35}S]Met-tRNA$_f$ bound (cpm)			
	None	Mg^{2+}	mRNA	Mg^{2+} + mRNA
eIF2	3100	3100	800	1200
+ Co-eIF2(C)	6400	6700	4600	5700
+ Co-eIF2(C) (12 S)	5900	5900	2200	3600
+ Co-eIF2(C) (15 S)	5700	5600	2700	3800

[a] Ternary complex formation was assayed in a one-stage reaction. Each incubation mixture (75 μl) contained 10 μg of bovine serum albumin, 0.92 μg of *Artemia* eIF2, 3 pmol of [^{35}S]Met-tRNA$_f$, 0.2 mM GTP, and where indicated, 1 mM Mg^{2+}, 1 μg of reticulocyte poly(A)-containing RNA, and 11 μg of the phosphocellulose 15-S or 12-S fraction of Co-eIF2(C). A blank in the absence of eIF2 (70 cpm) has been subtracted from each value.

TABLE VII
AUG-Dependent Met-tRNA$_f$ Binding to 40-Ribosomal Subunits[a]

Additions	[^{35}S]Met-tRNA$_f$ bound (cpm)	
	−AUG	+ AUG
eIF2	400	900
+ Co-eIF2(C)	3300	5000
+ mRNA binding-factor	500	1400
+ Co-eIF2(C) + mRNA binding factor	3200	4700
+ Co-eIF2(C) (12 S)	3000	3800
+ Co-eIF2(C) (12 S) + mRNA binding-factor	1900	5400
Co-eIF2(C) + mRNA binding-factor	100	300

[a] Reaction mixtures contained, where indicated, 1.6 µg of *Artemia* eIF2, 11 µg of phosphocellulose fraction Co-eIF2(C), 12 µg of mRNA binding-factor, and 8 µg of the sucrose-gradient 12-S Co-eIF2(C) fraction. A value of 350 cpm, obtained in the absence of eIF2, has been subtracted from each value. Each reaction contained 3 pmol of [^{35}S]Met-tRNA$_f$ (19,500 cpm/pmol).

than Co-eIF2(A) or Co-eIF2(B) in stabilizing the ternary complex in the presence of 9 mM ATA (Table III) (68).

iii. *Formation of 40-S initiation complex.* The eIF2-dependent binding of Met-tRNA$_f$ to 40-S ribosomal subunits in the presence of AUG (68, 84) is assayed by sucrose density-gradient centrifugation (Table VII). In the presence of eIF2, Met-tRNA$_f$ binding to 40-S ribosomal subunits is stimulated 2-fold by addition of AUG. Co-eIF2(C) stimulates Met-tRNA$_f$ binding 6- to 8-fold in the presence or the absence of AUG. The mRNA-binding fraction, which elutes from phosphocellulose between 0.4 and 1 M KCl, increases Met-tRNA$_f$ binding by 65% in the presence of AUG. Maximum stimulation of Met-tRNA$_f$ binding is observed in the presence of AUG, Co-eIF2(C), and the mRNA binding-factor. In the absence of eIF2, Co-eIF2(C) and the mRNA binding-factor have no significant activity, either alone or in combination, in binding Met-tRNA$_f$ to 40-S ribosomal subunits.

Co-eIF2(C) also stimulates 40-S initiation-complex formation with natural message as template. The binding of Met-tRNA$_f$ and ascites mRNA to 40-S ribosomal subunits is determined by pelleting the initiation complex through a sucrose gradient (Table VIII). In the absence of ancillary factors, no message and very little Met-tRNA$_f$ is bound.

In the absence of eIF2 addition of Co-eIF2(C) to the assay results in significant message binding, but no Met-tRNA$_f$ binding. However, in the presence of eIF2 and Co-eIF2(C), significant Met-tRNA$_f$, but

TABLE VIII
40-S INITIATION COMPLEX FORMATION WITH ^{3}H-LABELED ASCITES MRNA[a]

Additions	mRNA bound (cpm)		Met-tRNA$_f$ bound (cpm)	
	− eIF2	+ eIF2	− eIF2	+ eIF2
None	0	0	0	300
Co-eIF2(C)	2400	300	80	2300
mRNA binding-factor	1500	3300	90	300
Co-eIF2(C) + mRNA binding-factor	2900	4200	70	3400

[a] Each reaction contained 3 pmol of [^{35}S]Met-tRNA$_f$ (15,700 cpm/pmol), 0.4 μg of ^{3}H-labeled ascites mRNA (70,000 cpm/pmol) and, where indicated, 1.6 μg of eIF2, 11 μg of phosphocellulose fraction Co-eIF2(C), and 12 μg mRNA binding-factor. A blank of 540 cpm for ^{3}H-labeled mRNA and 375 cpm for [^{35}S]Met-tRNA$_f$ has been subtracted from each value.

no message binding, is obtained. It should be noted that the mRNA binding-factor binds mRNA but not Met-tRNA$_f$ to 40-S ribosomal subunits in the absence of eIF2, and that addition of eIF2 further increases this mRNA binding. Maximum stimulation of message and Met-tRNA$_f$ binding requires the addition of eIF2, Co-eIF2(C), and the mRNA binding-factor. No significant Met-tRNA$_f$ binding is obtained when either Co-eIF2(C) or the mRNA binding-factor is added. This is consistent with results obtained in the AUG-dependent assay (Table VII). Although Co-eIF2(C) stimulates ternary complex formation, its primary function appears to be in stabilizing the 40-S initiation complex.

3. EFFECT OF PHOSPHORYLATION ON eIF2

The level of eIF2 does not seem to be a limiting factor in regulating protein synthesis in *Artemia*, since both dormant and developing embryos contain equivalent amounts of it (83). However, the possibility existed that it could be regulated *in vivo* by covalent modification, such as by phosphorylation. As a first step, it was important to determine the effect of phosphorylation on eIF2 activity in various model reactions, including ternary- and 40-S initiation-complex formation and *in vitro* protein synthesis (68).

a. *Phosphorylation of Artemia eIF2.* The rabbit reticulocyte heme-controlled repressor (HCR) phosphorylates the smallest (α) subunit (M_r = 37,000) of reticulocyte eIF2 (75–80). This enzyme was purified from the postribosomal supernatant of rabbit reticulocytes through the phosphocellulose step (68, 78). In the presence of ATP and Mg^{2+}, it phosphorylated approximately 90% of the eIF2 from either dormant or developing embryos. The reaction products were assayed by electrophoresis in polyacrylamide gels containing sodium dodecyl sulfate. As illustrated in Fig. 4, the α-subunit (M_r = 41,000) of Artemia eIF2 and a 200,000-M_r protein were the only phosphopeptides detected (68).

After phosphorylation with HCR and [γ-^{32}]ATP, Artemia eIF2 was isolated by chromatography on a phosphocellulose column. The tryptic digest of the phosphorylated factor was chromatographed on Whatman 3MM paper, followed by high-voltage electrophoresis (68). Of

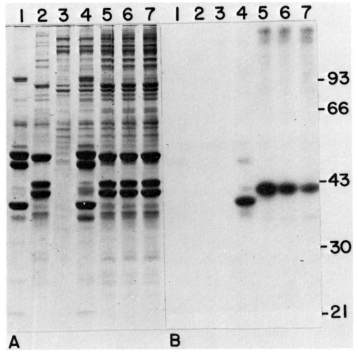

FIG. 4. Phosphorylation of reticulocyte and Artemia eIF2 by the heme-controlled repressor. Phosphorylation products were electrophoresed on 10% polyacrylamide gels containing sodium dodecyl sulfate and (A) stained with Coomassie Blue R-250, (B) autoradiographed. Lane 1, 4.8 μg of rabbit reticulocyte eIF2; lane 2, 5.6 μg of Artemia eIF2; lane 3, 3.5 μg of repressor, hydroxyapatite fraction; lane 4, reticulocyte eIF2 plus repressor; lane 5, Artemia eIF2 plus repressor; lanes 6 and 7 contain Artemia eIF2 and repressor plus 12 μM and 24 μM heme, respectively. The phosphorylation reaction was terminated with an equal volume of sodium-dodecyl-sulfate buffer (139). From Woodley et al. (68).

the five phosphopeptides obtained, a single, predominant peptide contained 54% of the total radioactivity. Acid hydrolysis of the tryptic peptides revealed only phosphoserine in the major and two minor phosphopeptides. Phosphothreonine was detected in two minor phosphopeptides. The net relative amounts of phosphoserine and phosphothreonine in HCR-phosphorylated eIF2 and 83% and 17%, respectively. The phosphothreonine may arise in part from contaminants in the eIF2 or repressor preparations.

b. Formation of Ternary Complex with Phosphorylated eIF2. Artemia eIF2 phosphorylated by reticulocyte HCR and reisolated by phosphocellulose chromatography has essentially the same activity in ternary-complex formation as mock-phosphorylated eIF2. The ability of reticulocyte eIF2 to interact with several ancillary proteins has been reported to be impaired by phosphorylation of the initiation factor (64–67). Therefore, the stimulation of ternary-complex formation with phosphorylated eIF2 in the presence of either Co-eIF2(A) or Co-eIF2(B) was determined. Addition of either cofactor produces the same degree of stimulation, independent of the phosphorylation state of eIF2. The specific activity of phosphorylated eIF2 is reduced approximately 15%. The presence of extra protein bands in the phosphorylated eIF2 preparation, as seen on gel electrophoresis, may in part explain the lower specific activity.

c. Phosphorylated eIF2 Activity in 40-S Initiation-Complex Formation. Three types of eIF2 preparations were assayed for AUG-dependent binding of Met-tRNA$_f$ to 40-S ribosomal subunits (68). The nonphosphorylated as well as the phosphorylated factor are active in 40-S initiation-complex formation (Table IX). However, the specific activity of the nonphosphorylated eIF2 is approximately 30% higher. If the factor, after treatment with HCR, is not reisolated by chromatography on phosphocellulose, it is not active in binding Met-tRNA$_f$ to 40-S ribosomal subunits. This suggests that some component in the HCR preparation, not phosphorylation of the eIF2 per se, is responsible for the inhibition of formation of the 40-S initiation complex.

d. Stimulation of Cell-Free Protein Synthesis by eIF2. The effect of phosphorylation on the ability of purified *Artemia* eIF2 to stimulate protein synthesis was assayed in two different translation systems. Purified *Artemia* eIF2 stimulates amino-acid incorporation in a partially fractionated system containing salt-washed ribosomes from dormant *Artemia* embryos, rabbit-reticulocyte-globin poly(A)-rich mRNA, and a limiting level of the 0.5 M KCl ribosomal wash fraction from rabbit reticulocytes (Table X). With 1.5 μg of mock-phosphorylated or phosphorylated eIF2, protein synthesis is stimulated approxi-

TABLE IX
ACTIVITY OF eIF2(α-P) IN 40-S INITIATION-COMPLEX FORMATION[a,b]

Expt. No.	Factor added (μg)		[^{35}S]Met-tRNA$_f$ bound	
			(cpm)	(pmol)
I			900	0.10
II	eIF2	1.0	2100	0.24
		2.5	6500	0.72
III	eIF2(α-P)	1.4	2600	0.28
		3.0	4800	0.53
IV	eIF2(α-P)	1.9	1000	0.10
		3.8	1000	0.11

[a] From Woodley et al. (68).
[b] No eIF2 was present in Expt. I. In Expt. II, eIF2 was mock-phosphorylated by omitting ATP, then reisolated by phosphocellulose chromatography. In Expt. III, after phosphorylation by HCR, eIF2 was chromatographed on phosphocellulose. This chromatography was not done in Expt. IV.

mately 1.8-fold. The increase in amino-acid incorporation with eIF2α is linear during the entire incubation period (68). During this time, a negligible amount of radioactivity is lost from the [^{32}P]eIF2[α-P], indicating that the stimulation does not result from dephosphorylation of the factor. Similar results were obtained when *Artemia* eIF2(α-P) was used in a reticulocyte lysate protein-synthesis system (68).

TABLE X
STIMULATION OF POLYPEPTIDE SYNTHESIS BY PHOSPHORYLATED eIF2α IN A FRACTIONATED CELL-FREE PROTEIN SYNTHESIS SYSTEM[a,b]

Additions	[^{3}H]Leucine incorporated	Total radioactivity (cpm) in [^{32}P]eIF2(α-P) at	
		0 min	60 min
None	1,500	—	—
Salt wash	55,500	—	—
Salt wash + eIF2	102,500	—	—
Salt wash + eIF2(α-P)	100,500	20,800	15,300

[a] From Woodley et al. (68).
[b] Each 50-μl reaction mixture contained 0.26 A_{260} unit of salt-washed *Artemia* ribosomes, 0.063 A_{260} unit of rabbit-hemoglobin poly(A)-rich mRNA, 36 μg of rabbit-reticulocyte ribosomal salt-wash, 10 μCi [^{3}H]leucine (50,000 cpm/pmol), and 1.5 μg of *Artemia* eIF2 or 1.4 μg of eIF2(α-P). Under these assay conditions, amino-acid incorporation and loss of radioactivity was linear for the assay period. The ^{32}P-labeled eIF2(α-P) had 19,280 cpm per microgram of protein.

D. Comparison of *Artemia* and Rabbit Reticulocyte eIF2

Both structurally and biochemically, there are differences as well as similarities between eIF2 isolated from *Artemia* embryos and from rabbit reticulocytes. Although both factors have three subunits, the M_r's of the individual polypeptides, as determined by electrophoresis in dodecyl sulfate-containing polyacrylamide gels, are different. The M_r's of the *Artemia* subunits are 52,000, 43,000, and 41,000, whereas those of the reticulocyte factor are 55,000, 52,000, and 37,000. However, the apparent M_r of the reticulocyte subunits, as determined by gel electrophoresis in dodecyl sulfate, may not be correct (87). Although the α-subunit of each factor may be phosphorylated by reticulocyte HCR, only the phosphorylated *Artemia* eIF2 remains active in protein synthesis. A major distinction between the *Artemia* and reticulocyte factors is the sensitivity to Mg^{2+}. Under conditions where Mg^{2+} inhibits Met-tRNA$_f$ and nucleotide binding to the reticulocyte factor, ternary and binary complex formation with *Artemia* eIF2 is not inhibited (39, 40, 67, 76, 88, 89). Evidence for the difference between *Artemia* and reticulocyte eIF2 is based on the following observations.

1. Effect of Mg^{2+} on Ternary- and Binary-Complex Formation

Concentrations of Mg^{2+} as low as 1 mM inhibit ternary-complex formation with reticulocyte eIF2 by approximately 75% (Figs. 5 and 6) (90). However, with *Artemia* eIF2, ternary-complex formation is stimulated by the addition of 3–4 mM Mg^{2+} (Fig. 5).

Reticulocyte and *Artemia* eIF2 preparations form a binary complex with GDP or GTP (68, 72). With *Artemia* eIF2, the K_D for GDP is 3×10^{-7} M and that for GTP about 1×10^{-5} M (68). Similar values were obtained for reticulocyte eIF2 (72). The K_D for GDP binding to reticulocyte eIF2 is 6-fold greater before chromatography of the preparation of the initiation factor on CM-Sephadex (91).

2. Effect of *Artemia* Co-eIF2(C) on Binary- and Ternary-Complex Formation

Protein factors that enhance eIF2 activity have been isolated from several sources, including rabbit reticulocytes, *Artemia*, ascites, and wheat germ (39, 40, 84, 85). Although the precise role of these factors in polypeptide chain initiation *in vivo* is not clear, they nevertheless stimulate eIF2 activity in model reactions, including binary, ternary, and 40-S initiation-complex formation. The interaction of eIF2 with its substrates, GDP, GTP, Met-tRNA$_f$, Mg^{2+}, and the eIF2 ancillary

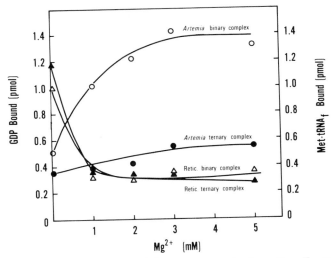

FIG. 5. Effect of Mg^{2+} on binary- and ternary-complex formation. For ternary-complex formation, each 75-μl reaction contained 0.8 μg of *Artemia* or 0.57 μg of reticulocyte (Retic.) eIF2, 0.2 mM GTP, and 3 pmol of Met-tRNA$_f$ (14,500 cpm/pmol). Assay of binary-complex formation was similar, except that no Met-tRNA$_f$ was present and unlabeled GTP was replaced by 0.53 μM [^{3}H]GDP (5000–6000 cpm/pmol). From Mehta *et al.* (90).

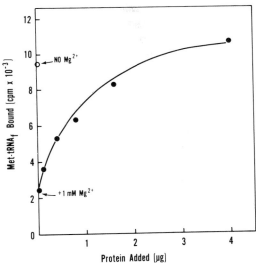

FIG. 6. Reversal of Mg^{2+}-induced inhibition of ternary-complex formation with reticulocyte eIF2. The standard ternary-complex assay in 75 μl contained 20 mM TrisC (pH 7.8), 100 mM KCl, 1 mM Mg^{2+} as indicated, 0.2 mM GTP, 1 mM dithiothreitol, 7! μg of bovine serum albumin, 3 pmol of [^{35}S]Met-tRNA$_f$ (11,350 cpm/pmol), 1.25 μg o reticulocyte eIF2, and amounts of the reticulocyte nucleotide exchange factor as indicated. From Mehta *et al.* (90).

factors have been examined most closely in the reticulocyte system. (For recent reviews, see 39 and 40.)

  a. *Ternary-Complex Formation.* The effect of *Artemia* Co-eIF2(C) and Mg^{2+} on ternary-complex formation with *Artemia* and reticulocyte eIF2 is illustrated in Table XI. As shown in Fig. 7, addition of 1 mM Mg^{2+} inhibits ternary-complex formation with reticulocyte but not with *Artemia* eIF2. The *Artemia* cofactor stimulates reticulocyte eIF2 activity 2-fold in the absence or the presence of Mg^{2+}, but does not effectively reverse the Mg^{2+} inhibition. Co-eIF2(C) stimulates ternary-complex formation 3-fold with *Artemia* eIF2, and there is no effect of added Mg^{2+}.

  b. *Binary-Complex Formation.* We examined the role of *Artemia* Co-eIF2(C) in promoting binding of GDP and GTP to *Artemia* and reticulocyte eIF2 preparations (Table XII). In the absence of Mg^{2+} or Co-eIF2(C), *Artemia* eIF2 binds GDP and GTP, although GDP is bound more efficiently. The reticulocyte factor also binds GDP, but fails to do so effectively with GTP. Addition of Mg^{2+} to reactions containing *Artemia* eIF2 increases the binding of either nucleotide, although that of GTP is selectively enhanced. GDP binding to reticulocyte eIF2 is reduced upon addition of Mg^{2+}, but there is nearly a 2-fold increase in GTP binding (86). In the absence of Mg^{2+}, the addition of Co-eIF2(C) stimulates both GDP and GTP binding to *Artemia* eIF2 2-fold, but only slightly increases nucleotide binding to the reticulocyte factor. However, in the presence of Mg^{2+}, Co-eIF2(C) preferentially stimulates GTP binding to both reticulocyte and *Artemia* eIF2.

TABLE XI
Effects of Mg^{2+} and Co-eIF2(C) on Reticulocyte and *Artemia* eIF2 Activities[a]

Factors added	[^{35}S]Met-tRNA$_f$ bound (cpm)	
	$-Mg^{2+}$	$+Mg^{2+}$
Reticulocyte eIF2	12,000	3,800
Reticulocyte eIF2 + Co-eIF2(C)	25,500	7,500
Artemia eIF2	6,000	5,700
Artemia eIF2 + Co-eIF2(C)	18,100	18,100

[a] Standard nitrocellulose membrane filtration assay conditions were used. Where indicated, reaction mixtures contained 1 mM magnesium acetate. Concentrations of the factors used were reticulocyte eIF2, 2 μg; *Artemia* eIF2, 2 μg; *Artemia* Co-eIF2(C) (phosphocellulose fraction), 8 μg. A filtration background of 150 cpm for [^{35}S]Met-tRNA$_f$ has been subtracted from each value. The specific activity of the [^{35}S]Met-tRNA$_f$ was 15,000 cpm/pmol. A blank containing *Artemia* Co-eIF2(C) was also subtracted from the appropriate data.

TABLE XII
Nucleotide Binding to eIF2 at 30°C[a]

Additions	[³H]Nucleotide bound (cpm)	
	GDP	GTP
Artemia eIF2	4,600	1140
+ Co-eIF2(C)	8,200	2600
+ Mg^{2+}	8,300	6900
+ Co-eIF2(C) + Mg^{2+}	9,500	8700
Reticulocyte eIF2	12,600	700
+ Co-eIF2(C)	18,000	1000
+ Mg^{2+}	5,900	1300
+ Co-eIF2(C) + Mg^{2+}	11,400	3200

[a] Reaction mixtures contained, where indicated, 1.7 µg Artemia eIF2 (for GDP binding) or 3.3 µg (for GTP binding), 2 µg of reticulocyte eIF2, 5.5 µg of step 4 Co-eIF2(C), and 1 mM Mg^{2+}. The concentrations of GDP and GTP were 0.35 and 27 µM, respectively. Appropriate blanks have been subtracted from each value. Nucleotide specific activities were [³H]GDP, 7000 cpm/pmol, and [³H]GTP, 3500 cpm/pmol.

3. Nucleotide Exchange

Gupta *et al.* (*39, 40*) isolated several factors that alter the Mg^{2+} sensitivity of eIF2 preparations from rabbit reticulocytes. Reticulocyte Co-eIF2(A) stimulates ternary-complex formation 2- to 3-fold, but does not reverse the inhibitory effect of Mg^{2+}; Co-eIF2(B) induces the Mg^{2+}-dependent dissociation of a preformed ternary complex at 0°C; Co-eIF2(C) overcomes the Mg^{2+} inhibition. Recently, a high-molecular-weight complex (Co-eIF2) that contains all the individual activities ascribed to Co-eIF2(A), Co-eIF2(B), and Co-eIF2(C) has been isolated (*67*). Apparently, this Co-eIF2 preparation is similar to GEF (RF), the nucleotide exchange factor described below.

Another factor, termed RF, that reverses HCR-induced inhibition of protein synthesis in heme-deficient reticulocytes has been isolated from the postribosomal supernatants of rabbit reticulocyte lysates (Fig. 6) (*88, 92, 93*). RF is apparently the same factor isolated from reticulocytes and ascites cells that reverses the Mg^{2+}-inhibition of ternary complex formation (*39, 40, 64, 91, 95*). The activities of both types of preparations are evaluated on the basis of the exchange of GTP for GDP with a stable eIF2·GDP·Mg^{2+} complex (*90, 91, 94, 95*). In the absence of Mg^{2+}, the eIF2·GDP complex readily exchanges bound GDP for free GTP, but in the presence of Mg^{2+} this complex is stabilized and both nucleotide exchange and ternary-complex formation are inhibited (64, 67, 90, 95). The nucleotide exchange factor is composed of five subunits of approximately M_r 82,000, 67,000, 55,000, 39,000, and 29,000. This factor can be isolated free or in a complex with eIF2 (88, 91, 93, 95).

   a. *Formation and Stability of eIF2·GDP Complexes.* The Artemia or reticulocyte eIF2·GDP complexes are formed at 30°C in the absence of Mg^{2+}. The binding reaction is then chilled to 0°C, the Mg^{2+} concentration is adjusted to 1 mM, and the binary complex is isolated by chromatography on phosphocellulose (90). The isolated eIF2·GDP complex thus formed with eIF2 from either Artemia or rabbit reticulocytes is stable at 30°C in the absence of added nucleotide (Fig. 7), (90). However, the Artemia binary complex readily exchanges bound GDP for free GTP (Fig. 7, top panel), and no dependence on added reticulocyte exchange factor is observed (90). A similar pattern with respect to stability and nucleotide exchange is observed at 15°C, except that the exchange occurs at a slower rate (unpublished observations). This is in contrast to the reticulocyte eIF·GDP complex, which is stable in the presence of added GTP. As seen in Fig. 7, bottom panel, nucleotide exchange is dependent on addition of the nucleotide exchange factor.

   b. *Role of Met-tRNA$_f$ in Nucleotide Exchange.* The kinetics of GDP release from the preformed eIF2·GDP complex, and subsequent ternary-complex formation with Artemia and reticulocyte eIF2 (90) are illustrated in Fig. 8. These reactions are carried out in the presence of 1 mM Mg^{2+}. As reported earlier (68, 90) and shown in Fig. 8A, Artemia eIF2 readily exchanges GTP for prebound GDP in the presence of 1 mM Mg^{2+}. Addition of the reticulocyte exchange factor has no effect on the rate or extent of nucleotide exchange. However, under the same conditions, there was a 3-fold stimulation of ternary-complex formation with Artemia eIF2. Clearly, in this instance, the factor has an effect on ternary-complex formation that is independent of GDP release. From these observations, it is not possible to conclude whether this stimulation is due to enhanced GTP binding or to a direct effect on initiator tRNA binding. Under the same reaction conditions, recitulocyte eIF2 required the addition of the nucleotide exchange factor for GDP release and subsequent ternary-complex formation (Fig. 8B). The addition of Met-tRNA$_f$ does not influence the rate of nucleotide exchange with eIF2 from either Artemia or reticulocytes (compare Figs. 7 and 8), and the amount of Met-tRNA$_f$ bound increases stoichiometrically with the release of GDP from the binary complex.

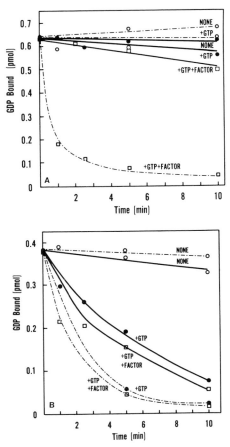

FIG. 7. Kinetics of GDP release from eIF2·GDP or eIF2(α-P)·GDP. The assay mixture in 75 μl contained 100 mM KCl, 20 mM TrisCl (pH 7.8), 1 mM Mg^{2+}, 0.2 mM GTP, and, as indicated, 42 ng of nucleotide exchange factor (90, 91), 0.6 pmol of reticulocyte eIF2-GDP or eIF2(α-P)·GDP (top panel), or 0.4 pmol Artemia eIF2·GDP or eIF2(α-P)·GDP (bottom panel). No addition (○); GTP (●); GTP and nucleotide-exchange factor (□); eIF2·GDP (----) and eIF2(α-P)·GDP (——).

c. Effect of the Nucleotide Exchange Factor on GDP/GTP Exchange with eIF2(α-P)·GDP Complexes. Recent observations indicate that recycling of eIF2 between successive initiation events in reticulocytes is impaired by phosphorylation of the smallest (α) subunit of eIF2 (64, 65, 91, 96). We reported earlier that Artemia eIF2 phosphorylated on this subunit is active in ternary-complex formation, in AUG-dependent binding of Met-RNA$_f$ to 40-S ribosomal subunits, and

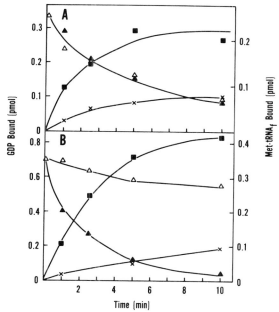

FIG. 8. Kinetics of GDP release and Met-tRNA$_f$ binding during ternary-complex formation with *Artemia* and reticulocyte binary complexes. The components of the assay mixture were as described in Fig. 6. Either *Artemia* eIF2·GDP (0.35 pmol) (panel A), or reticulocyte eIF2·GDP (0.7 pmol) (panel B) was used. GDP retained with factor (▲——▲) and without factor (△——△); Met-tRNA$_f$ bound with (■——■) or without (X——X) factor. From Mehta *et al.* (90).

in cell-free protein synthesis (68). The phosphorylated factor is also recognized by two eIF2 ancillary factors (68). With these observations, we initiated a comparative study on the effect of phosphorylation on *Artemia* and reticulocyte eIF2.

Preparations of eIF2 from *Artemia* and reticulocytes were phosphorylated with [^{32}P]ATP by reticulocyte HCR and reisolated by chromatography on phosphocellulose (68). The ^{32}P-labeled eIF2(α-P) was subjected to isoelectric focusing (97) and autoradiographed. In each case only one labeled polypeptide was detected. Based on precipitation of the factor in hot 5% trichloroacetic acid, approximately 1 mol of phosphate was bound per mole of eIF2.

The effect of the nucleotide exchange factor and Mg^{2+} on GDP binding and ternary-complex formation with phosphorylated eIF2 from *Artemia* and from reticulocytes was examined. It may be seen in Table XIII that, in the absence of Mg^{2+} isolated phosphorylated and

TABLE XIII
FORMATION OF BINARY (eIF2·GDP) AND TERNARY (eIF2·GTP·Met-tRNA) COMPLEX WITH PHOSPHORYLATED [eIF2(α-P)] AND NONPHOSPHORYLATED eIF2 FROM RABBIT RETICULOCYTES AND *Artemia*[a]

Additions	[^{3}H]GDP bound (cpm)		[^{35}S]Met-tRNA$_f$ bound (cpm)	
	$-Mg^{2+}$	$+1$ mM Mg^{2+}	$-Mg^{2+}$	$+1$ mM Mg^{2+}
Reticulocyte				
eIF2	11,200	3,300	8700	1700
eIF2 + factor	10,300	11,000	9900	7700
eIF2(α-P)	9,500	1,300	6900	1600
eIF2(α-P) + factor	9,000	1,400	8800	1600
Artemia				
eIF2	5,300	8,800	4200	6200
eIF2 + factor	8,200	8,500	7500	8400
eIF2(α-P)	4,900	8,200	4800	6400
eIF2(α-P) + factor	6,600	8,500	6300	7100

[a] Reaction mixtures contained, where indicated, 0.5 µg (3.8 pmol) of reticulocyte eIF2 or eIF2(α-P); 1 µg (7.1 pmol) of *Artemia* eIF2 or eIF2(α-P); nucleotide exchange factor, 0.2 µg (approximately 0.4 pmol assuming 70% purity); and 0.2 mM GTP. Filtration backgrounds for [^{3}H]GDP, [^{35}S]Met-tRNA$_f$, and ^{32}P-labeled eIF2 have been subtracted.

nonphosphorylated eIF2 from either source form stable binary or ternary complexes. However, Mg^{2+} inhibits both binary- and ternary-complex formation with reticulocyte eIF2 and eIF2(α-P), but not with either *Artemia* eIF2 or eIF2(α-P) (Table XIII). Addition of the nucleotide exchange factor to binary- or ternary-complex reactions readily reverses the Mg^{2+} inhibition with reticulocyte eIF2, but not with eIF2(α-P). On the other hand, there is an increase in GDP or Met-tRNA$_f$ binding with *Artemia* eIF2 or eIF2(α-P) upon addition of Mg^{2+} Maximum binding of Met-tRNA$_f$ with either eIF2 or eIF2(α-P) occurs in the presence of both Mg^{2+} and the nucleotide exchange factor. The ability of the nucleotide exchange factor to promote GDP/GTP exchange with these eIF2 preparations was also determined.

Preformed, isolated eIF2·GDP binary complexes were used to examine the kinetics of nucleotide exchange with *Artemia* and reticulocyte eIF2(α-P). With either reticulocyte eIF2·GDP or eIF2(α-P)·GDP, GDP release does not occur in either the presence or the absence of GTP (Fig. 7A). Only upon addition of the nucleotide exchange factor is GDP released from the eIF2·GDP complex; the exchange factor does not promote nucleotide release from eIF2(α-P)·GDP. The *Artemia* binary complex is also stable in the absence of free nucleotide

(Fig. 7B). However, in contrast to the reticulocyte binary complex, *Artemia* eIF2·GDP preparations readily release GDP in the presence of added GTP (Fig. 8B) (*68, 90*) and addition of the nucleotide exchange factor does not produce any significant effect on nucleotide exchange with eIF2·GDP or eIF2(α-P)·GDP. *Artemia* eIF2(α-P) exchanges GTP for GDP at approximately 50% the rate of the nonphosphorylated preparation.

The purification of *Artemia* eIF2 includes chromatography on CM-Sephadex, which should quantitatively remove any nucleotide-exchange-factor activity (*88, 95*). Recently we examined the possibility that catalytic levels of an *Artemia* nucleotide exchange factor might be present in *Artemia* eIF2 preparations. Addition of *Artemia* eIF2 to a reticulocyte eIF2·GDP binary complex did not promote nucleotide exchange (unpublished observations). Additional work is necessary to determine whether *Artemia* contains a factor corresponding to the reticulocyte nucleotide exchange factor and to determine the relationship between this presumptive exchange factor and Co-eIF-2(C).

II. Status of Message in *Artemia* Embryos

Eukaryotic mRNA has at least three components that may contribute to translational control: a modified 5' terminus ("cap"), a 3' poly(A) tract, and associated proteins.

A "cap" with the general structure $m^7G(5')ppp(5')N_1$, in which N_1 is the first nucleotide of the actual message, appears at the 5' end of most eukaryotic mRNAs (*43*) and its precursor, "heterogeneous nuclear" RNA (hnRNA), and does not occur in rRNA or tRNA. The $m^7GpppGm$ terminus (N_1 = Gm), as shown by ribosome-protection experiments, comprises part of the ribosomal binding site on the message (*44, 103–107*). However, the observation that translation of both capped (*98–102*) and uncapped mRNAs (*108–112*) is inhibited by cap analogs suggests that multiple structural features of the messenger (including cap, internal sequences, secondary structure (*46, 48, 49*), and position of the AUG codon) may influence the rate and extent of formation of a stable initiation complex (*104–112*).

Recently, the capping and uncapping of eukaryotic messages was proposed to be a mechanism for regulation of translation (*45, 113*). This type of control is unlikely in *Artemia* as the extent of message capping does not vary during development (*101, 114*).

The mRNA in eukaryotic cells is associated with proteins in ribonucleoprotein (mRNP) complexes. Although the reported number and M_r's of the proteins vary (*115–123*), many mRNPs contain two predominant proteins with approximate M_r's of 49,000–52,000 and 73,000–78,000. The larger protein may be associated with the poly(A) "tails" of the mRNA and may disappear as these poly(A) tails are degraded (119).

Functions proposed for the mRNA-associated proteins include selective binding of mRNA to the endoplasmic reticulum (116), protection against ribonuclease attack (119, 158), storage of mRNAs for use later in cellular growth or development (19, 50, 124–128), selection of specific messages for translation (115, 119, 121–123), and translation effectors (120–122). Thus, free and polysomal mRNPs may contain factors for initiation, elongation, termination, and regulation of the translation of the message (129). It has been suggested (123) that eIF2 may be associated with mRNA, but no initiation or elongation factors in mRNP complexes have been found. The differences in results may be due in part to the different experimental techniques utilized.

Initially it was thought that *Artemia* cysts were deficient in mRNA and one or more of the factors required for mRNA translation (27), but it has since been shown (15–17, 19, 20, 50, 124) that dormant embryos have both poly(A)-containing and poly(A)-deficient mRNAs, both of which can be actively translated in cell-free systems. The active message isolated from dormant *Artemia* embryos is found in the cytosolic and membrane fractions as mRNP particles (15–20, 50, 130).

A. mRNP Particles

Dormant *Artemia* embryos were ground according to the procedure of Grosfeld and Littauer (50). mRNPs sedimenting between 20 S and 55 S on linear sucrose gradients were pooled and pelleted by centrifugation (132). The pelleted mRNPs were then fractionated on isopycnic sucrose gradients containing 28 to 87% sucrose. The A_{260} peak banding at $\rho = 1.29$ g/cm³ was further fractionated by chromatography on CL-Sepharose-4B. The mRNP pools were fractionated (139) by electrophoresis in polyacrylamide gels containing dodecyl sulfate (Fig. 9). Molecular weight standards included *Artemia* eIF2, eEF-Tu, and a 27,000-M_r polypeptide that accumulates in dormant embryos (131). The six major polypeptides associated with the mRNPs have M_r's of 32,000 to 30,500, 27,000, 24,000, and 20,000, and are similar to the low-molecular-weight proteins that Huynh-Van-Tan and Schapira (133) found associated with rabbit reticulocyte mRNA.

One of the proteins associated with the mRNP particles comigrates with the 27,000 M_r polypeptide that accumulates in dormant embryos. To determine whether any mRNP proteins were antigenically related to eEF-Tu and the 27,000 M_r protein, antibodies prepared against

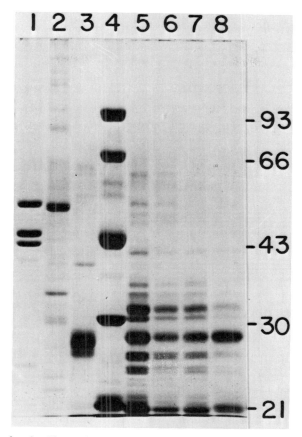

FIG. 9. Dodecyl sulfate/polyacrylamide gel electrophoresis of *Artemia* mRNP fractions. Lane 4 contained molecular weight standards (Bio-Rad) with $M_r \times 10^{-3}$, as indicated; lane 1, *Artemia* eIF2, 6 μg; lane 2, *Artemia* eEF-Tu, 3 μg; lane 3, the 27,000 M_r protein of *Artemia*; lane 5, *Artemia* mRNP from isopycnic gradients; lanes 6–8, mRNP fractions from the CL-Sepharose-4B chromatography step in the order in which they eluted, 12–13 μg each. From Woodley and Wahba (*132*).

these two proteins were tested for cross-reactivity in Ouchterlony preparations with mRNP fractions from the isopycnic gradients. A positive reaction (*132*) was obtained only with antibodies prepared against the 27,000 M_r protein (*131, 134*). Apparently the 27,000 M_r polypeptide that accumulates in dormant embryos, but not eEF-Tu, is a component of mRNP particles (Fig. 9).

The mRNP fractions from the isopycnic sucrose gradient, as well as the message extracted by phenol from these mRNPs, were examined for messenger activity in a wheat germ lysate system (*132*). Under conditions where poly(A)-containing globin mRNA stimulated translation 30-fold over background, amino-acid incorporation with the mRNP fraction and the phenol-extracted mRNA increased 2-fold and 10-fold, respectively. This suggests that the isolated RNA contains competent message, and the proteins associated with it serve to restrict its translation.

The pooled 20- to 55-S material from the sucrose rate-sedimentation gradient, as well as the subsequent isopycnic gradient fraction, were fractionated by electrophoresis in 2% agarose gels containing formaldehyde (*132, 135*). As shown in Fig. 10, the fraction from the sucrose rate sedimentation gradient contains RNA in the 8.5- to 18-S range. Since *Artemia* 28-S RNA is "nicked" (*136*), each fragment essentially comigrates with 18-S ribosomal RNA in denaturing gels. The 18-S RNA isolated from the rate-sedimentation gradient may be of ribosomal origin. However, RNA prepared from the isopycnic gradient is apparently free of ribosomal RNA.

B. mRNA

There is conflicting evidence concerning the properties of *Artemia* mRNA during development. Initially, RNA isolated from dormant embryos appeared to be poly(A)-deficient (*19, 20*), but later, levels comparable to those found in metabolically active cells were observed (*50*). Template activity of mRNA from dormant embryos was found associated with the poly(A)-minus RNA coding primarily for proteins below 30,000 M_r (*19, 20, 50*). These results were interpreted to mean that transcription of new gene products is required for further devel-

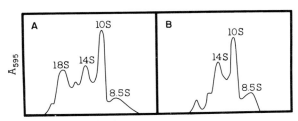

Fig. 10. Electrophoresis in formaldehyde/agarose gels of RNA extracted from *Artemia* mRNP fractions. Electrophoresis was 4.5 hours at 100 V in 2% agarose gels containing 2.2 M formaldehyde (*135*). Panel A, 0.8 A_{260} unit RNA from 40-S RNP isolated from the sucrose rate-sedimentation gradient; panel B, 0.4 A_{260} unit RNA from the isopycnic step. Gels were fixed with 15% trichloroacetic acid, stained with 0.2% methylene blue in 0.4 M potassium acetate (pH 4.3), destained, and scanned on a densitometer. *E. coli* rRNA and tRNA served as M_r standards.

opment. Therefore, a careful examination of template activity of message, from embryos at different stages of development, was undertaken to determine the relative contributions of stored and new mRNA transcripts to the pool of active message.

1. Poly(A) Content and Template Activity of RNA from *Artemia* Embryos

Since *Artemia* embryos contain high levels of ribonucleases, RNA extraction was performed in the presence of vanadyl-ribonucleoside complexes (*137*). The RNA pellets were treated to remove glycogen and residual proteins (*138*). The amounts of poly(A)-containing message at each stage of development was determined by chromatography of the extracted RNA on oligo(dT)-cellulose columns. Poly(A)-containing mRNA was isolated from dormant and 12-hour developing *Artemia* embryos, as well as from 48-hour nauplii. Equivalent amounts of total RNA and poly(A)-containing mRNA were obtained from all stages of development when vanadyl-ribonucleoside complexes were used in the isolation procedure. The extracted RNA was translated in a nuclease-treated rabbit reticulocyte lysate system, and the incorporation of methionine into polypeptides coded for by globin mRNA was used as a standard.

Message from each stage of the developing embryos and nauplii is similarly effective in supporting polypeptide synthesis and is approximately 65% as efficient as globin message (Fig. 11). Maximum translation of *Artemia* message is obtained at 20 μg of RNA per milliliter. Therefore, mRNAs from all stages of development appear to be equally competent as templates for polypeptide synthesis.

2. M_r of Polypeptides Synthesized *in Vitro*

The polypeptides obtained by translation of RNA from *Artemia* at three stages of development were fractionated on 10% polyacrylamide gels containing dodecyl sulfate (Fig. 12) (*139*). With message from dormant embryos, no significant polypeptides above 53,000 M_r are observed. The major ones synthesized are at M_r 53,000 and 27,000 to 29,000 and in a broad area from 14,000 to 24,000. At 12 hours of development, messages for many of the polypeptides in the M_r range of 15,000 to 29,000 are still present, but there is an increase in messages coding for polypeptides in the 30,000 to 50,000 range, as well as for numerous proteins above 55,000. As development proceeds from 12 hours to the naupliar stage, additional quantitative and qualitative changes are apparent. This includes the disappearance of message coding for several proteins in the 36,000 to 38,000 M_r range.

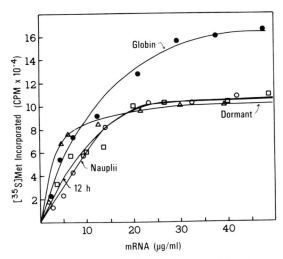

Fig. 11. Translation of *Artemia* and reticulocyte mRNA in an nuclease-treated reticulocyte lysate. Poly(A)-containing RNA was prepared from rabbit reticulocytes, from dormant and 12-hour developing *Artemia* embryos, and from nauplii. [^{35}S]Methionine incorporation into hot trichloroacetic acid-insoluble peptides was assayed in a nuclease-treated reticulocyte lysate (11 μl) containing mRNA as indicated. Globin (●——●); dormant *Artemia* embryos (△——△); 12-hour developing embryos (□——□); and nauplii (○——○).

3. Analysis of Translation Products by Two-Dimensional Gel Electrophoresis

Translation products obtained with *Artemia* message, as well as proteins extracted from *Artemia* embryos, were examined by two-dimensional gel electrophoresis at each stage of development (97). Figure 13 illustrates the autoradiograms of gels containing translation products of message from dormant, 12-hour developing embryos, and nauplii. A number of the changes that occur during resumption of development are summarized in Table XIV. There are proteins whose

Fig. 12. Autoradiograph of *Artemia* and globin mRNA translation products separated in dodecyl sulfate/polyacrylamide gels. A reticulocyte lysate system was programmed with mRNA from either dormant (0 hour) or 12-hour developing *Artemia* embryos, or from nauplii (48 hours). An additional incubation was programmed with globin mRNA. The translation products were processed as in Fig. 11 and were analyzed by electrophoresis in 10% polyacrylamide gels containing dodecyl sulfate. After staining with Coomassie Blue R-250, the gels were treated with EN³HANCE (NEN) and subjected to autoradiography. Each lane represents the radioactive translation products obtained with the indicated message. The markers are: A, eEF-Tu; B, tubulin; C, actin; and D, the 27,000 M_r major translation product of dormant embryos.

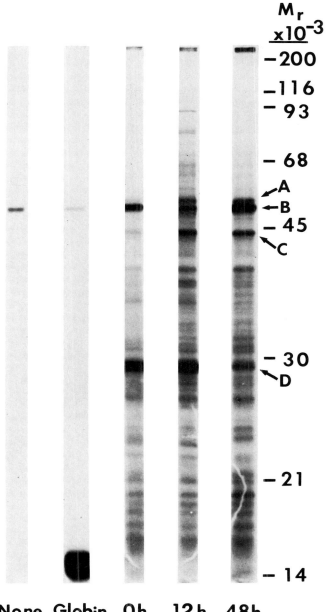

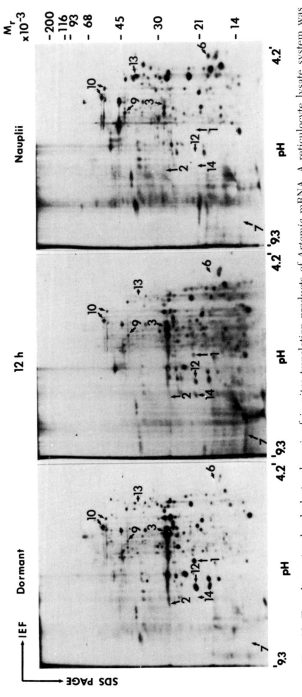

FIG. 13. Two-dimensional gel electrophoresis of *in vitro* translation products of *Artemia* mRNA. A reticulocyte lysate system was programmed with mRNA from either dormant or 12-hour developing embryos or from nauplii. The translation mixtures were lyophilized, dissolved in 9 M urea buffer, and applied to a 3% polyacrylamide gel containing 2% ampholytes (pH 5–9.5) (*95*). Electrophoresis in the second dimension was in 11% polyacrylamide gels containing dodecyl sulfate (*139*). Translation products were visualized by autoradiography as in Fig. 12.

TABLE XIV
POLYPEPTIDES SYNTHESIZED FROM ISOLATED Artemia mRNA

Protein	Characterization
1–3 and 12	Decrease or disappear with development
6, 13	Appear once the embryos reach the naupliar stage
7, 14	Appear transiently at 12 hours of development
9	Actin
10	α- and β-Tubulin

level of message remains nearly constant, increases, decreases, or transiently increases during development. The corresponding *in vivo* proteins, as visualized by silver staining of the polyacrylamide gels (*140*), are shown in Fig. 14 and summarized in Table XV. There are numerous polypeptides synthesized *in vitro* that correspond to *in vivo* proteins that change during the course of development.

4. STAGE-SPECIFIC EXPRESSION OF mRNAs

Based on the amino-acid sequence for eEF-Tu (*195*), a tetradecanucleotide complementary to a segment of the eEF-Tu gene was synthesized and was labeled at the 5′ end using [γ-^{32}P]ATP and T4 polynucleotide kinase. This ^{32}P-labeled oligomer was used to prime the reverse transcription of *Artemia* poly(A)-containing mRNA (*196, 197*). Specificity of priming and thus the purity of the reverse-transcribed eEF-Tu cDNA was established by sequencing the reverse transcription products (*193*).

Both total RNA and poly(A)-containing mRNA from the cytoplasm of *Artemia* embryos was purified from various stages of development (*138, 198*), fractionated by denaturing agarose gel electrophoresis (*199*), and blotted (*200*) onto nitrocellulose. Preliminary results of

TABLE XV
PROTEINS EXTRACTED FROM Artemia EMBRYOS

Protein	Characterization
1–4	Decrease or disappear with development
5	Elongation factor Tu
6–8	Appear after resumption of development
9	Actin
10	α- and β-Tubulin
11	Soybean trypsin inhibitor, included in nauplii preparation to inhibit protease activity

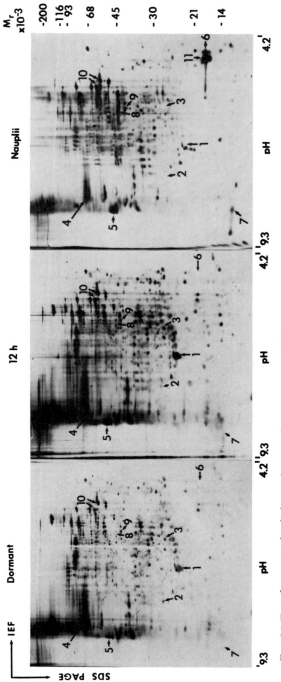

FIG. 14. Two-dimensional gel electrophoresis of proteins extracted from *Artemia*. Proteins were isolated from dormant, 12-hour developing embryos, and nauplii and subjected to two-dimensional electrophoresis as in Fig. 13 (95). Each gel contained approximately 20 μg of protein, which was visualized by silver staining (140).

hybridization (201, 202) with the eEF-Tu-specific cDNA probe demonstrated the presence at all stages of development of mRNA that hybridized with the eEF-Tu-specific cDNA. Furthermore, there was a significant increase in the amount of this mRNA appearing after emergence of embryos, indicating a substantial increase in transcription of the eEF-Tu gene(s) during development.

III. Questions for the Future

The mechanisms by which gene expression is regulated at the levels of transcription and translation during early embryo development of eukaryotic systems is a major, unanswered question in molecular biology. Indeed, the framework for transcriptional and translational controls is incomplete, and many details are lacking. Among the unanswered questions is the manner in which the synthesis of specific gene products required for differentiation is regulated. At the level of translational control, it is not clear how specific messages from the mRNA and mRNP pool are recruited. In addition, it is not known how initiation factor activities are modulated. For example, the roles of ancillary factors and protein kinases in the regulation of eIF2 have not been definitely determined.

Another area that requires clarification is the mechanism by which ribosomes recognize specific sequences and structural features of messages. Even the identification of all the polypeptide chain-initiation factors and their specific functions is uncertain.

Numerous factors associated with development in *Artemia* have been examined, but several questions must be answered before these can be related to development. Some of the pertinent observations are discussed briefly below.

1. Encysted dormant embryos of *Artemia* contain a peptidyl-tRNA hydrolase that converts N-acetylphenylalanyl-tRNA to N-acetylphenylalanine and tRNA, but does not hydrolyze other N-substituted aminoacyl-tRNAs (141). The level of this enzyme does not change during development. In contrast, an unspecific aminoacyl-tRNA hydrolase appears abruptly during development.

2. Little information is available regarding tRNA, aminoacyl-tRNA synthetases, or activators and inhibitors of these enzymes in *Artemia* embryos. There is an inhibitor of aminoacyl-tRNA synthetases in 3T3 and SV101 cells (143); however, details describing the nature of this inhibitor have not been published.

Examples of tRNA-mediated translational control of protein synthesis involving changes in isoaccepting species of tRNA during transition from one cellular growth phase to another have been reported (*144–149*). tRNA was prepared from *Artemia* cysts and nauplii (*26*), and quantitative differences in the degree of aminoacylation of isoaccepting species of tRNA for each of nine amino acids were observed.

3. Protease activity is low in dormant *Artemia* embryos, but for the first 2 days of development four proteases, termed A, B, C, and D, increase in activity (*150, 152*). The increase in protease B, which has a lytic effect on *Artemia* yolk platelets, parallels the decrease in yolk platelets during development. Protease B causes, *in vitro*, a modification of RNA polymerase I to Ia and perhaps also the loss of RNA polymerase III (*152*). There is a proteolytic activity in extracts of *Artemia* nauplii that produces, *in vitro*, the transformation of elongation factor 1 from heavy to a light form (*153*). Such a transformation appears to occur during development (*154*). The presence of proteases could contribute significantly to the morphological changes that occur during development as well as to the regulation of protein turnover. An awareness of the proteases in *Artemia* is essential if protein degradation is to be avoided during purification, especially in view of the inability of α-toluenesulfonyl fluoride to inhibit *Artemia* proteases B, C, and D (*151*).

4. Protein synthesis in rabbit reticulocytes is under the control of inhibitors that may be activated by double-stranded RNA (*154a*) (dsRNA), interferon (*154b*), or heme deprivation (*79*). These inhibitors are thought to be 3',5'-cyclic monophosphate-independent protein kinases that phosphorylate the α subunit of the eIF2 (*155–162*). Inhibitors of translation similar to the rabbit reticulocyte hemin-controlled repressor appear in Friend leukemia (*163*), ascites tumor (*164*), and rat liver cells (*165*).

The role of phosphorylation of *Artemia* eIF2 is not understood. How this phosphorylation is regulated in a system such as *Artemia*, which lacks heme, is not clear. Although phosphorylation of *Artemia* eIF2 by reticulocyte heme-controlled repressor reduces, but does not block, nucleotide exchange, the significance of this observation cannot be fully understood until it is determined whether *Artemia* contains a factor analogous to the reticulocyte nucleotide exchange factor (*88, 92, 93*). Furthermore, we have routinely observed that eIF2 preparations contain a polypeptide of 90,000 M_r that copurifies with the initiation factor. Phosphorylation of eIF2(α) as well as of the 90,000 M_r polypeptide is observed when the eIF2 preparation is incubated with [γ-^{32}P]ATP.

Dormant embryos of *Artemia* contain an inhibitor of polypeptide chain elongation that is found both in the cytosol and associated with the 60-S ribosomal subunits (166, 167). This inhibitor lacks protease and nuclease activity and is heat-labile. It apparently resembles EF2 in function and inhibits poly(U) translation. Whether this inhibitor, which disappears during embryo development, has a role in the translational control of protein synthesis remains to be elucidated.

The presence of inhibitors in Artemia may explain why extracts of dormant embryos are not capable of translating exogenous natural mRNA (27). Regulation of inhibitor activity during early development may control mRNA transcription and subsequent translation of the message. Even if these inhibitors lack true physiological functions, their *in vitro* activities must still be understood to evaluate experimental results.

5. The efficiency of translation may be influenced by the regulation of elongation or termination, but information on these aspects is limited. Elongation factor 1 (EF1) from Artemia cysts (23, 154, 168, 169) and other organisms (25, 170) is an aggregate of subunits. The heavy form of EF1 occurs in dormant embryos, but it is replaced by light EF1 as development progresses. Free-swimming nauplii contain only the light form of EF1 (154). The heavy form of EF1 consists of three polypeptides, α, β, and γ, with M_r's of 53,000, 51,000, and 30,000, respectively, whereas the light form has a M_r weight of 53,000 and lacks the β- and γ-subunits (23, 163). The γ-polypeptide may maintain EF1 activity during periods of biological dormancy and be an important part of the developmental program of Artemia (23).

Artemia EF2 is associated with a high molecular weight nonribosomal complex, perhaps analogous to the heavy form of EF1 isolated from dormant Artemia embryos (24). Upon resumption of development, as with EF1, EF2 activity shifts from the high-molecular-weight aggregate to a low-molecular-weight (96,000 M_r) form.

6. eIF2 may play a key role in the regulation of protein synthesis. However, the basic functions of eIF2 and its interactions with other components of the initiation apparatus are not understood. eIF2 is a three-subunit factor in most eukaryotic systems, although a two-subunit form has been observed (171, 172, 174). We have isolated a two-subunit form of Artemia eIF2 that is as active in ternary complex formation as the usual factor. Treatment of reticulocyte eIF2 with trypsin produced a β-subunit-depleted form of the initiation factor (173). Loss of ternary complex formation after this trypsin treatment appeared to parallel damage to the α-subunit rather than to loss of the β-subunit. The role of eIF2 subunits in nucleotide, Met-tRNA$_f$, and mRNA binding, as well as their interactions with ancillary factors and ribosomal subunits, remains to be elucidated.

7. An important step in binding message to 40-S ribosomal subunits during the initiation cycle is the recognition of the 5'-terminal cap structure. Several proteins have been implicated in the recognition of the cap and the secondary structure of the 5' terminus (175–178). Recently, a protein complex, termed eIF4F, containing polypeptides of 24,000, 50,000, and 200,000 M_r was isolated and found to be essential for cap recognition (179). The 50,000 M_r polypeptide was identified as eIF4A.

The proteins in *Artemia* responsible for message binding to ribosomes have not been determined. The 15-S form of Co-eIF2(C), which facilitates Met-tRNA$_f$ and, to a lesser extent, message binding to 40-S ribosomal subunits, contains a 24,000 M_r subunit (86). The relationship of this protein to the one isolated from rabbit reticulocyte lysates is not known (176, 177). The mRNA binding factor from the high-salt wash of *Artemia* ribosomes greatly stimulates VSV-mRNA binding to 40-S ribosomes. The specificity of this protein for different messages has not been determined. Other initiation factors isolated from rabbit reticulocytes have not been characterized in *Artemia*. These factors include eIF3, eIF5, and eIF6. How these and the other factors participate in the formation of the 80-S initiation complex is yet to be determined.

8. Dormant *Artemia* embryos contain mainly 80-S ribosomes that are devoid of message (180). At present, we are not sure whether the polypeptide-chain-initiation factors are high-molecular-weight aggregates that pellet with 80-S ribosomes during isolation. Dormant embryos offer a source of eukaryotic 80-S ribosomes that may be suitable for studies of ribosome structure, organization, and function.

Purified 80-S ribosomes from HeLa cells contain both loosely and tightly bound protein kinases (181). These kinases phosphorylate primarily 60-S ribosomal proteins, although several 40-S ribosomal polypeptides are also phosphorylated. At least one of these kinases appears to be associated with the 40-S/60-S interface. The role of protein kinases associated with *Artemia* ribosomes is yet to be determined.

9. The purpose and fate of stored mRNP particles in dormant *Artemia* embryos has yet to be elucidated. That there may be more than one mechanism for storage and activation of message was recently suggested (125). DNA-dependent RNA polymerase activities and the amount of polymerase bound to templates increase during development in *Artemia* as well as other developing systems (182–185). The impermeability of cysts makes it difficult to assess which RNAs are increasing during development, and at which stage of development the newly transcribed mRNA may influence development.

The storage of mRNA in mRNP particles in *Artemia* has analogies in sea urchin oocytes. Maternal mRNA is stored until oocytes are fertilized, at which time active protein synthesis occurs (127, 128). During early development of the sea urchin, as in cyst embryogenesis, there is a progressive increase in the number of ribosomes attached to the mRNA (185). The mRNA from oocytes and developing embryos can both be translated, but the mRNA from developing sea urchin embryos is somewhat more active (186, 187). Again, this is analogous to the brine shrimp system.

The bulk of the free mRNP particles in developing sea urchin embryos may not serve as a store of stable mRNA, but may represent transcripts present in excess of what can be translated (188, 189). The proteins attached to mRNA may play a role in translational control by influencing which mRNAs may attach to ribosomes. They may also aid in competition between maternal and newly synthesized mRNA.

Although the sea urchin is the best-studied example of stored mRNA, the relationship between mRNA sequestration and its mobilization on polyribosomes is, as for *Artemia*, obscure. There are many opportunities for further study of this problem in *Artemia*.

10. When analyzed by gel electrophoresis, *in vitro* translation products of *Artemia* mRNA isolated at different stages of development showed differences in a 50,000 M_r polypeptide, which comigrated with eukaryotic elongation factor Tu (eEF-Tu). We therefore began to investigate the expression of the eEF-Tu gene(s) to explain these differences. Based on the amino-acid sequence of eEF-Tu, a mixed oligonucleotide complementary to eEF-Tu mRNA was synthesized and used as a primer for reverse transcription of eEF-Tu mRNA. The primary reverse transcript was identified as cDNA for eEF-Tu by sequencing (193) as well as by hybridization to *Artemia* mRNA (202). The nucleotide sequence data are in close agreement with the partial sequence published by van Hemert *et al.* (203, 204) and extends the sequence 50 nucleotides. *Artemia* RNA isolated from various developmental stages was hybridized to the ^{32}P-labeled reverse-transcription products (202), revealing different steady-state levels of the eEF-Tu transcript during development. Hybridization of reverse-transcription products to *Artemia* chromosomal DNA (200) indicated the possible presence of multiple eEF-Tu genes. Different eEF-Tu clones will be prepared to determine whether true multiple genes or pseudogenes exist in *Artemia* for eEF-Tu. Changes in expression of these genes during development, as well as the nature of the differences between the genes, need to be determined.

Acknowledgments

We thank H. Mehta and H. Daum from our laboratory and other colleagues in the Department of Biochemistry for helpful discussions. This work was supported in part by National Institutes of Health Grant GM-25451.

References

1. F. J. Finamore and J. S. Clegg, in "The Cell Cycle: Gene–Enzyme Interactions" (G. M. Padilla, I. L. Cameron, and C. L. Whitson, eds.), p. 249, Academic Press, New York, 1974.
2. C. C. Hentschel and J. R. Tata, *Trends Biochem. Sci.* **5**, 97 (1976).
3. J. S. Clegg and F. P. Conte, *Brine Shrimp Artemia, Proc. Int. Symp.* Vol. 2, p. 11 (1980).
4. J. S. Clegg, *Nature (London)* **212**, 517 (1966).
5. A. Golub and J. S. Clegg, *Dev. Biol.* **17**, 644 (1968).
6. C. S. Olson and J. S. Clegg, *Wilhelm Roux' Arch. Dev. Biol.* **184**, 1 (1978).
7. D. Epel, *PNAS* **57**, 899 (1967).
8. A. M. Rinaldi and A. Monroy, *Dev. Biol.* **19**, 73 (1968).
9. T. Humphreys, *Dev. Biol.* **26**, 201 (1971).
10. M. Zalohar, *Dev. Biol.* **49**, 425 (1976).
11. J. A. Lovett and E. S. Goldstein, *Dev. Biol.* **61**, 70 (1977).
12. E. S. Goldstein, *Dev. Biol.* **63**, 59 (1978).
13. L. M. Paglia, S. J. Berry, and W. H. Kastern. *Dev. Biol.* **51**, 173 (1976).
14. L. M. Paglia, W. H. Kastern, and S. J. Berry, *Dev. Biol.* **51**, 182 (1976).
15. M. O. Nilsson and T. Hultin, *Dev. Biol.* **38**, 138 (1974).
16. M. O. Nilsson and T. Hultin, *FEBS Lett.* **52**, 269 (1975).
17. H. Grosfeld and U. Z. Littauer, *EJB* **70**, 589 (1976).
18. H. Grosfeld, H. Soreq, and U. Z. Littauer, *NARes* **4**, 2109 (1977).
19. P. O. Amaldi, L. Felicetti, and N. Campioni, *Dev. Biol.* **59**, 49 (1977).
20. J. M. Sierra, W. Filipowicz, and S. Ochoa, *BBRC* **69**, 181 (1976).
21. A. Golub and J. S. Clegg, *Dev. Biol.* **17**, 644 (1968).
22. T. Hultin and J. E. Morris, *Dev. Biol.* **17**, 143 (1968).
23. L. I. Slobin and W. Moller, *EJB* **69**, 351 (1976).
24. Z. Yablonka-Reuveni and A. H. Warner, *Biochem. Artemia Dev. (Proc. Symp.)*, 233 (1979).
25. M. A. Reddington, A. P. Fong, and W. P. Tate, *Dev. Biol.* **63**, 402 (1978).
26. J. C. Bagshaw, F. J. Finamore, and G. D. Novelli, *Dev. Biol.* **23**, 23 (1970).
27. J. M. Sierra, D. Meier, and S. Ochoa, *PNAS* **71**, 2693 (1974).
28. T. Hultin and J. E. Morris, *Dev. Biol.* **17**, 143 (1968).
29. J. S. Clegg and A. L. Golub, *Dev. Biol.* **19**, 644 (1968).
30. S. C. Hand and F. P. Conte, *J. Exp. Zool.* **219**, 17 (1982).
31. J. C. Bagshaw, *Dev. Genet.* **3**, 41 (1982).
32. J. Sebastian, J. Cruces, C. Osuna, and J. Renart, *Brine Shrimp Artemia, Proc. Int. Symp.* Vol. 2, p. 335 (1980).
33. T. Hultin and M. O. Nilsson, *Brine Shrimp Artemia, Proc. Int. Symp.* Vol. 2, p. 83 (1980).
34. Y. C. Chen, C. L. Woodley, K. K. Bose, and N. K. Gupta, *BBRC* **48**, 1 (1972).
35. M. H. Schreier and T. Staehelin, *Nature NB* **242**, 25 (1973).
36. G. L. Dettman and W. M. Stanley, Jr., *BBA* **287**, 123 (1972).

37. D. H. Levin, D. Kyner, and G. Acs, *PNAS* **70**, 41 (1973).
38. U. Maitra, E. A. Stringer, and A. Chaudhuri, *ARB* **51**, 869 (1982).
39. N. K. Gupta, M. Grace, A. C. Banerjee, and M. Bagchi, *in* "Interaction of Translational and Transcriptional Controls in the Regulation of Gene Expression" (M. Grunberg-Managò and B. Safer, eds.), p. 339. Elsevier, Amsterdam, 1982.
40. N. K. Gupta, *Curr. Top. Cell. Regul.* **21**, 1 (1982).
41. M. H. Schreier and T. Staehelin, *Nature NB* **242**, 35 (1973).
42. R. Benne and J. W. B. Hershey, *JBC* **253**, 3078 (1978).
43. A. J. Shatkin, *Cell* **9**, 645 (1976). (Also, This series, Vol. 19)
44. S. Muthukrishnan, M. Morgan, A. K. Banerjee, and A. J. Shatkin, *Bchem* **15**, 5761 (1976).
45. D. L. Nuss, Y. Furuichi, G. Koch, and A. J. Shatkin, *Cell* **6**, 21 (1975).
46. N. Sonenberg, D. Guertin, and K. A. Lee, *MCBiol* **2**, 1633 (1982).
47. N. Sonenberg, D. Guertin, D. Cleveland, and H. Trachsel, *Cell* **27**, 563 (1981).
48. J. C. Chang, G. F. Temple, R. Poon, K. H. Neumann, and Y. W. Kan, *PNAS* **74**, 5145 (1977).
49. F. E. Baralle, *Nature (London)* **267**, 279 (1977).
50. H. Grosfeld, and U. Z. Littauer, *BBRC* **67**, 76 (1975).
51. S. M. Tahara, M. A. Morgan, and A. J. Shatkin, *JBC* **256**, 7691 (1981).
52. J. L. Hansen, D. O. Etchison, J. W. B. Hershey, and E. Ehrenfeld, *MCBiol* **2**, 1639 (1982).
53. W. Filipowicz, J. M. Sierra, C. Nonbela, S. Ochoa, W. C. Merrick, and W. F. Anderson, *PNAS* **73**, 44 (1976).
54. H. Trachsel, B. Erni, M. H. Schreier, and T. Staehelin, *JMB* **116**, 755 (1977).
55. M. Kozak, *Cell* **22**, 459 (1980).
56. A. Marcus, *JBC* **245**, 962 (1970).
57. G. Kramer, D. Konecki, J. M. Cimadevilla, and B. Hardesty, *ABB* **174**, 355 (1976).
58. J. A. Grifo, S. M. Tahara, J. P. Leis, M. A. Morgan, A. J. Shatkin, and W. C. Merrick, *JBC* **257**, 5246 (1982).
59. R. Benne, M. L. Brown-Luedi, and J. W. B. Hershey, *JBC* **253** 3070 (1978).
60. D. T. Peterson, B. Safer, and W. C. Merrick, *JBC* **254**, 7730 (1979).
61. H. Trachsel and T. Staehelin, *PNAS* **75**, 204 (1978).
62. W. C. Merrick, *JBC* **254**, 3708 (1979).
63. M. J. Clemens, V. M. Pain, S. Wong, and E. Henshaw, *Nature (London)* **297**, 93 (1982).
64. J. Siekierka, L. Mauser, and S. Ochoa, *PNAS* **79**, 2537 (1982).
65. H. O. Voorma and H. Amesz, *in* "Interaction of Translational and Transcriptional Controls in the Regulation of Gene Expression" (M. Grunberg-Managò and B. Safer, eds.), p. 297. Elsevier, New York, 1982.
66. B. Safer, R. Jagus, A. Konieczny, and D. Crouch, *in* "Interaction of Translational and Transcriptional Controls in the Regulation of Gene Expression" (M. Grunberg-Managò and B. Safer, eds.), p. 311. Elsevier, New York, 1982.
67. M. K. Bagchi, A. C. Banerjee, R. Roy, I. Chakrabarty, and N. K. Gupta, *NARes* **10**, 6501 (1982).
68. C. L. Woodley, M. Roychowdhury, T. H. MacRae, K. W. Olsen, and A. J. Wahba, *EJB* **117**, 543 (1981).
69. D. M. Valenzuela, A. Chadhuri, and U. Maitra, *JBC* **257**, 7712 (1982).
70. D. W. Russell and L. L. Spremulli, *JBC* **254**, 8796 (1979).
71. S. A. Austin and M. J. Clemens, *FEBS Lett.* **110**, 1 (1980).
72. G. M. Walton and G. N. Gill, *BBA* **390**, 231 (1975).

73. M. J. Clemens, C. O. Echetebu, V. J. Tilleray, and V. M. Pain, *BBRC* **92**, 60 (1980).
74. B. Safer and R. Jagus, *PNAS* **76**, 1094 (1979).
75. R. S. Ranu, *BBRC* **109**, 872 (1982).
76. R. S. Ranu and I. M. London, *PNAS* **76**, 1079 (1979).
77. T. Hunt, in "Recently Discovered Systems of Enzymes Regulated by Reversible Phosphorylation" (P. Cohen, ed.), p. 175. Elsevier, New York, 1980.
78. G. Kramer, M. Cimadevilla, and B. Hardesty, *PNAS* **73**, 3078 (1976).
79. P. J. Farrell, K. Balkow, T. Hunt, R. J. Jackson, and H. Trachsel, *Cell* **11**, 187 (1977).
80. S. Ochoa and S. de Haro, *ARB* **48**, 549 (1979).
81. G. Almis-Kanigur, B. Kan, S. Kospancali, and E. Bermek, *FEBS Lett.* **145**, 143 (1982).
82. A. H. Warner, T. H. MacRae, and A. J. Wahba, *Methods Enzymol.* **60**, 298 (1979).
83. T. H. MacRae, M. Roychowdhury, K. J. Houston, C. L. Woodley, and A. J. Wahba, *EJB* **100**, 67 (1979).
84. S. R. Lax, J. J. Osterhout, and J. M. Ravel, *JBC* **257**, 8233 (1982).
85. J. J. Osterhout, S. R. Lax, and J. M. Ravel, *JBC* **258**, 8285 (1983).
86. C. L. Woodley, H. B. Mehta, D. M. Hunt, and A. J. Wahba, *FP* **41**, 1039 (1982).
87. M. A. Lloyd, J. C. Osborne, Jr., B. Safer, G. M. Powell, and W. C. Merrick, *JBC* **255**, 1189 (1980).
88. J. Siekierka, K.-I. Mitsui, and S. Ochoa, *PNAS* **78**, 220 (1981).
89. A. Das, R. O. Ralston, M. Grace, R. Roy, P. Ghosh-Dastider, H. K. Das, B. Yagmai, S. Palmieri, and N. K. Gupta, *PNAS* **76**, 5076 (1979).
90. H. B. Mehta, C. L. Woodley, and A. J. Wahba, *JBC* **258**, 3438 (1983).
91. A. Konieczny and B. Safer, *JBC* **258**, 3402 (1983).
92. R. O. Ralston, A. Das, M. Grace, H. Das, and N. K. Gupta, *PNAS* **76**, 5490 (1979).
93. H. Amesz, H. Goumans, T. Haubrich-Morree, H. O. Voorma, and R. Benne, *EJB* **98**, 513 (1979).
94. J. Siekierka, V. Manne, L. Mauser, and S. Ochoa, *PNAS* **80**, 1232 (1983).
95. R. Panniers and E. C. Henshaw, *JBC* **258**, 7928 (1983).
96. V. M. Pain and M. J. Clemens, *Bchem* **22**, 726 (1983).
97. R. Z. O'Farrell, H. M. Goodman, and P. H. O'Farrell, *Cell* **12**, 1133 (1977).
98. B. L. Adams, M. Morgan, S. Muthukrishnan, S. M. Hecht, and A. J. Shatkin, *JBC* **253**, 2589 (1978).
99. D. Canaani, M. Revel, and Y. Groner, *FEBS Lett.* **64**, 326 (1976).
100. J. Suzuki, *FEBS Lett.* **79**, 11 (1977).
101. U. Groner, H. Grosfeld, and U. Z. Littauer, *EJB* **71**, 281 (1976).
102. L. A. Weber, E. D. Hickey, and C. Baglioni, *JBC* **253**, 178 (1978).
103. S. G. Lazaravitz and H. D. Robertson, *JBC* **252**, 7842 (1977).
104. M. Kozak and A. J. Shatkin, *Cell* **13**, 201 (1978).
105. M. Kozak, *Nature (London)* **269**, 390 (1977).
106. M. Kozak and A. J. Shatkin, *JBC* **252**, 6895 (1977).
107. S. Muthukrishan, B. Moss, J. A. Cooper, and E. S. Maxwell, *JBC* **253**, 1710 (1978).
108. C. E. Samuel, D. A. Farris, and K. H. Levin, *Virology* **81**, 476 (1977).
109. J. Booker and A. Marcus, *FEBS Lett.* **83**, 118 (1977).
110. W. A. Held, K. West, and J. K. Gallagher, *JBC* **252**, 8489 (1977).
111. S. N. Seal, A. Schmidt, M. Tomeszeuski, and A. Marcus, *BBRC* **82**, 553 (1978).
112. H. F. Lodish and J. K. Rose, *JBC* **252**, 1181 (1977).
113. H. Shinshi, M. Miura, T. Sugimura, K. Shimotohno, and K. I. Miura, *FEBS Lett.* **65**, 254 (1976).

114. S. Muthukrishnan, W. Filipowicz, J. M. Sierra, G. W. Both, A. J. Shatkin, and S. Ochoa, *JBC* **250**, 9336 (1975).
115. P. E. Mirkes, *J. Bact.* **131**, 240 (1977).
116. J. Cardelli and H. C. Pitot, *Bchem* **16**, 5127 (1977).
117. R. U. Mueller, V. Chow, and E. S. Gander, *EJB* **77**, 285 (1977).
118. W. R. Jeffery, *JBC* **252**, 3525 (1977).
119. W. J. Van Venrooij, C. A. G. van Eckelen, R. T. P. Jansen, and J. M. G. Princen, *Nature (London)* **270**, 189 (1977).
120. A. S. Spirin, *FEBS Lett.* **88**, 15 (1978).
121. T. N. Vlasik, L. P. Ovchinnikov, K. M. Radjabov, and A. S. Spirin, *FEBS Lett.* **88**, 18 (1978).
122. L. P. Ovchinnikov, A. S. Spirin, B. Erni, and T. Staehelin, *FEBS Lett.* **88**, 21 (1978).
123. J. G. Hellerman and D. A. Shafritz, *PNAS* **72**, 1021 (1975).
124. L. Felicetti, P. P. Amaldi, S. Moretti, N. Campioni, and C. Urbani, *Cell Differ.* **4**, 339 (1975).
125. J. Simons, E. De Herdt, M. Kondo, and H. Slegers, *FEBS Lett.* **91**, 53 (1978).
126. H. Slegers and M. Kondo, *NARes* **4**, 625 (1977).
127. J. K. Kaumeyer, N. A. Jenkins, and R. A. Raff, *Dev. Biol.* **63**, 266 (1978).
128. N. A. Jenkins, J. K. Kaumeyer, E. M. Young, and R. A. Raff, *Dev. Biol.* **63**, 279 (1978).
129. L. Moens and M. Kondo, *Dev. Biol.* **49**, 457 (1976).
130. H. Slegers, E. De Herdt, and M. Kondo, *EJB* **117**, 111 (1981).
131. L. I. Slobin, *Brine Shrimp Artemia, Proc. Int. Symp.* Vol. 2, p. 557 (1980).
132. C. L. Woodley and A. J. Wahba, *Brine Shrimp Artemia, Proc. Int. Symp.* Vol. 2, p. 591 (1980).
133. Huynh-Van-Tan and G. Schapira, *EJB* **85**, 271 (1978).
134. E. De Herdt, F. De Voeght, J. Clauwert, M. Kondo, and H. Slegers, *BJ* **194**, 9 (1981).
135. H. LeBrach, D. Diamond, J. M. Woznoey, and H. Boedtker, *Bchem* **16**, 4743 (1977).
136. M. P. Roberts, and J. C. Vaughn, *BBA* **697**, 148 (1982).
137. G. E. Lienhard, I. I. Secemski, K. A. Koehler, and R. N. Lindquist, *CSHSQB* **36**, 45 (1971).
138. A. R. Bellamy and R. K. Ralph, *Methods Enzymol.* **12B**, 156 (1968).
139. U. K. Laemmli, *Nature (London)* **227**, 680 (1970).
140. B. R. Oakley, D. R. Kirsch, and N. R. Morris, *Anal. Biochem.* **105**, 361 (1980).
141. J. Miralles, J. Sebastian, and C. F. Heredia, *BBA* **518**, 326 (1978).
142. A. J. Wahba, M. J. Miller, A Niveleau, T. A. Landers, G. G. Carmichael, K. Weber, D. A. Hawley, and L. I. Slobin, *JBC* **249**, 3314 (1974).
143. V. G. Malathi and R Mazumder, *BBA* **517**, 228 (1978).
144. L. Keliman, J. Woodward-Jack, R. J. Cedergren, and R. Dion, *NARes* **5**, 851 (1978).
145. A. Carpousis, P. Christner, and J. Rosenbloom, *JBC* **252**, 8023 (1977).
146. A. Ziberstein, B. Dudock, H. Berissi, and M. Revel, *JMB* **108**, 43 (1976).
147. R. E. Law, A. J. Ferro, M. R. Cummings, and S. K. Shapiro, *Dev. Biol.* **54**, 304 (1976).
148. O. K. Sharma, L. A. Loeb, and E. Borek, *BBA* **240**, 558 (1971).
149. P. Paradiso and P. Schofield, *Exp. Cell Res.* **100**, 9 (1976).
150. A. Olalla, C. Osuna, J. Sebastian, A. Sillero, and M. A. G. Sillero, *BBA* **523**, 181 (1978).

151. C. Osuna, A. Olalla, A. Sillero, M. A. G. Sillero, and J. Sebastian, *Dev. Biol.* **61**, 94 (1977).
152. C. Osuna, J. Renart, and J. Sebastian, *BBRC* **78**, 1390 (1977).
153. T. Twardowski, J. M. Hill, and H. Weissbach, *BBRC* **71**, 826 (1976).
154. L. I. Slobin and W. Möller, *Nature (London)* **258**, 452 (1975).
154a. T. Hunter, T. Hunt, R. T. Jackson, and H. D. Robertson, *JBC* **250**, 409 (1975).
154b. W. K. Roberts, A. Hovanessian, R. E. Brown, M. J. Clemens, and I. M. Kerr, *Nature (London)* **264**, 477 (1976).
155. R. S. Ranu and I. M. London, *PNAS* **73**, 4349 (1976).
156. D. H. Levin, R. S. Ranu, V. Ernst, and I. M. London, *PNAS* **73**, 3112 (1976).
157. R. S. Ranu, D. H. Levin, J. Delauray, V. Ernst, and I. M. London, *PNAS* **73**, 2720 (1976).
158. M. Gross and J. Mendelewski, *BBRC* **74**, 559 (1977).
159. G. Kramer, A. B. Henderson, P. Pinphanichakarn, M. H. Wallis, and B. Hardesty, *PNAS* **74**, 1445 (1977).
160. P. Pinphanichakarn, G. Kramer, and B. Hardesty, *BBRC* **73**, 625 (1976).
161. M. J. Clemens, *Bchem.* **66**, 413 (1976).
162. S. M. Tahara, J. A. Traugh, S. B. Sharp, T. S. Lundak, B. Safer, and W. C. Merrick, *PNAS* **75**, 787 (1978).
163. P. Pinphanichakarn, G. Kramer, and B. Hardesty, *JBC* **252**, 2106 (1977).
164. M. J. Clemens, V. M. Pain, E. C. Henshaw, and I. M. London, *BBRC* **72**, 768 (1976).
165. J. Delaunay, R. S. Ranu, D. H. Levin, V. Ernst, and I. M. London, *PNAS* **74**, 2264 (1977).
166. A. H. Warner, V. Shridhar, and F. J. Finamore, *Can. J. Biochem.* **55**, 965 (1977).
167. F. L. Huang and A. H. Warner, *ABB* **163**, 716 (1974).
168. C. Nombela, B. Redfield, S. Ochoa, and H. Weissbach, *EJB* **65**, 395 (1976).
169. L. I. Slobin and W. Möller, *EJB* **69**, 367 (1976).
170. H. Grasmuk, R. D. Nolan, and J. Drews, *EJB* **67**, 421 (1976).
171. L. L. Spremulli, B. J. Walthall, S. Lax, and J. M. Ravel, *ABB* **178**, 565 (1977).
172. A. Barrieux and M. Rosenfeld, *JBC* **252**, 3843 (1977).
173. G. Zardeneta, G. Kramer, and B. Hardesty, *PNAS* **79**, 3158 (1982).
174. A. Das, M. K. Bagchi, P. Ghosh-Dastidar, and N. K. Gupta, *JBC* **257**, 1282 (1982).
175. K. A. Lee, D. Guertin, and N. Sonenberg, *JBC* **258**, 707 (1983).
176. N. Sonenberg, K. M. Rupprecht, S. M. Hecht, and A. J. Shatkin, *PNAS* **76**, 4345 (1979).
177. K. M. Rupprecht, N. Sonenberg, A. J. Shatkin, and S. M. Hecht, *Bchem.* **20**, 6570 (1981).
178. N. Sonenberg, *NARes* **9**, 1643 (1981).
179. J. A. Grifo, S. M. Tahara, M. A. Morgan, A. J. Shatkin, and W. C. Merrick, *JBC* **258**, 5804 (1983).
180. P. Nieuwenhuysen and J. Clauwaert, *JBC* **256**, 9626 (1981).
181. I. Horak and D. Schiffmann, *EJB* **79**, 375 (1977).
182. J. M. Alession and J. C. Bagshaw, *Differentiation (Berlin)* **8**, 53 (1977).
183. J. Renart and J. Sebastian, *Cell Differ.* **5**, 97 (1976).
184. C. C. Hentschel and J. R. Tata, *Dev. Biol.* **57**, 293, (1977).
185. T. Humphreys, *Dev. Biol.* **26**, 201 (1971).
186. A. M. Pirrone, G. Spinelli, P. Acierno, M. Errera and G. Guidice, *Cell Differ.* **5**, 335 (1977).

187. G. Sconzo, M. C. Roccheri, M. Di Liberto, and G. Guidice, *Cell Differ.* **5**, 323 (1977).
188. M. B. Divorkin and A. A. Infante, *Dev. Biol.* **53**, 73 (1976).
189. M. B. Divorkin, L. M. Rudensey, and A. A. Infante, *PNAS* **74**, 2231 (1977).
190. D. Ish-Horowicz and J. F. Burke, *NARes* **9**, 2989 (1981).
191. D. Alexandraki and J. V. Ruderman, *MC Biol.* **1**, 1125 (1981).
192. D. Filer and A. V. Furano, *JBC* **255**, 728 (1980).
193. A. M. Maxam and W. Gilbert, *PNAS* **74**, 560 (1977).
194. M. L. Goldberg, R. P. Lefton, G. R. Stark, and J. G. Williams, *Methods Enzymol.* **68**, 206 (1979).
195. R. Amons, W. Pluijms, K. Roobol, and W. Möller, *FEBS Lett.* **153**, 37 (1983).
196. G. N. Buell, M. P. Wickens, F. Paybar, and R. T. Schimke, *JBC* **253**, 2471 (1978).
197. E. F. Retzel, M. S. Collet, and A. J. Faras, *Bchem* **19**, 513 (1980).
198. H. Aviv and P. Leder, *PNAS* **69**, 1408 (1972).
199. J. M. Bailey and N. Davidson, *Anal. Biochem.* **70**, 75 (1976).
200. E. Southern, *JMB* **98**, 503 (1975).
201. T. Maniatis, E. F. Fritsch, and J. Sambrook, "Molecular Cloning, A Laboratory Manual." CSH Lab, 1982.
202. P. S. Thomas, *PNAS* **77**, 5201 (1980).
203. F. J. van Hemert, H. van Ormondt, and W. Möller, *FEBS* **157**, 289 (1983).
204. F. J. van Hemert, J. A. Lenastra and W. Möller, *FEBS* **157**, 295 (1983).

Translational Control Involving A Novel Cytoplasmic RNA and Ribonucleoprotein

Satyapriya Sarkar

Department of Muscle Research
Boston Biomedical Research
Institute and
Department of Neurology
Harvard Medical School
Boston, Massachusetts

I.	Eukaryotic RNAs with Translation Modulator Activities	269
II.	The Translation-Inhibiting Cytoplasmic 10-S Ribonucleoprotein of Chick Embryo Muscle	273
III.	The Secondary Structure of iRNA	277
IV.	Dissociation and Reassociation of iRNP	278
V.	Differential Effects of iRNA and iRNP on Endogenous and Exogenous mRNA Translation in Reticulocyte Lysate	281
VI.	Effect of iRNA on the Different Intermediates in Polypeptide Chain Initiation	283
VII.	Concluding Remarks	290
	References	291

In many eukaryotes, where the mRNAs are relatively stable in contrast to short-lived prokaryotic mRNAs, posttranscriptional controls of many cellular events operate as subtle modes of regulation of gene expression. One important aspect of eukaryotic mRNA metabolism that has relevance to translational control is the fact that the primary transcripts containing mRNA sequences in the nucleus as well as the fully processed mRNAs in the cytoplasm are present as ribonucleoproteins (RNP).[1] The RNP particles may play important roles in various cellular processes, such as processing of the primary transcripts, nucleocytoplasmic transport of mRNAs, *in vivo* stability of mRNAs, and regulation of translation. However, our knowledge about the structure and function of various classes of RNP particles is

---

[1] Abbreviations: RNP, ribonucleoprotein; tcRNA, translational control RNA; iRNA and iRNP, translation-inhibiting RNA and ribonucleoprotein particle, respectively; mRNP, messenger ribonucleoprotein; NaDodSO$_4$, sodium dodecyl sulfate; dsRNA protein kinase, protein kinase sensitive to double-stranded RNA; snRNA and snRNP, small nuclear RNA and small nuclear ribonucleoprotein, respectively; ATA, aurin tricarboxylic acid; tRNA$_f$, formylatable methionine tRNA.

quite limited. This has been due mainly to difficulties in obtaining the RNP particles in purified and undenatured forms and to the lack of a well-defined biological assay for the protein moieties of RNP particles.

Among the well-known examples of cytoplasmic regulation of mRNA translation are the heme-regulated protein-synthesis inhibitor that is activated during heme deficiency in reticulocyte lysate (1–3) and the double-stranded RNA-activated inhibitor found in interferon-treated cells and also in reticulocytes (4–9). The activated inhibitors act as protein kinases that specifically phosphorylate the α-subunit of the initiation factor eIF2 (10–17; for reviews, see 18–20). This ultimately leads to inhibition of the polypeptide chain-initiation step, specifically 43-S initiation-complex formation (10, 21–24).

In addition to the above-mentioned eIF2 phosphorylation pathway for translational control, certain small RNA species recently isolated from a wide variety of eukaryotic cells have been implicated in cytoplasmic regulation of mRNA translation (25–43). Some of these RNA species stimulate the *in vitro* translation of mRNAs in a nonspecific manner (26, 27), while others inhibit (31–43). Among the inhibitory RNA species, there is a class of oligo(U)-rich RNA (referred to as translational-control RNA, or tcRNA (34–37) from chick embryo muscle that blocks the translation of specific mRNA in a selective manner (34, 36, 37, 44, 45). Recent work from this laboratory shows that a heterogeneous small RNA species in the 70–90 nucleotide size range (referred to as inhibitory RNA or iRNA) is present in chick embryonic muscle as a novel cytoplasmic 10-S RNP particle (39, 40, 46). Both the 10-S RNP (iRNP) and the deproteinized iRNA are potent inhibitors of *in vitro* translation of a variety of mRNAs (39, 40, 46), suggesting a role of the 10-S iRNP in modulation of mRNA translation. The various small translation modulator RNAs described in the literature appear to be quite diverse with respect to their biochemical nature and mechanism of action.

In order to investigate the role of cellular regulatory mechanisms involving iRNA and iRNP, we have undertaken a detailed study on the characterization and mechanism of action of these macromolecules. It is the purpose of this article to outline recent progress in this area. The results presented strongly imply that iRNA and iRNP are genuine cytoplasmic macromolecular entities involved in modulation of mRNA translation in a subtle manner at a posttranscriptional level.

I. Eukaryotic RNAs with Translation Modulator Activities

Numerous reports in the literature imply that RNAs of low molecular weight ("small" RNAs, M_r = 6000–20,000), isolated from many different types of eukaryotic cells, modulate protein synthesis in cell-free systems. It was first observed that the activity of crude initiation factors, isolated from the 0.5 M KCl wash of rabbit reticulocyte ribosomes, is lost upon dialysis, but is restored by the addition of an RNA fraction isolated from the dialyzate (26). This RNA (M_r 11,000) is purine-rich (46% A, 33% U, 15% C, and 7% G) and stimulates the *in vitro* translation of a variety of mRNAs in a nonspecific manner (27). Subsequently, it was shown that two types of stimulatory factors are present in this dialyzate, an RNA fraction and the polyamines spermine and spermidine (29). Both factors stimulate globin synthesis in an additive manner, and it was suggested (29) that a transitory complex, mRNA·polyamines·RNA, is responsible for the stimulation.

Small U-rich oligoribonucleotides (tcRNA), isolated from the dialyzate of crude initiation factor eIF3 of chick embryonic muscle and chick erythroblasts, inhibit translation in cell-free systems (34, 35). The tcRNA from muscle has an M_r about 6500, is pyrimidine-rich (48% U, 28% C, 12% A, and 12% G), and inhibits specifically the translation of heterologous mRNA, but not of homologous mRNAs (35). The isolation of two types of tcRNA from the cytoplasm of chick embryonic muscle, one form from the polysomes (referred to as polysomal tcRNA) and the other from the cytoplasmic free messenger ribonucleoprotein (mRNP) particles (referred to as mRNP-tcRNA) has also been described (34). The mRNP-tcRNAs are larger than the polysomal tcRNAs (34, 44). They have a higher U content and presumably an oligo(U) tract capable of binding to oligo(dA)-cellulose, and they inhibit homologous mRNA translation in a discriminatory manner (34, 44, 45). In contrast, the polysomal tcRNAs, which have biological properties very similar to those of the tcRNA initially isolated from the dialyzate of eIF3 (35), stimulate the translation of homologous mRNA to a small extent (34, 35).

One of the mRNP-tcRNA species has been isolated from the 90- to 120-S sucrose gradient fractions of postpolysomal particles (37, 44), the same fractions in which the free myosin heavy-chain mRNP is located (47). This tcRNA has been referred to as myosin heavy-chain mRNP-tcRNA (37, 44, 45, 48). Highly purified preparations of this tcRNA, as initially reported, have molecular weights of 10,000 and specifically inhibit the translation of myosin heavy-chain mRNA. The translation of mRNAs coding for other muscle proteins, such as actin and myoglobin, as well as heterologous mRNA, such as globin mRNA, is not inhibited by this tcRNA (34, 44, 45, 48). The tcRNA reacts stoichiometrically with myosin heavy-chain mRNA. At a 1:1 mole ratio, tcRNA inhibits maximally the translation of myosin heavy-chain mRNA and increases the resistance of the mRNA to digestion with RNase T1 and RNase T2, suggesting an interaction of mRNA with tcRNA (44, 45). The presence of a 3'-poly(A) segment in mRNA is necessary for the inhibition by tcRNA, as deadenylated myosin heavy-chain mRNA is not inhibited by mRNP-tcRNA (44, 48). According to a model presented (34, 44) to explain the inhibitory activity of tcRNA, the oligo(U) tract of tcRNA interacts with the 3'-poly(A) "tail" of the mRNA. Presumably, the tcRNA also contains a region or domain that specifically interacts with the 5' end of the mRNA, causing a structural change, the "cirularization" of mRNA (34).

The association of an mRNA present in the free mRNP particles with its specific mRNP-tcRNA has been suggested as the basis of the translation-inhibiting role of tcRNA (44, 45, 48). In support of this view, tcRNA isolated from muscle mRNP particles of different size classes, which does not contain myosin heavy-chain mRNA, does not inhibit the translation of myosin heavy-chain mRNA (44, 48). The tcRNA of mRNP-tcRNA has a nucleotide length of 102 (M_r 36,000 in contrast to the previous value of 10,000), is rich in uracil and guanine residues (37), and differs in many biochemical properties from the previously purified tcRNA (44, 45). Although tcRNA has been isolated from the same mRNP particles from embryonic chick leg muscle, no explanation of these differences has been offered.

A number of points related to the biological effect of tcRNA remain to be answered. The relationship of multiple subspecies of polysomal and mRNP-tcRNA and the nature of the interaction between tcRNA and mRNA is not clearly understood. Further characterization of tcRNA is necessary for understanding the mechanism of its action and its reported ability to inhibit the translation of specific homologous mRNA in a discriminatory manner.

Other small eukaryotic RNA species that show modulating effects in *in vitro* translation systems exist in extracts of *Artemia salina* embryos (30–32), neonatal rat calvaria (33), rabbit reticulocytes (42), duck erythroblasts (43), rat liver (41, 49), and human placenta (50). The properties of some of these RNA species is briefly reviewed here. An RNA from *A. salina*, which is similar to tcRNA isolated from the dialyzate of chick muscle eIF3 (34, 35) in size and base composition

(M_r = 6000; pyrimidine-rich), inhibits the *in vitro* translation of mRNAs in a nonspecific manner (30). At a level of two molecules of RNA per ribosome, 50% inhibition of GTP- and eIF1-dependent binding of aminoacyl-tRNA to 80-S ribosomes was observed. Only at higher concentrations of the RNA was polypeptide chain initiation partially blocked, suggesting that the chain-elongation step is more sensitive to the inhibitor. Interestingly, the inhibitor RNA was active on both eukaryotic and prokaryotic systems.

The amount of the inhibitory RNA was unaltered in dormant or developing stage of the *A. salina* embryos. In contrast, another RNA species stimulating *in vitro* protein synthesis is present only in developing embryos. Like the activator RNA of reticulocytes (26), this RNA is also purine-rich (M_r 9000; 51% G, 10% A, 6% C, and 33% U). However, these two RNA species are not identical in size and base composition (26, 30). The activator RNA from *A. salina* counteracts the effect of the inhibitory RNA from the same source, presumably by complexing with the latter. Since RNase activity in *A. salina* embryos is significantly increased during development, it was suggested that these two RNA species are produced by cleavage of embryo RNA (30). A dual role in the regulation of translation during the development of *A. salina* embryos has been suggested (30), but this remains to be confirmed.

Artemia salina embryos contain stored messenger RNAs as free nonpolysomal mRNP particles in the cryptobiotic stage. A translation-inhibiting RNA species has been reported to be specifically associated with the free poly(A)-containing mRNP particles of *A. salina* embryos, and this association seems to be correlated with the inability of the mRNP particles to be translated *in vitro*. Interestingly, the mRNP particles of *A. salina* embryos lacking poly(A) do not contain the inhibitory RNA species and are translated efficiently in cell-free systems. The inhibitory RNA does not discriminate between homologous and heterologous mRNAs in cell-free translation. Since the RNA has not been characterized, its mechanism of action and relationship to the U-rich translation-inhibiting oligonucleotide isolated from the same source (30) remains to be defined. Small translation-inhibiting RNAs that presumably contain tracts of oligo(U) have also been isolated from neonatal rat calvaria by molecular-sieve and oligo(dA)-cellulose chromatography (33). This RNA inhibited the translations of globin mRNA and of a 26-S procollagen mRNA of rat calvaria in a nonselective manner (33). An RNA species of about 3 S, isolated from the postpolysomal supernatant of rabbit reticulocytes, inhibits the *in vitro* translation of mRNA owing to an effect on the initiation step (42). A similar inhibitory RNA species, about 4 S, which affects primarily the initiation step of polypeptide synthesis, has also been isolated from the free mRNP particles of human placenta (50).

A common parameter related to many of these inhibitory RNA species of diverse cell types is that they are associated with free mRNP particles. This has been reported for tcRNA of chick embryonic muscle (37, 44, 45), the inhibitory RNAs of human placenta (50), rat liver (41, 49), and duck erythroblasts (43). It has been postulated that these RNA species are involved in maintaining sequestered mRNAs in a translationally repressed form as free mRNP particles (37, 43, 44, 49, 50). However, it should be noted that the information in the literature is only suggestive. Any role of inhibitory RNAs in translational repression of mRNAs can be established only by well-documented reconstitution of free mRNP particles, which has not yet been reported.

A poly(A)-containing 7-S RNA isolated from chick embryonic heart promotes differentiation of the cardiac cells *in vitro* (51). This RNA (M_r 80,000) has a 3'-poly(A) tract of about 80 residues and strongly inhibits mRNA translation *in vitro* (52). Since poly(A) itself strongly inhibits under similar conditions, it was concluded that inhibition by 7-S RNA was due to its poly(A) fragment. The small nuclear RNAs, (snRNAs) found in eukaryotic cells have drawn considerable interest because of their involvement in the splicing of primary transcripts containing mRNA sequences (reviewed in 53). Among, these, one species, U_1-snRNA, which has been sequenced, has a 5' cap and an initiator AUG codon and strongly inhibits mRNA translation *in vitro* (54, 55). The inhibitory action is presumably due to the cap structure and the AUG codon present in the snRNA, which competes with mRNA for binding to ribosomes.

It appears from the above survey that small RNAs isolated from a wide variety of eukaryotic cells show translational effects *in vitro*. Some of these RNAs stimulate whereas others inhibit mRNA translation. Some of the observed inhibitions, e.g., that by 7-S cardiac differentiation-promoting RNA and the U_1-snRNA, appear to be due to specific structural features of the RNA molecules and may not necessarily be related to their *in vivo* function. Some of the inhibitory RNA species may be involved in sequestering the mRNAs as free mRNP particles in the cytoplasm.

The question of mRNA specificity in the inhibitory activity of the various RNA species appears to be quite controversial and unresolved. Only for tcRNA has an mRNA discriminatory activity been postulated; all other reports seem to agree with a nonselective mode of inhibition by the various inhibitory RNA species. Thus, an understanding of the structure, function, and mechanism of action of the translation-modulating eukaryotic RNAs will require more information than what is currently available.

II. The Translation-Inhibiting Cytoplasmic 10-S Ribonucleoprotein of Chick Embryo Muscle

The translation-inhibiting cytoplasmic small RNAs (iRNA) of chick embryo muscle (38–40, 56, 57), which appears to be quite distinct from the various eukaryotic RNAs with modulatory activities in cell-free translation systems, was initially isolated in this laboratory from the dialyzate of the 0.5-M KCl wash of chick embryo muscle ribosomes (100,000 g pellet) by a combination of DEAE-cellulose and hydroxyapatite chromatography (38, 58). The iRNA sediments as 4-S RNA in sucrose gradients, shows multiple electrophoretic bands in the 70–90-nucleotide size range, and is a potent inhibitor of translation of both homologous and heterologous mRNAs reticulocyte lysates is treated with micrococcal nuclease. The RNA is purine-rich. It does not bind to oligo(dT)- or oligo(dA)-cellulose, lacks detectable oligo(U) sequences, and is thus different from the oligo(U)-rich tcRNA of chick embryonic muscle, reviewed in the preceding section. Although the purified RNA is similar in size to tRNA, it is clearly distinct by a number of criteria, such as differences in elution profiles from hydroxyapatite columns (38, 58) and its inability to act either as a substrate or inhibitor of *in vitro* aminoacylation using a pH-5 enzyme preparation (38). The latter property indicates that the observed inhibition of mRNA translation by the RNA species (iRNA) is not caused by a nonspecific blocking of the tRNA aminoacylation step.

A cytoplasmic 10-S RNP particle (iRNP) that contains a translation-inhibiting 4-S RNA (iRNA) has been isolated by us (39, 40) and others (57) from chick embryo muscle. The iRNP was purified in this laboratory by a combination of sucrose-gradient fractionation of postpolysomal particles, gel filtration of the 10-S fraction, and ultrafiltration of iRNP on membrane filters (40). The nucleoprotein nature of iRNP was established by biochemical criteria that distinguish RNP particles from deproteinized RNA (40).

Electrophoresis of iRNP on polyacrylamide gels in the presence of sodium dodecyl sulfate shows a complex pattern of about 30 proteins of M_r 12,000 to 150,000 and consisting of both major and minor species (Fig. 1, panel A). The differences in the staining intensity of some of the protein bands imply that iRNP is a family of heterogeneous particles, a situation analogous to the well-known nuclear snRNP particles.

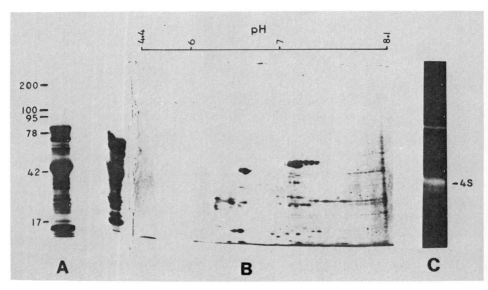

FIG. 1. Polyacrylamide gel electrophoresis of proteins and RNA components of iRNP. For details, see also Sarkar et al. (40). (A) Electrophoretic pattern of proteins in 5 to 15% linear gradient slab gel run in NaDodSO$_4$. The gels were stained with Coomassie blue. Known molecular weights (indicated by the numbers shown × 10^3) are represented at the left. (B) Two-dimensional gel electrophoresis of proteins, stained with Coomassie blue. A sample of proteins of iRNP was run in a separate slot in the second dimension, and the resulting band pattern is shown on the left side of the gel. (C) Electrophoresis of 4 μg of iRNA in 99% formamide. The mobility of a marker sample of yeast tyrosine transfer RNA of 78 nucleotides is indicated. Reproduced from Sarkar et al. (40).

Two-dimensional electrophoresis, using isoelectric focusing in the first dimension and resolution in the presence of NaDodSO$_4$ in the second dimension (59), revealed additional complexities in the protein patterns of iRNP (40). About 45 distinct spots of M_r 14,000 to 80,000, distributed in the pI range of 5.9 to 8.1 (Fig. 1, panel B), were observed. Also, several proteins of similar M_r (e.g., 75,000 to 80,000 and 45,000 to 50,000) were resolved into multiple distinct spots of different pI values. The proteins of iRNP were also compared with those of highly purified preparations of free and polysomal mRNP particles of chick embryonic muscle (60–62) using both one- and two-dimensional electrophoresis. Most, if not all, of the mRNA-associated proteins are absent from iRNP (39, 40). It should be pointed out that both iRNP and the free mRNPs of chick embryo muscle are protein-rich particles [RNA : protein ratio of about 1 : 4 (40, 63)]. Nevertheless, the absence of iRNP proteins in mRNP particles indicates that iRNP is a novel class of cytoplasmic nucleoprotein.

Electrophoresis of iRNA in 99% formamide shows three major bands in the 70–80 nucleotide size range constituting about 85% of the total iRNA (Fig. 1, panel C). In addition, there were two RNA bands in the 90 to 100-nucleotide size range. This heterogeneity of iRNA is also consistent with the view that iRNP represents a family of macromolecules.

Both purified iRNP and iRNA are potent inhibitors of *in vitro* translation of a variety of mRNAs, such as chick muscle poly(A)$^+$ mRNA, rabbit globin mRNA, uncapped mRNA such as EMC virus RNA, and poly(A)$^-$ mRNA of rat liver (Fig. 2). The effect of the inhibitors was tested using optimized conditions for translation of each mRNA species. Both iRNA and iRNP inhibited the translation of mRNAs in a concentration-dependent manner, complete inhibition (about 90–95%) being obtained with 2 to 4 µg/ml of iRNA or equivalent amount of iRNP (40). The iRNA did not interfere with the aminoacylation of tRNA, nor did tRNA mimic the activity of iRNA. With the exception of poly(A)$^-$ mRNA of rat liver, iRNA and iRNP were equally effective as inhibitors of translation for each mRNA tested. As judged by the concentration required to produce 50% inhibition, translation of globin mRNA was inhibited somewhat less than that of chick muscle poly(A)$^+$ mRNA (Fig. 2, A and B). Since *in vitro* translation assays are strongly influenced by numerous variables, this observed difference should not be taken as evidence for a discriminating effect of iRNA and iRNP on homologous and heterologous mRNA translation. In incubations programmed with chick muscle poly(A)$^+$ mRNA, iRNA and iRNP inhibited the linear phase of amino-acid incorporation without any lag, and the degree of inhibition was dependent on the concentration of iRNA and iRNP (40). Fluorography of the *in vitro* translation products in incubations containing iRNA and iRNP indicated that the synthesis of the authentic polypeptides coded by rabbit globin mRNA and chick muscle poly(A)$^+$ mRNAs is inhibited in a concentration-dependent manner by the inhibitors (46, 58). The possibility that iRNA and iRNP may block selectively the translation of homologous mRNAs (i.e., coding for specific myofibrillar proteins such as myosin heavy chain) in a manner similar to that reported for tcRNA (37, 44, 45, 48) was also tested. Comparing the relative intensities of the fluorogram bands corresponding to different marker myofibrillar proteins as a function of increasing concentration of iRNA or iRNP, it was observed that the synthesis of all myofibrillar proteins is inhibited in a nonselective manner (38, 39, 58).

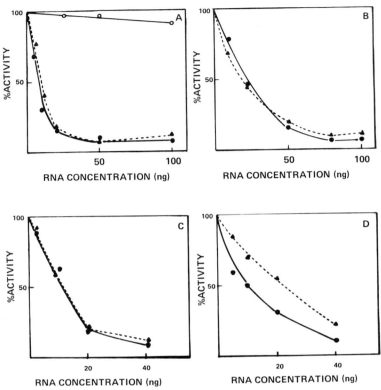

FIG. 2. Inhibition of translation of various mRNAs by iRNA and iRNP in micrococcal nuclease-treated rabbit reticulocyte lysate. For details, see also Sarkar et al. (40). Incubations (volume 25µl) contained (A) 0.4 µg of muscle poly(A)$^+$ mRNA; (B) 0.4 µg of globin mRNA; (C) 0.3 µg of EMC virus mRNA; (D) 0.4 µg of poly(A)$^-$ mRNA of rat liver. ● = iRNA; ▲ = iRNP; ○ = embryonic chick muscle tRNA (panel A). The abscissas represent the concentrations of inhibitors used per assay as the equivalent RNA content. The radioactivities incorporated into proteins in 5-µl portions of the reaction mixtures were nuclease-treated lysate, 98 cpm; lysate + muscle poly(A)$^+$ mRNA, 2500 cpm; lysate + globin mRNA, 2850 cpm; lysate + EMC virus mRNA, 1330 cpm; lysate + poly(A)$^-$ mRNA, 1150 cpm. Reproduced from Sarkar et al. (40).

Several possibilities that may account for the observed biological activity of iRNA and iRNP were also tested. The possibility of a nuclease activity associated with both iRNA and iRNP was eliminated by showing that radiolabeled synthetic polyribonucleotides such as poly(A) and poly(A, C) as well as *B. brevis* rRNAs, were not cleaved after incubation with iRNA or with iRNP (40). The possibility that mRNA is specifically degraded in the presence of iRNA was also eliminated by the fact that both the sedimentation profile and the electrophoretic pattern of 9-S globin mRNA remained unaltered after incubation with iRNA. (40). Also, inhibition by iRNA is not accounted for by irreversible inactivation of mRNA due to interaction with iRNA. This possibility was tested by incubating a sample of 9-S globin mRNA with iRNA and analyzing the reaction mixture by gel filtration on an Ultro-Gel ACA-34 column capable of resolving 4-S iRNA and 9-S globin mRNA. The incubated reaction mixture was quantitatively resolved into two RNA peaks by gel filtration. Also, both the iRNA and the 9-S globin mRNA fractions, thus obtained, retained full biological activities (40).

III. The Secondary Structure of iRNA

The possibility that iRNA behaves like double-stranded RNA in stimulating the phosphorylation of the α-subunit (M_r = 38,000) of eIF2 in reticulocyte lysates should be carefully considered, as this would account for the potent translation inhibitory activity of iRNA. Three lines of evidence obtained during the initial phase of our studies strongly suggest that the biological activity of iRNA does not arise from its extensive double-stranded nature or the presence of a double-stranded RNA in the preparations. These are that (a) treatment of iRNA with pancreatic RNase, which does not cleave double-stranded RNA, resulted in complete loss of biological activity (40); (b) heating of iRNA at 75° for 3 min followed by quick cooling, which should destroy the double-stranded structure, caused no change in biological activity of the iRNA (39, 40), (c) hyperchromicity measurements of iRNA with increasing temperature showed that iRNA melts over a wide temperature range (30–80°C) in a progressive and noncooperative manner (38–40). A net hyperchromicity of about 0.35 was obtained for iRNA; this value was indistinguishable from that obtained with a sample of poly(A), which is known to melt like a single-stranded homopolyribonucleotide. Comparative studies carried out with chick embryo muscle tRNA showed that tRNA melts differently over the range of 70 to 80° with about 20% larger hyperchromicity than iRNA (39). This agrees with the view that tRNA contains considerable secondary structure (64). It was concluded that the hyperchromicity profile of iRNA reflects primarily the unstacking of bases.

The average number of base-paired nucleotides in iRNA was also estimated in this laboratory by a sensitive Tb^{3+}-induced fluorescence method (65). Tb^{3+} complexed only with the single-stranded bases of nucleic acids gives characteristic fluorescence enhancements at 490 and 540 nm. Double-stranded nucleic acids do not show this fluorescence enhancement. By comparing the fluorescence increments of Tb^{3+}–iRNA, and Tb^{3+} complexes with chick muscle tRNA and yeast tyrosine tRNA (78 nucleotides long with about 50% base-pairing on the average; gift from U. L. RajBhandary), it was estimated that the base-paired nucleotides in iRNA fall within the range of 12–14 nucleotides (66 and unpublished results). The effect of iRNA on eIF2 phosphorylation, using an *in vitro* assay system for the latent double-stranded RNA-sensitive protein kinase of rabbit reticulocytes (ds protein kinase) (18–20) is presented in Section VI.

The translation-inhibiting activity of iRNP remains unaltered after treatment with insoluble pancreatic RNase (39, 40). This is because iRNP is RNase-resistant whereas deproteinized iRNA is RNase-sensitive (40).

IV. Dissociation and Reassociation of iRNP

The possibility that a nonspecific association of cytoplasmic proteins with iRNA leads to the formation of a 10-S RNP-like complex must be eliminated in order to establish any regulatory role in iRNP in cellular processes. Therefore, the dissociation of iRNP and its reassociation from the dissociated components were studied in order to probe the nature of the interaction of protein and nucleic acid moieties in the process. Native iRNP, which is RNase-resistant, becomes RNase-sensitive when the particle is exposed to ionic conditions in which K^+ and Mg^{2+} are absent (46). Furthermore, the RNase-sensitivity of the particle thus obtained is quantitatively similar to that of deproteinized iRNA (40, 46). These results imply that both K^+ and Mg^{2+} are required for maintaining the nucleoprotein structure of iRNP. Based on this observation, we developed a convenient method for the dissociation of iRNP into protein and RNA components using mild conditions.

Samples of 10-S iRNP were applied to DEAE-cellulose in "dissociation buffer" (10 mM TrisCl, pH 7.6, and 5 mM EDTA), and the bound material (about 90–95%) was eluted with a linear KCl gradient (46). Two well-resolved peaks were obtained. The first peak, eluted at 0.18 M KCl, contained the proteins of 10-S iRNP. The second peak, eluted at 0.42 M KCl, contained iRNA. Electrophoresis of the material present in the two peaks indicated that the full complement of the protein and RNA components of iRNP were recovered in the two resolved peaks (46).

The protein and RNA components resolved by DEAE-cellulose

were tested separately for biological activity in nuclease-treated reticulocyte lysates programmed with globin and muscle poly(A)$^+$ mRNA. The RNA inhibited the translation of both mRNA species in a concentration-dependent manner (46). Also, the observed inhibition was almost identical to that obtained with deproteinized iRNA isolated from iRNP by extraction with phenol/CHCl$_3$/isoamyl alcohol (40). In contrast, the proteins resolved by DEAE-cellulose showed no inhibition of translation of either mRNA (46).

When samples of native 10-S iRNP and the DEAE-cellulose-resolved RNA and protein moieties were subjected to gel filtration separately on an Ultogel ACA-34 column, each species was eluted as a characteristic single peak (positions indicated in Fig. 3, panel A). We then tested for the reassociation of the dissociated protein and RNA

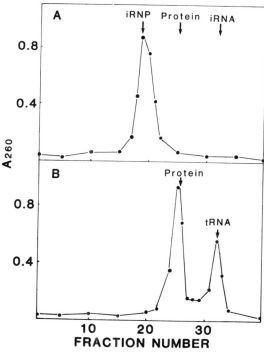

FIG. 3. Specificity of the reassociation of dissociated components of iRNP. For details, see also Mukherjee and Sarkar (46). (A) Reassociation of the protein and RNA components of iRNP. The arrows indicate the positions where the protein and RNA components appeared in separate runs on the same column. (B) Elution profile of incubation mixture containing proteins of iRNP and chick muscle tRNA. Reproduced from Mukherjee and Sarkar (46).

moieties by a method used for the reconstitution of 30-S ribosomal subunits from 16-S rRNA and 30-S proteins (67). The protein and RNA samples in a ratio of 4:1 were incubated for 30 min at 35°C in the "reassociation buffer" (25 mM TrisCl, pH 7.6, 10 mM $MgCl_2$, 0.3 M KCl, 2 mM dithiothreitol) and then analyzed by gel filtration on an Ultrogel column. The material was quantitatively recovered in a single peak whose position corresponded to that of 10-S iRNP (Fig. 3, panel A). Reassociation did not occur to a significant degree when the reaction mixtures were incubated in the absence of K^+ or Mg^{2+}, or at 0°C. The specificity of the reassociation process was indicated by the inability of the proteins of iRNP to form a 10-S complex with chick embryo muscle tRNA under identical conditions. Gel filtration of the reaction mixture gave quantitative recovery of the tRNA and protein samples as two well-resolved peaks at positions corresponding to those of the DEAE-cellulose-dissociated protein and RNA components (Fig. 3, panel B). Similarly, the proteins of iRNP showed no detectable degree of reassociation with globin mRNA (46).

The reassociated 10-S RNP particle was indistinguishable from the native iRNP in properties such as UV spectrum, buoyant density, RNase-resistance, and ability to inhibit translation of both homologous and heterologous mRNAs (46).

Since iRNP is RNase-resistant and its protein moieties per se do not inhibit mRNA translation, it is quite likely that an *in vivo* role of the protein moieties is to provide a RNase-resistant structure to iRNA, which would otherwise, because of its predominantly single-stranded nature, be cleaved by intracellular nucleases. It should be noted that not only are the proteins of iRNP electrophoretically distinct from those of the cytoplasmic mRNP particles (39, 40), but they are also distinct from the relatively simple protein patterns of eukaryotic snRNP particles as reported in the literature (53, 68, 69). Although iRNP particles are similar in size to snRNP particles, iRNA differs from snRNA in both size and base composition (53, 55, 68, 69). It has been observed by us that the proteins of iRNP show no detectable immunopositive reaction with a number of monospecific autoimmune antibodies (e.g., Sm, RNP, PM-1, SSB, MA, and ScL-70; reviewed in 53, 70) to nuclear antigens. These results of iRNP, considered together with the specificity of the reassociation of iRNP from the dissociated RNA and protein components (Fig. 3) lead to the conclusion that iRNP represents a true cellular entity unrelated to other classes of nuclear and cytoplasmic RNP particles (46).

V. Differential Effects of iRNA and iRNP on Endogenous and Exogenous mRNA Translation in Reticulocyte Lysate

The effects of iRNA and iRNP on endogenous mRNA translation in reticulocyte lysates not treated with micrococcal nuclease reveal some interesting differences from those observed with exogenous mRNA translation in nuclease-treated lysates (66). In the presence of levels of iRNP and iRNA sufficient to inhibit quantitatively exogenous globin mRNA translation (Fig. 2, panel B), only about 15% inhibition of amino-acid incorporation was observed in untreated lysate. Furthermore, in the presence of 0.1 mM aurin tricarboxylic acid, a specific inhibitor of peptide chain initiation, this small degree of inhibition by iRNP disappeared. It was also observed that high levels of iRNA and iRNP (4 μg/ml) caused only a 15% inhibition of amino-acid incorporation in nuclease-treated lysate programmed with 0.5 M KCl-washed muscle polysomes. Since amino-acid incorporation in polysome-directed incubations represents primarily the elongation phase of peptide synthesis, the above results are consistent with the view that the translation of exogenous and endogenous mRNAs in reticulocyte lysate is inhibited in a contrasting manner by iRNA and iRNP.

The effects of iRNA and iRNP on exogenous mRNA translation were also compared with those obtained with cycloheximide and aurin tricarboxylate, known inhibitors of the elongation and initiation phase of peptide synthesis, respectively. The inhibitors were added at different times after amino-acid incorporation was started in globin-mRNA-directed incubations. The time course of amino-acid incorporation shows that cycloheximide caused an immediate cessation due to blocking of the elongation step (Fig. 4, panel A). In contrast, after the addition of aurin tricarboxylate, amino-acid incorporation continued at a slightly diminished rate for a measurable time before it finally stopped (Fig. 4, panel A). The duration of the continued incorporation after the addition of the inhibitor reflects approximately the time required for the completion of nascent chains, since reinitiation does not occur in such incubations. The kinetics of incorporation after the addition of 3 μg of iRNA per milliliter at two different time points is strikingly similar to that obtained with aurin tricarboxylate (Fig. 4, panel B), and is quite distinct from that obtained with cycloheximide.

The effect of iRNA was also tested in another *in vitro* assay system in which the rate of peptide-chain elongation can be directly monitored. Nuclease-treated reticulocyte lystate programmed with exoge-

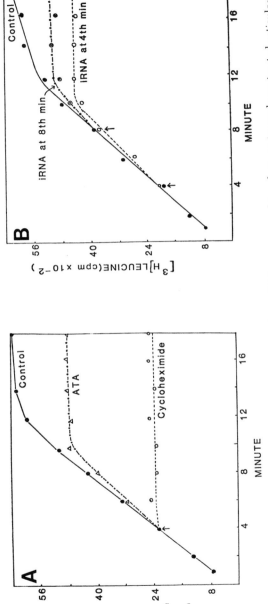

FIG. 4. Time course of amino-acid incorporation in globin mRNA-programmed incubations in nuclease-treated reticulocyte lysate containing cycloheximide, aurin tricarboxylate (ATA), and iRNA or iRNP. For details see references 40 and 58. Conditions for cell-free protein synthesis were the same as previously described (40). The arrows indicate the time points where the various inhibitors were added separately in identical incubations after the reaction was started. Panel A: Effect of cycloheximide and ATA. ●——●, Control incubation globin mRNA; ○--○, 0.2 μg of cycloheximide was added 4 minutes after incubation was started (indicated by the arrow); △--△, 0.1 mM ATA was added after 4 minutes of incubation. Panel B: effect of iRNA. ●——●, control incubation containing globin mRNA; ○--○, 75 ng of iRNA was added 4 min after incubation was started; ○--○, 75 ng of iRNA was added 8 min after incubation was started. Essentially similar results were obtained when iRNP containing 70 ng of iRNA were used. Two-mitroliter samples were withdrawn at different times for the estimation of radioactivities incorporated into proteins, as previously described (40).

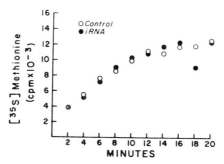

FIG. 5. Effect of iRNA on elongation rate of polypeptide synthesis. TMV mRNA (30 μg/ml) was incubated for 10 minutes in 100 μl of a nuclease-treated reticulocyte lysate. Edeine (10^{-4}M) and [^{35}S]methionine were added, and the lysate was divided into two equal aliquots. Either water or iRNA (10 μg/ml) was added, and 2.5-μl aliquots were removed at 2-minute intervals for determination of [^{35}S]methionine protein-bound counts. Reproduced from Winkler et al. (56).

nous mRNA, such as TMV RNA, was incubated for 10 minutes; during this time the mRNA became loaded with ribosomes. Then, edeine, a potent and specific inhibitor of peptide-chain initiation (70a), was added together with [^{35}S]methionine. The lysate was divided into two equal portions, one of which served as the control; to the other iRNA, 10 μg/ml was added. Incorporation of labeled methionine into nascent chains under these conditions is an indicator of the elongation phase, since edeine blocks reinitiation on the exogenous mRNA. As shown in Fig. 5, the rate of methionine incorporation in the presence of iRNA was virtually identical to that obtained in the control incubation. These results confirm that iRNA has no detectable effect on the elongation phase of protein synthesis (56).

VI. Effect of iRNA on the Different Intermediates in Polypeptide Chain Initiation

Multiple experimental approaches were used by us to gain insight into the mechanism of inhibition of peptide chain initiation by iRNA. First, the possibility that the limited base-paired domain(s) in iRNA, about 12–14 nucleotides, estimated from Tb^{3+} fluorescence measurements (66), may serve as an activator of the dsRNA-protein kinase, leading to eIF2 phosphorylation was considered by testing directly the effect of iRNA on eIF2 phosphorylation using an in vitro dsRNA-protein-kinase assay system (71). As shown by autoradiograms in Fig. 6, under conditions where double-stranded reovirus RNA was fully

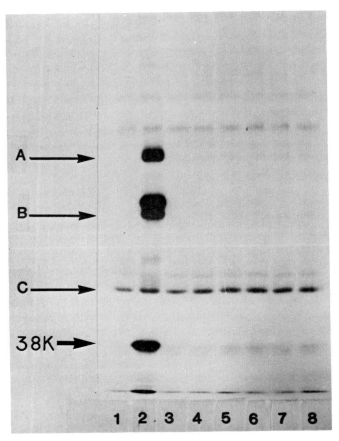

Fig. 6. Effect of iRNA on dsRNA-protein-kinase activity. The 25-µl incubation mixture contained 3 µg of purified latent dsRNA protein kinase, 0.5 µg of purified eIF2 of rabbit reticulocytes (gifts from I. M. London), 10 µCi of [γ-^{32}P]ATP (20 µM), and indicated amounts of the various RNA samples. Protein-kinase assay, slab gel electrophoresis of the incubation mixture, and autoradiography were carried out as previously described (71). Autoradiograms of gel runs of the incubation mixtures are shown. Lane 1, minus RNA; lane 2, reovirus RNA (dsRNA), 40 ng/ml; lane 3, iRNA 25 ng/ml; lane 4, iRNA 100 ng/ml; lane 5, chick muscle rRNA, 29 ng/ml; lane 6, chick muscle rRNA, 116 ng/ml; lane 7, chick muscle tRNA, 26.5 ng/ml; lane 8, chick muscle tRNA, 106 ng/ml. A, B, and C indicate positions of labeled protein bands (M_r = 94,000, 68,000, 50,000) autophosphorylated during activation of dsRNA protein-kinase (71). 38K is the position of the α-subunit of eIF2.

active in stimulating that latent dsRNA-protein kinase and gave pronounced phosphorylation of the α-subunit of eIF2 ($M_r = 38,000$), samples of iRNa, chick muscle rRNA, and tRNA gave no detectable activity. Thus, it appears that translation-inhibition by iRNA is not due to nonspecific phosphorylation of the α-subunit of eIF2.

The effect of iRNA on different steps in polypeptide chain initiation, using an unfractionated reticulocyte lysate as described earlier (72) was also tested. According to this assay (the "shift assay"), the formation of various "intermediates" in the peptide chain-initiation process, such as Met-tRNA$_f$·0-S (the 43-S complex) and the complete 80-S initiation complex, can be conveniently monitored in a sequential manner. Elongation is inhibited by the addition of diphtheria toxin to the lysate. Then [^{35}S]Met-tRNA$_f$ is added, which exchanges with unlabeled initiator tRNA in the 43-S initiation complexes. If the reaction mixture is fractionated at this point on sucrose gradients, a prominent radioactive peak associated with the 40-S ribosomal subunit is observed (Fig. 7, panel A). If mRNA is now added, it binds to the labeled 43-S complex, a 60-S subunit is joined, and the labeled complexes are shifted to 80 S on sucrose gradients (Fig. 7, panel B). The elongation inhibitor prevents any further change from taking place. With globin mRNA as the message for the shift assay, iRNA has no effect on the formation of Met-tRNA$_f$·40-S complex (Fig. 7, panels A and C). Rather, it shows a reproducible stabilization of the complex, presumably by affecting the dissociation of Met-tRNA$_f$ from the 43-S. The iRNA strongly inhibits the subsequent shift of the Met-tRNA$_f$· 40-S complex to the complete 80-S initiation complex (Fig. 7, panels B and D). These results strongly suggest that the mRNA-dependent conjugation of the 60-S subunit to the 43-S complex is specifically blocked by iRNA (56).

The iRNA effect on peptide chain initiation also appears to be quite distinct from those of other initiation inhibitors, such as the heme-regulated repressor and double-stranded RNA, both of which cause a reduction in the level of 43-S initiation complexes (72, 73).

Another possible site of action of iRNA in the initiation process is the conjugation of 60-S subunits following the binding of mRNA to the 43-S initiation complex. This seems unlikely since [^{35}S]Met-tRNA$_f$ is expected to sediment at 48-S as an mRNA-initiation complex (74). However, the possibility that Met-tRNA$_f$ is hydrolyzed in iRNA-inhibited lysates with a concomitant loss of radioactivity from the 48-S complex should not be ruled out. The addition of iRNA after the shift reaction had no effect on the 80-S complex.

We also carried out the shift assay with labeled mRNA in order to

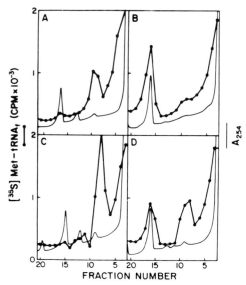

FIG. 7. Effect of iRNa on the shift assay. Untreated reticulocyte lysate (200 μl) was incubated for 3 minutes, and then elongation was inhibited by the addition of NAD (150 μM) and diphtheria toxin (50 μg/ml). After 5 minutes, approximately 1.1×10^6 cpm of [^{35}S]Met-tRNA$_f$ was added per milliliter. After 2 minutes the following additions (10 μl) were made: (A) water; (B) globin mRNA (20 μg/ml); (C) iRNA (8 μg/ml): or (D) iRNA followed after 2 minutes by globin mRNA. After another 2 minutes 25-μl aliquots were diluted with 125 μl of gradient buffer and centrifuged for 2.5 hours at 50,000 rpm on 15 to 40% sucrose gradients in an SW-56 Beckman rotor at 4°C. Gradients were examined for absorbance at 254 nm (———) and for acid-insoluble (0°C) radioactivities (●———●). Sedimentation was from right to left. Reproduced from Winkler et al. (56).

sort out whether iRNA inhibits the mRNA binding to the 43-S complex or the conjugation of the 60-S subunit. Reovirus mRNA was selected as the message for two reasons. Considerable difficulty was encountered in preparing biologically active globin mRNA of high specific activity by *in vitro* labeling. We had shown previously that iRNA strongly inhibits the *in vitro* translation of reovirus mRNA (75). Also, control experiments indicate that reovirus mRNA induces a shift reaction that is also blocked by iRNA. The effect of iRNA on the binding of labeled reovirus mRNA to 80-S ribosomes is shown in Fig. 8. A large reduction in the amount of labeled reovirus mRNA bound to 80-S ribosomes and a decreased level of radioactivity sedimenting in the 40-S region were observed in the presence of iRNA. These results indicate that iRNA blocks mRNA binding to the 43-S initiation complex. If the conjugation of 60-S were inhibited by iRNA, the amount of

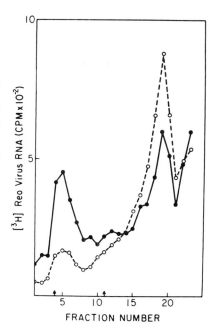

FIG. 8. Effect of iRNA on reovirus RNA binding in the shift assay. (For details, see also the legend to Fig. 7.) Untreated reticulocyte lysate (200 μl) was treated with NAD (150 μM) and diphtheria toxin (50 μg/ml) and incubated for 5 minutes. Then 8 μg of iRNA (○---○) or water (●——●) as a control was added. After 2 minutes, 10,000 cpm of reovirus [^{3}H]RNA was added. After 2 more minutes, 25-μl aliquots were diluted with 125 μl of gradient buffer and centrifuged for 170 minutes at 50,000 rpm on 15 to 40% gradients in an SW50.1 Beckman rotor at 4°C. Gradients were analyzed for acid-insoluble (0°C) radioactivities. Arrows indicate the peaks of radioactivity of 43-S and 80-S particles (sedimenting from right to left) from incubation mixtures labeled with [^{35}S]Met-tRNA$_f$ and run on an identical gradient. Reproduced from Winkler et al. (56).

radioactive RNA sedimenting at the 40-S region would increase. The large radioactive peak sedimenting at the top region of the gradient (Fig. 8) is due to unbound reovirus mRNA.

The inhibition of protein synthesis by iRNA in the nuclease-treated lysate was compared with the inhibition of shift assay in the untreated lysate in order to study whether these two parameters are correlated. The dose response to inhibition by iRNA was found to be similar in both assays (Fig. 9), suggesting that the inhibition of polypeptide chain initiation in the shift assay can account for the inhibition of protein synthesis in the mRNA translation assay. In agreement with this view, we also observed that iRNA inhibits the shift assay in both nuclease-treated and untreated lysate in a similar manner. However, iRNA acts as a more potent inhibitor in nuclease-treated lysate than in untreated lysate (38, 40). A possible interpretation of these results is that the shift assay analyzes the interaction of the first initiation event of mRNA with the ribosome. We speculate that subsequent initiation events may be less sensitive to iRNA. Thus, it is quite likely that the same inhibitory action of iRNA is involved in both the shift assay in the untreated lysate and the protein synthesis assay in the nuclease-treated lysate.

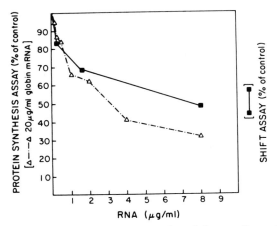

FIG. 9. Comparison of the effect of iRNA on the inhibition of protein synthesis and the inhibition of the shift assay. Protein synthesis was carried out in nuclease-treated reticulocyte lysate programmed with 20 µg of globin mRNA per milliliter as previously described (40, 56). (For details see also legend to Fig. 2). Values are the average of duplicate 2.5-µl samples taken 60 minutes after the start of the experiment. The shift assay in untreated lysate was performed as described in legends to Figs. 7 and 8. The left ordinate (△----△) represents the percentage of protein synthesis activities of incubation mixtures containing iRNA. The right ordinate (■——■) represents the percentage of shift assay activities. The counts per minute in the 80-S region of the gradient runs of incubations without iRNA was used as the control (100%). The abscissa represents the iRNA concentration used in both assays.

There are a number of possibilities that can account for the observed inhibition of peptide chain initiation by iRNA. (i) The iRNA may form a complex with mRNA, presumably at or near the ribosome binding site, and this complex is inactive. If this is true, iRNA should contain some domain that recognizes common nucleotide sequence(s) present in a variety of eukaryotic mRNAs (76). (ii) The iRNA may bind to the 40-S ribosomal subunit, presumably as an analog of mRNA, and this iRNA·40-S complex is incapable of binding to mRNA. (iii) The iRNA may bind to one or more of the initiation factors involved in mRNA binding. These possibilities need not be mutually exclusive. If iRNA binds to mRNA, then increasing levels of mRNA should overcome the degree of inhibition. The stimulation of protein synthesis in the nuclease-treated lysate by the addition of increasing levels of globin mRNA in the presence and the absence of 2.5 µg of iRNA per milliliter was measured. At high levels of added globin mRNA the inhibition of protein synthesis by iRNA is markedly reduced (Fig. 10). However, the shape of the observed curve is more consistent with the

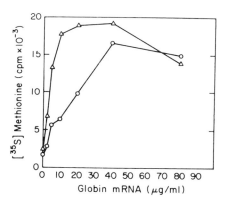

FIG. 10. Effect of iRNA on the stimulation of protein synthesis in the nuclease-treated reticulocyte lysate programmed with varying amounts of globin mRNA. Incubations were carried out in a total volume of 25 μl as previously described (40). (For details, see also legend to Fig. 2.) The assay values are the average of duplicate 2.5-μl samples taken 60 minutes after the start of the experiment. △, control; ○, iRNA (2.5 μg/ml).

view that iRNA competes with mRNA for a common binding site. If iRNA were binding strictly to mRNA, then the level of protein synthesis should remain low until the iRNA is titrated by the added mRNA. It appears that iRNA causes a reduction of the concentration of globin mRNA in the translation system. These results are also in agreement with our previous observation that radiolabeled iRNA neither hybridized with globin mRNA or muscle poly(A)$^+$ mRNA in "Northern blots" (unpublished results), nor formed a stable complex with mRNA, which could be detected by gel filtration (40).

The effect of iRNA on the shift assay programmed with the trinucleotide AUG was also studied. It was observed that AUG was capable of inducing the shift of labeled Met-tRNA$_f$ to 80-S (Fig. 11, panel A),

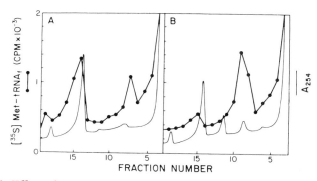

FIG. 11. Effect of iRNA on the AUG-induced shift reaction. The shift assay was carried out as described in the legend to Fig. 7 except that the shift reaction was induced by AUG instead of globin mRNA. Gradient fractions were analyzed for absorbance at 254 nm (———) and [^{35}S]Met-tRNA$_f$ bound radioactivity at 0°C (●———●). Sedimentation is from right to left. (A) Incubation containing 33 μM AUG; (B) incubation containing 8 μg of iRNA per milliliter followed after 2 minutes by 33 μM AUG.

and iRNA inhibited this shift (panel B). This result strongly supports the view that iRNA does not bind to mRNA, since it is highly unlikely that iRNA could recognize a binding site on AUG. Also, the AUG was present in about 100-fold molar excess over the iRNA in the assay system, their approximate levels being 0.27 μM for iRNA and 33 μM for AUG.

In order to study a possible interaction of iRNA with initiation factors, we have tested the effect of addition of purified eIF2, eIF3, eIF4A, eIF4B, and eIF5 as well as crude salt-wash fractions to iRNA-inhibited lysates. None of these preparations stimulated protein synthesis in the iRNA inhibited lysates, suggesting that an interaction of iRNA with initiation factors is unlikely. However, the possibility still exists that iRNA interacts with an initiation factor present in the lysate but absent from the preparations tested. We have used iRNA either labeled with iodine-125 or 3'-end-labeled with [^{32}P]pCp as a probe to study the specific binding of iRNA with ribosomal subunits. These studies did not show any significant binding of iRNA to the 40-S subunit. However, the possibility of a labile interaction of iRNA with the 40-S subunit cannot be excluded by these results. Thus, although iRNA inhibits mRNA binding to ribosomes or to the 43-S initiation complexes, the biochemical processes through which this effect is exerted remain to be understood.

VII. Concluding Remarks

The results presented in this article document that a translation-inhibiting heterogeneous low-molecular-weight RNA species in the 70–90 nucleotide size range exists as a 10-S RNP particle in the cytoplasm of chick embryo muscle. The iRNA and iRNP are quite distinct from the various eukaryotic RNA species that have modulating effects on *in vitro* translation systems. The potent inhibitory effect of iRNA and iRNP on mRNA translation is not exerted in an absolutely discriminatory manner. Rather, they act as specific inhibitors of mRNA binding to the 43-S complex. This mode of inhibition distinguishes iRNA and iRNP from other cellular inhibitors that act through phosphorylation of the eIF2 α-subunit. Since iRNA inhibits mRNA binding to ribosomes, it is quite likely that the relative affinities of mRNA and iRNA for this rate-limiting step in polypeptide synthesis may function as a subtle parameter in cytoplasmic regulation of translation. Thus, quantitative modulation of translation of various mRNAs by iRNA rather than a specific discriminatory effect can lead to striking effects on cellular translation patterns. It is conceivable that the selection of a specific mRNA for ribosome loading may be influenced by the level of iRNA in the cell. The iRNA and iRNP may be considered as a family of regulatory cytoplasmic macromolecules in analogy to the well-known nuclear entities snRNAs and snRNP particles.

Acknowledgments

The author wishes to thank M. G. Pluskal, C. Guha, A. K. Mukherjee, S. Dasgupta, M. M. Winkler, J. W. B. Hershey, D. Chakraborty, P. F. Agris, H. Stedman, and A. Boak for research contributions; R. Petryshyn for his generous help in the ds-protein kinase assay; H. Paulus and J. Gergely for helpful discussion; and Ms Swantana Mukherjee for excellent technical assistance.

Research in this laboratory has been supported in part by the USPHS Grant AM 13238 and a grant from Muscular Dystrophy Associations of America, Inc.

References

1. C. A. Howard, S. D. Aadamson, and E. Herbert, *BBA* **213**, 273 (1970).
2. T. Hunt, G. Vanderhoff, and I. M. London, *JMB* **66**, 471 (1972).
3. M. Rabinovitz, M. L. Freedman, J. M. Fisher, and C. R. Maxwell, *CSHSQB* **34**, 567 (1969).
4. E. Ehrenfeld and T. Hunt, *PNAS* **68**, 1075 (1976).
5. B. Lebleu, G. C. Sen, S. Shaila, B. Carbrer, and P. Lengeyl, *PNAS* **73**, 3107 (1976).
6. J. R. Lenz and C. Baglioni, *JBC* **253**, 4219 (1978).
7. W. K. Roberts, H. C. Hovanessian, R. E. Brown, M. J. Clemens, and I. M. Kerr, *Nature (London)* **264**, 477 (1976).
8. G. C. Sen, H. Taira, and P. Lengeyl, *JBC* **253**, 5915 (1978).
9. A. Silberstein, P. Fefermann, L. Shulman, and M. Revel, *FEBS Lett.* **68**, 119 (1976).
10. H. K. Das, A. Das, P. Ghosh-Dastidar, R. O. Ralston, B. Yaghmai, R. Roy, and N. K. Gupta, *JBC* **256**, 6491 (1981).
11. P. J. Farrell, K. Balkow, T. Hunt, R. J. Jackson, and H. Trachsel, *Cell* **11**, 187 (1977).
12. H. Grosfeld and S. Ochoa, *PNAS* **77**, 6526 (1980).
13. M. Gross and J. Mendelenski, *BBRC* **74**, 559 (1977).
14. G. Kramer, J. M. Cimadevilla, and B. Hardesty, *PNAS* **73**, 3078 (1976).
15. D. Levin and I. M. London, *PNAS* **75**, 1121 (1978).
16. D. Levin, R. S. Ranu, V. Ernst, and I. M. London, *PNAS* **73**, 3112 (1976).
17. R. S. Ranu, *BBRC* **97**, 252 (1980).
18. M. Revel and Y. Groner, *ARB* **47**, 1079 (1978).
19. S. Ochoa and C. deHaro, *ARB* **48**, 459 (1979).
20. S. A. Austin and M. J. Clemens, *FEBS Lett.* **110**, 1 (1980).
21. A. Das, R. O. Ralston, M. Grace, R. Roy, P. Ghosh-Dastidar, H. K. Das, B. A. Yaghmai, S. Palmieri, and N. K. Gupta, *PNAS* **76**, 5076 (1979).
22. C. deHaro, A. Datta, and S. Ochoa, *PNAS* **75**, 243 (1978).
23. G. Kramer, A. B. Henderson, P. Pinphanichakran, M. H. Wallis, and B. Hardesty, *PNAS* **74**, 1445 (1977).
24. R. S. Ranu, I. M. London, A. Das, A. Dasgupta, A. Majumdar, R. Ralston, R. Roy, and N. K. Gupta, *PNAS* **75**, 749 (1978).
25. J. E. Fuhr and C. Natta, *Nature NB* **240**, 274 (1972).
26. D. Bogdanovsky, W. Hermann, and G. Schapira, *BBRC* **54**, 25 (1973).

27. A. Berns, M. Salden, D. Bogdanovsky, M. Raymondjean, G. Schapira, and H. Bloemendal, *PNAS* **72**, 714 (1975).
28. J. E. Fuhr and M. Overton, *BBRC* **63**, 742 (1975).
29. M. Raymondjean, D. Bogdanovsky, L. Bachner, B. Kneip, and G. Schapira, *FEBS Lett.* **76**, 311 (1976).
30. S. Lee-Huang, J. M. Sierra, R. Naranjo, W. Filipowicz, and S. Ochoa, *ARB* **180**, 276 (1977).
31. H. Slegers, R. Mettrie, and M. Kondo, *FEBS Lett.* **80**, 390 (1977).
32. E. DeHerdt, H. Slegers, E. Piot, and M. Kondo, *NARes* **7**, 1363 (1979).
33. M. Zeichner and D. Breitkreutz, *ABB* **188**, 410 (1978).
34. A. J. Bester, D. S. Kennedy, and S. M. Heywood, *PNAS* **72**, 1523 (1975).
35. S. M. Heywood, D. S. Kennedy, and A. J. Bester, *PNAS* **71**, 2428 (1974).
36. D. S. Kennedy, E. Siegel, and S. M. Heywood, *FEBS Lett.* **90**, 209 (1978).
37. T. L. McCarthy, E. Siegel, B. Mroczkowski, and S. M. Heywood, *Bchem* **22**, 935 (1983).
38. M. G. Pluskal and S. Sarkar, *Bchem* **20**, 2048 (1981).
39. A. K. Mukherjee, C. Guha, and S. Sarkar, *FEBS Lett.* **127**, 133 (1981).
40. S. Sarkar, A. K. Mukherjee, and C. Guha, *JBC* **256**, 5077 (1981).
41. B. Kuhn, A. Villringer, H. Falk, and P. C. Heinrich, *EJB* **126**, 181 (1982).
42. C. A. Dionne, G. B. Stearns, G. Kramer, and B. Hardesty, *JBC* **257**, 12373 (1982).
43. A. Vincent, O. Civelli, K. Maundrell, and K. Scherrer, *EJB* **112**, 617 (1980).
44. S. M. Heywood and D. S. Kennedy, This Series **19**, 477 (1976).
45. S. M. Heywood and D. S. Kennedy, *Bchem* **15**, 3314 (1976).
46. A. K. Mukherjee and S. Sarkar, *JBC* **256**, 11301 (1981).
47. J. Bag and S. Sarkar, *JBC* **251**, 7600 (1976).
48. S. M. Heywood, D. S. Kennedy, and A. J. Bester, *EJB* **58**, 587 (1975).
49. W. Northeman, E. Schmelzer, and P. C. Heinrich, *EJB* **112**, 451 (1980).
50. M. S. Eller, R. E. Cullinan, and P. M. McGuire, *FP* **41**, 1457 (1982).
51. A. K. Deshpande, S. B. Jakowlew, H. H. Arnold, P. A. Crawford, and M. A. Q. Siddiqui, *JBC* **252**, 6521 (1977).
52. H. H. Arnold, M. A. Innis, and M. A. Q. Siddiqui, *Bchem* **17**, 2050 (1978).
53. H. Busch, R. Reddy, L. Rothblum, and C. Y. Choi, *ARB* **51**, 617 (1982).
54. R. Reddy, T. S. Ro-choi, D. Henning, and H. Busch, *JBC* **249**, 6486 (1974).
55. M. S. Rao, M. Blackstone, and H. Busch, *Bchem* **16**, 1756 (1977).
56. M. M. Winkler, C. Lashbrook, J. W. B. Hershey, A. K. Mukherjee, and S. Sarkar, *JBC* **258**, 15141 (1983).
57. J. Bag, M. Hubley, and B. H. Sells, *JBC* **255**, 7055 (1980).
58. A. K. Mukherjee and S. Sarkar, *Mol. Biol. Rep.* **8**, 51 (1981).
59. P. H. O'Farrell, *JBC* **250**, 4007 (1975).
60. S. K. Jain, M. G. Pluskal, and S. Sarkar, *FEBS Lett.* **97**, 84 (1979).
61. S. K. Jain, R. K. Roy, M. G. Pluskal, D. E. Croall, C. Guha, and S. Sarkar, *Mol. Biol. Rep.* **5**, 79 (1979).
62. R. K. Roy, A. S. Lau, H. N. Munro, B. S. Baliga, and S. Sarkar, *PNAS* **76**, 1751 (1979).
63. S. K. Jain and S. Sarkar, *Bchem* **18**, 745 (1979).
64. V. A. Bloomfield, D. M. Crothers, and I. Tinoco, in "Physical Chemistry of Nucleic Acids" (V. A. Bloomfield, D. M. Crothers, and I. Tinoco, eds.), p. 293. Harper, New York, 1974.
65. M. D. Topal and J. R. Fresco, *Bchem* **19**, 5531 (1980).
66. S. Sarkar, H. Stedman, and A. K. Mukherjee, *FP* **41**, 1456 (1982).
67. P. Traub and M. Nomura, *PNAS* **59**, 777 (1968).

68. M. R. Lerner and J. A. Steitz, *PNAS* **76**, 5496 (1979).
69. C. Brunnel, S. J. Widada, M. Lelay, P. Jeanteur, and J. Liautard, *NARes* **9**, 815 (1981).
70. M. Takano, S. S. Golden, G. C. Sharp, and P. F. Agris, *Bchem* **21**, 5929 (1981).
70a. S. Pestka, *ARB* **40**, 697 (1971).
71. D. H. Levin, R. Petryshyn, and I. M. London, *JBC* **256**, 7638 (1981).
72. C. Darnbrough, S. Legon, T. Hunt, and R. J. Jackson, *JMB* **76**, 379 (1973).
73. R. Jagus, W. F. Anderson, and B. Safer, This Series **25**, 127 (1981).
74. B. Safer, W. Kemper, and R. Jagus, *JBC* **253**, 3384 (1978).
75. M. G. Pluskal, R. K. Roy, and S. Sarkar, *Biochem. Soc. Trans.* **7**, 1091 (1979).
76. M. Kozak, *in* "Protein Biosynthesis in Eukaryotes" (R. Perez-Bercoff, ed.), p. 167. Plenum, New York, 1982.

The Hypoxanthine Phosphoribosyltransferase Gene: A Model for the Study of Mutation in Mammalian Cells

A. Craig Chinault and
C. Thomas Caskey

Howard Hughes Medical Institute
Laboratories and Departments
of Biochemistry and Medicine
Baylor College of Medicine
Houston, Texas

I. General Aspects: Enzyme and Genetic Properties	296
II. Comparison of Normal and Mutant Hypoxanthine Phosphoribosyltransferase Proteins	298
III. Molecular Cloning and Analysis of Hypoxanthine Phosphoribosyltransferase Sequences	301
IV. Normal and Mutant Gene Structures	305
V. Future Prospects	309
References	310

Maintenance of the integrity of the inheritable information encoded in DNA is crucial for survival. Given the strong correlations between mutagenic and carcinogenic activities of many environmental compounds (1, 2) and the profound effects that seemingly minor changes in base sequence may have on the expression of genetic information [as dramatically shown, for example, in the activation of a human proto-oncogene (3–7)], an understanding of the mechanisms of mutational alteration of DNA is obviously important. In mammalian cells, the principal genetic locus that has proved to be amenable to mutation studies is the one coding for hypoxanthine phosphoribosyltransferase (HPRT; EC 2.4.2.8).[1] Along with adenine phosphoribosyltransferase (APRT), this enzyme is considered to be primarily responsible for the recycling of purine bases into cellular nucleotide pools (Fig. 1). The lack of HPRT activity in humans has been corre-

---

[1] Abbreviations: HPRT, hypoxanthine phosphoribosyltransferse (EC 2.4.2.8); HAT, hypoxanthine + aminopterin + thymidine (medium); EMS, ethyl methanesulfonate; MNNG, N-methyl-N'-nitro-N-nitrosoguanidine; cDNA, complementary DNA; CRM, cross-reacting material.

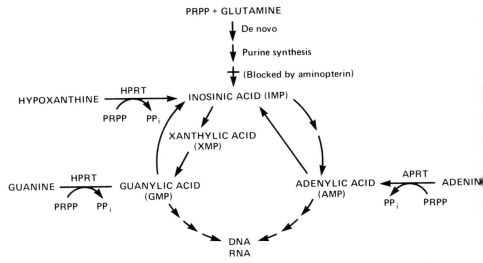

FIG. 1. Pathways of purine interconversion in mammalian cells showing the reactions catalyzed by the salvage enzymes HPRT and APRT.

lated with the debilitating neurological disorder known as Lesch–Nyhan syndrome (8, 9), and partial enzyme deficiency is frequently found in patients with gouty arthritis (10, 11). These associations, together with the discovery of powerful selective systems for the identification of mutants and revertants in cultured somatic cells, have made the HPRT locus an excellent system for study. This essay reviews results obtained from the various approaches that have been applied to the study of mutation at the HPRT locus, and summarizes recent advances in the molecular cloning of HPRT genes, which will now allow molecular analysis of both spontaneous and induced mutational events.

Earlier reviews of certain somatic cell genetic and biochemical aspects of this system have appeared (12, 13). A recent summary of studies on the molecular basis of the clinical syndromes associated with HPRT deficiency in humans, based on protein sequence analysis, is recommended (14).

I. General Aspects: Enzyme and Genetic Properties

Hypoxanthine phosphoribosyltransferase (HPRT) is a cytoplasmic multimeric enzyme that catalyzes the formation of 5'-IMP and 5'-

GMP by condensation of 5-phosphoribosyl diphosphate (PRPP) and the bases hypoxanthine and guanine. Apparent K_m values for hypoxanthine, guanine, and PRPP are 1.7×10^{-5} M, 5.0×10^{-6} M, and 2.5×10^{-4} M, respectively, for HPRT isolated from human erythrocytes, and similar values have been found for HPRT isolated from other sources (9). Early estimates of the subunit molecular weights of purified human and rodent proteins, based on electrophoretic mobilities in denaturing polyacrylamide gels, ranged from 24,000 to 27,000 (15–20); molecular weights of 24,500 have now been established from amino-acid sequences. Although it is agreed that the active form of the enzyme is a multimer of identical subunits, the exact configuration has been controversial. Native molecular weights of 68,000 (human), 80,000 (mouse), and 78,000–85,000 (hamster) have been determined by Sephadex column chromatography, and the latter two values were interpreted as indicating a trimeric structure (15, 17, 21). However, more recent results, based on cross-linking experiments (20) and on isoelectric focusing studies of human/mouse heteropolymers (22), indicate that the native enzyme is probably a tetramer under physiological conditions, and that an equilibrium mixture of dimeric and tetrameric forms at lower ionic strength was responsible for the lower estimates arrived at previously. Human HPRT isolated from erythrocytes exhibits substantial electrophoretic heterogeneity (23–26). In at least one case, this has been attributed to deamidation of an asparagine residue, which is believed to occur *in vivo* (27). The possibility remains that there are additional posttranslational modifications, and, in fact, it has been suggested that such events may play a role in the modulation of protein metabolism in erythrocytes as a result of cell aging (28).

The HPRT deficiency is inherited as a recessive X-linked trait (29, 30); the human gene has been regionally located at Xq26-Xq27 using human/mouse cell hybrids containing various portions of the human X (31, 32). The HPRT structural gene has also been located in the X chromosome in a variety of other mammals (33), including hamsters [in the distal Xp (34–36)] and mice [in the Xcen-XD region (37, 38)]. Location of the gene on the X chromosome has been particularly helpful for genetic studies, as not only is it hemizygotic in males, but, as a consequence of X-chromosome inactivation, the locus is also functionally hemizygotic in females.

In addition to the normal substrates, hypoxanthine and guanine, HPRT can utilize a number of base analogs, such as 6-thioguanine and 8-azaguanine, to produce the corresponding ribonucleotides, which are toxic to the cell. Phenotypic resistance to the effects of these compounds has, therefore, been the criterion generally used for identification of HPRT-deficient cells in culture (see 39, 40). Spontaneous mutation rates at the HPRT locus in cultured cells by this criterion have generally been estimated to be in the range of 10^{-6} to 10^{-7}. However, considerable variation has been noted, which can probably be attributed to differences in selection protocols and/or the cell lines used (39–41). The apparent mutation rate is dramatically enhanced in a dosage-dependent manner by prior treatment with ultraviolet light, X-rays, and a variety of chemical mutagens, such as ethylmethanesulfonate (EMS) and N-methyl-N'-nitro-N-nitrosoguanidine (MNNG) (see 41, 42). Although HPRT is normally not essential for the growth of cultured cells, the enzyme is required for the growth of cells in hypoxanthine/aminopterin/thymidine (HAT) medium, in which aminopterin blocks de novo purine and thymidine synthesis, while exogenous hypoxanthine and thymidine are supplied as purine and pyrimidine sources (43). This property is the basis for selection of HPRT$^+$ revertants and has also been widely exploited in the construction of interspecies somatic cell hybrids (44), which have proved to be valuable for genetic mapping (45).

II. Comparison of Normal and Mutant Hypoxanthine Phosphoribosyltransferase Proteins

Since the discovery of the correlations between HPRT deficiency and human disease, a large number of patients have been examined whose cells exhibit a wide range of HPRT enzymatic activity. Immunological studies with anti-HPRT antibodies have shown that there is generally a reasonable correlation between the levels of cross-reacting material (CRM) and enzyme activity (46–48), although there are examples in which nearly normal levels of CRM were found even when enzyme activity was low or undetectable (46, 47, 49–51). The properties of mutant enzymes have been examined in those situations where there was sufficient activity, usually with unfractionated hemolysates or fibroblast lysates. The basic conclusion that can be drawn from these studies is that there is much genetic heterogeneity in the mutant alleles, leading to the production of enzymes with a variety of changes in primary sequence that are reflected in aberrant enzymatic and physical parameters. These changes have included increased thermolability, which can be demonstrated *in vivo* (49, 52) and *in vitro* (10, 52, 53), increased K_m values for PRPP or the base substrates

(51, 54–58), differential sensitivity to product inhibition (51, 52, 59), or altered electrophoretic mobility (46, 51, 55, 58, 60–64).

Acquired resistance to purine analogs by somatic cells in culture has been adequately proved to be a result of mutation at the HPRT locus in a variety of rodent and human lines (12, 13, 65–72). Our laboratory has used a male Chinese hamster cell line (V79) to examine forward and reverse mutation at the HPRT locus; a summary of some of the results obtained is presented in Table I (73). Twenty-one of 79 independent 8-azaguanine-resistant isolates had the capacity to incorporate hypoxanthine into nucleic acid under culture conditions where de novo purine biosynthesis was inhibited with aminopterin. The remaining 58 isolates had no detectable HPRT enzymatic activity; however, 5 of these 58 isolates were CRM$^+$ by radioimmunoassay.

As with naturally occurring variations in humans, genetic heterogeneity has been the hallmark of spontaneous and mutagen-induced mutation in cultured cells. Biochemical studies of mutant HPRT proteins have demonstrated substantial differences in thermal stability (69, 74, 75), kinetic constants (69, 75–78), immunological reactivity (79, 80), and apparent molecular weight (76, 77, 81). Molecular mechanisms responsible for these changes remain largely unknown.

Several investigators have attempted to apply comparative peptide analysis to the problem of defining, or at least categorizing, mutational changes in variant HPRT proteins (82–84). They showed specific alterations in the tryptic peptide patterns obtained by high-pressure chromatography on cation-exchangers with mutant human and rodent proteins, and, in at least one case, they found genetic revertants of a mutant protein with an altered pattern that produced an apparently

TABLE I
CHARACTERIZATION OF CHINESE HAMSTER CELLS RESISTANT TO 8-AZAGUANINE

Mutagen	Frequency ($\times 10^5$)	Number of isolates			
		Selected	HPRT$^+$	HPRT$^-$/CRM$^+$	HPRT$^-$/CRM$^-$
None	1	10	0	0	10
MNNG	70	23	14	2	7
EMS	70	18	5	2	11
UV light	20	10	1	0	9
ICR-191	8–10	18	1	1	16
		79	21	5	53

normal enzyme at this level of analysis (84). However, precise amino-acid changes responsible for the observed changes were not identified in any of these cases, and, in retrospect, some misleading conclusions were reached.

Recent studies (27, 85) have greatly clarified the situation regarding HPRT and also serve to emphasize the hazards inherent in this type of analytical approach. Reverse-phase liquid chromatography was used to purify the tryptic peptides of human erythrocyte HPRT, and the complete amino-acid sequence of the normal protein was established by Edman degradation analysis. This knowledge was used to isolate altered peptides from erythrocytes of several patients with gout or Lesch–Nyhan syndrome and to determine the exact amino-acid substitutions that account for the altered enzymatic phenotype (14, 86–88). These results are summarized in Table II. In addition, these studies identified some reproducible artifacts. In several examples, trypsin treated with tosylphenylalanylchloromethane (TPCK) either failed to cleave at expected sites or gave spurious cleavage at positions other than the normal ones (i.e., following lysine and arginine residues) (27). Of particular interest is the finding that the carboxyl-terminal peptide from human HPRT is Tyr-Lys-Ala (85); based on identity of amino-acid sequences at these positions (see Section III,C), one would expect the same result for mouse and hamster HPRT. Previously, it had been assumed that the carboxyl-terminal tryptic peptide could be identified as one not containing lysine or arginine; in fact, the loss of such a fragment has been used as the criterion for trying to identify nonsense mutations in the HPRT gene that result in premature chain termination (73, 83, 84). Such mutations are particularly important in establishing whether tRNA-mediated suppression mechanisms are utilized to correct mutations in mammalian cells; previous conclusions in this area should be reanalyzed in light of the information now available.

TABLE II
AMINO ACID SUBSTITUTIONS IN MUTANT FORMS OF HUMAN HPRT

Mutant enzyme	Disorder	Amino-acid change	Position[a]
HPRT$_{Toronto}$	Gout	Arg to Gly	50
HPRT$_{Munich}$	Gout	Ser to Arg	103
HPRT$_{London}$	Gout	Ser to Leu	109
HPRT$_{Kingston}$	Lesch–Nyhan	Asp to Asn	193

[a] See Fig. 2.

III. Molecular Cloning and Analysis of Hypoxanthine Phosphoribosyltransferase Sequences

A. Isolation of HPRT cDNA Sequences

Precise knowledge of the structure of any normal gene is a prerequisite for understanding the effects of mutation at the molecular level, and this need served as the motivation for applying recombinant DNA technology to the problem of cloning and characterizing HPRT gene sequences. Initial efforts by the now classical methods of cDNA cloning were inhibited by the fact that HPRT is constitutively expressed at low levels; thus, significant enrichment of the mRNA and subsequent identification of complementary sequences present difficult problems. Enzyme activity studies had indicated that there is some variation in relative levels of expression in mammalian tissues, with brain showing the highest levels (9). Corroboration of this finding was obtained by *in vitro* translation studies on mRNA isolated from brain, liver, and testes of Chinese hamsters; although the level was approximately 7-fold higher in brain than in the other tissues, the translatable message coding for HPRT still represented only about 0.04% of the total mRNA (89). Similar studies showed only about a 0.01% level of HPRT mRNA in cultured hamster and mouse cell lines (89).

The problem of low-level expression of the HPRT gene was partially alleviated by the discovery of a mouse neuroblastoma cell line, NBR4, which overproduces a mutant form of HPRT (90). This cell line arose as a spontaneous revertant of an HPRT-deficient line (NB⁻) during cell-fusion experiments. The apparent level of *in vivo* enzyme activity measured in NBR4 cells was approximately 10 times that in the parental cells, and immunoprecipitation assays with anti-HPRT antibody showed a similar increase in precipitable protein. These conclusions were qualified by the fact that the HPRT enzyme produced in NBR4 appeared to have altered kinetic and stability properties. In addition, there was no information concerning the ability of the subunits to assemble into the active multimeric form. Indeed, cell-free translation assays carried out on mRNA isolated from this cell line showed a 20- to 50-fold overproduction of translatable mRNA compared to mRNA from wild-type neuroblastoma cells, giving an estimated HPRT mRNA level of about 0.3–0.5% (89).

The significant overproduction of HPRT mRNA in NBR4 led to selecting this cell line as a source of mRNA for cDNA cloning experiments. First, sucrose-gradient-enriched mRNA was converted to double-stranded DNA by reverse transcriptase; this material was then inserted into the *Pst*I site of pBR322 by the "(dG-dC)-tailing" method. After transformation into competent bacterial cells, recombinants containing HPRT-gene sequences were identified by differential hybridization followed by a positive hybridization-selection/*in vitro* translation procedure (*91*). The largest of the initial isolates, pHPT2, which contained an insert of 1050 base-pairs, was isolated and used as a probe in subsequent experiments.

RNA- and DNA-blot hybridization analyses (*92*, *93*) with this probe immediately led to several important conclusions (*91*).

1. The size of the HPRT mRNA was about 1550 nucleotides, indicating that the pHPT2 cDNA was still significantly shorter than full-length. Screening of larger cDNA libraries constructed from NBR4 mRNA enabled the subsequent isolation of a recombinant, pHPT5, that appeared to be nearly full-length, although the length of the poly(A) tail and the precise identification of the mRNA initiation site remain uncertain.

2. NBR4 had at least a 20-fold higher level of HPRT mRNA than did NB^-, although, interestingly, the NB^- line did have levels of apparently normal-size HPRT mRNA comparable to those in the $HPRT^+$ neuroblastoma cell line (NB^+) from which the mutant was derived. This was consistent with the conclusions reached earlier on the basis of the *in vitro* translation assays. This analysis of mRNA levels, together with the earlier information on the enzyme properties, led to the hypothesis that NB^- is phenotypically $HPRT^-$ under selective conditions, not because of failure to produce functional enzyme, but because the level of activity of the mutant enzyme produced is insufficient for survival. Thus, this problem is circumvented in NBR4 by overproduction of the enzyme at a level allowing growth rather than by a true genetic reversion event (*94*).

3. The NBR4 cell line showed an estimated 20- to 50-fold amplification of HPRT gene sequences leading to the conclusion that gene dosage, rather than increased transcription from a single gene, was the most likely explanation for the overproduction of HPRT mRNA in the cell line.

4. Finally, cross-hybridization experiments with the mouse cDNA probe indicated excellent sequence homology with mRNA and DNA from Chinese hamster and human cell lines, which greatly simplified subsequent isolation of HPRT sequences from these sources.

The discovery that the Chinese hamster cell line RJK159 overproduces HPRT mRNA at least 20-fold (Section IV,C) was exploited for the cloning of hamster cDNA sequences (95). Messenger RNA isolated from this cell line was used to construct a cDNA library of 3200 recombinants as described above. Colony hybridization screening with the mouse cDNA probe identified seven HPRT recombinants, which were characterized by restriction endonuclease mapping and nucleotide sequence analysis. A cDNA recombinant carrying human HPRT sequences has also been isolated in our laboratory by means of the mouse cDNA probe to screen a cDNA library (96). This plasmid has an insert of about 950 base-pairs and appears to extend from about 100 nucleotides before the protein initiation codon to a position about 200 nucleotides downstream from the termination codon.

An alternative approach to the isolation of human HPRT gene sequences is that of DNA-mediated gene transfer to transfect HPRT-deficient mouse cells with total human DNA and select cells producing the human enzyme by growth in HAT medium (97, 98). Human DNA sequences are then rescued from the transfectants by identifying plasmids from a genomic library carrying human-specific repetitive sequences. A subcloned unique sequence from one such plasmid has been used to screen a human cDNA library constructed in a vector designed to give expression of insert sequences in mammalian cells (98, 99). These colony hybridization studies identified a nearly full-length HPRT cDNA recombinant (1350 base-pairs) that functioned efficiently in gene transfer experiments (98).

B. Nucleotide Sequence Analysis

The nucleotide sequences of mouse, hamster, and human cDNA recombinants described in the preceding section have been determined and lead to the following picture of HPRT mRNA structure. In each case, there is an open reading frame of 654 nucleotides, beginning with an AUG initiation codon and followed by a UAA termination codon, corresponding to the protein coding sequence (see Section III,C). Although the precise cap site of HPRT mRNA has not been defined, the cDNA sequences indicate a minimum of 100 nucleotides of untranslated sequence precede the initiator. There are relatively long 3'-untranslated sequences of about 550 and 590 nucleotides in the mouse and human sequences, respectively. Two independent hamster cDNA recombinants gave 3'-untranslated stretches of different lengths (470 and 550 nucleotides), suggesting the possibility that more than one class of processed mRNA are produced in the cell. In every case, the sequences contain the putative processing signal AAUAAA (100) at 10–20 nucleotides preceding the poly(A) tail. Comparison of the three sequences reveals a high 96% level of nucleotide homology within the coding regions; this drops off to approximately 80% in both the 5' and 3' untranslated regions.

C. Comparison of HPRT Protein Sequences

Amino-acid sequences for HPRT can be derived from the nucleotide sequences; the results are summarized in Fig. 2. Results obtained from the human cDNA (98) are in exact agreement with the sequence determined for normal human erythrocyte enzyme by protein-sequencing techniques (85). However, the initiator methionine residue in the human protein is apparently removed by processing, as the

```
                       1                      10                       20
Mouse      (Met) ProThrArgSerProSerValValIleSerAspAspGluProGlyTyrAspLeuAspLeu
Hamster    (Met) Ala  —  —  —  —  —  —  —  —  —  —  —  —  —  —  —  —  —  —
Human      (Met) Ala  —  —  —  —  Gly —  —  —  —  —  —  —  —  —  —  —  —  —

                            30                          40
  PheCysIleProAsnHisTyrAlaGluAspLeuGluLysValPheIleProHisGlyLeuIleMetAspArgThr
   —  —  —  —  —  —  —  Val  —  —  —  —  —  —  —  —  Val  —  —  —  —  —
   —  —  —  —  —  —  —  —  —  —  Arg  —  —  —  —  —  —  —  —  —  —  —  —

             50                           60                          70
  GluArgLeuAlaArgAspValMetLysGluMetGlyGlyHisHisIleValAlaLeuCysValLeuLysGlyGly
   —  —  —  —  —  —  —  —  —  —  —  —  —  —  —  —  —  —  —  —  —  —  —
   —  —  —  —  —  —  —  —  —  —  —  —  —  —  —  —  —  —  —  —  —  —  —

                           80                          90
  TyrLysPhePheAlaAspLeuLeuAspTyrIleLysAlaLeuAsnArgAsnSerAspArgSerIleProMetThr
   —  —  —  —  —  —  —  —  —  —  —  —  —  —  —  —  —  —  —  —  —  —
   —  —  —  —  —  —  —  —  —  —  —  —  —  —  —  —  —  —  —  —  —  —

                    100                         110                        120
  ValAspPheIleArgLeuLysSerTyrCysAsnAspGlnSerThrGlyAspIleLysValIleGlyGlyAspAsp
   —  —  —  —  —  —  —  —  —  —  —  —  —  —  —  —  —  —  —  —  —  —  —
   —  —  —  —  —  —  —  —  —  —  —  —  —  —  —  —  —  —  —  —  —  —  —

                              130                         140
  LeuSerThrLeuThrGlyLysAsnValLeuIleValGluAspIleIleAspThrGlyLysThrMetGlnThrLeu
   —  —  —  —  —  —  —  —  —  —  —  —  —  —  —  —  —  —  —  —  —  —  —
   —  —  —  —  —  —  —  —  —  —  —  —  —  —  —  —  —  —  —  —  —  —  —

                  150                          160                        170
  LeuSerLeuValLysGlnTyrSerProLysMetValLysValAlaSerLeuLeuValLysArgThrSerArgSer
   —  —  —  —  Arg  — AsnLeu —  —  —  —  —  —  —  —  —  —  —  —  —  —
   —  —  —  — Arg  —  Asn  —  —  —  —  —  —  —  —  —  —  —  — Pro  —  —

                             180                         190
  ValGlyTyrArgProAspPheValGlyPheGluIleProAspLysPheValValGlyTyrAlaLeuAspTyrAsn
   —  —  —  —  —  —  —  —  —  —  —  —  —  —  —  —  —  —  —  —  —  —
   —  —  —Lys —  —  —  —  —  —  —  —  —  —  —  —  —  —  —  —  —  —  —

                  200                         210
  GluTyrPheArgAspLeuAsnHisValCysValIleSerGluThrGlyLysAlaLysTyrLysAla
   —  —  —  —  —  —  —  — Ile —  —  —  —  —  —  —  —  —  —  —
   —  —  —  —  —  —  —  —  —  —  —  —  —  —  —  —  —  —  —  —
```

FIG. 2. Amino acid sequences for normal mouse and human HPRT and a mutant form of Chinese hamster HPRT (RJK159).

amino terminus is an acetylated alanine (85). On the basis of previous results indicating that the amino termini of mouse and hamster proteins are also blocked, it is likely than an analogous posttranslational modification occurs in these proteins as well. Both the mouse and hamster cell lines used as sources of mRNA for cloning produce mutant proteins; however, it has been shown directly by DNA sequencing that the mouse HPRT in NBR4 differs from wild-type by only a single amino-acid change, namely the replacement of the normal aspartic acid with an asparagine at position 200 (101), and it may be anticipated that a similar missense mutation will be found in the hamster sequence.

Comparisons between the protein sequences of human HPRT, wild-type mouse HPRT, and the RJK159 hamster HPRT show a remarkably high level of sequence conservation with 7 amino-acid differences out of 217 between the mouse and hamster proteins, and also only 7 differences between the mouse and human sequences. Nearly all these differences are the result of single nucleotide changes in the corresponding DNA sequences, and they represent fairly neutral changes with respect to amino-acid class. There is a region of substantial similarity in amino-acid sequence between HPRT and two functionally related enzymes, *Salmonella typhimurium* ATP-phosphoribosyltransferase (EC 2.4.2.17) and *Escherichia coli* amidophosphoribosyltransferase (EC 2.4.2.14), which may represent the PRPP binding domain (102). As more information becomes available on related eukaryotic genes, it may be possible to identify the functional domains of the enzyme and test the possible correlation with exon blocks in the gene structure, which has been suggested to play a role in the evolution of related enzymes (103).

IV. Normal and Mutant Gene Structures

A. HPRT Genomic Organization

The cDNA clones described above have been used as probes to identify lambda recombinants in genomic libraries constructed from mouse and human DNAs to characterize the genomic organization of HPRT genes. Work on the mouse gene structure is now complete, and the results are summarized in Fig. 3 (101). Like most eukaryotic genes that have been examined, the HPRT gene is a mosaic of coding sequences (exons) separated by intervening sequences (introns); extensive posttranscriptional processing must occur to produce the mature mRNA. In the mouse HPRT gene, the coding sequence is split into

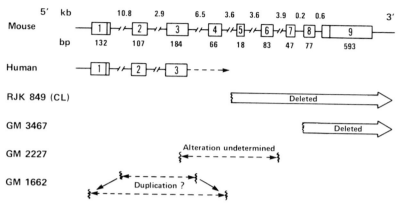

FIG. 3. Summary of the gene structure of HPRT and alterations found in Lesch–Nyhan cell lines. The boxes represent exon elements with the lengths shown underneath; numbers at the top are intron lengths. The vertical lines in exons 1 and 9 indicate boundaries of the protein coding region. kb = kilobase (pairs); bp = base-pairs.

nine exon blocks that encompass a total of 33,000 base-pairs of DNA. This gives an inordinately high ratio of intron/exon sequences of more than 20 to 1. The 3'-terminal half of the mRNA, consisting mostly of untranslated sequences, is coded for in a single exon; thus, the actual protein-coding portion (654 nucleotides) is dispersed over a very large region. Although the analysis of the human gene is less complete, it appears to have a very similar, if not identical, structure. The large size of these genes probably contributed to the difficulties noted previously in achieving high efficiency transfection with HPRT in DNA-mediated gene-transfer experiments (104–106).

In addition to the functional locus on the X chromosome, other homologous sequences, believed to represent pseudogenes, have been identified. At least one such sequence has been detected in both the hamster and mouse genomes, and, in each case, it appears to be unlinked to the HPRT locus. With both mouse and hamster cell lines, amplification of the functional gene failed to give coamplification of these sequences (101, 107); in one hamster cell line, with an apparent deletion of the entire HPRT locus, the putative pseudogene sequence remained intact (107). At least three apparent pseudogene sequences have been found in the human genome; they have been mapped to chromosomes 3, 5, and 11 (P. Patel and A. C. Chinault, unpublished). In contrast to the functional gene, the pseudogenes appear to be relatively small and uninterrupted, and presumably they represent examples of dispersed, processed genes.

A great deal of information concerning gene structure can be gained by DNA hybridization analysis with appropriate probes. Although minor changes may not be detectable at this level of analysis, identification of events such as deletion, insertion, rearrangement, and amplification should be possible. Therefore, as the first step toward examining mutation and reversion events, such studies have been carried out on DNA isolated from human cells and from Chinese hamster cells.

B. Lesch–Nyhan Patients

Analysis of mutation at the human HPRT locus has proved to be difficult because of (i) the presence of homologous non-X sequences that create a complex pattern in DNA-blot experiments; and (ii) the discovery of natural variations in DNA sequences at or near the HPRT gene that are not directly relevant to gene defects, but, rather, represent restriction-fragment-length polymorphisms (108). Of the Lesch–Nyhan patients examined to date, approximately 20% have shown restriction-fragment patterns indicating major gene-alterations based on hybridization analysis with cDNA probes. To further delineate the regions of change in these genes, hybridizations were carried out using cDNA subfragments chosen to represent specific exon regions of the normal gene. Preliminary results from this approach are summarized in Fig. 3. RJK849 is missing exons 7, 8, and 9 and most restriction fragments associated with exons 4, 5, and 6, while the first three exons are present. Since no new DNA fragments were detected, this patient apparently has a partial deletion extending from near the middle of the gene through the 3′ end. Similar analysis of GM3467 reveals a partial gene-deletion extending from a position around exons 7 and 8 and extending past the last exon. An additional patient with a total deletion of the HPRT locus has been identified. GM2227 represents a mutant with structural alterations associated with exons 3, 4, 5, and possibly 6, but the exact nature of this alteration is unclear. GM1662 represents an unusual alteration at the HPRT locus that has been tentatively interpreted as representing a partial gene duplication involving exons 2 and 3.

The analysis of Lesch–Nyhan mutant-gene expression has also included the examination of mRNA size and amount by RNA ("northern") blots. As anticipated, none of the deletion mutants produced HPRT mRNA. Two patients with normal DNA patterns also produce a normal-sized mRNA and are presumed to carry point mutations, whereas a third patient with a normal DNA pattern produces no detectable mRNA. The mutant GM1662 produces mRNA larger than normal, which is consistent with the proposed partial gene duplication.

In one case, it has been possible to demonstrate directly a single-base mutation in an HPRT allele by restriction endonuclease analysis (109). As indicated in Table II, $HPRT_{Toronto}$ was shown to differ from normal HPRT by an arginine-to-glycine substitution at position 50, a variation that can be explained by a single C-to-G change. From the cDNA sequence, it was predicted that such a change would result in the loss of a TaqI site, and indeed Southern analysis indicated that a normal 2000-nucleotide fragment was lost in the mutant DNA while a new 4000-nucleotide fragment appeared. This fragment variation was used to detect the $HPRT_{Toronto}$ allele in a heterozygote that was otherwise normal with respect to the classical techniques used for heterozygote detection.

C. Chinese Hamster HPRT Mutants and Revertants

A set of 10 spontaneous and 9 UV-induced subclones of V79 Chinese hamster cells, originally selected as being HPRT deficient on the basis of resistance to 8-azaguanine and failure to grow in HAT medium, was chosen for the initial study of mutation in cultured cells (107). Eighteen of these failed to show *in vivo* HPRT enzymatic activity or production of immunoprecipitable protein, while 1 showed detectable levels of an enzyme with abnormal kinetic properties. DNA isolated from these cell lines was digested with *Pst*I and subjected to electrophoresis in agarose gels. Subsequent DNA-blot analysis, using radiolabeled hamster cDNA as a probe, revealed that 16 of the isolates gave hybridization patterns identical to that obtained with DNA from wild-type cells, consisting of 7 fragments of 8.8, 7.8, 6.0, 5.0, 3.2, 2.6, and 0.9×10^3 base-pairs. However, in two of the cell lines (RJK71, a UV-induced mutant, and RJK88, a spontaneous mutant), five of the seven bands were absent, amounting to deletion of a minimum of 27,000 base-pairs at the HPRT locus. The two remaining fragments were shown to be homologous pseudogene sequences unlinked to the functional gene locus. These conclusions were supported by similar studies using other restriction enzymes. Interestingly, cytogenetic analysis of both of the cell lines with deletions showed translocation events involving the X chromosome, while the remaining 17 mutants appeared to have normal karotypes. Further support for the conclusion that most of the mutations occurring at the HPRT locus do not involve major alterations was obtained by analysis of mRNA levels (J. C. Fuscoe and C. T. Caskey, unpublished). Twenty HPRT$^-$ CRM$^-$ hamster mutant cell lines, with no detectable DNA alterations, were examined. Of these, 18 showed essentially normal levels of HPRT mRNA, 1 had a reduced level, and 1 showed no detectable hybridization signal.

Based on the reasoning that phenotypic reversion from HPRT⁻ to HPRT⁺ could result from amplification of a gene coding for an unstable or kinetically defective, albeit functional, enzyme, experiments were undertaken to screen Chinese hamster cell lines for overproduction of HPRT (107, 110). Previous studies had shown that both spontaneous and mutagen-induced revertants producing antigenically altered HPRT could be obtained from RJK10, a nitrosoguanidine-induced mutant of RJK0 that does not produce detectable levels of HPRT protein but does produce HPRT mRNA of normal size. Four of five spontaneous revertants examined appeared to have 10- to 20-fold increases in HPRT gene-copy-number, and two of these (RJK159 and 160) had comparable increases in mRNA levels. On the other hand, HPRT gene amplification was not found in three UV-induced revertants or in a MNNG-induced revertant. These mutagens enhanced the frequency of reversion of RJK10 5- to 40-fold. Although the sample size is small, the results imply that spontaneous and mutagen-induced reversion of RJK10 occur predominantly by different mechanisms.

Similar analysis was used to examine HPRT⁺ revertants arising from the Chinese hamster cell line RJK526. This line produces a temperature-sensitive HPRT, resulting in reduced plating efficiency in HAT medium at 39°C. By selecting for ability to grow under these conditions, seven phenotypic revertants were isolated. These cell lines produced 2- to 13-fold higher levels of HPRT activity at 33°C than did the parental RJK526 cells, and six of the seven expressed a thermolabile enzyme. HPRT overproduction in these cell lines was correlated with 2- to 7-fold levels of gene amplification, and subsequent selection for 6-thioguanine-resistant cells demonstrated that loss of these amplified sequences occurred concomitantly with loss of HPRT enzyme activity. These results, coupled with similar findings on amplification of defective dihydrofolate reductase (111) and thymidine kinase (112) genes, imply that this method of phenotypic reversion in cultured animal cells is a relatively common occurrence.

V. Future Prospects

The ultimate goal in the study of genetic mutation is the elucidation of specific molecular alterations in the DNA structure that result in the elimination or modification of gene expression. This information may then be used to infer the mechanisms by which particular agents, such as carcinogenic compounds or radiation, induce these changes. Such changes may involve major deletions or rearrangements in the gene or may be quite subtle differences in nucleotide sequence that result in missense, frameshift, or nonsense mutations affecting the enzyme product directly or lead to defects in regulation of transcription, mRNA processing, etc. The availability of cloned probes will lead to detailed molecular study of both experimentally induced and naturally occurring mutations at the HPRT locus. Although the gene is very large, a number of changes that involve substantial deletions or rearrangements have already been identified, and even smaller changes have been examined either by protein-sequencing techniques or selective restriction-enzyme cleavage studies. Further development of methods for using specific primers to examine mRNA sequences directly (*113, 114*) or the application of recombination "rescue" approaches (*115*) to isolate defined regions for detailed analysis may increase the feasibility of routine examination of minor changes in nucleotide sequences. The fact that small expressing-vectors constructed from HPRT cDNA sequences can be used in gene-transfer experiments (*98, 116, 117*) allows the potential for site-directed mutagenesis *in vitro* followed by examination of the *in vivo* effects and may well provide a good system for future studies on gene correction.

References

1. J. McCann, E. Choi, E. Yamasaki, and B. N. Ames, *PNAS* **72**, 5135 (1975).
2. B. N. Ames, *Science (Washington, D. C.)* **204**, 587 (1979).
3. C. J. Tobin, S. M. Bradley, C. I. Bargmann, R. A. Weinberg, A. G. Papageorge, E. M. Scolnick, R. Dhar, D. R. Lowy, and E. H. Chang, *Nature (London)* **300**, 143 (1982).
4. E. P. Reddy, R. K. Reynolds, E. Santos, and M. Barbacid, *Nature (London)* **300**, 149 (1982).
5. D. J. Capon, E. Y. Chen, A. D. Levinson, P. H. Seeburg, and D. V. Goeddel, *Nature (London)* **302**, 33 (1983).
6. E. P. Reddy, *Science (Washington, D. C.)* **220**, 1061 (1983).
7. E. Santos, E. P. Reddy, S. Pulciani, R. J. Feldmann, and M. Barbacid, *PNAS* **80**, 4679 (1983).
8. J. E. Seegmiller, F. M. Rosenbloom, and W. N. Kelley, *Science (Washington, D. C.)* **155**, 1682 (1967).
9. W. N. Kelley and J. B. Wyngaarden, *in* "The Metabolic Basis of Inherited Disease" (J. B. Stanbury, J. B. Wyngaarden, D. S. Fredrickson, J. L. Goldstein, and M. S. Brown, eds.), p. 115. McGraw-Hill, New York, 1983.
10. W. N. Kelley, F. M. Rosenbloom, J. F. Henderson, J. E. Seegmiller, *PNAS* **57**, 1735 (1967).
11. W. N. Kelley, M. L. Green, F. M. Rosenbloom, J. F. Henderson, and J. E. Seegmiller, *Ann. Intern. Med.* **70**, 155 (1969).
12. C. T. Caskey and G. D. Kruh, *Cell* **16**, 1 (1979).

13. R. G. Fenwick, Jr., D. S. Konecki, and C. T. Caskey, in "Somatic Cell Genetics" (C. T. Caskey and D. C. Robbins, eds.), p. 19. Plenum, New York, 1982.
14. J. M. Wilson, A. B. Young, and W. N. Kelley, *N. Engl. J. Med.* **309**, 900 (1983).
15. A. S. Olsen and G. Milman, *JBC* **249**, 4030 (1974).
16. A. S. Olsen and G. Milman, *JBC* **249**, 4038 (1974).
17. S. H. Hughes, G. M. Wahl, and M. R. Capecchi, *JBC* **250**, 120 (1975).
18. A. S. Olsen and G. Milman, *Bchem* **16**, 2501 (1977).
19. H. Muensch and A. Yoshida, *EJB* **76**, 107 (1977).
20. J. A. Holden and W. N. Kelley, *JBC* **253**, 4459 (1978).
21. W. J. Arnold and W. N. Kelley, *JBC* **246**, 7398 (1971).
22. G. G. Johnson, L. R. Eisenberg, and B. R. Migeon, *Science (Washington, D. C.)* **203**, 174 (1979).
23. W. J. Arnold and W. N. Kelley, in "Isozymes" (C. L. Markert, ed.), Vol. 1 ("Molecular Structure"), p. 213. Academic Press, New York, 1975.
24. G. S. Ghangas and G. Milman, *Science (Washington, D. C.)* **196**, 1119 (1977).
25. V. I. Zannis, L. J. Gudas, and D. W. Martin, Jr., *Biochem. Genet.* **18**, 1 (1980).
26. J. M. Wilson, B. W. Baugher, L. Landa, and W. N. Kelley, *JBC* **256**, 10306 (1981).
27. J. M. Wilson, L. E. Landa, R. Kobayashi, and W. N. Kelley, *JBC* **257**, 14830 (1982).
28. G. G. Johnson, A. L. Ramage, J. W. Littlefield, and H. H. Kazazian, Jr., *Bchem* **21**, 960 (1982).
29. S. L. Shapiro, G. L. Sheppard, Jr., F. E. Dreifuss, and D. S. Newcombe, *PSEBM* **122**, 609 (1966).
30. W. L. Nyhan, J. Pesek, L. Sweetman, D. G. Carpenter, and C. H. Carter, *Pediatr. Res.* **1**, 5 (1967).
31. M. A. Becker, R. C. K. Yen, P. Itkin, S. J. Goss, J. E. Seegmiller, and B. Bakay, *Science (Washington, D. C.)* **203**, 1016 (1979).
32. G. S. Pai, J. A. Sprenkle, T. T. Do, C. E. Mareni, and B. R. Migeon, *PNAS* **77**, 2810 (1980).
33. P. L. Pearson and T. H. Roderick, *Cytogenet. Cell Genet.* **25**, 82 (1979).
34. A. Westerveld, R. P. L. S. Visser, M. A. Freeke, and D. Bootsma, *Biochem. Genet.* **7**, 33 (1972).
35. S. A. Farrell and R. G. Worton, *Somatic Cell Genet.* **3**, 539 (1977).
36. R. G. Fenwick, Jr., *Somatic Cell Genet.* **6**, 477 (1980).
37. V. M. Chapman and T. B. Shows, *Nature (London)* **259**, 665 (1976).
38. U. Franke and R. T. Taggart, *PNAS* **77**, 3595 (1980).
39. L. H. Thompson and R. M. Baker, in "Methods in Cell Biology" (D. M. Prescott, ed.), Vol. 6, p. 209. Academic Press, New York, 1973.
40. J. Morrow, "Eukaryotic Cell Genetics," Chapter 2. Academic Press, New York, 1983.
41. A. W. Hsie, J. P. O'Neill, J. R. San Sebastian, and P. A. Brimer, in "Mammalian Cell Mutagenesis: The Maturation of Test Systems" (A. W. Hsie, J. P. O'Neill, and V. K. McElheny, eds.), p. 407. CSHLab, CSH, New York, 1979.
42. J. Morrow, "Eukaryotic Cell Genetics," Chapter 12. Academic Press, New York, 1983.
43. W. Szybalski and E. H. Szybalska, *Univ. Mich. Med. Bull.* **28**, 277 (1962).
44. J. W. Littlefield, *Science* **145**, 709 (1964).
45. V. A. McKusick and F. H. Ruddle, *Science (Washington, D. C.)* **196**, 390 (1977).
46. J. M. Wilson, B. W. Baugher, P. M. Mattes, P. E. Daddona, and W. N. Kelly, *J. Clin. Invest.* **69**, 706 (1982).
47. G. S. Ghangas and G. Milman, *PNAS* **72**, 4147 (1975).

48. K. S. Upchurch, A. Leyva, W. J. Arnold, E. W. Holmes, and W. N. Kelley, PNAS 72, 4142 (1975).
49. W. J. Arnold, J. C. Meade, and W. N. Kelley, J. Clin. Invest. 51, 1805 (1972).
50. M. Strauss, L. Lübbe, and E. Geissler, Hum. Genet. 57, 185 (1981).
51. W. Gutensohn and H. Jahn, Eur. J. Clin. Invest. 9, 43 (1979).
52. W. N. Kelley and J. C. Meade, JBC 246, 2953 (1971).
53. M. P. Uitendaal, C. H. M. M. de Bruyn, T. L. Oei, and P. Hösli, Biochem. Genet. 16, 1187 (1978).
54. J. A. McDonald and W. N. Kelley, Science (Washington, D. C.) 171, 689 (1971).
55. B. Bakay, W. L. Nyhan, N. Fawcett, and M. D. Kogut, Biochem. Genet. 7, 73 (1972).
56. P. J. Benke, N. Herrick, and A. Hebert, J. Clin. Invest. 52, 2234 (1973).
57. J. F. Henderson, J. B. Dossetor, M. K. Dasgupta, and A. S. Russell, Clin. Biochem. 9, 4 (1976).
58. L. Sweetman, M. A. Hoch, B. Bakay, M. Borden, P. Lesh, and W. L. Nyhan, J. Pediatr. (St. Louis) 92, 385 (1978).
59. I. H. Fox, I. L. Dwosh, P. J. Marchant, S. Lacroix, M. R. Moore, S. Omura, and V. Wyhofsky, J. Clin. Invest. 56, 1239 (1975).
60. B. Bakay and W. L. Nyhan, Biochem. Genet. 6, 139 (1972).
61. G. Milman, E. Lee, G. S. Ghangas, J. R. McLaughlin, and M. George, Jr., PNAS 73, 4589 (1976).
62. I. H. Fox and S. Lacroix, J. Lab. Clin. Med. 90, 25 (1977).
63. G. S. Ghangas and G. Milman, Science (Washington, D. C.) 196, 1119 (1977).
64. C. R. Merril, D. Goldman, and M. Ebert, PNAS 78, 6471 (1981).
65. E. H. Y. Chu and H. V. Malling, PNAS 61, 1306 (1968).
66. E. H. Y. Chu, P. Brimer, K. B. Jacobson, and E. V. Merriam, Genetics 62, 359 (1969).
67. F. D. Gillin, D. J. Roufa, A. L. Beaudet, and C. T. Caskey, Genetics 72, 239 (1972).
68. A. L. Beaudet, D. J. Roufa, and C. T. Caskey, PNAS 70, 320 (1973).
69. J. D. Sharp, N. E. Capecchi, and M. R. Capecchi, PNAS 70, 3145 (1973).
70. K. H. Astrin and C. T. Caskey, ABB 176, 397 (1976).
71. M. Fox, J. M. Boyle, and B. W. Fox, Mutat. Res. 35, 289 (1976).
72. M. B. Meyers, O. P. van Diggelen, M. van Diggelen and S. Shin, Somatic Cell Genet. 6, 299 (1980).
73. C. T. Caskey, R. G. Fenwick, G. Kruh, and D. Konecki, in "Nonsense Mutations and tRNA Suppressors" (J. E. Celis and J. D. Smith, eds.), p. 235. Academic Press, New York, 1979.
74. R. G. Fenwick, Jr. and C. T. Caskey, Cell 5, 115 (1975).
75. U. Friedrich and P. Coffino, BBA 483, 70 (1977).
76. R. G. Fenwick, Jr., T. H. Sawyer, G. D. Kruh, K. H. Astrin, and C. T. Caskey, Cell 12, 383 (1977).
77. J. Epstein, G. S. Ghangas, A. Leyva, G. Milman, and J. W. Littlefield, Somatic Cell Genet. 5, 809 (1979).
78. L. A. Chasin and G. Urlaub, Somatic Cell Genet. 2, 453 (1976).
79. R. G. Fenwick, Jr., J. J. Wasmuth, and C. T. Caskey, Somatic Cell Genet. 3, 207 (1977).
80. G. M. Wahl, S. H. Hughes, and M. R. Capecchi, J. Cell. Physiol. 85, 307 (1974).
81. R. G. Fenwick, Jr., Somatic Cell Genet. 6, 477 (1980).
82. G. Milman, S. W. Krauss, and A. S. Olsen, PNAS 74, 926 (1977).
83. M. R. Capecchi, R. A. Vonder Haar, N. E. Capecchi, and M. M. Sveda, Cell 12, 371 (1977).

84. G. D. Kruh, R. G. Fenwick, Jr., and C. T. Caskey, *JBC* **256**, 2878 (1981).
85. J. M. Wilson, G. E. Tarr, W. C. Mahoney, and W. N. Kelley, *JBC* **257**, 10978 (1982).
86. J. M. Wilson, R. Kobayashi, I. H. Fox, and W. N. Kelley, *JBC* **258**, 6458 (1983).
87. J. M. Wilson, G. E. Tarr, and W. N. Kelley, *PNAS* **80**, 870 (1983).
88. J. M. Wilson and W. N. Kelley, *J. Clin. Invest.* **71**, 1331 (1983).
89. D. W. Melton, D. S. Konecki, D. H. Ledbetter, J. F. Hejtmancik, and C. T. Caskey, *PNAS* **78**, 6977 (1981).
90. D. W. Melton, *Somatic Cell Genet.* **7**, 331 (1981).
91. J. Brennand, A. C. Chinault, D. S. Konecki, D. W. Melton, and C. T. Caskey, *PNAS* **79**, 1950 (1982).
92. P. S. Thomas, *PNAS* **9**, 5201 (1980).
93. E. M. Southern, *JMB* **98**, 503 (1975).
94. D. W. Melton, D. S. Konecki, D. H. Ledbetter, J. Brennand, A. C. Chinault, and C. T. Caskey, *in* "Gene Amplification" (R. T. Schimke and V. K. McElheny, eds.), p. 59. CSHLab, CSH, New York, 1982.
95. D. S. Konecki, J. Brennand, J. C. Fuscoe, C. T. Caskey, and A. C. Chinault, *NARes* **10**, 6763 (1982).
96. A. M. Michelson, A. F. Markham, and S. H. Orkin, *PNAS* **80**, 472 (1983).
97. D. Jolly, A. Esty, U. Bernard, and T. Friedmann, *PNAS* **79**, 5038 (1982).
98. D. Jolly, H. Okayama, P. Berg, A. C. Esty, D. Filpula, P. Bohlen, G. G. Johnson, J. E. Shively, T. Hunkapillar, and T. Friedmann, *PNAS* **80**, 477 (1983).
99. H. Okayama, and P. Berg, *MCBiol* **2**, 161, (1982).
100. N. J. Proudfoot and G. G. Brownlee, *Nature (London)* **263**, 211 (1976).
101. D. W. Melton, D. S. Konecki, J. Brennand, and C. T. Caskey, *PNAS* **81**, 2147 (1984).
102. P. Argos, M. Hanei, J. M. Wilson, and W. N. Kelley, *JBC* **258**, 6450 (1983).
103. W. Gilbert, *Nature (London)* **271**, 501 (1978).
104. K. Willecke, M. Klomfass, R. Mierau, and J. Doehmer, *MGG* **170**, 179 (1979).
105. L. H. Graf, Jr., G. Urlaub, and L. A. Chasin, *Somatic Cell Genet.* **5**, 1031 (1979).
106. S. C. Lester, S. K. LeVan, C. Steglich, and R. DeMars, *Somatic Cell Genet.* **6**, 241 (1980).
107. J. C. Fuscoe, R. G. Fenwick, Jr., D. H. Ledbetter, and C. T. Caskey, *MCBiol* **3**, 1086 (1983).
108. R. L. Nussbaum, W. E. Crowder, W. L. Nyhan, and C. T. Caskey, *PNAS* **80**, 4035 (1983).
109. J. M. Wilson, P. Frossard, R. L. Nussbaum, C. T. Caskey, and W. N. Kelley, *J. Clin. Invest.* **72**, 767 (1983).
110. R. G. Fenwick, Jr., J. C. Fuscoe, and C. T. Caskey, *Somatic Cell Mol. Genet.* **10**, 71 (1984).
111. L. A. Chasin, L. Graf, N. Ellis, M. Landzberg, and G. Urlaub, *in* "Gene Amplification" (R. T. Schimke and V. K. McElheny, eds.), p. 161. CSHLab CSH, New York, 1982.
112. J. M. Roberts and R. Axel, *Cell* **29**, 109 (1982).
113. P. H. Hamlyn, M. J. Gait, and C. Milstein, *NARes* **9**, 4485 (1981).
114. M. Kaartinen, G. M. Griffiths, P. H. Hamlyn, A. F. Markhan, K. Karjalainen, J. L. T. Pelkonen, O. Makela, and C. Milstein, *J. Immunol.* **130**, 937 (1983).
115. B. Seed, *NARes* **11**, 2427 (1982).
116. A. D. Miller, D. J. Jolly, T. Friedmann, and I. M. Verma, *PNAS* **80**, 4709 (1983).
117. J. Brennand, D. S. Konecki, and C. T. Caskey, *JBC* **258**, 9593 (1983).

The Molecular Genetics of Human Hemoglobin

FRANCIS S. COLLINS* AND
SHERMAN M. WEISSMAN

Departments of Human Genetics,
Medicine, and Molecular
Biophysics and Biochemistry
Yale University School of
Medicine
New Haven, Connecticut

I. Structure of Human Globin Genes.............................. 317
   A. General 317
   B. Organization of Exons and Introns......................... 319
   C. Codon Utilization... 321
   D. 5' and 3' Untranslated Regions............................ 324
II. Approaches for Studying Expression of Globin Genes 325
   A. Cellular Systems.. 325
   B. Systems for Introducing Globin Genes into Cells............... 326
III. Physiological Globin Gene Expression 330
   A. Developmental Regulation 330
   B. Transcriptional Unit...................................... 332
   C. Promoter Function 333
   D. Splicing.. 343
   E. DNase Hypersensitivity................................... 348
   F. Methylation.. 354
IV. Structure of Human Globin Gene Clusters 359
   A. The α-Globin Cluster..................................... 359
   B. The β-Globin Cluster...................................... 363
   C. Evidence for Gene Conversion of Repeated Genes............ 373
V. Evolutionary Considerations................................... 379
   A. Early Dispersion of the Globin Gene Family 379
   B. Mammalian Comparisons................................. 381
VI. Naturally Occurring Mutations 390
   A. General Comments....................................... 390
   B. The Thalassemias... 390
   C. Deletions in the β-Globin Complex......................... 402
   D. Polymorphisms without Known Functional Correlates 411
   E. Prenatal Diagnosis 416
VII. Summary and Prospects...................................... 419
   References... 421

* Present address: Department of Medicine, University of Michigan Medical School, Ann Arbor, Michigan 48109.

Hemoglobin constitutes approximately 90% of the soluble protein in circulating red cells and essentially the same percentage of soluble protein extractable from the total cellular mass present in blood.[1] The major physiological functions of hemoglobin were elucidated decades ago, and the protein itself was one of the earlier proteins purified and characterized by sequence analysis. The X-ray-diffraction determination of the crystalline structure of hemoglobin is one of the highlights in structural molecular biology and represents the first successful analysis of the three-dimensional structure of a protein. Many chemical derivatives and structurally mutant hemoglobins as well as species variants are known. These have further provided an extensive body of knowledge concerning structure–function relationships in the hemoglobin protein.

In addition to the protein, globin messenger RNA (mRNA) may be readily isolated in a very enriched form from nucleated red cells or reticulocytes. Interest in the pathophysiology of the various disorders of hemoglobin, information about a large number of hemoglobin mutants, and the availability of mRNA created a favorable situation for the application of recombinant DNA technology. Isolation of complementary DNAs (cDNAs) and genomic segments from hemoglobin genes were among the first successes in eukaryotic cloning. By the time of this writing (October, 1983), there had been an accumulation of knowledge of the comparative structure and linkage of the globin genes in several species exceeding that available on any other eukaryotic gene system. Extensive data on the effects both of natural mutations and those produced by surrogate genetics has been derived, and

[1] Abbreviations and terms: bp, base-pair; kb, kilobase; IVS, intervening sequence, intron; Southern blotting, a technique whereby DNA cut with a particular restriction enzyme is gel-electrophoresed to separate the fragments by size, and the fragments are then transferred to nitrocellulose paper and hybridized with a labeled DNA or RNA probe to reveal the position of complementary DNA sequences. If the starting material was total genomic DNA, this is also denoted as genomic blotting; Northern blotting, technique similar in principle to Southern blotting, except that the starting material is uncut RNA. The hybridization then reveals the size and abundance of the message(s) corresponding to the probe used; MEL, mouse erythroleukemia cells, also called Friend cells; HEL, K562, two different human continuous erythroid cell lines; TK, thymidine kinase; HAT, hypoxanthine+aminopterin+thymidine (selective medium); HPRT, hypoxanthine phosphoribosyltransferase (EC 2.4.2.8), sometimes also denoted HGPRT; the bacterial form is often denoted XGPRT; APRT, adenine phosphoribosyltransferase; HMBA, hexamethylene bisacetamide; BPV, bovine papillomavirus; LTR, long-terminal-repeat of a retrovirus; SV40, simian virus 40; E1A, an early protein of adenovirus; COS cells, a monkey kidney cell line containing a stably integrated SV40 genome and expressing functional SV40 early proteins; *cos* site, sequence at the ends of λ bacteriophage recognized by the apparatus that packages λ DNA into phage heads.

has led to important conclusions about the signals necessary for transcription and processing of globin mRNA that seem to be broadly applicable to other eukaryotic genes. Application of this information in the field of human pathophysiology has revealed an enormous degree of genetic heterogeneity in hereditary disorders of production of specific proteins and has provided a testing field for the applications of modern DNA technology to antenatal diagnosis.

The present review is intended principally to summarize the state of current knowledge about the structure–function relationships of the globin gene system as derived from studies at the level of DNA and RNA, to discuss some of the sequence comparisons that relate to the evolution of this gene system, and to review briefly the current status of applications of this molecular genetic knowledge to the understanding of the human hemoglobinopathies. The bulk of literature on globin (more than 10,000 references are listed in *Index Medicus* in the last four years) precludes anything approaching exhaustive coverage. We will attempt to concentrate on developments of the last three years. There are several excellent reviews and texts that present further details and background (*1–12*).

We apologize to the many colleagues whose important work we may have failed to quote adequately. In particular, we have omitted a discussion of the cellular basis of erythropoiesis, in spite of the great importance of this topic. *In vitro* analysis of clonal erythroid cultures has already yielded important insights, most notably that adult erythroid precursors appear to follow a program of maturation that includes a shift in the capacity to synthesize globin from fetal hemoglobin (HbF) in earlier precursors to adult hemoglobin (HbA) in more mature precursors (*13–15a*). The observation of a humoral factor in fetal sheep serum that influences this program (*16, 16a*) is an interesting recent piece in the puzzle of hemoglobin switching, but space precludes our considering this area any further. We also regret that we are not able to include a discussion of the various retroviruses that induce erythroid proliferation of mouse (*17–22*) or birds (*24*), although studies in this area may provide important insights into erythroid differentiation and host control of virus susceptibility.

I. Structure of Human Globin Genes

A. General

Hemoglobin is a metalloprotein tetramer consisting of two α-like globin polypeptides and two β-like globin chains. These form a com-

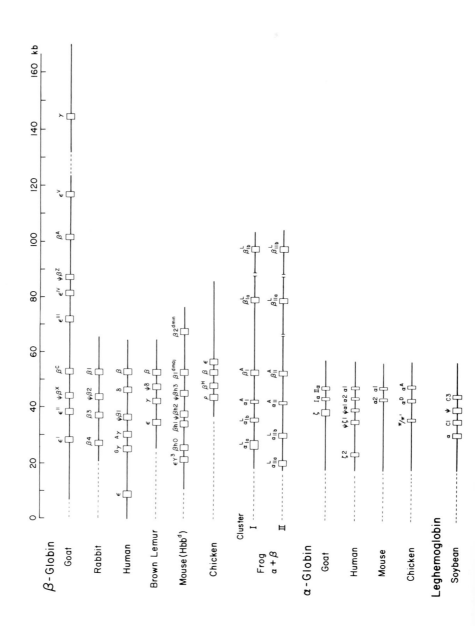

plex with four molecules of heme, one for each peptide chain. The structure of the human α and β gene loci, as well as the globin loci for several other species for which detailed mapping information is available, is presented in Fig. 1 (25–41). We have much more to say about the structure of the human clusters (Section IV) and about evolutionary relationships (Section V). The developmental regulation of the α-like and β-like genes is considered in Section IIIA.

Figure 2 shows the basic structure of the human α- and β-globin genes to scale, including the untranslated sequences, exons, and introns. Complete sequences are available for all the expressed human genes: ε (42), $^G\gamma$ and $^A\gamma$ (43, 44), δ (45), β (46), ζ (47), and α (48, 49).

In this section we consider the basic structure of globin genes. A more functional analysis is given in Section III, after presentation of some of the approaches (Section II) used to deduce the nature of globin gene expression.

B. Organization of Exons and Introns

Throughout the vertebrates, the basic intron–exon structure of the globin genes remains constant. The 5' leader sequences and approximately the first 100 nucleotides of the globin gene are encoded in a single exon, followed by intron 1, a second exon domain extending to approximately codon 100, a second intron, and a third exon domain comprising the rest of the transcribed DNA. In mammals, the first intron [110–130 base-pairs (bp)] of β-like genes is considerably smaller than the second intron (600–900 bp). In the α cluster, the second intron is only slightly larger than the first, and considerably smaller than in the β-gene cluster (50). There are exceptions; in the ζ and ψζ genes in man, for example, the first intron contains a tandemly repeated element (see Section IV,A) and is much larger than the usual IVS-1 (47). The introns of *Xenopus laevis* are somewhat larger, up to 1500 bp (40).

The myoglobin gene of the seal has been cloned recently and shows similar intron and exon structure, but with remarkably long introns; intron 1 is about 4800 bp and intron 2 about 3400 bp (50). This finding casts some doubt on the suggestion that intron length per se is important for globin gene expression.

FIG. 1. Structure of the α- and β-globin loci of several species. The scale is indicated at the top of the figure. Dashed lines between ε^V and γ in the goat and dashed lines in the frog indicate remaining uncertainties until overlapping genomic clones can be identified (see Addendum 2, p. 459). Linkage arrangements are taken from the following references: goat (25–27a), rabbit (28, 29), human (30, 31), lemur (32), mouse (33–36), chicken (37–39), frog (40), and leghemoglobin (41).

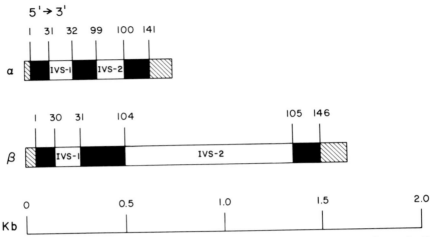

FIG. 2. Structure of the human α- and β-globin genes. Hatched boxes indicate the 5' and 3' untranslated regions, open boxes represent the first and second introns of each gene, and solid boxes represent the coding regions. The codon numbers of the boundaries of each exon are shown above the gene. Reproduced by permission of the publisher from B. G. Forget (Normal and Abnormal Globin Genes), in "The Regulation of Hemoglobin Biosynthesis" (E. Goldwasser, ed.), p. 27, Elsevier Science Publishing Co., Inc., New York, 1983. IVS = intervening sequence.

The amino acid residues of globin can be divided into four groups, with members of each group located near or in contact with one another in native globin (51). The first intervening sequences of the globin genes lie between the coding regions for groups one and two, and the second lies between the coding regions for groups three and four. A most interesting and remarkable correlation with this analysis emerged when the leghemoglobin gene was isolated from plants and found to have three intervening sequences, the first two of which were at positions analogous to those of animal hemoglobin genes (52, 53, 53a). The third intervening sequence was located so as to separate exons coding for the second and third of the four regions postulated (51). Thus, the locations of intervening sequences correlates with coding regions previously defined on the basis of the structure of the folded protein.

This correlation between location of introns and the structure of proteins, while not universal, has been seen remarkably often. It is sometimes very striking, as in the genes for immunoglobulins (54) and the major histocompatibility antigens (55–57). The numerous varieties of thalassemia caused by mutations within intervening sequences show that there is some evolutionary disadvantage associated with the presence of intervening sequences. Arguments have been advanced for the evolutionary advantages of having the intervening sequences located between coding regions for separate protein domains, because of the possibility of exon exchange during evolution (58, 59, 59a). The globin family has a long evolutionary history without evidence of a major role for exon exchange with any other protein family. It is not clear that these evolutionary considerations alone provide adequate selection pressure to maintain and locate intervening sequences, or whether more specific selection mechanisms have evolved to maintain intervening sequences, such as alternate splicing events.

To maintain the relationship between sequence location and protein structure, there would apparently have to be feedback of differently processed forms of RNA into events that affect biological fitness over periods short enough to provide for continuing evolutionary selection. For example, the correlation of exon position with protein tertiary structure could occur by translation of incompletely spliced mRNA or mRNA in which there are "aberrant" splices excluding or creating additional protein segments. If such proteins were produced and if their structures affected any cellular functions, they would provide additional constraints on the location of intervening sequences. Sensitive methods have commonly failed to detect aberrantly spliced cytoplasmic RNA from normal globin genes, but there may be exceptions, and the putative RNAs could be short-lived compared to normal globin mRNA.

A variation on the exon exchange model that has been proposed recently is that properly located intervening sequences would limit homogenizing effects of gene conversion within a gene family (60). As we see in Section IV,C, introns may provide favorable targets for initiation of the gene-conversion process and often (but not always) seem to act as barriers for its propagation. This could provide more selection for properly placed intervening sequences if gene conversion is a more frequent event than unequal recombination. On the other hand, this explanation would not pertain to single-copy genes.

C. Codon Utilization

Most prokaryotes display a relatively narrow range or unimodal distribution of base composition along their chromosomal DNA. Nevertheless, there are unexplained divergences even in prokaryotic systems. For example, λ bacteriophage has segments of DNA that exhibit (G + C):(A + T) ratios deviating from the mean considerably more than would be expected on a statistical basis (61). Eukaryotic DNA displays a wider dispersion of average base composition over long

α

F TTT	0	S TCT	3	Y TAT	1	C TGT	0
F TTC	7	S TCC	4	Y TAC	2	C TGC	1
L TTA	0	S TCA	0	* TAA	0	* TGA	0
L TTG	0	S TCG	0	* TAG	0	W TGG	1
L CTT	1	P CCT	2	H CAT	0	R CGT	1
L CTC	2	P CCC	3	H CAC	10	R CGC	0
L CTA	1	P CCA	0	Q CAA	0	R CGA	0
L CTG	14	P CCG	0	Q CAG	0	R CGG	1
I ATT	0	T ACT	0	N AAT	0	S AGT	0
I ATC	0	T ACC	9	N AAC	4	S AGC	4
I ATA	0	T ACA	0	K AAA	0	R AGA	0
M ATG	3	T ACG	0	K AAG	10	R AGG	1
V GTT	1	A GCT	3	D GAT	0	G GGT	0
V GTC	3	A GCC	11	D GAC	8	G GGC	8
V GTA	0	A GCA	0	E GAA	0	G GGA	0
V GTG	9	A GCG	7	E GAG	4	G GGG	0

β

F TTT	5	S TCT	1	Y TAT	2	C TGT	2
F TTC	3	S TCC	2	Y TAC	1	C TGC	0
L TTA	0	S TCA	0	* TAA	0	* TGA	0
L TTG	0	S TCG	0	* TAG	0	W TGG	2
L CTT	0	P CCT	5	H CAT	2	R CGT	0
L CTC	3	P CCC	0	H CAC	7	R CGC	0
L CTA	0	P CCA	2	Q CAA	0	R CGA	0
L CTG	15	P CCG	0	Q CAG	3	R CGG	0
I ATT	0	T ACT	3	N AAT	1	S AGT	2
I ATC	0	T ACC	3	N AAC	5	S AGC	0
I ATA	0	T ACA	1	K AAA	3	R AGA	0
M ATG	2	T ACG	0	K AAG	8	R AGG	3
V GTT	3	A GCT	6	D GAT	5	G GGT	4
V GTC	2	A GCC	2	D GAC	2	G GGC	2
V GTA	0	A GCA	2	E GAA	2	G GGA	0
V GTG	13	A GCG	6	E GAG	6	G GGG	1

ζ

F TTT	0	S TCT	2	Y TAT	2	C TGT	0
F TTC	7	S TCC	7	Y TAC	7	C TGC	1
L TTA	0	S TCA	0	* TAA	0	* TGA	0
L TTG	1	S TCG	1	* TAG	0	W TGG	2
L CTT	0	P CCT	0	H CAT	0	R CGT	0
L CTC	3	P CCC	1	H CAC	1	R CGC	7
L CTA	1	P CCA	0	Q CAA	0	R CGA	2
L CTG	12	P CCG	4	Q CAG	3	R CGG	0
I ATT	2	T ACT	2	N AAT	1	S AGT	0
I ATC	8	T ACC	8	N AAC	8	S AGC	3
I ATA	0	T ACA	0	K AAA	0	R AGA	0
M ATG	9	T ACG	2	K AAG	9	R AGG	0
V GTT	0	A GCT	0	D GAT	0	G GGT	5
V GTC	4	A GCC	12	D GAC	8	G GGC	0
V GTA	1	A GCA	0	E GAA	0	G GGA	0
V GTG	6	A GCG	4	E GAG	6	G GGG	1

ε

F TTT	7	S TCT	3	Y TAT	0	C TGT	1
F TTC	2	S TCC	1	Y TAC	2	C TGC	0
L TTA	0	S TCA	0	* TAA	0	* TGA	1
L TTG	1	S TCG	1	* TAG	0	W TGG	3
L CTT	0	P CCT	2	H CAT	4	R CGT	0
L CTC	4	P CCC	4	H CAC	3	R CGC	0
L CTA	0	P CCA	0	Q CAA	0	R CGA	0
L CTG	11	P CCG	0	Q CAG	3	R CGG	0
I ATT	4	T ACT	4	N AAT	1	S AGT	1
I ATC	1	T ACC	2	N AAC	6	S AGC	6
I ATA	0	T ACA	0	K AAA	0	R AGA	2
M ATG	4	T ACG	0	K AAG	13	R AGG	0
V GTT	2	A GCT	8	D GAT	2	G GGT	2
V GTC	3	A GCC	9	D GAC	3	G GGC	4
V GTA	0	A GCA	0	E GAA	3	G GGA	3
V GTG	8	A GCG	0	E GAG	6	G GGG	0

segments than does prokaryotic DNA. It has been known for some years that the globin system provides a puzzling example of bias in base composition. Thus, even though α- and β-globin mRNAs encode proteins of very similar peptide length that are formed at essentially identical rates in the same cell, the base compositions of the α-related globin DNA and the β-related DNA are very different. This is reflected in codon utilization, as shown in Fig. 3. For example, the dinucleotide C-G is remarkably rare in β, γ, and ε mRNA compared with α and ζ mRNA. In addition, the α-globin mRNAs have a substantial preference for C over U in the third position of a number of codons. This is not true for the β-globin mRNAs. Some of the intervening sequences of the α-globin cluster are also markedly rich in G and C. A striking example of this is provided by the large intron of ζ-globin, which contains large stretches of imperfect repeats of a tandem pentanucleotide CGGGG (47). It is difficult to understand how such biases could arise on the basis of codon selection, since the same transfer RNAs are used for translating the (G + C)-rich and the (G + C)-poor mRNAs in red cells. The presence of tandem repeats of short (G + C)-rich sequences suggests that they may have arisen by slippage and/or unequal crossing-over mechanisms. Alternate and unelucidated aspects of chromatin structure must be considered as possible factors. If some segments of chromosomal DNA were under higher torsional stress, for example, there might be evolutionary advantages to maintaining them at a high (G + C)-content. This could have been accomplished either by evolutionary selection at the nucleotide level, or on the basis of more general mechanisms such as the use of different enzyme systems for repair or replication.

Other biases may exist in globin codon usage. For example, it has been observed (62) that globin-coding regions are relatively deficient in codons that could mutate by a single-base change to a chain-terminator codon. Also the rate of change to or from T is low (62), and this would tend to preserve the hydrophilic or hydrophobic nature of amino-acid residues. Conversely, the G to A substitution rate may be

---

FIG. 3. Codon utilization patterns in the α-, β-, ζ-, and ε-globin genes of man. The number of times each codon appears is shown. Standard one-letter amino-acid symbols are also included; * indicates a termination codon. Initiation codons (ATG) are included. Note the rarity of codons containing C-G in the β and ε genes (0 and 1, respectively), while these are more common in α and ζ (11 and 15, respectively). Two examples illustrate this bias: (i) β and ε use only the AGA and AGG codons for arginine, whereas α and ζ also use the CGN codons; (ii) the codon used for alanine in α and ζ is GCG in 11 out of 37 instances, whereas this codon is not used at all in 32 appearances of alanine in the β and ε genes.

increased, but this substitution is relatively conservative with respect to hydrophobicity of encoded amino acids.

D. 5′ and 3′ Untranslated Regions

The mammalian globin genes are remarkable for the similarity in length of the 5′ untranslated sequences of the β-globin genes, which vary from 50 to 56 bp in length. The 5′ untranslated sequences of α-globin are somewhat more variable in length. The sequence homologies among various untranslated 5′ regions of β-like globin have been compared previously (63). There are two prominently conserved sequences in the various globin genes (both α and β). One is the sequence of the form GCTTCTGR (R = a purine nucleoside), found 6 through 13 nucleotides downstream from the cap site of goat, human, rabbit, and mouse β-like genes, and near the 5′ end of several α-like genes. In chicken genes, a similar sequence (GCTCTR) is found. It has been suggested that these sequences are involved in ribosomal binding by formation of base-pairs with a purine-rich sequence at the 3′ end of 18-S rRNA (64–66). A second conserved sequence present in both birds and mammals is the sequence CAYCATG (Y = a pyrimidine nucleoside), in which the last three residues correspond to the codon initiating translation (Table I).

TABLE I
Conserved Sequences in 5′ Untranslated Portion of mRNA[a]

A. Presumed ribosomal binding site

Position relative to cap site								
	+6	+7	+8	+9	+10	+11	+12	+13
Base:	G	C	T	T	C	T	G	R
Frequency:	14/18	17/18	17/18	17/18	16/18	17/18	18/18	17/18

B. Initiation sequence

Base:	C	A	Y	C	ATG
Frequency:	20/21	18/21	20/21	19/21	21/21

[a] The numerator in each frequency fraction is the number of times the consensus base occurs; the denominator is the number of sequences. Highly homologous sequences, such as the two human fetal globin genes, were not compared. Y = pyrimidine (nucleoside), R = purine (nucleoside).

Sequences considered are α-like and β-like genes for man, mouse, goat, and rabbit; chicken genes are considered in the initiation sequence comparisons. References for the β genes are in the legend to Fig. 6, and for the α genes in Section III,D.

The untranslated 3' sequences of the globin genes vary somewhat more in length, but most fall in the range of 80 to 150 nucleotides. The only obvious conserved sequence is the core polyadenylation signal AAUAAA located approximately 20 nucleotides upstream from the site of polyadenylate addition. No other specific functional sequences have yet been demonstrated in this region of the gene. As with the intervening sequences, variation between 3'-untranslated sequences of different genes may occur by single-base changes, but also by duplications and deletions. In primates, both 5' and 3' sequences of globin mRNA are relatively conserved. It has been suggested that this reflects some functions that restrain their divergence (67). However, the rate of nucleotide substitution as a whole may be reduced in higher primates (see Section V,B) and this may be an additional important factor accounting for the similarities.

II. Approaches for Studying Expression of Globin Genes

A. Cellular Systems

Because erythropoietic cells are intermingled with many other cell types in the marrow, cloned cell lines that can be propagated *in vitro* are highly desirable for studying intracellular globin gene expression and regulation. The oldest and most extensively studied of the systems for studying long-term expression of globin genes *in vitro* is the virus-induced murine erythroleukemia (MEL) cell system originally developed by Friend (67a). There are a number of variant strains of MEL cell lines originally derived from mice exposed to the complexes of spleen focus-forming virus (SFFV) and helper virus. These cell lines can be propagated indefinitely and may be induced to form relatively large amounts of hemoglobin by a variety of compounds, such as dimethyl sulfoxide (Me_2SO) or hexamethylene bisacetamide (HMBA). The specifics of the variation in the properties of various MEL cell lines, and the nature and property of their inducers, have been extensively discussed (68) and are beyond the scope of the present review. In brief, the MEL cells produce large amounts of adult globin chains and may produce smaller amounts of embryonic chains (69). The amount of globin produced may increase up to a hundredfold on induction. Much, but not all (70, 71), of the control of globin production appears to be mediated by increased transcription of the globin genes.

Attempts have been made to derive cell lines representing earlier stages of erythroid differentiation. Recently, Mak *et al.* derived cell lines that appear to be arrested at various stages of erythroid proliferation, but are inducible to varying degrees (*18, 19, 21, 72, 73*). These interesting lines have not yet been employed for studies of globin gene control.

In addition to the Friend erythroleukemia cell itself, fusion hybrids that express human as well as murine globin chains have been produced between Friend erythroleukemia cells and human bone marrow cells (*74–76*), lymphoblasts (*77*), or even fibroblasts (*78, 79*). Cell lines that produce adult human β- but not γ- or ε-globin mRNA as well as the murine mRNA have been obtained suggesting the existence of *trans*-active factors that operate across species in the regulation or stabilization of globin production.

Human erythroleukemic cell lines that also continue to produce variable amounts of globin have been obtained. The K562 cell line (*80*) was the first such line derived and the one most extensively studied, but other lines, such as the human erythroleukemia (HEL) line, have been derived (*81*). These cells contain genes for all the hemoglobins, and at the level of Southern blotting there is no evidence for any deletion or alteration of any globin gene (*82*). They produce all globin chains except for the adult β-globin (*83*). Even though small amounts of δ-globin mRNA may be produced, repeated studies have failed to detect any evidence of mature β-globin mRNA, and the predominant globin mRNA produced is embryonic (*85*). Globin production (*86*) and transcription of globin genes (*85, 87, 88*) are regulated in response to external inducers. Because both the murine and human cell lines are neoplastic, there is some concern as to whether they completely reflect the structure and regulation of globin chromatin in normal cells.

B. Systems for Introducing Globin Genes into Cells

A variety of methods have been developed in recent years for introducing DNA, including globin genes, into various cell lines. The first to be applied successfully, and one still widely used, depends upon the observation that precipitates of calcium phosphate bind DNA and promote its uptake into cells. The mouse L cell line has been studied extensively in this regard. L cells deficient in thymidine kinase (TK) have been obtained and can be readily transfected with herpesvirus TK genes. Selection of these cells in HAT media (hypoxanthine + aminopterin + thymidine) permits growth only of those cells that have a functional TK gene, as the aminopterin blocks endogenous synthesis of thymidine. A large fraction of cells that are cotransfected with the TK gene and another gene, such as a globin gene, will take up and express both. It is further possible to select against TK-expressing cells by the use of drugs, such as bromodeoxyuridine, that depend upon thymidine kinase for activation. In this way, for example, a study (89) of the expression of γ-globin genes linked to the thymidine kinase gene of herpes virus shows that the γ-globin gene is expressed in stably transformed cells when they are expressing the linked TK gene. However, cell lines that can, by apparently epigenetic means, turn off the production of TK have been obtained. Selection for such cells shows that there has been a simultaneous loss of β-globin expression, and that β-globin expression can be recovered in cells reselected for expression of thymidine kinase, presumably demonstrating activation and inactivation phenomena extending over substantial regions (at least 20 kb) of DNA (89a).

DNA in transfected cells early becomes ligated into large masses of DNA (called pekelasomes) (90). After varying periods of time in culture, the extrachromosomal DNA may become integrated in chromosomal sites. Although this method may be modified to introduce varying amounts of DNA, commonly multiple copies of DNA are retained and integrated. Sometimes large fractions of the DNA are not rearranged and may be in tandem copies, but some rearranged copies also occur. The disadvantages of the system are that (a) it requires relatively long periods in culture to recover and expand transfected clones; (b) the DNA may be partially rearranged and subject to further selection before its expression is studied; and (c) as initially employed, one needs TK$^-$ cell lines for transfection, and the efficiency of transfection varies considerably with cell lines.

A number of improvements have been made in the procedures for transfection. New selective markers, particularly the neomycin derivative G418 (91) and the use of mycophenolic acid and the bacterial gene for hypoxanthine phosphoribosyltransferase (HPRT) (EC 2.4.2.8) (92), provide dominant cotransfectants that can be used to select cells without requiring preexisting mutations in the cell line. A number of improvements to increase the efficiency of DNA transfection have been proposed. It is, again, beyond the scope of the present review to discuss the relative merits of these procedures. A particularly attractive procedure appears to be the electropulse induction of DNA uptake (93, 93a). This procedure has the advantage that it makes it possible to transfect rounded cells, such as lymphocytes or Friend erythroleukemia cells in suspension.

Globin genes introduced into MEL cells retain responsiveness to certain inducing agents, but do not exhibit all features of normal regulation. For example, transfected human embryonic (94), fetal, and adult human globin genes are expressed in MEL cells. Wright et al. (95), transfected MEL cells with each of several cosmids containing portions of the human β-globin complex and studied globin induction by HMBA. At least 75% of the cell clones carrying a β gene showed a 4 to 100-fold induction of the β genes. Only two of 19 clones with γ-globin genes showed induction, and this was less than 6-fold. The embryonic gene was also variably inducible (96). The sequences responsible for induction are being sought. Curiously, when the 3' end of the fetal or embryonic gene is fused to the herpes TK gene, inducible RNA that presumably initiates in an anti-sense direction within the TK gene is found. A fused gene containing the 5' portion of a mouse globin gene is also inducible in MEL cells (97), but the hybrid mouse-human genes are 10-fold less inducible than the chromosomal mouse gene. This contrasts with the observations when a single human chromosome is introduced into MEL cells by cell fusion (77, 78). In these experiments, human globin is produced at a level roughly comparable to that of mouse globin. The hybrid MEL cells are also inducible for γ-globin mRNA, although the levels before and after induction are substantially lower than the levels of β-globin mRNA in similar cells.

A modified protocol considerably expedites the analyses of globin gene expression. This is the use of transient expression systems, in which the RNA or protein products are studied 30–48 hours after exposure of the cell population to DNA. This procedure requires no selective markers or integration of DNA into stably transmitted forms. Although the methods used have in general transfected only a fraction of the cells, sufficient RNA is produced to permit primer extension, S1-nuclease mapping, and Northern blotting of RNA products (see Section VI,B,2). As discussed in Section III,C, the β-like globin genes appear to require a *cis*-linked enhancer element (98) for expression in these systems, although the α gene does not. Very recently it has been observed that the adenovirus E1a product may provide *trans* effects that permit the β genes to be expressed in the absence of enhancers (98a,b), although this appears to operate through a different mechanism, since only the TATA box is required in *cis*.

Several vector systems have been developed to generate encapsidated DNA that can be used to infect cells by means of modified virus systems (reviewed in 99, 100). Among these are SV40 systems, some of which permit substitution of late or early regions of SV40 with up to more than 2000 bases of DNA, after which a molecule may be propagated in appropriate cells, or with appropriate helper virus, and encapsidated. The advantage of this and other virus systems is that every cell in the culture can be transfected, and stable stocks can be obtained from which repeated transfections can be readily performed. The more recently developed adenovirus vectors offer in principle the advantage of being able to introduce much larger segments of DNA. Retrovirus-like vectors have the advantage that they promote direct integration of the DNA into the chromosome, where it can be stably transmitted from cell to cell; single copies of DNA will be incorporated with retroviral long-terminal-repeats at their ends, and without rearrangements (100a–f). A potential disadvantage is that the retroviral agents are propagated through RNA intermediates, and splicing events may alter the sequence of the inserted DNA. Nevertheless, the systems are very attractive and have not yet had their full potential explored. The recent report of successful genetic transformation of human Lesch–Nyhan (HPRT$^-$) cells by a retrovirus bearing the HPRT coding region is an important step (100g).

An additional attractive vector that has been used is the bovine papilloma virus (BPV) system (100h), in which a portion of the late-region viral DNA may be substituted by the gene of interest. Transfected cells transformed by the BPV-early region will grow up as colonies that will contain and often express the linked DNA. In principle, the system could accept large amounts of DNA, but rearrangements may be a serious problem. An advantage of the system is that the BPV DNA exists largely or solely as extrachromosomal elements and might permit reisolation of the chromatin containing reintroduced DNA by physical means.

Recently, a modified cosmid vector that includes a dominant genetic marker, an SV40 origin of replication, and an enhancer sequence was developed (101). It contains the DNA sequence signal (cos) for packaging of λ-bacteriophage DNA; thus a stably integrated cosmid can be rescued by packaging of whole-cell DNA with a λ packaging system, providing a means of cloning any gene whose expression can be selected for. Such a cosmid containing both ζ- and α-globin genes expressed the adult α-globin gene substantially more efficiently than the embryonic ζ gene (101) in both transient expression systems and stably transformed mouse or hamster cells.

Direct techniques for injection of DNA into cells by means of glass capillaries have been developed and applied to several cell systems (102, and references therein), including teratocarcinoma cells. Perhaps its greatest interest and potential value lie in the introduction of DNA molecules into fertilized mouse eggs. Improvements in technique make it possible to obtain viable mice from a significant fraction, perhaps up to 10–25%, of all eggs isolated. Initial results show that the injected DNA can become incorporated into murine chromosomes (*103, 104*) and propagated as genetic material. The expression of herpes thymidine kinase in somatic cells has been demonstrated (*105–107*). Very recent studies have shown increasing success: The growth hormone gene linked to the metallothionein promoter was overexpressed *in vivo* (*108, 108a*). Appropriate expression of transferrin (*109*) genes injected into fertilized murine eggs has also been reported, and a functional immunoglobulin K was expressed in spleens, but not in livers, of transgenic mice (*109a*).

Unfortunately, the experience with the β-globin genes to date has been less successful. There are several reports dealing with the introduction of β-globin genes into fertilized mouse eggs, and their subsequent incorporation into the DNA of intact mice (*103, 104, 110–114*), but it appears that a relatively small fraction of the mice so injected produced any β-globin mRNA. In the thoroughly studied cases, the very limited expression has been ectopic, e.g., in testes (*111, 112*). Interestingly, the pattern of expression was heritable, as was the chromosomal location of the injected DNA by *in situ* hybridization of metaphase spreads of the progeny. This observation suggests that the chromosomal domain into which exogenous DNA is integrated plays a crucial role in determining its pattern of expression. Unlike mice developing from injection of fertilized eggs, most mice developing from blastocyst–teratocarcinoma cell chimeras do not transmit the acquired DNA to their progeny. Further exploration of these systems should lead to a much more profound understanding of the sequences determining the tissue specificity and developmental timing of the expression of a variety of genes.

III. Physiological Globin Gene Expression

A. Developmental Regulation

The human globin genes undergo an orderly ontogenetic program of expression during prenatal and postnatal life. Figure 4 is a schematic of the relative levels of the various globin chains at various stages of development (*3*). At all times, the amounts of α-like and β-like chains are in a 1:1 ratio. In early embryonic development, the primary hemoglobin is $\zeta_2\varepsilon_2$ (hemoglobin Gower I). As development progresses, the α and γ chains begin to be expressed and the embryonic genes are shut off. During the time of transition, hemoglobins Portland ($\zeta_2\gamma_2$) and Gower II ($\alpha_2\varepsilon_2$) can be found, and a start of produc-

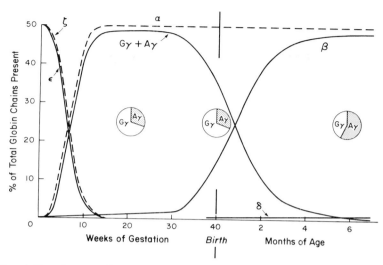

FIG. 4. Developmental control of globin gene expression in man. The time course of expression of the various human globin genes is schematically depicted (3). In early fetal life, ε and ζ predominate, but are then replaced by expression of α and γ genes. Beginning somewhat before birth and continuing to completion at about 6 months of age, there is an orderly switch of production of γ-globin to β-globin, as well as a coordinate induction of synthesis of the minor adult δ-globin. The circles represent the relative proportion of $^G\gamma$ and $^A\gamma$ chains in the fetal hemoglobin of normal Caucasian individuals. At 20 weeks of fetal life this ratio is approximately 7:3, and it does not change at birth, even though production of fetal globin chains has already begun to drop at that time (115, 116). The adult ratio of approximately 4:6 is achieved somewhat later. Differences in these ratios in normal black individuals have been noted (117, 118).

tion of fetal hemoglobin (HbF, $\alpha_2\gamma_2$), which remains the primary hemoglobin throughout most of the rest of gestation. Fetal hemoglobin has a higher oxygen affinity than HbA ($\alpha_2\beta_2$) and hence allows more efficient extraction of oxygen from the maternal circulation. In the third trimester, production of γ decreases and the synthesis of β chains increases coordinately. The total amount of γ + β is kept relatively constant. At the time of birth, the ratio of γ to β is approximately 2:1; the orderly shutoff of γ continues after birth, so that by 6 months of age the adult level of less than 1% hemoglobin F has been achieved, and β-chain synthesis has reached its appropriate level, together with a small amount of the minor adult hemoglobin δ which makes up hemoglobin A_2 ($\alpha_2\delta_2$).

The sequential expression of these genes therefore follows their order on the chromosome in the 5' to 3' direction in man, although this is not the case in many other vertebrate species (see Section V,B). In this context, it is interesting to note that the γ chains made during fetal life have a ratio of approximately 7:3 of $^G\gamma$ to $^A\gamma$, whereas the small amount of fetal hemoglobin made in adult life has a ratio of about 4:6 (*115*, *116*) (Fig. 4). Thus, like overall expression from the β-globin complex, the 5' gene is expressed predominantly early in development and the 3' gene more so in adulthood. However, the timing of the $^G\gamma$ to $^A\gamma$ switch is not quite the same as the γ to β switch, as there appears to be *no* decrease in the $^G\gamma$ to $^A\gamma$ ratio until γ-chain production has dropped below the level of β-chain production.

Certain individuals in the American black population have higher $^G\gamma$ to $^A\gamma$ ratios (*117*, *118*). Analysis suggested the existence of $^A\gamma^+$-thalassemia, manifested by the production of 80–90% $^G\gamma$ at birth. One presumed homozygote still had small amounts of $^A\gamma$ present, leading to the conclusion that the basis is not a $^G\gamma$-$^G\gamma$ chromosome. A rarer situation was the occurrence of reduced (30–50%) $^G\gamma$ at birth, which was thought to be due to heterozygosity for an $^A\gamma$-$^A\gamma$ chromosome. Such a chromosome has recently been demonstrated (*119*).

We will have more to say about the fetal to adult globin switch and the various theories of its regulation in Section VI,C, which concerns deletions in the β-globin complex that alter this normal program. Switching also occurs in goats (*25*) and sheep (*120*). Understanding the mechanism is a high-priority subject of current research, both because of its potential as a model system for differential control of gene expression, and because of its possible therapeutic significance. Because even small amounts of fetal hemoglobin, if uniformly distributed, are protective against the β-thalassemias (*121*) and sickle-cell disease (*122*), the ability to reverse this switch in adults would have enormous therapeutic implications.

B. Transcriptional Unit

In Section I,D we described briefly the conserved portions of the 5'- and 3'-untranslated sequences. An abundance of evidence suggests that "cap" sites for mRNA actually correspond to transcription initiation sites. On this assumption, major transcription initiation sites for globin genes correspond to the upstream end of the 5'-untranslated sequences of the major mRNA, located approximately 50 nucleotides upstream from the initiation codon. However, as described in Section III,C, there is increasing evidence that additional transcription may be initiated still farther upstream. In contrast to the cap sites, a polyadenylation site is believed not to be a site at which transcription terminates, but a position at which RNA is cleaved and subsequently polyadenylated in a posttranscriptional process. Pulse-labeling indicates that the transcription termination sites from mouse β-globin genes probably lie from 700 to 1700 nucleotides downstream from the polyadenylation site for the gene (*123–124*). There has not been a demonstration of a similar phenomenon for the α genes. It is not known whether the extra untranslated sequences at the 3' end present in some mRNAs or their precursors have any functional significance. Detailed studies of the 5' transcription initiation starts or the extension of the transcript 3' to the gene have been performed in only a very few systems. Therefore, it is not clear whether these properties are evolutionarily conserved or subject to random variation between species. Overlapping of extended transcription units could create some relatively complex interactions in the expression of different globin genes.

C. Promoter Function

The control of expression of globin, and probably most other genes, is exerted in major part at the level of initiation of transcription (for review, see *124a*). The levels of globin gene expression may be modulated by small molecules such as heme and steroids, but the (nearly) absolute decision as to whether a cell expresses a globin gene, and which globin gene is expressed is set by processes of determination and developmental regulation that are at present quite mysterious. Unfortunately, the globin genes cannot yet be manipulated *in situ* in the chromosome, and all cell-free polymerase II transcription systems are still rather crude, reflecting only partially the control of transcription *in vivo*. The most extensive results on transcription signals of the globin system have come from the application of the techniques of recombinant DNA to mutate the cloned gene, followed by introduction into cells in an unphysiological chromosomal or extrachromosomal environment. Within these limitations, the definition and comparison of the sequences necessary for the initiation of globin gene transcription *in vivo* and *in vitro* are well under way.

In both prokaryotes and eukaryotes, the sequences necessary for transcription of mRNA lie upstream from the translation initiation site, and principally upstream from the site of transcription initiation (reviewed in *125*). In prokaryotes, there is a single RNA polymerase responsible for all RNA transcription, and promoter sequences that specify the initiation of transcription lie within 40 nucleotides upstream from the site of transcription initiation. Two principal conserved components of the prokaryotic promoter have been noted: the "Pribnow box" located between 5 and 12 nucleotides upstream from the site of transcription initiation and most commonly of the form

ATAAT; and a sequence of the form TTGACA located approximately 35 nucleotides upstream from the transcription initiation site. The principal site of initiation of transcription usually deviates by no more than 1 nucleotide with respect to the Pribnow box, and the TTGACA sequence is usually 17 nucleotides upstream from the Pribnow box, although promoters in which the −35 sequence is displaced by 1 nucleotide in either direction can function moderately effectively. In addition, in various prokaryotic promoters, either the Pribnow box or the −35 box may vary to a substantial extent from the usual form so that a better −35 box may form a functional promoter in association with a weaker Pribnow box, and vice versa. The conclusion that these sequences are the only essential ones for prokaryotic promoters is consistent with a body of mutational data and has recently received considerable support from the demonstration that chemically synthesized promoter sequences function effectively in bacterial cells.

In comparison to the prokaryotic promoters, eukaryotic polymerase II promoters show a major divergence in transcription signals, in respect to both their exact sequence and this location relative to the transcription initiation site. We discuss first promoter features in general (Fig. 5), then make a few comments based on comparisons of the β-like globin promoters (Fig. 6).

Promoter elements have been defined in two general ways: by sequence comparisons of different eukaryotic genes looking for conserved elements, and by examination of mutations in the promoter regions. Such mutations may be either naturally occurring, as in several of the β-thalassemias, or created by deletional or site-specific mutagenesis with subsequent functional analysis. This analysis may be carried out either in an *in vitro* transcription assay or, more commonly, in an *in vivo* expression system using a plasmid vector and DNA-mediated gene transfer into eukaryotic cultured cells.

There appear to be at least four, and probably five, definable elements contributing in varying degrees to one or another eukaryotic promoter (see Fig. 5). The first element of the polymerase II promoter recognized originally by sequence homology among promoters is the "TATA," or "ATA," or "Hogness–Goldberg box." This sequence is located 25 to 30 nucleotides upstream from the transcription initiation site and is an (A+T)-rich sequence often resembling somewhat the Pribnow box of prokaryotes. In both the β-like (Fig. 6) and α-like globin genes, the core of the sequences is an ATA located at the same position upstream from the initiation site in each gene studied. The "ATA box" is necessary, and often sufficient, to direct globin transcription in cell-free systems (*125*) and also in cells expressing certain "immortal-

HERPES TK GACACAAACCCCGCCCAGCGTCTTGTCATTGGCGAATTCGAACACGCAGATGCAGTCGGGGCGGCGCGGTCCGAGGTCCACTTCGCATATTAAGGTGACGCGTGTGGCCTCGA

HUMAN α-GLOBIN AGGCCGCGCCCCGGGCTCCGCGCCAGCCAATGAGCGCCGCCCGGCCGGGCGTGCCCCCGCGCCCCAAGCATAAACCCTGGCGCGCTCGCGGCCCGGC

HUMAN β-GLOBIN ACCTCACCCTGTGGAGCCACACCCTAGGGTTGGCCAATCTACTCCCAGGAGCAGGGAGGGCAGGAGCCAGGGCTGGGCATAAAAGTCAGGGCAGAGCCATCTATTGCTT

RABBIT β-GLOBIN CCTCACCCTGCAGAGCCACACCCTGGTGTTGGCCAATCTACACACGGGGTAGGGATTACATAGTTCAGGACTTGGGCATAAAAGGCAGAGCAGGGCAGCTGCTGCTT

SV40

EARLY PROMOTER CATAGTCCCGCCCCTAACTCCGCCCATCCCGCCCCTAACTCCGCCCAGTTCCGCCCATTCTCCGCCCCATGGCTGACTAATTTTTTTTATTTATGCAGAGGCCGAGGCCGCCTCGGCCTCT

FIG. 5. Comparison of promoter sequences. The figure compares five promoters whose essential components have been analyzed by mutagenesis. The TATA boxes of the promoters are aligned, and the rightmost nucleotide immediately precedes the cap site. The ψ 11-bp repeat refers to sequences that could be considered to resemble the 11-bp halves of a single 21-bp repeat in the SV40 promoter region. In the herpes TK promoter, these largely overlap the two essential regions (*126*). Further discussion of the role of these sequences in promoting globin transcription is presented in the text. The herpes TK sequence is from reference *126*, SV40 early promoter from *127*, human α from *48*, human β from *46*, and rabbit β from *128*. ·····, CCAAT box; -----, Hogness–Goldberg box; ——, ψ 11-bp repeat, or "3rd box"; -·-·, ψ 11-bp repeat (inverted).

```
              -130       -120      -110       -100        -90        -80        -70        -60        -50        -40        -30        -20        -10       CAP
HUMAN β       TCACTTAGACCTCACCC-TGTGGAG--CCACACCTAGGGTTGGCCAATCT--ACTCCCAGGAGGAGGAGGCAGGAG--CCA--------GGGC--TGGGCATAAAAGTGAGGGCAGAGC-CATCTATTGCTTACA
RABBIT β1     TCACCCAGACCTCACCC--TGCAGAG--CCACACCCTGGTGTTGGCCAATCT--ACACAC-GGGGTAGGGATTACATAGT-TCA--------GGACTTGGGCATAAAAGGCAGAGCAGG--CAGCTGCTGCTTACA
GOAT βA       CCATTCAAGCCTCACCC--TGTGGAA--CCACAACTTGGCACGAGCCAATCTGCTCACA--GAAGCAGGGAGGGCAGGAG--GCA--------GGGC--TGGGCATAAAAGGAAGAGGCCGGGCCAGCTGCTGCTTACA
GOAT γ        TCATTCAAGCCTCACCC--TGTGGAA--CCACACCTTGGCCTCACCCTTGAGCCAATCTGCTCACA--GAAGCAGGGAGGGCAGGAG--GCA--------GGGC--TAAGCATAAAAGGAAGAGGCCGGGCCAGCTGCTGCTTACA
MOUSE βmaj    TCACCGAAGCCTGATTC--CGTAGAG--CCACACCCTAAGGGCCAATCTGGTAGGGCCAATC--TGCTCACA-TAGAGAGGGCAGGAG--CCA--------GGGC--AGAGCATATAAGGTGAGGTAGGATCAGTTGCTCCTCACA
MOUSE βmin    TCTCTGAAGCCTCACCC--TGCAAGG--TAACACCCTGGCATTGGCCAATCTGCTCAGAGGA-CAGAGTGGGCAGGAG--CCA--------GCAT-TGGGTATATAAAGCTGAGCAGGGTCAGTTGCTTCTTACA

HUMAN δ       TCATTC-----TCAAACTAATGAAACCCTGTATCCTTTAAACCAACCTGCTCACT---GAAGCAGGGAGGACAGGA-------------GCATAAAAGGCAGGGCAGAGTGACTGTTGCTTACA
                                                   TATC
                                                  C
RABBIT ψβ2    CTCTTCTGGCCTCACCCTGGCATTG-----------------------------GTGACTCAC-CA-GGGTAGGCAGTGCAGGGG-CCA-----------CTACTGGGCATAAAAGGCAGAGTGGAA-CAGCTGCTGCTTATG
HUMAN ψβ      CTGCCCAAACCCACCCCACCCCCTGGAGT---CACAAGCACCCCCTGAT CAATGATTCATTT-CACTG-GGAGAGGCAAAGGGCTGGGG-CCAGAGAGG-AGAAT AAAAAGCCACACATG--AAGCAGCAATGCAGG
                                                        CCAATAGATTCATTTTCACTGAGG-GAGGCAAAGGGCTGG                                G
LEMUR ψβ      C----CCAAACTCCACCCCCTTGGAT---CACAAACCCGCCCTTGAACAATAGC--CTCATTTCATTAGGAGAGACAAACGGCTGGGGG--CCAGAGAGGCAAAGGCCATGGAGAGAAG-CAGCAGTACAGG
HUMAN Aγ      C-----TAAACTCCACCCATGGGTTGGCCA--GCCTTGCCTGACCAATAGT--CTTAGAGATCCAGTGAGGCCA--GGGGCCGGGCCTGGCTAGGATGAGGATGAAGAATAAAAGCAAGCACCCT-CAGCAGTTCCACACA
                                                          TGACCAATAGCCTTGACAAGGCAAACT                CG
RABBIT β3     C-----TAAACCTCACCCCTGCGCTGACCA---GCCTTGCCTGACCAATAGC---CTCAGAGAACACGGCGAAACAAGGGCCAGATGTCCAGGGAGGAAGAATAAAAGGACGAGCCTTA-GAGCAGTTTCACATA
                                                          TGACCAATAGTCGTTACACAAAAACAC
GOAT εII      TTGCCCAAGTTCCACCCTGGCAGTGAC-CA--CCTAGCTTGCCTGACCTTTGAC-TCTTCATTTTATTGGGGAA-GGAAGGGCCTGGGG--CAGCAGATGAGGAATAAAAGGCCATGAGTGAAGCAG--CGGCACAGA
MOUSE βh0     CCCACTGGACCCCACCCCTGGCCTTG--CC-CAGACTC--TCTTGACCAATAGC---CTCAGAGTCTTGAAAAGGGTAAGGGAGCAAGGTCTTAGGGTGCAAGAATCTTGAAGAATAAAAGAACAGGTCTT-CAGCCTCTTGAACATT
MOUSE βh1     CCCACTGGACCCCACCCCTGTCTTG--CC-CAGACTC--TCTTGACCAATAGC---CTCAGAGTCCTGGGAGGGGTAAGGGAGCAAGGTCTAAGGGAGCAAGAATAAAAGAACAGGTCTT-CAGCCTCTTGAACATT
HUMAN ε       CGGACCTGACTCCACCCCTGAGGA---CACAGGTCAGCTTTTAAGTACC--ATGGAGAA-CAGGGGGGCCAGAACTTCGCAGTAGAATAAAGAATAAAAGCCAGTGGGAATAAAAGGCCAGTGAAGCAGCACATA
RABBIT β4     GGTGTCTTG-CTCCACCCATGAGGA---CACATCCAG-TCTTGACCACAAGCTTC-AAGTAT--GAAGAGAA-CAAGGGCCAGATCAGCAGTGGGAATAAAAGGCCAAGCAGCAGCACAAA
GOAT εI       CTGACCTGACTCCACCCCTGAGGGA---CACAGCCTAACCCTGACCAATGACTTGACCAAGGAC-AAGGGGGAGCAGAGTTCAGCAGTAAAGGAATAAAAGCCACAGCAGCAGACAGA
MOUSE εY      TGACCAATGGCTTCAAAGATAA-ATGCAGAATCAAAGGXCAGAACATTGTCTGCGAAGAATAAAAGGCCACCACTTC-TAGCAGCAGTACGTA
```

ization" genes, such as the adenovirus E1a protein. The ATA box is thought to play an important role in locating the site at which transcription initiates.

The second readily identifiable component in human globin gene promoters (also commonly in other polymerase-II promoters) is the "CCAAT box" located approximately 80 nucleotides upstream from the transcription initiation site. However, the position of this sequence relative to transcription initiation is somewhat more variable than that of the ATA box, and it may be duplicated (e.g., in the γ-globin genes of man). The CCAAT sequence is found in all the actively functioning globin genes studied to date (e.g., see Fig. 6 and ref. 63). The only substantially deviant sequence is that noted for the human δ-globin gene, which is transcribed less efficiently than the genes for the major forms of globin. There are three CCAAT-like sequences upstream of the δ-globin gene, but the copy in the usual CCAAT position appears degenerate. It is possible that the differences from the usual CCAAT box may contribute to the lack of efficiency of transcription of the δ-globin gene (see below).

A third component of polymerase-II promoters is a set of sequences of the general form GGGGYG, or the inverted form CRCCCC. These have been identified as critical components for transcription *in vivo* and *in vitro* of the SV40 early-region. Similar sequences are critically important for transcription of the herpes virus thymidine kinase gene in oocytes. In the case of SV40, there are six copies of similar sequences located in tandem upstream from the TATA box region and extending through the region of the putative CCAAT box, although a sequence of 21 nucleotides including only two copies is sufficient for transcription initiation *in vivo* and *in vitro* (*139–141*). In the thymidine-kinase-promoter region, there are two copies of the (G+C)-rich sequence in inverted orientation, one upstream and one

---

FIG. 6. Comparison of 5′ flanking sequences of β cluster genes. This is an expansion of compilations previously presented (*63, 129, 130*). Sequences were aligned to maximize homology; we have followed the alignments of Hardison (*130*) for the sequences included in that comparison and have added to that the sequences of human $\psi\beta$, goat ε^I and ε^{II}, and mouse βh0 and βh1, which have been aligned by inspection. References for the sequences presented are human β (*46*), rabbit β1 (*128*), goat β^A and γ (*129*), mouse β maj and β min (*131*), human δ (*45*), rabbit $\psi\beta$2 (*132*), human $\psi\beta$ (*133*), lemur $\psi\beta$ (*134*), human $^A\gamma$ (*44*), rabbit β3 (*135*), goat ε^I and ε^{II} (*136*), mouse βh0 and βh1 (*60*), human ε (*42, 137*), rabbit β4 (*130*), and mouse εY (*138*). Areas of homology are easily apparent and are outlined by boxes, corresponding to the repeated "CACA" boxes, the "CCAAT" box, the "ATA" box, and the region of the cap site. A further area of weak homology between −50 and −60 has been suggested (*130*) but is not marked here.

downstream from the CCAAT box site and both upstream of the TATA box. Modification of either sequence decreases transcription of the TK gene (*142, 143*), and it has been suggested that the two sequences might base-pair during activation of the promoter. A potential analog of this sequence in the β-globin system is the repeated sequence CACCC (see Fig. 6 and below). A transcription factor that binds specifically to the SV40 21-bp region and stimulates transcription from the early promoter in HeLa cells has recently been described (*143a*).

A fourth component contributing to polymerase-II transcription is the sequence located at the site of transcription initiation. Usually the first two nucleotides transcribed are a purine followed by a pyrimidine, but additional neighboring bases may play an ancillary role in favoring transcription initiation (discussed in Section I,D).

Fifth and most intriguing of the elements that may be involved in enabling transcription to occur *in vivo* are the so-called enhancing elements (reviewed in *98*). The enhancement properties of sequences near the SV40 origin of replication were initially recognized because of their ability to increase the efficiency of transformation of cells with the thymidine kinase gene after microinjection (*144*) or calcium phosphate transfection (*145*), regardless of the orientation of these sequences relative to the thymidine kinase gene, or even whether the sequences were located 3' or 5' to the thymidine kinase gene. Separately it was noted that sequences upstream from a histone gene could enhance the expression of the gene in oocytes regardless of their orientation or exact position with respect to the gene (*146*). Deletion of a portion of the SV40 91-bp repeats (72-bp repeat in the small-plaque mutant virus used in many laboratories) impaired markedly the expression of the SV40 early-region genes after transfection of cells. Sequences with properties similar to the SV40 enhancer occur in polyoma virus, and mutation within these sequences endows the viruses with an ability to express early regions in otherwise nonpermissive undifferentiated teratocarcinoma cells.

A limited number of enhancing sequences in genomic DNA have been identified. For example, an enhancer-like function has been located within an intron in the immunoglobulin gene system (*147–150*). These sequences are much more effective in plasma cells than in other cell types. This raises the intriguing possibility that tissue-specific enhancers might be found in other host genes. There is a sequence 3' to the β-globin gene that enhances transformation of mouse cells when linked to bovine papillomavirus DNA (*151*), but its relation to other enhancers is uncertain. The β-globin gene also is linked to sequences that can act in a *cis* manner to depress transcription of a linked protamine gene (152).

Several workers (153–155) have used *in vitro* mutagenesis to modify the rabbit β-globin gene, have linked the gene to papovavirus replication origin and enhancer sequences, and have then studied the expression of the modified gene 24–48 hours after transfection of mouse "L" cells or HeLa cells, respectively. Grosveld *et al.* (155) noted that deletion of the TATA box of the rabbit globin gene led to greater heterogeneity of transcription initiation sites. However, Dierks *et al.* (153) reported that deletion of the TATA sequences reduced the level of correct transcription of the globin gene in transfected cells, but that transcription began at the normal site. Both reports noted that deletion of sequences downstream from the TATA box of the rabbit β-globin leads to transcription initiation beginning in plasmid sequences at approximately the correct position relative to the TATA box. Changing the rabbit β-globin TATA box from GCATAAAA to GCATGTAA led to a 5-fold reduction in transcription initiation efficiency and produced heterogeneous transcription initiation. There are several forms of β-thalassemia with single mutations in the TATA box that are associated with decreased level of transcription (see Section VI,B,2). Therefore, the TATA box of the β-globin gene plays an important role in promoting efficient transcription as well as in defining the site at which transcription initiates.

Talkington and Leder (156) analyzed the role of sequences about the cap site for the mouse α gene using an *in vitro* transcription system, and they noticed that a 44-base stretch of DNA from −55 to −11 relative to the cap site was sufficient to direct RNA transcription *in vitro*, but transcription efficiency was increased if the sequence from −10 to +7 including the cap site was retained in the promoter segment. They also studied transcription of an α-globin pseudogene that differed from the true gene in several nucleotides about the cap site (normal AGACACT, mutant GCAAATT). Replacement of the nucleotides from −9 to +7 of the pseudogene with the corresponding nucleotides from the active gene partly restored the pseudogene promoter *in vitro*, and replacement of the sequence from −7 to +10 of the normal gene with pBR322 DNA remarkably decreased the efficiency of transcription initiation. In contrast, the principal determinant of *in vitro* transcription of rabbit β-globin is the stretch of nucleotides from −34 to −20 with respect to the transcription initiation site, and there seems to be no effect of the sequences at the cap site in promoting transcription (154).

The CCAAT box sequences appear to be important *in vivo* but not *in vitro* for the transcription of the β-globin gene. For example, when a series of deletions extending from an upstream site downstream toward the rabbit β-globin gene was used to transfect mouse L cells, expression was dramatically reduced when the deletion removed the CCAAT box (located at −75 to −71 (*153*). A single-base substitution (the C at −75 by a T, creating TCAAT) reduced transcription to less than 25% of normal. Substitution of the T at position −78 by a C and the A at −73 by G combined to reduce transcription to 12% of the normal level while substitution of the T at −78 alone produced no effect on transcription.

Constructs in which the SV40 enhancing sequences were ligated in the same plasmid with either the δ- or the β-globin gene or both together help to localize the basis for the relatively low rate of production of δ as compared with β hemoglobin in man (*158*). In these experiments, the δ-globin gene gave rise to about 1/40th as many transcripts as the β-globin gene after transfection into HeLa cells. Furthermore, substitution of the 3′ portion of the δ-globin gene by the corresponding portion of the β-globin gene did not increase the level of transcripts, while the composite gene (with the 5′ portion of the β-globin including its promoter and the 3′ portion of the δ-globin gene) was expressed as well as the intact β-globin gene. This localized a defect in the δ-globin gene in the 5′ region, probably the promoter, of the gene, and is consistent with the earlier suggestion that the deviation of the CCAAT box region from the forms noted in the major globin genes may lead to less effective *in vivo* action and underproduction of the δ-globin message. However, sequences at the 3′ end of the δ-globin gene (or the β-globin gene) may also influence the production of mRNA (*159, 160*), as a fusion gene (anti-Lepore) with the 5′ end of β and the 3′ end of δ produces less mRNA than normal β-globin. Hemoglobin Parchman (*159, 159a*), on the other hand, is derived by a double crossover and has 5′ and 3′ ends of δ-globin but internal β-globin sequences, probably including the small intron. It is produced at a lower rate than either hemoglobin Lepore or anti-Lepore and only slightly greater than that of δ-globin. The turnover rates of (labeled) δ- and β-globin RNA in cultured marrow cells show average half-lives for β mRNA to be 16.5 hours and for δ, 4.5 hours, indicating that mRNA instability is another significant factor in establishing the normal 40-fold excess of β- over δ-globin in red cells.

A sequence upstream from the CCAAT box (−75 to −71) has been identified as important for rabbit β-globin transcription after transfection (153). A deletion removing sequences upstream of residue −109 transcribes normally. Extension of the deletion to −95 only decreases transcription to 35% of the control level, but a deletion extending to −84 decreases transcription to 10% of the normal level. Substitution of C at −91 by a T reduces transcription in transfected cells to 25%, and C at −89 by T to 10% of the normal level. The sequence from −91 to −85 is CACACCC; there are three copies of the sequence (T/A)CACCC in this region of the β-globin gene, and the possible homology to the SV40 sequence CGCCC (see above) that is part of the 21-bp sequences necessary for SV40 transcription has been pointed out (153). Deletions of all three copies of the globin sequence reduces transcription to 10% of normal, but an internal deletion leaving one of the three copies intact transcribes at 4% of the normal rate. The internal deletions moved a CACC sequence to a position near residue −82, presumably compensating for the removal of the normal sequence at this point and also explaining why a single base change at this site had a greater effect than did a deletion. Also, in man a nucleotide transversion from C to G at position −87 (corresponding to the −86 site of the rabbit gene) produces a form of the β-thalassemia that is presumably a consequence of decreased transcription of the β-globin gene (161, 162). This thalassemia mutation is particularly informative, since it shows the relative importance of these sequences in a physiological chromosomal environment.

Figure 6 presents a comparison of the 5' flanking sequences of the β-like globin genes from several species, including adult, fetal, embryonic, and pseudogenes. This is an extension of previous comparisons, and the alignments are basically those of Hardison (130). The conserved regions are outlined and include the cap site, the ATA box, the CCAAT box, and two CAC (C/A) sequences. The correlation of these sites with regions shown to have functional significance as described above is a strong confirmation that evolutionary conservation reflects function. The various promoters have been arranged in accordance with their evolutionary origins (see Section V). Searching for specific sequences that might affect the timing of developmental expression reveals two possible but inconclusive candidates: (i) Shapiro et al. (136) have previously suggested that in the −120 to −110 region the sequence CACCCTG is specific for embryonic expression; however, the recent addition of the rabbit β4 sequence (130) contradicts this generalization. (ii) Comparison of the goat $β^A$ (adult) and γ (fetal) genes just upstream from the ATA box raises the question whether the change in GGGC to AAGC might be involved in altering timing of expression, since AAGA appears in most fetal or embryonic genes (Fig. 6). However, the mouse β-major and rabbit β4 genes contradict this rule.

There is a contrast between the homologies among the untranscribed 5' sequences of the β-globin gene family in various species (Fig. 6 and references therein) and the homologies among the α globin gene sequences that have been published to date. These include man (47–49), goat (26), mouse (35), chicken (164), and duck (165). The sequences of the α-globin genes of different species are strongly homologous only from a few nucleotides before the CCAAT site up through the transcription initiation site. In the goat there are two adult α-globin genes whose sequences are very similar for about 85 bases upstream from the cap site (26). Farther upstream they show only random homology. Both genes are transcribed; only one has the −85 sequence CACCC, and this gene is transcribed at several times the rate of the other. Also, the sequence GGGCGT is found between the CCAAT box and the TATA box in man, mouse, goat, and chicken α genes. This sequence is reminiscent of a complement to the SV40 21-bp repeats and also of the sequence in the herpes virus TK promoter between the CCAAT and TATA boxes (see Fig. 5). Although the sequences between the CCAAT and ATA box have not yet been demonstrated to be important for transcription of transfected genes, it is tempting to consider that they may yet prove to be important for regulation of normal transcription of the genes in the chromosome *in situ*. Somewhat similar sequences of the forms GGGCG are found in the analogous region of a number of β-globin-like genes (Fig. 6); as mentioned, there is a potential analogy between the sequence CACCC noted upstream from the CCAAT box and the sequence GGGC noted downstream, as compared with the inverted (G+C)-rich sequences located on either side of the CCAAT box in the herpesvirus thymidine-kinase gene (see above). Further study of transcription *in vivo* of deletion and substitution mutants affecting the globin gene sequences between the CCAAT and ATA boxes would be of interest.

A feature of transcription of the α- and β-globin genes that is as yet not correlated with their nucleotide sequence is the observation (158) that α-globin genes are expressed efficiently on transfection into HeLa cells whereas β-globin genes are not expressed unless they are linked to viral enhancer sequences. The possibility exists that the α-globin genes have sequences that either correspond to viral enhancers or bypass the need for them. Interpretation is difficult because the mechanism of action of enhancer sequences is unknown and their requirement or effectiveness may vary between species or cell types.

No enhancer sequences have yet been described in either the α- or β-globin gene complex. Very recently it has been observed (98a,b) that adenovirus E1a protein acting in *trans* can circumvent the requirement for a *cis* enhancer sequence.

There is some suggestion that more upstream alternate promoters may exist for several globin genes. Between 10 and 15% of mouse β-globin RNA initiates in the 1000 nucleotides upstream from the principal cap site (123). There is also an apparent duplication of a promoter-like region located about 100 nucleotides farther upstream from the cap site. In human bone marrow cells, about 10% of the RNA is derived from this upstream promoter (166). This suggests that this promoter is active *in vivo* and that the RNA derived from it is also fairly stable. Several upstream 5' ends for β-globin RNA in both marrow cells and reticulocytes have been detected (166a). Approximately 90% of this RNA is polyadenylated. A major 5' end occurs at -172, relative to the cap site, but minor 5' ends were found at -196, -213, and -235. RNA transcription initiates at similar positions *in vitro* although the ratio of the abundance of the various transcripts relative to one another is considerably different than *in vivo*. Most of the *in vitro* transcription initiations are resistant to low levels of amanitin, suggesting that they are polymerase III transcripts (166a). It would be of interest to know the time course, both during erythroid maturation and during organismal development, of expression of the β-globin RNA from the upstream promoter.

Multiple forms of ε gene transcripts with 5' ends at -65 to -250, -900, -1480, and -4500 upstream of the conventional cap site have been found (167, 168). These transcripts were polysome associated and were capped at least some of the time. In total, these transcripts amounted to 10–15% of ε-globin transcripts in K562 cells or purified first-trimester erythroblasts. In some established nonerythroid cell lines, transcripts apparently initiating at these upstream sites were found even though "normal" ε-globin mRNA was absent. The upstream transcripts sometimes were present at levels comparable to those seen in K562 cells. The presence of such additional transcription initiation sites in the globin gene cluster would add new dimensions to the potential complexity of regulation mechanisms affecting globin gene expression and chromatin structure.

D. Splicing (See Addendum 2)

As in most genes of animal cells, the coding regions of globin genes are interrupted by sequences of unknown function, called introns. As summarized in Section I,B, seal myoglobin and all functional vertebrate hemoglobin genes have two intervening sequences. The still unanswered question of the function of introns in the globin genes is also discussed in Section I,B. A second category of incompletely explained phenomena related to mRNA splicing is the observation that splicing seems to expedite, or in some cases to be obligatory, for mRNA stabilization and or export from the nucleus. This rule is clearly not invariable. There are genes lacking splice signals that encode mRNA precursors, yet these genes are expressed adequately. In other cases, an mRNA precursor may be exported from the nucleus partly in spliced and partly in unspliced form.

The globin system initially appeared to provide an extreme example of the requirement of splicing for mRNA export. Both the small intron and flanking sequences and the large intron and flanking sequences of the β-globin gene have been separately incorporated in the late region of SV40 (*169, 170*). In both cases, when transcription proceeded over the intron so as to produce sense-strand RNA, the transcript was spliced and abundant cytoplasmic RNA was formed. When the orientation of the intron was reversed relative to the promoter so that the antisense-strand RNA was transcribed from the intron and the flanking sequences, only small amounts of heterogeneous RNA was seen. A number of mutants that diminish or abolish the function of one or another of the splice sites or create abnormal splice sites in globin are observed in the human thalassemias (see Section VI,B) and have been created by *in vitro* mutagenesis of the rabbit β-globin gene (*171*). These mutations may lead to aberrantly spliced RNAs, but also often to marked reduction in the total amount of RNA. The reduction in the amount of RNA in thalassemias might well prove to be the result of a combination of causes including failure to complete splicing reactions and selective destabilization of aberrantly spliced RNAs at early times subsequent to their formation.

In contrast to the above, recent results show that splicing is not always an obligatory event for expression of mRNA containing globin introns. The herpesvirus thymidine-kinase gene produces an unspliced mRNA. Greenspan (*172*) created a recombinant plasmid in which the small globin intron is incorporated either in the sense or antisense orientation into the thymidine kinase gene downstream from the termination codon but preceding the polyadenylation site. Cells transfected with this thymidine-kinase gene express similar amounts of TK mRNA containing globin intron sequences within the untranslated 3' sequences whether the orientation was in sense or antisense strand. Little splicing of the small intron was detected even though all the splice site signals immediately surrounding the intron were retained. That experiment is perhaps an extreme example illustrating that sequences remote from the splice junctions may play critical but poorly understood roles in specifying the utilization of splice sites.

Until recently all effective splices known to occur in animal cells removed an intron whose first two nucleotides were G and U and whose last two were A and G (the consensus splice signals defined by Breathnach and Chambon *173*). Other nucleotides near the 5' splice junction tend to follow more or less strictly a limited range of patterns consistent with models postulating that U1-RNA pairs with the 5' splice junction (reviewed in *174*). Mutations near the globin splice sites that affect normal splicing commonly alter one of the consensus nucleotides (Fig. 7) and are the cause of certain β-thalassemias. A more surprising and less anticipated effect of mutations, observed in several cases, is the creation of a new splice site by a single-base change in the globin gene. This has been observed to occur with base changes either in introns or in the coding region of the first exon (Fig. 8). In some of the cases in which a single base change creates a new splice site, the base change has itself generated a dinucleotide consistent with the Breathnach–Chambon rule, either an A-G for the 3' splice site, or a G-T for a 5' splice site. On the other hand, there is one case in which a nucleotide change one base upstream from a G-T created a new splice site that competed effectively with a normal splice site. In addition, in two cases—one a natural splice site in chicken α-globin (*175*), the second a splice site that appears in a rabbit β-globin *in vitro* mutated (*177*)—the 5' dinucleotide, normally expected to be a G-T, is a G-C.

In several cases where the normal splice site is abolished or a new splice site is created, additional "cryptic" splice sites are uncovered (*161, 171, 177–180*). For example, inactivation of the 5' splice site of the first exon of human β-globin by a base change of the first nucleo-

	5' ↓ 3'	
Normal I-VS1	CAGGTTGGT	
IVS-1 Pos1 G→A	CAGATTGGT	⎫ Splice Site Inactivated - β° Thalassemia
IVS-1 Pos1 G→T	CAGTTTGGT	⎭
IVS-1 Pos5 G→C	CAGGTTGCT	β⁺ Thalassemia - More Severe
IVS-1 Pos6 T→C	CAGGTTGGC	β⁺ Thalassemia - Less Severe
U1 RNA	$_{3'}$GUCCAΨΨCAU^{m}A$^m_{ppp}$G$^m_3$$_{5'}$	

FIG. 7. Mutations in the 5' splice site of intron 1 of the human β-globin gene. Thalassemic mutations shown are referenced in Table II. Complementarity to U1-RNA is discussed in reference *174*.

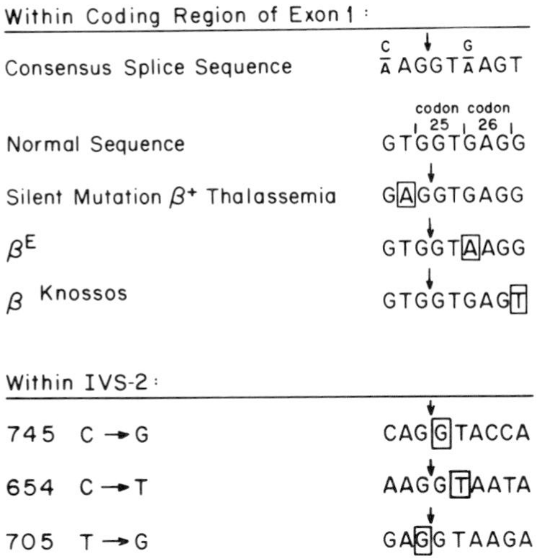

Within Coding Region of Exon 1:	
Consensus Splice Sequence	$\overset{C\ \ \ \ \ \ \ \ \ \ \ G}{\underset{A}{}AGGT\underset{A}{}AGT}$
Normal Sequence	codon 25, codon 26 GTGGTGAGG
Silent Mutation β^+ Thalassemia	G[A]GGTGAGG
β^E	GTGGT[A]AGG
β Knossos	GTGGTGAG[T]

Within IVS-2:	
745 C → G	CAG[G]TACCA
654 C → T	AAGG[T]AATA
705 T → G	GA[G]GTAAGA

FIG. 8. Thalassemic mutations creating a new 5′ splice site. The sequences shown are sites where a single base change has generated a new splice site in an adult human globin gene. Of note is that β^E and $\beta^{Knossos}$ are both mutations that change the amino-acid sequence of globin, as well as creating a new splice site. References for these mutations are in Table II.

tide of intron 1 (IVS-1) leads to the utilization of G-T dinucleotides at positions 105 and 127 of exon 1 and IVS-1 position 13 as 5′ splice sites (Fig. 9). The same splice sites are activated by base changes at IVS-1 position 5 or 6, even though these latter mutations can still produce substantial amounts of normally spliced RNA (161). Yet another form of β-thalassemia is due to a base change at IVS-2 position 745 that creates a new 5′ splice site. This new splice site is joined to the normal 3′ splice site of IVS-2, and the normal 5′ splice site of IVS-2 is joined to a cryptic 3′ splice site at position 579, creating a new exon. Yet another thalassemic syndrome is due to a change of T to G at IVS-2 position 705, creating a sequence GAGGTAAGA somewhat resembling a 5′ splice site. In this case, the cryptic 3′ splice site at 579 is used as in the preceding case. However, the pseudo-splice site at 705 is not used. In effect, the mutation at 705 activates the 579 site without itself forming a new splice, as though it inactivated the normal IVS-2 3′ splice site (181). In summary, both 5′ and 3′ cryptic splice sites have been uncovered, and the absence of an effective 3′ or 5′ splice site forces the utilization of latent complementary 5′ or 3′ sites that are not otherwise used.

5' SPLICE SITES

Consensus 5' Splice Sequence		$\frac{C}{A}$AG↓GT$\frac{G}{A}$AGT
Human β	Exon I residue 105	AAG↓GTGAAC
	Exon I residue 127	GTG↓GTGAGG
	IVS-1 residue 13	AAG↓GTTACA
	IVS-2 residue 48	ATG↓GTTAAG
Rabbit β	Exon 2 residue 42	TAA↓GCTGAG
	Exon 2 residue 85	AAG↓GTGAAG
	IVS-2 residue 4	TGA↓GTTTGG
Human α	Exon I residue 83	GGG↓GTAAGG

3' SPLICE SITE

Consensus 3' Splice Sequence		YYYYYYNCAG↓G$\frac{G}{T}$
Human β	IVS-2 residue 579	TTTCTTTCAG↓GG

FIG. 9. Cryptic splice sites in human and rabbit globin genes. The splice sites shown are those used when a mutation inactivates the nearest normal splice site. The splice sites in rabbit IVS-2 are from reference 177. Those in the human gene are from the β-thalassemias referenced in Table II, except for the α mutation, which is from reference 179. Consensus sequences are from reference 173; Y represents a pyrimidine, and N any nucleotide.

Normal splicing always joins the 5' splice site of intron 1 to the 3' splice site of intron 1, but apparently never to the 3' splice site of intron 2. There are genes such as those for the procollagens with as many as 50 intervening sequences that are spliced pairwise in the correct fashion. To explain this precision of the splicing process, several workers have considered various local scanning models. These models suggest that if a 5' or 3' splice site is recognized, the splicing mechanism then scans sequentially along the RNA molecule downstream or upstream, respectively, until it comes to the nearest effective 3' or 5' splice site and completes splicing.

Several observations test predictions of the scanning model. Two groups have created globin genes in which there is duplication of either the 5'-most splice site or the 3'-most splice site, and have concurrent results in that when they duplicate the 5'-most splice site in a globin gene, the splice site nearer the 5' end is used preferentially (177–178a). However, they obtained opposite results when they duplicated the 3'-splice site of the large exon. Spritz (182) finds that it is the upstream 3' splice site that is used preferentially, and Wieringa (177, 178) finds the downstream splice site used. Spritz's results, but not Wieringa's would be consistent with a model in which the splicing apparatus scans from the 5' end toward the 3' end of the globin gene. The reasons for the differences are not entirely clear. It should be noted, however, that different globin genes were used for the construct. In addition, studies of splicing of transfected genes may produce quantitatively or qualitatively variable results with the same gene depending on the experimental conditions. The results on the use of cryptic splice sites quoted above show that splice sites either 5' or 3' to the normal splice site may become active when the normal site is inactivated, which cannot be fully explained by a 5'-to-3' scanning model. Finally, in a thalassemia mutation deleting 25 nucleotides including the IVS-1 3'-splice site (*183*), the 5'-splice site of IVS-1 was not used for splicing, and instead the undeleted portion of IVS-1 was retained in mRNA, which is also inconsistent with such a simple model. Much more needs to be learned about the actual nature of the direct substrates for splicing actions before this phenomenon can be understood; Kuhne *et al.* (*178a*) proposed a model of splice site selection based on RNA secondary structure.

E. DNase Hypersensitivity

In recent years it has become apparent that the structure of chromatin containing expressed genes is such that the DNA corresponding to the gene is more sensitive to cleavage by a variety of double- or single-stranded endonucleases or restriction endonucleases than is chromatin containing unexpressed genes. While the entire transcribed region of the gene appears to have an increased sensitivity to DNase as compared to nontranscribed regions, the most striking feature is the development of particular hypersensitive sites often located in the 5' region of the gene, but sometimes 3' to the gene (see *184*). Among the most thoroughly studied hypersensitive regions in the globin system are those that occur between 10 and 200 nucleotides upstream from the chicken adult β-globin transcription initiation sites (Fig. 10) (*185–188*). DNase sensitivity and hypomethylation are not seen in the globin region in 20-hour chicken erythroid precursor cells, but develop at 35 hours concomitant with the onset of globin synthesis (*188a*). Sites hypersensitive to nuclease have also been studied in the β-globin complex of 5-day chicken embryo erythroblasts that express principally the embryonic rather than the adult β-globin gene and have been compared with the sites in 12- to 14-day erythrocytes that actively express the adult, but not the embryonic, gene. In 5-day cells, both embryonic and adult genes are sensitive to DNase-I digestion, while in adult cells the embryonic gene is resistant. Relative sensitivity extends 6–7 kb 5' and 8 kb 3' to the β-gene cluster. The more mature erythrocytes develop sites 5' to the gene that are more sensitive to DNase I, DNase II, micrococcal nuclease, several restriction enzymes, the single-strand-specific nuclease S1 (*186*) and the single-strand-specific DNA-modifying reagent bromoacetaldehyde (*189*). The last two reagents appear to attack in part the same sites in chromatin and in "naked" supercoiled DNA, and these sites are slightly different from those mapped by the other endonucleases. Treatment of "active" chicken β-globin chromatin with the endonuclease *Msp*I releases in high yield a 115-bp fragment corresponding to residues −224 to −110. A substantial portion of the released fragment behaves on gel electrophoresis as though it were naked DNA (*185*). Thus in this case there is a close correlation among demethylation, development of hypersensitive and single-strand-like sites, and loss of nucleosomes at the 5' end of the globin gene with activation of the gene.

DNase-sensitive and DNase-hypersensitive sites in the mouse globin genes have been studied in murine erythroleukemia cells before and after induction of globin genes. Some DNase sensitivity of globin genes is present prior to induction and is also detected in lymphoblasts but not in liver cells (*71, 190*). A DNase-hypersensitive site is located in a small region near the 5' terminus of the β^{major} globin, and shows a 5- to 10-fold increase in sensitivity after induction. Curiously, a hypersensitive site is detectable prior to induction in the 5' portion of the large intron, but it disappears on induction. In a cell line resistant to the inducer HMBA, the intron becomes less sensitive after induction, but the site upstream from the cap site remains largely resistant.

The DNase sensitivity of human globin genes has been studied in fetal and adult erythropoietic tissues and in the erythroblastic cell lines K562 and HEL (Section II,A). These two cell lines produce γ-globin mRNA but not β-globin mRNA. Hypersensitive sites 5' to the δ and β genes are present in HEL cells as well as fetal and adult erythroid cells, but no such hypersensitive structures are found around the nontranscribed β-globin gene in K562 cells (*191*), even though the β-globin gene region is sensitive to the DNase (*193*). No difference was noted in the sensitivity of γ-globin genes in stimulated as compared to unstimulated K562 cells in spite of the difference in mRNA production. There was also no hypersensitive site around the 5' end of the γ-globin genes in adult erythroid tissues (*191*). In the β-globin gene, there are hypersensitive sites 200 bp in the 5' direction from the

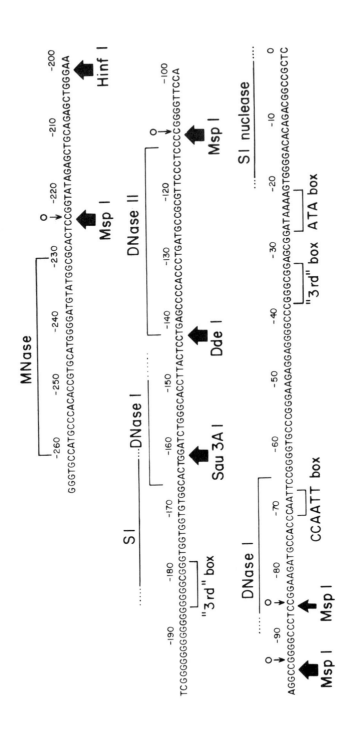

transcription initiation site, in the large intron, and approximately 800 bp 3' to the polyadenylation site (*191*). These sites were not detected in K562 cells.

DNase-hypersensitive sites do not seem to be simply a consequence of active transcription, since they can sometimes be found in cells not producing transcripts. They can be induced in chicken fibroblasts by salt treatment, without induction of globin synthesis. Chicken bone marrow cells expressing active avian-erythroblastosis-virus transforming genes fail to produce globin transcripts or exhibit DNase hypersensitive sites in the β-globin genes. Transcripts and new hypersensitive sites appear after thermal inactivation of temperature-sensitive viral products. The hypersensitive sites, but not the transcripts, persist in these cells after the virus product is reactivated by lowering the temperature (*194*). In this last case, the hypersensitivity seems to be propagated for a number of generations without detectable transcription of the globin genes; it is a potential model for the study of determination and differentiation events producing an activated gene that can be inherited by progeny cells prior to evocation of gene expression by differentiation.

Regions hypersensitive to DNase in the SV40 virus genome have also been extensively studied. Here, they lie at or immediately upstream from the sites of early and late DNA transcription (*194a*), and the appearance of some of the sites is correlated with the onset of late mRNA production (*195*). In this case, it has been possible to visualize the DNase hypersensitive regions under the electron microscope, and they appear to be free of detectable nucleosomes (*196, 196a*). Some, but not all, of the DNase hypersensitivity sites of SV40 are located in

---

FIG. 10. Comparison of nuclease-sensitive sites in the promoter region of the chicken adult β-globin gene. This figure is adapted from that of McGhee and Felsenfeld (*185*) with inclusion of the data of Weintraub and colleagues (*186*). Not shown is the recently reported fine mapping of the S1-sensitive sites (*186a,b*). MNase stands for micrococcal nuclease, S1 for the single-strand-specific nuclease from *A. oryzae*; *Sau*3A, *Dde*I, *Msp*I, and *Hin*fI are the restriction endonucleases cleaving at the site of the arrow. "3rd" box refers to sequences resembling the SV40 11-bp repeat (see Fig. 5). CCAATT box and ATA box refer to other promoter components described in the text. Cap indicates the site where transcription initiates. The width of arrows indicates relative sensitivity to cleavage of DNA in adult erythroid chromatin. Horizontal lines show approximate location of preferred cleavage site by nonspecific nucleases digesting active chromatin. The nucleotides are numbered relative to the transcription initiation site. The CCGG sequences indicated by $\overset{\circ}{\mid}$ are unmethylated in adult red cell chromatin, but methylated in embryonic red cells. The CCGG at −40 is unmethylated in both embryonic and adult red cells. A reciprocal relationship is seen in the methylation of CCGG and GCGC sites surrounding the 5' end of the embryonic β genes.

the 72-bp enhancer element and adjacent DNA, even when this element is translocated to new positions in the viral DNA (*197, 197a*). It has been argued that regions of left-handed "Z" DNA may be involved (*198*), with hypersensitive sites occurring at the boundaries between left-and right-handed DNA.

As mentioned above, sites hypersensitive to S1 nuclease may appear also in regions of the DNA very close to those in chromatin when naked supercoiled DNA is studied in plasmids. However, the exact pattern of S1 cleavage may vary considerably with experimental conditions, and additional S1-hypersensitive sites appear that are not detected in chromatin. S1-sensitive sites are also present in reconstituted chromatin, even though these molecules are topologically relaxed. The specific suggestion has been made that supercoiling of the DNA molecule may be concentrated at the region of S1-hypersensitive sites and might even be associated with alternate conformations of DNA (cruciform structures, etc.), that will tend to "pop off" single nucleosomes (*199*).

Fine mapping of the S1-sensitive sites in the chicken adult β-globin gene has been reported (*186a,b*), and in supercoiled plasmids neither methylation nor a "Z"-DNA conformation could account for the S1 sensitivity of a major cutting site at the $(G)_{16}$ sequence (see Fig. 10). Some features of the S1 cutting, particularly the lack of temperature dependence, have been interpreted to suggest that the sensitive sites may not have single-stranded character in the conventional sense (*186b*).

Sites in supercoiled human globin DNA that are sensitive to S1 nuclease have been mapped in detail in two instances. A cluster of sites in the human α1-α2 intergenic region occurs in a long sequence of pyrimidines containing multiple imperfect repeats of the trinucleotide CCT (*199a*). The other cluster consists of several sites in the long pyrimidine stretch upstream from the human δ-globin gene (*199b*). In both sites, there are prominent tandem duplicated sequences (Fig. 11). These observations indicate that at least a portion of the S1-sensitive sites may arise from slippage and mispairing of repeat sequences and suggest that this slippage might occur preferentially in DNA, where there is a strong bias for purines on one strand. Such slippage could promote rapid mutation by deletion-duplication events and possibly favor development of pyrimidine stretches in DNA. All this suggests that the relative single-stranded character of stretches of 5′ flanking DNA may be a consequence of promoter activation, but may also occur at sites not directly related to transcription initiation,

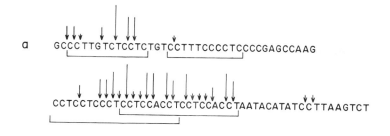

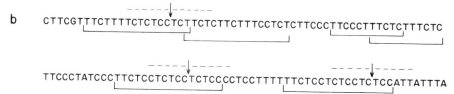

FIG. 11. Sequences embedding sites hypersensitive to S1 nuclease in globin DNA. The hypersensitive sites are detected in globin DNA cloned into supercoiled plasmids. (a) Sequence upstream from human α-globin gene (*199a*); (b) sequence about 1000 bases upstream from human δ-globin gene (*199b*). (a) Arrows indicated preferred sites of cleavage, with the height of the arrow proportional to the probability of cleavage. (b) Arrow indicates site of cleavage ± about 10 bp. Horizontal brackets under sequences show imperfect repeats that could be involved in generating S1 sensitivities.

possibly revealed because of the generally less protected nature of DNA in active chromatin.

In summary, nuclease hypersensitivity in the region around the 5' end of the gene in the globin system is a common concomitant of gene expression, but is not itself sufficient to ensure active expression. Hypersensitivity at other sites in and around the genes occurs. These additional sites have not yet been correlated with alternate transcription initiation sites (see Sections III,B and III,C) and specific features of the DNA sequences.

Before leaving this section, it is appropriate to make a few comments on the elusive search for *trans*-active factors responsible for differential control of gene expression, which presumably interact with the *cis*-active DNA sequences we have been discussing.

The HMG (high-mobility group) (*200*) nonhistone proteins 14 and 17 are abundant in nuclei. They bind preferentially to single-stranded DNA and to active nucleosomes and may be necessary for exhibition of sensitivity to DNase (*201, 202*). However, the potential for specific

DNase sensitivity rests with other properties of the chromatin rather than the HMGs per se. For example, HMGs 14 and 17 from erythrocyte chromatin fail to induce DNase sensitivity in globin genes in brain chromatin, but HMGs 14 and 17 from brain nuclei do induce such sensitivity in the globin gene of erythroid chromatin.

The nuclei of erythroid precursors contain many other peptides of unknown function, including tissue-specific proteins and proteins that may modulate β-globin DNA transcription *in vitro* (*202a*) or bind preferentially to β-globin as compared to λ bacteriophage DNA (*202b*). Among such proteins, one might hope to find the products of genes functioning to establish or maintain tissue or stage-specific differentiation. An initial approach to the sorting of avian erythrocyte nuclear proteins involves using the powerful tool of monoclonal antibodies (*203*). Interestingly, a large fraction of the antibodies raised to erythroid nuclear proteins were erythrocyte specific. Many, but not all, of these reacted only with adult red cells, and two reacted also with those of *Xenopus*. Some antibodies reacted with several peptides of different molecular weights. These early results are intriguing and reveal considerable structural complexity in search of functional correlates.

An exciting development is the recent report of partial purification of a factor from chicken erythrocyte nuclei that can induce hypersensitivity in the 5' flanking region of the chicken adult (β^A) gene when combined with plasmid clones of this gene and histones (*203a*). This factor is not present in extracts from cells not expressing the β^A gene, suggesting that it may play an important role *in vivo* in the specific activation of the gene. Furthermore, this factor, which contains protein and is distinct from RNA polymerase II, binds specifically to DNA sequences 5' to the gene, where DNase hypersensitivity has been mapped.

F. Methylation

A small fraction of the bases in DNA are modified by methylation, which occurs after incorporation of the nucleotides into the DNA chain. The predominant modification in animal cells is methylation at carbon-5 of deoxycytidylic acid. About 4% of the dC residues in animal DNA are methylated. Most of the m^{5}dC is in the dinucleotide dC-dG, and about 70% of these in human DNA are so methylated. If DNA is methylated *in vitro* at these residues and reintroduced into cells, the progeny DNA also is often methylated. This is a sort of epigenetic inheritance and provides an attractive model for one of the determination events during cellular differentiation. This model cannot account for differentiation in all organisms since, for example, *Drosophila* DNA lacks methylated bases (see *204* for review and references). Also, methylation may not be the primary event, as methylation may not occur until several days after viral DNA is introduced into nonpermissive cells (*204*).

DNA methyltransferases may form complexes with DNA-synthesizing enzymes (*204b*). Two DNA methyltransferases, extensively purified from MEL cells (*204c*), methylate dC-dG in a wide variety of sequence contexts; they are 50- to 100-fold more active with hemimethylated than with unmethylated DNA and shows a strong DNA-chain-length dependence of methylation activity. These properties of the enzymes *in vitro* indicate that other factors in the cell must limit their activity and provide the sequence specificities seen *in vivo*.

The differential specificity of restriction endonucleases toward DNA sequences containing methylated and unmethylated deoxycytidylate has been a useful tool for analysis of methylation of unique-copy DNA *in vivo*. In particular, with the exception of the sequence GGC^{m}CGG (*205, 205a*) the endonuclease *Msp*I cleaves the sequence CCGG regardless of whether the second C is methylated or not, provided the first C is not methylated (*206*). Conversely, the enzyme *Hpa*II cleaves the same sequence only if the second C is not methylated, but will cleave regardless of whether the first is methylated or not (*207*). Successive and combined application of these two enzymes to genomic DNA followed by analysis of the resulting restriction fragments makes it possible to determine the extent of C-G methylation at the different occurrences of CCGG within a region of unique DNA, and to study this as a function of the state of differentiation of the cells from which the DNA was extracted. Unfortunately, only a small fraction of C-G residues occurs in this tetranucleotide. A second sequence, GCGC, is cleavable by *Hha*I only if the internal C is not methylated; this provides an additional partial test for methylation, but the state of methylation of C in isolated C-G dinucleotides remains unmeasurable by these techniques.

The state of various globin DNAs has been investigated extensively by use of the restriction endonucleases *Hpa*II, *Hha*I, and *Msp*I. Certain generalizations (*208*) are worthy of repetition here. (i) Sperm DNA appears to be completely methylated throughout the globin gene clusters. (ii) In general, CCGG sequences and in some cases GCGC sequences flanking embryonic globin genes are undermethylated in embryonic erythroid tissue cells relative to adult erythroid tissues. (iii) The CCGG or GCGC sequences surrounding the adult globin genes are extensively methylated in nonerythroid tissues.

(iv) The methylation of sequences containing C-G and flanking adult globin gene regions is decreased in adult erythroid cells. (v) Only a few sites show dramatic differences in the extent of methylation between expressing tissues and nonexpressing tissues, and most of these sites are located near the 5' end of the genes.

These conclusions (208) for the globin system are similar to those drawn from observations on other active genes, but there are noteworthy exceptions and complexities. With certain other genes, demethylation may not accompany active expression of the genes. In the Friend MEL cells (190, 209), there do not appear to be substantial changes in the methylation in the region of the β-major and αI-globin genes during HMBA-mediated differentiation (210, 211). The adult β-globin gene of the chicken is a well-studied example that illustrates the complexities of the system. The CCGG sites between residues -110 and -80 are unmethylated in both adult and embryonic red cells, whereas sites farther upstream are completely methylated in embryonic cells and completely unmethylated in adult cells. Another site at the 3' end of the adult β gene is completely unmethylated in both embryonic and adult cells, whereas a site within the large intron is partially methylated in both embryonic and adult cells (185). Chicken erythrocyte precursors transformed by a temperature-sensitive avian-erythroblastosis virus begin synthesis of globins when placed at temperatures nonpermissive for the virus. Some but not all C-G sites become less methylated at the elevated temperature, but the extent of those changes is small compared to those during normal erythropoeisis (212). However, in no case does increased methylation correlate with gene expression. Very recently, an apparent exception to this rule has been reported. The major histocompatibility complex (H-2) genes of mouse are not expressed in undifferentiated teratocarcinoma cells, but are expressed once differentiation is induced *in vitro*. In at least one case, induction has been associated with *increased* methylation of the H-2 gene (212a). However, methylation of specific sequences at the 5' end of the gene were not studied, and these might yet prove to follow the conventional pattern.

An additional feature of globin DNA methylation is a region of about 5 kb just 5' to the chicken π (embryonic α) gene in all tissues that have fully methylated *Hha* and *Hpa*II sites and also four blocked *Msp* sites (213, 214). This last could be due to a C^{m}CGG or to local sequences that make it resistant to *Msp* (215). Also, the methylation of some sequences as far as 10 kb upstream from the α-globin cluster correlates with globin gene expression (214). Similarly, hypomethylation of sites around the ε gene is seen at 6 weeks when ε is actively expressed, but not in later stages of gestation or after birth (216). In man, there is an inverse correlation between methylation of CCGG sequences 5' to the fetal globin genes and fetal globin synthesis. These sequences are undermethylated in the fetus but completely methylated in adults (216a).

In another approach (217), DNA molecules fully methylated on the C's of one of the two strands within localized regions are prepared by use of a single-stranded DNA templates, suitable DNA primers, and *in vitro* DNA synthesis employing m^{5}dCTP rather than dCTP. The results (218) with the fetal globin gene are dramatic. Methylation of the 3' half of the DNA does not impair subsequent expression *in vivo*, whereas methylation of the sequences from −700 through +100 abolishes expression of the genes following transfection. The CCGG sites remain methylated in the stable cell lines that express or do not express the globin gene. These results are consistent with earlier studies of other genes, such as the early region of adenovirus 2 and the APRT (adenine phosphoribosyltransferase) gene, where methylation markedly decreased or prevented expression (reviewed in 204). On the other hand, transcription of methylated DNA *in vitro* proceeds to an extent not detectably different from that seen with unmethylated templates, so that the methylation restriction of gene expression is lost in the *in vitro* system.

In summary, extensive methylation of DNA sequences *in vitro*, particularly those of the 5' positions of the gene, impairs the subsequent expression of the DNA in transfected cells. This correlates with a tendency for demethylation of 5' sequences of expressed chromosomal genes. However, the correlation is imperfect, and gene expressions may vary without detectable concurrent methylation changes. More precise knowledge of the state of methylation of individual residues could show whether there are key methylation sites and what part of the control of transcription could be directly related to DNA methylation.

The chemical sequencing method of Maxam and Gilbert (218a) distinguishes between methylated and unmethylated dC, as the former is not cleaved and does not appear as a dC on the gel. Potentially, this method can be applied to genomic DNA by adoption of the indirect end-labeling method of Wu and Gilbert (219). While the experiment may be somewhat tedious if enrichment of globin sequences is technically necessary prior to running sequencing gels, it offers the valuable prospect of displaying the methylation state of each dC-dG.

The nucleoside analog 5-azacytidine cannot be methylated, and it may inhibit methylation of other cytidylates *in vivo* (220, 220a). Under appropriate conditions, it can induce differentiation in fibroblasts and MEL cells (221). Its effect on globin gene expression in baboon and man has been studied in an effort to induce expression of fetal globin genes in adults. DeSimone et al. noted (222) that marked elevations in HbF occurred in baboons recovering from hemolysis or hypoxia, and that the levels reached were under genetic control (223). Individual animals could be distinguished as low, medium, or high responders with familial clustering, but the precise pattern of inheritance was not determined. Reasoning that demethylation of γ genes might also produce this effect, they treated anemic baboons with 5-azacytidine and observed a striking increase in HbF, to levels as high as 80% (224, 225). This increased γ-gene expression, whether induced by 5-azacytidine or by phenylhydrazine-induced hemolytic anemia, is accompanied by hypomethylation of the γ genes (226), but to a lesser degree in hemolytic anemia.

A bold and exciting step was taken with the application of this finding to patients severely ill with β-thalassemia (227) or sickle-cell anemia (228, 229). Administration of intravenous 5-azacytidine resulted in an increase in fetal globin synthesis after as little as 48–72 hours, leading to HbF levels of 8–20%. Bone marrow analyses demonstrated that hypomethylation of the CCGG sites 5' to the γ genes accompanied the increased HbF production. Interestingly, and raising concern over potential deleterious effects, hypomethylation of total genomic DNA, a Y-chromosome-specific DNA fragment, and the ε gene were also noted, although there was no apparent initiation of ε synthesis (227) and no rise in serum α-fetoprotein (228).

Some debate has ensued as to whether the effects of 5-azacytidine are indeed due to hypomethylation of the γ gene or whether the drug in some other way leads to selection of earlier erythroid precursors that have a greater capacity to make HbF (229a). This latter possibility is made more real by the observation that hydroxyurea and arabinosylcytosine drugs, which are cytotoxic but not known to affect methylation, also result in increased HbF in primate models (229a).

The use of 5-azacytidine has quite naturally been greeted with some excitement (230, 231), as it represents an attempt at a new therapy based on an understanding of molecular genetics. However, the therapeutic efficiency of this particular approach remains to be demonstrated, and the potential carcinogenicity of long-term use of a drug that may inappropriately activate other cellular genes restricts its use at present to a research setting, applied to patients with shortened life expectancy or severe disability.

IV. Structure of Human Globin Gene Clusters

A. The α-Globin Cluster

The human α gene is located on the short arm of chromosome 16 between 16p12 and 16pter (232), and occupies a region of approximately 30 kb. Figure 12 shows the relative position of the five α-like genes (31). The locus consists of an embryonic gene denoted as ζ at the 5' end of the cluster, a ψζ gene approximately 11 kb 3' to this, a ψα gene, and the duplicated adult genes denoted α2 and α1. In this cluster, as in the β cluster, the genes in man are arranged in the 5' to 3' direction in order of their activation.

The ζ and ψζ genes have recently been sequenced (47). There is extremely close homology between them extending from 140 bases on the 5' side of the cap site to a point about 50 bases on the 5' side of the poly (A) addition signal, with marked divergence of the two sequences outside these boundaries. This observation suggests the possibility of a recent gene conversion (see Section IV,C). Specifically, except for the variable length of the introns, there are only six single-base differences between ζ and ψζ, three of which are in noncoding sequences, two of which result in amino-acid replacements, and one that results in termination at codon 6 of ψζ. Thus, ψζ is unlike other

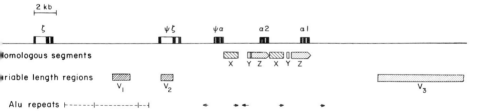

FIG. 12. Map of the human α cluster. The embryonic and adult α genes on chromosome 16 are drawn to scale, with the coding strand arranged in a 5' → 3' direction from left to right. Filled-in areas of the genes are coding regions, and open areas are introns. The cluster consists of an embryonic gene ζ, a ψζ gene, ψα gene, and the two adult α genes denoted as α2 and α1. Blocks of sequence homology designated as X, Y, and Z, which presumably arose by duplication of the α segment, are shown. Regions of DNA of variable length in different individuals are shown as V_1, V_2, and V_3. The molecular basis of V_1 has been determined, and stems from a variable length of a tandemly repeated element (see text). It is very likely that V_2 has a similar basis, and V_3, although less is known about it, may also. The location of *Alu* repetitive sequences and their orientation is shown. Between ζ and ψζ at least three *Alu* elements are known to be present, but the exact location of these elements has not yet been determined by sequencing. References are given in the text.

globin pseudogenes (233) in that it contains apparently all the necessary sequences for transcription initiation, splicing, and polyadenylation; one would expect that some mature cytoplasmic message may be produced from this gene. Another interesting feature of the ζ and $\psi\zeta$ genes is their introns, which are unusually large. Intron I of the ζ gene is 886 bases long, and that of $\psi\zeta$ is fully 1264. A major portion of the first intron in both genes is made up of a simple repeat of a 14-bp sequence that has the consensus sequence ACAGTGGGGAGGG. This sequence is repeated 12 times in ζ and 39 times in $\psi\zeta$, accounting for the difference in the size of the introns. This sequence is similar to others, both in this cluster and elsewhere, that seem to have the ability to form regions of variable length (see below). The second intron also contains a simple repeat, this time of the sequence CGGGG, with 35 repeats in ζ and 52 repeats in $\psi\zeta$. This results in a very large number of potential methylation sites, the significance of which is unknown.

The $\psi\alpha$ gene is approximately 73% homologous to $\alpha 2$ and $\alpha 1$, with many single-base changes and small frameshifts, as well as a 20-bp deletion from codons 38–45 that results in UGA terminations at codons 75, 79, and 83 (234).

The transcriptional activity of $\psi\alpha$ has recently been studied, both *in vitro* and in a transient expression system (234a). The promoter, which contains a deletion bringing the CCAAT and ATA boxes unusually close together (17 bp) was functional at about 10% of the efficiency of α. The mutated poly(A) addition signal in the $\psi\alpha$ (AATGAA), however, appears to be nonfunctional, which may explain the absence of $\psi\alpha$ transcripts in human reticulocytes.

If the possibility of ancient gene conversion is ignored, a comparison of silent-site versus replacement-site substitutions in $\psi\alpha$ versus $\alpha 2$ suggests that a duplication event occurred about 60 million years (My) ago to give rise to $\psi\alpha$ and α (which presumably much later duplicated to $\alpha 2$ and $\alpha 1$; see below), but $\psi\alpha$ became inactivated about 45 My ago. In accordance with this estimate is the observation that the chimpanzee $\psi\alpha$ has the same 20-bp deletion seen in man (235); the divergence of chimp and man is much more recent than 45 My. However, it is difficult to explain by this model why no homology has been detected in the 5' flanking regions of $\psi\alpha$ and $\alpha 2$ (though dot-matrix analysis to look for subtle homologies has not yet been reported); a simple duplication model would predict about the same homology between these flanking regions and the $\psi\alpha$ and $\alpha 2$ introns. It has been proposed (235) that the flanking regions are particularly prone to expansion of simple sequences and thus may diverge more rapidly. An alternate explanation, however, and more in line with analyses of the β-globin cluster, is that the $\psi\alpha$ and α genes actually diverged much more than 60 My ago, but that the gene sequences were corrected against each other at about this time; the gene conversion would presumably have extended from the region of the promoter to about the poly(A) signal. A well-described example of such a conversion with a 5' end precisely at the CCAAT box occurs in the human δ gene (see Section V,B).

The $\alpha 2$ and $\alpha 1$ genes are very similar, differing only by two base-substitutions and a seven base-pair insertion in the second intron (48, 49, 236). However, there is a sudden divergence in the 3' untranslated region that enables separate quantitation of $\alpha 2$ and $\alpha 1$ mRNA from human bone marrow. Such studies have demonstrated an approximately 2:1 to 3:1 ratio of $\alpha 2$ to $\alpha 1$ mRNA (237, 238). Interestingly, the translational efficiency of $\alpha 1$ is higher, so that the total protein production from $\alpha 2$ and $\alpha 1$ is very nearly equal (238a). Since $\alpha 2$ and $\alpha 1$ mRNAs differ only in their 3' untranslated regions, this provides incontrovertible evidence that this region contributes to translation efficiency.

Heteroduplex mapping of cloned genes from the α-globin complex demonstrated from the outset that there are interesting regions of internal homology interrupted by short stretches of nonhomology in the region of the $\alpha 1$- and $\alpha 2$-globin genes (31). More recently, most of this region has been sequenced, and the basis for the homology has become clear (239). Apparently, at the time of the duplication of the α locus, three regions of DNA (denoted as XYZ in Fig. 12) were contiguously arranged in the 5' and 3' direction, with the α locus in the 3' portion of the Z block. The XYZ unit was then duplicated. The triply repeated sequence GCCTGTGTGTGCCTG appears at the 5' ends of the X blocks and 3' ends of the Z blocks, and presumably bounded the original XYZ block prior to duplication. It is tempting to speculate that this was the 3' border of the gene conversion between $\psi\alpha$ and α mentioned above. This conversion, by homogenizing the $\psi\alpha$ and α sequences, may have promoted a subsequent unequal crossover that gave rise to $\alpha 2$ and $\alpha 1$. The X and Y blocks have slowly diverged, now containing about 10–15% base substitutions, but the Z blocks are highly homologous, suggesting that an additional gene conversion has occurred (239, 265). The sequences of the chimpanzee $\alpha 1$ and $\alpha 2$ cDNAs lend support to such a conversion in the human lineage after divergence from the chimpanzee (239a).

Other interesting sequences have also appeared. Nonhomologous insertions are found between the X and Y blocks; neither of these is flanked by direct repeats, and so the mechanism of insertion remains open to question. The insertion in the α2 unit contains a simple repeat of the sequence $(C-A)_{15}$. It is interesting that such a sequence also appears in the large intron of the γ genes (43) and has been considered to be a possible "hot spot" for gene conversion (see Section IV,C below). There is also a 224-base sequence present between the Y and Z blocks in the α1 unit, which is not flanked by direct repeats. This could represent either an insertion or the ancestral sequence that has been deleted in the α2 unit.

A very interesting feature of this repeated block is the location of Alu sequences, which are discussed in greater detail in Section IV,B,2. The 3' end of the X block (Fig. 12) contains a typical representative of the Alu family. In the X block for α2, this Alu member is seen to be flanked by a typical direct repeat 15 bp long. The identical direct repeat appears at the 5' end of the homologous Alu sequence in the α1 X block, although the 3' copy of this repeat has been lost, presumably in the same event that led to the insertion of nonhomologous material between X and Y. The identity of the 5' direct repeats leads to the conclusion that this Alu sequence must have been inserted prior to the duplication of the XYZ unit. There is another Alu sequence in the duplicated region, however, as shown in Fig. 12. This is a dimeric Alu inserted just inside the 5' end of the Y block in the α2 unit and flanked by a 13-bp repeat. This 13-bp sequence can be found intact and uninterrupted at the 5' end of the α1 Y block; this suggests that a staggered break occurred in the α2 Y block after the time of duplication. This therefore provides clear evidence for dispersion of Alu sequences into the α locus at two different times in evolution (239).

A further interesting feature of the α-globin locus is the presence of variable length regions, as denoted in Fig. 12. In a search for restriction-fragment-length polymorphisms in the α-globin locus, it was noted that several restriction enzymes that cleave fragments containing the regions V1, V2, or V3 gave fragments with a range of sizes from different individuals (240, 241). The fact that any enzyme that cut the sequences flanking these variable regions showed such behavior, and that at least three, and possibly many more, alleles seemed to be present in the general population and were inherited in a Mendelian fashion, led to the conclusion that these polymorphisms did not result from single-base changes that abolished or created a particular restriction site, but were the result of regions of DNA that by some special property had developed a range of alleles of variable length in the normal population.

It was therefore of particular interest to see what the sequences of such regions might be. The explanation is now at hand for the region denoted as V1, which contains a tandemly repeated 36-bp sequence that, in a "short" allele, was present in 32 copies, but in a "long" allele was present in approximately 58 copies (242). This 36-bp repeat is seen on close analysis to consist of an internal 9-bp piece flanked on either side by 13- or 14-bp domains, which are themselves homologous. These domains are also noted to be homologous to the simple repeat element present in the first intron of the ζ and $\psi\zeta$ genes. Although it has not yet been confirmed by sequencing, the variable region denoted as V2 in Fig. 12 is in the first intron of the $\psi\zeta$ gene and very likely represents a variable copy-number of this simple repeat, especially since the number of repeats between ζ and $\psi\zeta$ is known to differ. It is also of considerable interest that the 13- or 14-bp domains of the repeat are homologous to a similar sequence 5' to the human insulin gene, which is also present there in variable copy-number, leading to a similar-length polymorphism (243). The fact that these two length-polymorphisms from different parts of the human genome have their basis in a similar sequence strongly suggests that there is something about this sequence that promotes its tandem expansion. Once expanded, variable lengths could presumably be further enhanced by unequal crossing-over. One intriguing possibility is that this sequence plays a role in DNA replication, and that somehow it is tandemly replicated during this process. It has been proposed that the single-stranded palindromic nature of the insulin repeat TGGGGA-CAGGGGT may somehow be tied to its function (244). However, this palindrome is largely destroyed in the versions present in the α-globin locus, which tends to reduce the likelihood of this explanation. The molecular basis of the variable-length region 3' to the α1 locus denoted as V3 remains to be determined, but it is likely to have a similar explanation.

We do not include sequence data from the α-globin locus in this review, though much of this has been determined. Sequences of the ζ (47), V1 (242), $\psi\zeta$ (47), $\psi\alpha$ (234, 235), α2 and α1 (48, 49, 236), X, Y, and Z blocks and their nonhomologous inserts (239, 365) and 3' α1 sequence (245) can be found in the appropriate primary references.

B. The β-Globin Cluster

1. GENERAL

The human β-globin cluster occupies a region of approximately 70 kb on the short arm of chromosome 11. (This cluster is also referred to by some as the non-α cluster, but this seems to be an unnecessarily negative designation and should be abandoned.) Using radiation-in-

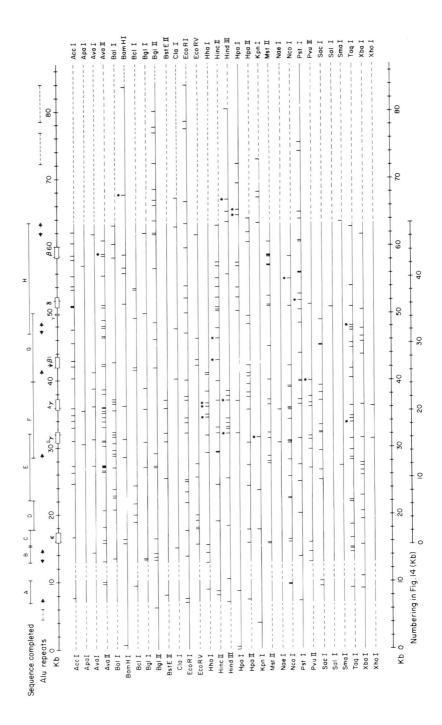

duced deletions of 11p in somatic cell hybrids, this cluster was originally mapped close to the centromere in band 11p11→11p1208 (246–248). However, more recent data suggest that the β gene may, in fact, be located in the distal third of 11p along with the insulin gene and the cellular Harvey ras gene (c-Ha-ras) (249–249b). The β-globin cluster is larger, more complex, contains more expressed genes, and in some ways is more exhaustively studied than the α cluster (30, 250–253). Figure 13 presents a map of the cluster together with a restriction map of most of the commonly used restriction enzymes that recognize a 6-bp site, as well as a few of the 4-bp recognizing enzymes. Figure 13 also shows the location of Alu repetitive sequences and a long sequence of pyrimidines.

Figure 14 (see Addendum 1, p. 439) presents the entire nucleotide sequence extending from a point approximately 400 bp 5' from the ε gene to a point about 3 kb 3' to the β gene. As the legend denotes, this sequence comes from several sources, including a considerable amount of previously unpublished data from this laboratory. In addition to the sequence presented in Fig. 14, published sequences from the area 5' to the ε gene are available, including a region that was thought to contain a β pseudogene (254). Sequencing of this area, however, revealed no β-like structure, and the basis for assignment of this region as a ψβ gene appears to be several runs of poly(dT), which

---

FIG. 13. Restriction map of the β-globin cluster. The available published sequences in the region of the β-globin cluster, together with sequences presented for the first time in this review, were used to generate by computer a restriction map of most of the common restriction enzymes that recognize a 6-bp site as well as a few of the enzymes recognizing 4-bp sites, which cut relatively infrequently. Shown above the β-globin cluster map are the regions that have been completely sequenced. Restriction sites that fall outside of these bars have been determined by genomic blotting or restriction digests of cloned genomic segments, and errors as much as 10% in the size of these fragments are possible. Regions where no information is available for a given restriction enzyme are shown as dashed lines. Polymorphic sites are marked with an asterisk and are further referenced in Fig. 24. No sites were found for MluI, NarI, or PvuI.

The location of Alu repetitive elements are marked with arrows, except in a few instances where the exact location is unknown and a dashed line indicates the boundaries of a fragment in which an Alu sequence is known to reside. The region marked Y is a 275-bp polypyrimidine sequence. Kilobase scales have been placed at both the top and the bottom of the figure to facilitate alignment of sites with the β-globin map. The scale at the very bottom corresponds to the sequence in Fig. 14. While every effort has been made to be accurate, users would be well advised to consult the primary references if discrepancies arise. The references for the sequenced areas are A (254), B (255), C (42), D (255a), E (256), F (44), G (257), H (258). A great many sources were utilized to construct sites in areas that have not been sequenced directly. Particularly helpful are references 253, 259–262.

may have hybridized with the poly(dA) sequence of the cDNA probe used.

2. Repetitive Sequences: the *Alu* Family

Two categories of repetitive DNA in the eukaryotic genome may be distinguished. One consists of multiple tandem repeats of very similar or identical runs of nucleotides; the second, of interspersed sequences present in multiple copies in the genome. The latter are not necessarily internally repetitive and are not commonly tandemly located with respect to one another. The former type of sequence may arise by a series of slippage events during DNA replication or unequal crossing-over events (265). Short stretches of simple, repetitive DNA have been noted in the β-globin gene cluster. They include runs of the dinucleotide T-G, and short blocks of tandem repeats of a sequence of four to five nucleotides (see Section VI,D). They are among the most variable sequences and are good candidates for mutation by slippage, either solely based on the DNA sequence or potentially promoted by sequence-specific protein/DNA interactions.

The second category of long stretches of interspersed, repetitive DNA is of considerably more current interest (266, 267, *et op. cit.*). The evidence, although largely inferential, is very strong that these interspersed DNA elements occur in multiple sites as a result of DNA insertion or transposition events that have disseminated these sequences through the genome. As yet no specific function for these interspersed repetitive elements has been described, and they are considered to be possible examples of "selfish DNA" (268, 269). It would be surprising if further detailed analysis did not show that some of these interspersion events played a significant role in the evolution of gene clusters and their regulation.

The most extensively studied of these interspersed sequences in man are the *Alu* sequences (270, 271, 271a). These are repetitive DNA sequences approximately 300 nucleotides in length that are widely interspersed through all chromosomes of human DNA. They constitute approximately 5% of human DNA, or 300,000–500,000 copies per genome. The sequences of individual *Alu* family representatives differ from each other by about 20%, and from a consensus sequence by about 13% (271a). They are relatively randomly distributed in intergenic DNA and have also been found in introns and immediately preceding the poly(A) tails in the untranslated 3′ portion of cytoplasmic RNA. There is no universal pattern of arrangement of *Alu* sequences with respect to one another; in particular, direct tandem repeat copies of complete *Alu* sequences are uncommon.

The *Alu* sequence itself consists of an internal imperfect direct repeat as though it had been generated by fusion of two monomers. One end of the *Alu* sequence contains oligodeoxyadenylate. Near the other end are internal sequences that specify initiation of transcription by RNA polymerase III (272–275). These sequences are so located that the transcript begins at the border of the repetitive element and extends across it through the poly(A) tract and into flanking unique-sequence DNA until it reaches a thymidylate stretch that serves as a termination signal for the enzyme (276). On the basis of these observations, it was suggested that the *Alu* sequences are transposable elements whose transposition is mediated by their polymerase-III transcript. The uridylate sequences at the end of the polymerase-III transcript have been postulated to serve as a primer for synthesis of a DNA strand beginning in the poly(A) stretch and going up to the 5' end of the RNA. The resulting cDNA would be composed of sequences corresponding to the repetitive unit in the genome. Since polymerase-III promoters are internal, this would generate a segment of DNA in the genome that could repeat this type of propagative event, initiating transcription, allowing the transcription to proceed into flanking DNA until it reached a thymidylate stretch, and generating another RNA that would be self-priming for DNA synthesis (277, 278).

Consistent with the suggestion that this DNA disperses by insertion into preexisting genomic sequences, each *Alu* sequence in the genome is flanked by a direct repeat (279) of several to over 20 nucleotides. In at least two instances where the prior sequence can be inferred (239, 280), a staggered break seems to have occurred at the time of insertion. The direct repeats are different in sequence for different insertions of the *Alu* DNA (57), and little if any base specificity can be discerned with regard to the site of insertion of *Alu* sequences. It is difficult to envision a biological role that would cause direct repeat sequences to be retained, and therefore the rate of deviation between pairs of repeats may represent the random drift of unselected sequence. An estimate of the average age for an *Alu* sequence can be derived by comparing the 5' direct repeat in each *Alu* to the corresponding 3' direct repeat. While the *Alu* sequences themselves exhibit an average of perhaps 80% homology, the 5' and 3' direct repeats of each *Alu* are more than 96% homologous. This suggests, first, that multiple different fertile *Alu* elements may have given rise to the present *Alu* insertions and thus account for their extent of divergence, and, second, that the average insertion time was such that there was 3% probability that any single base would be changed over this period of evolution. The evolutionary clock (discussed in Section V) in terms of silent base-substitution seems to run more slowly in primates. Nevertheless, the 3% figure suggests that the *Alu* sequences were interspersed in the human DNA subsequent to evolution of mammalian precursors and possibly even subsequent to the earliest emergence of primates.

In the β-globin cluster the intergenic DNA sizes and location of restriction sites suggest that at least some of the *Alu* sequences exist in the evolutionary precursor common to the Old-World monkeys and the great apes. There is evidence (discussed above) (*239*) that in the α-globin cluster *Alu* sequences were dispersed at more than one point in evolution. The similarities between *Alu* sequences raises the question whether, in addition to initial insertion events, gene-conversion-like events operate to maintain the homogeneity of the sequences. However, the *Alu* sequences that have developed in association with the α-globin gene differ in sequence by 14%. This is at least as large a drift in base sequence as is seen in the surrounding unique sequences.

There are short repetitive sequences interspersed in the genomes of other mammals and even birds. In the case of mammals, some of these sequences, including the terminal adenylate sequence, appear to have some homologies to *Alu* sequences. However, not all the short interspersed segments in the DNA of other species are clearly homologous to *Alu* sequences, and some but not all (e.g., mouse B2 but not B1 repeats) contain polymerase-III promoters (*267*).

The origin of *Alu* sequences is not known, but there is a nearly universal well-preserved 7-S RNA in eukaryotes that is localized in the cytoplasm and appears to be part of the apparatus involved in the export of proteins into and through cell membranes. The 5' and 3' sequences of this 7-S RNA resemble half of an *Alu* sequence, and it has been suggested that the 7-S RNA may be an evolutionary precursor of the *Alu* sequences (*281*). If so, one might imagine that on more than one occasion during evolution an event might have occurred to give rise to *Alu* sequences from 7-S genes. On the other hand, the multiplicity of forms of short interspersed DNA noted in other species, such as mouse and ungulate, and the absence of detectable polymerase-III promoter activity, leave open the possibility that other novel mechanisms may operate to generate some of these short interspersed DNAs.

An important recent observation is that *Alu* elements in a prosimian, the galago, fall into two classes (*281a*). One type is closely analogous to the human; the other has a right half that is analogous to man, but a left half that appears to be homologous only in the regions that participate as RNA polymerase-III promoters. This appears to be of more recent evolutionary origin, and could be the result of the replacement of sequences within the left half of a particular *Alu*, followed by its propagation throughout the genome (*281a*).

Several *Alu* sequences have been localized in the human β-globin clusters (Fig. 13) (*282–284*), including single *Alus* upstream from the fetal globin genes and upstream from the ψβ gene. Inverted pairs of *Alu* sequences appear upstream from ε, upstream from δ, and downstream from the β-globin gene. Interestingly, the three inverted pairs in the β cluster are all arranged tail-to-tail with about 800 bp of nonrepetitive DNA between them. The *Alu* sequences that have attracted most interest are the inverted pair located approximately 3 kb upstream from the δ-globin gene. Further discussion of these *Alu* sequences and their polymorphisms is presented in connection with the section on deletions in the β-globin complex. (Section VI,C)

3. REPETITIVE SEQUENCES: THE *Kpn* FAMILY

This second major family of interspersed repetitive DNA sequences in the human genome (*267, 285*) is known as the *Kpn* family because of eponymic *Kpn*I restriction endonuclease cleavage sites located at analogous positions in many copies of the sequences (*286, 287a*). These sequences have been independently investigated by a number of groups who approached them from different points of view (*267, 286–295*). It is now clear that several short, repetitive-sequence segments released by specific endonucleases are derived from internal regions of long repeat-sequences, called LINES (*267*), of which the major component is the *Kpn* sequence. Since individual segments of the *Kpn* family of length 1.0 to 1.5 kb occur in more than 10^4 copies per genome, the total *Kpn* sequence may constitute several percent of the genome in man and at least approach in abundance the amount of *Alu* sequences.

Major portions of *Kpn* sequences occur within stretches of repetitive DNA up to 6.4 kb long. Other shorter *Kpn*-related inserts have been studied (referred to in Fig. 15). In some cases these could have resulted from deletions in a longer insert. However, the short *Kpn* segments may be flanked by direct repeats suggesting that the 5' and 3' boundaries of the original insertion event are intact. In one example, a *Kpn* element was inserted within the tandem-repeated α-satellite repetitive-DNA of monkey (*294, 302*). The *Kpn* element was flanked by 14-bp direct repeats of a sequence that occurs only once elsewhere in each copy of the α-DNA element (*294*), consistent with a

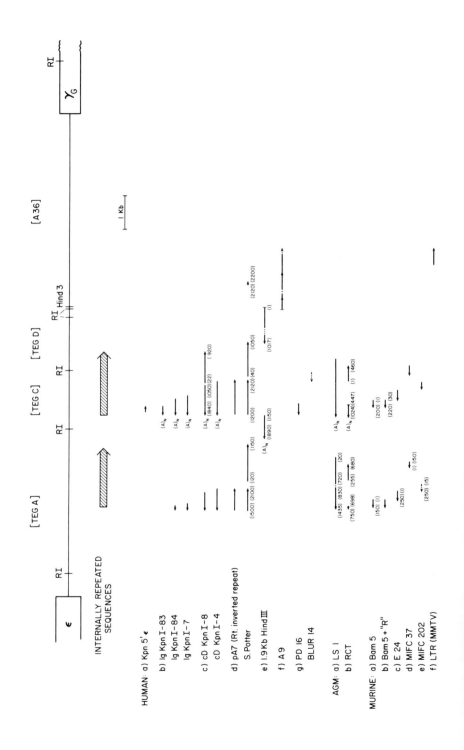

transpositional model for *Kpn* sequence insertion. One end of the *Kpn* element terminated in a run of deoxyadenylates.

Both the human *Kpn* family and the mouse *Bam*HI family sequences may be transcribed. A large portion of the RNA (*292, 301, 303, 303a*) that is sequence-homologous to *Kpn* DNA is found as heterogeneous nuclear RNA (hnRNA), but discrete polyadenylated cytoplasmic transcripts containing *Kpn* sequences have been described (*301*). Several workers have suggested that transposition of *Kpn*-like sequences may occur by way of a polyadenylated RNA intermediate (*292, 301, 303*) and raised the possibility that certain of the internal deletions found in some *Kpn* sequences could be a consequence of splicing. An internal segment of DNA has been found in different copies of *Kpn* repeats at the same relative location, but in opposite orientations, as though it had been inverted.

Two types of *Kpn* sequences are apparently widely dispersed. One abundant form releases a 1.9-kb fragment on digestion with the restriction endonuclease *Hind* III (*291, 293*). In another abundant subset of *Kpn* sequences (see Fig. 15), this fragment is interrupted by large blocks of additional sequence as though insertion (or deletion) had occurred in a single parental sequence that had then sired a substantial subset of current *Kpn* sequences. The detailed structures of complete *Kpn* family sequences are only beginning to emerge, although multiple sequences are published for segments of the *Kpn* family and for portions of the homologous *Bam*HI-R sequence family of the mouse (reviewed in *304a* and *304c*) (Fig. 15).

Kpn sequences have been clearly identified in the human globin cluster (*289, 290, 304*). At least one copy lies downstream from the β-globin gene; another lies between the ε and fetal genes and is over 6 kb in length. The latter has been sequenced. At the end nearer the

---

FIG. 15. Comparison of "*Kpn*I" sequences of man and the "*Bam*HI" sequences of mouse. Sequenced repeated DNA segments are compared to the "full-length" *Kpn* family sequence located between the human ε and $^G\gamma$ genes. The top horizontal line represents the DNA of the human β globin cluster beginning within the ε gene and extending to the 5' end of the $^G\gamma$ gene. *Eco*RI cleavage sites are indicated. Each DNA sequence is numbered continuously 5' to 3' in the direction presented in the original report. Sequences in (c) are derived from cDNA clones from cellular RNA complementary to *Kpn* genomic DNA. The mouse sequences indicated are regions of strong homology between segments of longer mouse repetitive sequences and the interglobin human *Kpn* sequence. AGM is African green monkey. References for the sequences are as follows: Human: (a) Short region of homology taken from 5' to ε gene; (b) *301*; (c) *301*; (d) *291g*; (e) *293*; (f) P. A. Biro and S. M. Weissman, unpublished data; (g) *291a*, *299*, *300*. AGM: (a) *292*; (b) *294*. Murine: (a) *295*; (b) *296*; (c) *297*; (d) *298*; (e) *297*, (f) *298a*.

fetal globin gene are some structural analogies to a retrovirus long-terminal-repeat (LTR) (305). This consists of a stretch of several hundred nucleotides bounded by imperfect inverted d(G + C)-rich repeats. However, the longer *Kpn* elements have not yet been found to be flanked by long direct repeats and no pseudo LTR-like element has been noted at the other end of the 6.4-kb *Kpn* insert between the ε and γ genes. If *Kpn* sequences are transcribed in the direction indicated by the cDNA clones, then the transcription would proceed from fetal toward ε-globin genes in the interglobin *Kpn* family, and the transcripts would be very rich in short runs of adenylic acid. DNA "blotting" studies of the sequences downstream from the β-globin gene suggest that there has been some scrambling of the order of homologous stretches of sequence of the downstream *Kpn* sequences as compared with that between the ε and fetal globin genes. There are additional suggestions that more than one type of scrambling of blocks of internal sequence of *Kpn* family may occur. One of Singer's monkey *Kpn* sequences contains a segment of several hundred base-pairs that is inverted relative to other members of the *Kpn* family. Potter (291b) has sequenced a 3000-bp *Kpn* element in which a block of 900 bp and a second adjacent block of 1100 bp occur in the same transcription sense, but in reverse order relative to the globin *Kpn*, without inversion (Fig. 15). He has also observed an open reading frame of over 650 bases near the 3' end of this *Kpn* element and has located a 5' end common to several *Kpn* sequences. The very intriguing possibility exists that some or many *Kpn* copies may encode expressed peptides. In this regard, there are repetitive sequences in mRNA specific for early embryonic and transformed cells (291c,d), and other such sequences occur in multiple brain-specific mRNAs (291e) and in a lymphocyte mRNA directing synthesis of a surface antigen (291f). There is insufficient evidence to decide whether any of these repeated sequences are related to *Kpn* elements, or to any other sequences present in the globin clusters.

No general mechanism for transposition of *Kpn* sequences has so far been proposed, and it appears quite possible that more than one mechanism is involved. For example, the sequence data available are consistent with the possibility that internal segments of the *Kpn* sequence may be independently transposed. A mechanism analogous to that suggested for the *Alu* family could operate for *Kpn* sequences (292, 301) and mouse *Bam*-H1-R sequences (304b), for example, with uridylate-rich termini of RNA transcripts folding back and pairing with internal adenylates. There are several internal oligoadenylate stretches that could act as sites for initiation of such reverse transcription. The 3' end of the 860-bp *Kpn* sequence (294), *Kpn* family cDNAs, and three genomic *Kpn* sequences (*301*) have a common boundary of homology at one end marked by dA-rich sequences. Long polyadenylated transcripts could hypothetically be generated by polymerase III, although the apparent termination of pol-III transcription by short oligo(U) segments would put constraints on the transcribed sequence. Although the classical components of polymerase-II promoters occur upstream from the transcription initiation site and therefore would not be transposable by such a simple mechanism, genes containing internal enhancers have been described. The structure surrounding the 5' terminals of other *Kpn* and *Bam*HI families would be of considerable interest in this regard. Speculatively, one might imagine a two(or more)-step transposition process for *Kpn* sequences. An initial step would be RNA-mediated transposition of transcribed portions of the sequence, after self-primed synthesis of cDNA. If a sequence near the 5' end of the transcript created a favorable site, LTR-like promoter sequences might then be mobilized into this site, presumably by DNA-mediated transposition. Other invertible, internal DNA sequences might have been acquired by transposition into parental *Kpn* sequences and transmitted as part of the complex unit. In this regard, there is evidence for probable insertion events into mouse R sequences (*304a*).

An important functional question is whether the *Kpn* insertions may alter the expression of other genes when they are inserted near them. Sufficient analogy exists for these effects in prokaryotic and simpler eukaryotic systems. In particular, the point in evolution at which a *Kpn* family element was inserted between the fetal and embryonic globin genes is not known. The distance between embryonic and fetal genes in lemur is considerably smaller than that in man (*32*), and gene mapping studies show that the equivalent region between analogous genes of mouse or rabbit is considerably shorter than that of man. This raises the possibility that the *Kpn* family may have been inserted after the separation of primates from rodents and lagomorphs, and perhaps after separation of the lemur linkage from that of higher primates. It would be interesting to know whether insertion of the *Kpn* element is correlated in any way with the altered timing of expression of fetal globin genes in man. Further discussion of this issue is presented in Section V.

C. Evidence for Gene Conversion of Repeated Genes

We have already alluded to the evidence for gene conversion in both the α- and β-globin clusters, and in the section on evolutionary comparisons it will become apparent that this process has been spread widely over evolutionary time. However, the actual mechanisms remain unclear, and there has been a tendency for the exact meaning of the term to become obscured. By strict definition, based on the original observations in yeast, gene conversion is the *non*reciprocal transfer of sequence information from one DNA duplex to another (*307–310*). Confusion has often arisen between this phenomenon and unequal crossing-over. Figure 16 shows a comparison of the two, demonstrating that in gene conversion the donor final sequence remains unaltered, whereas in double unequal crossing-over the same final sequence can be created in one duplex, but a reciprocal change occurs in the other. Obviously, one cannot distinguish between these two possibilities unless all the products of recombination can be assessed; this has not been the case in globin system analyses thus far. Thus,

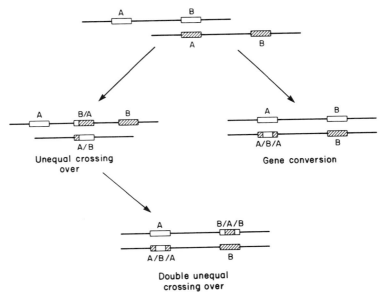

FIG. 16. Schematic comparison of the results of unequal crossing-over and gene conversion. A pair of homologous genes *A* and *B* are arranged in tandem. Alignment of *A* with *B* followed by a crossover results in two products, one of which contains three genes and other one. A second unequal crossover between the *B/A* and *A/B* genes can then result in the re-creation of a two-gene arrangement, but with part of one gene replaced by its homolog. In gene conversion, on the other hand, the event is nonreciprocal. In this example, there is a direct copying of part of gene B into the middle of gene A residing on the other DNA duplex. Note that the resulting lower haplotype is identical for a double unequal crossover or for gene conversion, so these mechanisms can be distinguished only if both products of the recombination event can be ascertained, which is difficult if not impossible to achieve in higher eukaryotes.

suggestions of gene conversion in mammals at present rest on inferential data. As we shall see, however, some observations can be explained otherwise only by double unequal crossing-over at extremely narrow intervals, which is unlikely.

It is possible to ascertain all the products of a given meiosis in yeast because the resulting spores are arranged sequentially in an ascus, allowing the genotype of each spore to be ascertained. Such studies have shown that intrachromosomal gene conversion between repeated adjacent genes may occur at a frequency as high as 4% in meiosis (311), although this is probably reduced by about two orders of magnitude in a mitotic division (312). Gene conversion between distant similar genes has also been demonstrated, but at a reduced frequency (313). The mechanism for such a nonreciprocal transfer of information is not yet completely clarified. The Meselson–Radding model (314) proposes that the initial event is a single-stranded nick in one DNA strand followed by its displacement as repair occurs. That single strand invades a homologous region of another DNA duplex, resulting in the displacement of a D loop. This D loop is then excised, the single-strand connection between the two duplexes is resolved in one of several possible ways, and the heteroduplex region of the recipient DNA is either resolved by replication or by mismatch repair. An alternative model (315), the so-called double-strand break-repair model, proposes the actual loss of double-stranded DNA in one of the two homologs, which is then filled in and replaced by information from the donor gene. Each of these models has strong points and weaknesses, which have been discussed elsewhere (315) and are not further elaborated here.

The best-studied example of gene conversion in the globin system, and indeed one of the earliest indications that this mechanism must operate in higher eukaryotes, is that involving the duplicated fetal globin genes (43, 44, 316). Homology between the flanking regions of the $^G\gamma$ and $^A\gamma$ globin genes shows approximately 14% differences, which suggests a duplication of the 5-kb region containing the fetal gene approximately 35 million years ago. However, sequencing of two different $^A\gamma$ genes showed that while one ($^A\gamma$-A in Fig. 17) displays approximately the divergence from $^G\gamma$ expected for this time of duplication, the other ($^A\gamma$-B) shows a sharply bordered region extending from approximately 500 bp 5' to the cap site to the middle of the second intervening sequence, within which interval the homology between the $^G\gamma$ and $^A\gamma$ genes of this chromosome is about 99%. This $^A\gamma$ gene is therefore much more similar to the $^G\gamma$ gene on the same chromosome than it was to an allelic $^A\gamma$ gene. This strongly suggests the

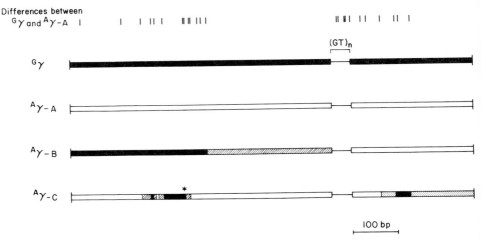

FIG. 17. Different gene conversion events in $^A\gamma$ IVS-2. Schematic depiction of IVS-2 from the different human fetal genes that have been sequenced. The sequence of the $^G\gamma$ gene from Smithies' clone 165.24 and the $^A\gamma$ gene from clone 51.1 (43, 44), which appear not to have undergone recent conversion, are represented as filled and open bars, respectively. Sites of differences between these genes, presumably representing divergence since the fetal gene duplication, are marked at the top of the figure. The $(G-T)_n$ region is slightly different in sequence between all four genes (43).

The $^A\gamma$-B gene is $^A\gamma$ from Smithies' clone 165.24 (44). The 5' portion of IVS-2 from this gene matches $^G\gamma$, whereas the 3' end matches $^A\gamma$-A, suggesting that a gene conversion has altered the 5' portion. The hatched bar represents the uncertainty of the boundary of the postulated conversion, which may include $(G-T)_n$ but cannot be further narrowed down because of the absence of differences between $^G\gamma$ and $^A\gamma$-A in this region.

The $^A\gamma$-C gene, sequenced by Stoeckert et al. (317) is more complex. There appear to be three separate regions where the $^G\gamma$ sequence has been imposed, separated by unconverted "$^A\gamma$" regions. The 5' most converted region cannot be more than 28 bp long, although there is some chance that this difference from $^A\gamma$-A might represent the occurrence of a "private" polymorphism in $^A\gamma$-A that makes it unrepresentative of the true ancestor of $^A\gamma$-C. The second area of conversion is unequivocal, however, and must be less than 76 bp in length. The asterisk marks a base change not seen in the potential donor ($^G\gamma$), suggesting the possibility of error-prone mismatch repair at the time of the resolution of the heteroduplex between $^G\gamma$ and $^A\gamma$, postulated to be intermediate in the process of conversion (see text).

possibility of a recent gene conversion of the $^A\gamma$-B gene. Analysis of the boundaries of this postulated conversion discovers no unusual sequence at the 5' end, but the 3' end is in the region of a repeated dinucleotide of the sequence $(T-G)_{9-18}(C-G)_{3-5}(T-G)_{8-10}$.

The sequence $d(T-G)_n$, for which the opposite strand is $(C-A)_n$, has attracted attention because of its tendency to form left-handed

Z-DNA structures when exposed to supercoil strain, perhaps even at physiological ionic strength (*318–321a*). This sequence or variations on it is sometimes flanked by direct repeats in the genome, and it has been found in immunoglobulin genes, a C-*myc* gene, near actin genes, in H-2 genes, and in the region of telomeres (*318*). Other examples of such repeats in globin are found in one of the nonhomologous regions at the 5' flanking side of the human α genes, between δ and β (*322*), and in the rabbit genes, where $(G-T)_n$ is found 5' to β4 and in β3 intron 2 (*323*). Interestingly, these specific repeats in rabbit are not found in the human analogs, consistent with the suggestion that they are prone to recombination or deletion. In the human fetal globin gene situation, one might speculate that the unusual physical properties of such a repeated dinucleotide create the possibility of single-stranded DNA stretches, perhaps in the borders between left-handed and right-handed DNA. The single-stranded regions might be prone to nicking, initiating the process of gene conversion as described in the Meselson–Radding model; or such regions might prevent further migration of branch points, halting a conversion at such a sequence.

We have recently sequenced an interesting $^A\gamma$ gene from a globin cosmid bearing the sickle (HbS) mutation (*317*). In this $^A\gamma$ gene ($^A\gamma$-C in Fig. 17) the large intron contains three regions that appear to represent gene conversion by the $^G\gamma$ gene from the same chromosome, and interspersed between these are *un*converted short regions. There is also at least one mutation not present in either the potential donor or the potential recipient. This might represent sequential gene conversion of the $^A\gamma$-C gene with different boundaries. Alternatively, a single conversion event involving the formation of a heteroduplex between the entire IVS-2 of $^G\gamma$ and $^A\gamma$-A might have been responsible; repair of the mismatches in such a heteroduplex might skip from strand to strand, resulting in the patchwork result observed (*309*). The new mutation observed (marked with an asterisk in the figure) might represent an error occurring in the process of mismatch repair.

Smithies and co-workers have recently studied a chromosome where both fetal globin genes produce an $^A\gamma$ chain (*119*). Interestingly, the only abnormality seen in the 5' γ gene from this chromosome is a single-base change at codon 136 that results in an amino-acid change from glycine to alanine; none of the surrounding regions of this 5' γ gene have been converted by the presumed $^A\gamma$ donor. This suggests that the region of conversion must have been quite short, in this example less than 213 bp. A similar result was found for a $^G\gamma$-$^G\gamma$ chromosome.

It has been proposed (*324*) that some δ-globin variants may have arisen by a conversion event with the β gene acting as a donor. This hypothesis could be tested by sequencing informative variant δ genes to see whether the mutant sequence represents a point mutation or a short conversion.

In another gene family, a particularly convincing example of such a short gene conversion has been described in the mouse H-2 system. A spontaneous mutant of the K^b gene that involves the change of three amino acids in a short stretch of the protein has been described (325). Cloning and sequencing of this mutant gene has revealed seven different base changes in 13 residues; a potential donor sequence was found in a different class I H-2 gene (326, 327, 327a). This event cannot, therefore, be explained by simple point mutation, but seems likely to represent a short (less than 20-bp) gene conversion of the H-$2K^b$ gene by a homologous class I gene. Similar evidence for conversion of repeated genes has been described in the mouse immunoglobulin system, where the γ2a gene has apparently been converted by the γ2b gene just 5′ to it in multiple small segments (328).

We have already alluded to the evidence for at least three gene conversions in the α-globin locus. The close homology between the ψζ and ζ genes suggests that a recent conversion has homogenized their sequences (47). The increased homology between the ψα and α genes relative to their flanking DNA suggests that a gene conversion may have occurred between these elements approximately 60 million years ago (see Section IV,A). The close correspondence between the Z-blocks containing the α2 and α1 genes, which is considerably greater than the homology of flanking sequences, is further evidence for a very recent conversion event (236, 365) or rapid cycles of unequal crossing-over (329) between α2 and α1. Finally, the 36-bp repeated element between ψζ and ζ mentioned above shows a pattern of base changes that is nonrandom and consistent with the suggestion of multiple short conversions between copies of the repeat (242).

As we shall see in Section V, evolutionary analyses show that gene conversion appears to be a widespread process in the globin gene systems of various vertebrates. It is somewhat puzzling that conversions often seem to be limited to the genes themselves, especially the coding regions; 5′ or 3′ flanking regions affected by the conversion are usually relatively short, and introns may be left untouched. This may reflect the fact that the coding regions are under selection and thus likely to be more homologous between duplicated genes some time after divergence than the flanking and intron regions. The loss of homology at these boundaries might be sufficient to halt branch migration and terminate a conversion. Alternatively, mRNA might act as an intermediate of this process, although this would not account for conversion of promoter regions, as has occurred with the human δ gene, unless an alternate cap site is used.

Further understanding of this phenomenon in higher eukaryotes may be forthcoming based on gene-transfer experiments such as those in which two tandemly arranged selectable thymidine kinase (TK) genes, each with a different mutation, are transfected into TK$^-$ mouse cells and HAT selection is applied (*330*). TK$^+$ segregants occur at a frequency of 10^{-4} to 10^{-5}, with at least 50% of the recombinants attributable to nonreciprocal exchange. This sort of analysis, though limited to mitotic conversions, allows determination of the actual frequency of such events, the sequence specificity of the borders of conversion, and the dependence on the distance between the pair of repeated genes. In some ways the effects of gene conversion are paradoxical: it has the potential both to prevent divergence of repeated genes and to increase the heterogeneity of a repeated gene family by allowing multiple transfers of small genetic segments within the members of that family. This may well be a powerful means of generating diversity in an evolutionary sense.

V. Evolutionary Considerations

A. Early Dispersion of the Globin Gene Family

Globins are found in insects and all extant families of vertebrates. Curiously, a protein like hemoglobin (leghemoglobin) occurs in the root nodules of leguminous plants and is encoded in the plant genome (*41, 53*). Recent analyses of leghemoglobin genes confirm the homology of these proteins to animal hemoglobins, but suggest that they have a relatively remote evolutionary relationship to vertebrate hemoglobin genes. The earliest forms of life to have evolved hemoglobin proteins are not known, but their occurrence in plants does not absolutely indicate that they had already evolved in some form of life that was a precursor to both plants and animals. Aside from remarkable convergent evolution, the hypothetical possibility remains that the hemoglobin gene was transferred from an insect or other form of life to the plant. This possibility might be better evaluated when various invertebrate hemoglobin genes are sequenced.

Myoglobin and hemoglobin are similar in structure and function and presumably evolved by gene duplication and specialization at some early stage in evolution. Since all jawed fish appeared to have both myoglobin and hemoglobin genes, the duplication preceded the appearance of the gnathostomes (*331, 332*). Comparisons of partial amino-acid sequences from myoglobin and hemoglobin of lower fish such as the lamprey has shown stronger homologies than are seen between the corresponding proteins of higher animals, suggesting that the duplication may have occurred just prior to the evolution of the jawed fish, or even that an independent and more recent duplication event gave rise to myoglobin in lamprey (332). Gnathostomes have both α and β chains in their hemoglobin, again indicating that a second gene-duplication producing α and β chains preceded the common evolutionary precursor for the jawed fish and might well represent an adaptation to the larger body size and/or greater activity of more modern vertebrates.

The frog is the lowest vertebrate whose globin genes have been isolated and characterized. Frog α- and β-globin genes are closely linked to one another in the chromosome (40). In *Xenopus laevis* there appear to be at least two clusters of globin genes (Fig. 1). The duplicated clusters may have arisen by tetraploidization (332a). In the direction of transcription, there are two larval α genes, one adult α-chain gene, an adult β-chain gene, and then two larval β-chain genes in the better studied cluster. There is probably a similar arrangement in the second cluster. All genes in a cluster appear to be oriented in the same direction relative to transcription.

In birds and mammals the α- and β-chain genes are found on separate chromosomes (Fig. 1). Presumably, the linkage of α- and β-globin in the frog represents the earlier arrangement. Separation of the α- and β-globin genes could have arisen either by loss of α genes from one cluster and β from the other, or by some chromosome break point occurring between the adult α- and β-globin genes.

In mammals, the arrangements of globin genes for both the α and β clusters are such that in the direction of transcription embryonic genes are 5′ and adult genes are 3′ (Fig. 1). In chicken, the same general arrangement holds except that there is a second embryonic gene 3′ to the adult β-globin gene (333–335). It is difficult to see how these gene arrangements could have arisen economically from the arrangement of the clusters that is seen in the frog. The gene arrangements would be consistent with models in which repeated events of gene duplication and change in regulation of the genes had occurred. In this case the larval genes of the tadpole might have no direct descendants in mammals. These comparisons support the concept of the fluidity of the globin gene family, in contrast to the constancy of the structure of the individual gene.

The consistent occurrence of a pseudogene between fetal and adult β-globin genes is curious and could represent either a consequence of the location of important positive controlling sequences at the 5' and 3' boundaries of the globin gene cluster or an advantage of unknown nature that is provided by a wide separation of differently regulated functional genes within the same family (342). As we shall see in Section V,B, the pseudogenes may have different evolutionary origins in different mammalian species.

B. Mammalian Comparisons

In this section we review a very interesting body of evidence being generated from detailed comparisons of globin gene sequences of various mammalian species. Our main emphasis in this review is on the human hemoglobin system; but a correct analysis of human globin gene origins is possible only by comparison with other mammalian species. It is not our purpose to provide a complete overview of the methods used in establishing evolutionary relationships, for that topic would be far too ambitious for a review such as this. However, it is appropriate to make a few comments about the methods being used and their limitations.

A pair of genes to be compared may be either orthologous or paralogous (336). Orthologous genes are those in different species that are derived from the same ancestral gene; paralogous genes are those within a species that arose by gene duplication from a common ancestor. If two genes being analyzed are relatively closely related, the number of differences between them will be small, and sophisticated methods may not be necessary to discern homology and count differences. However, for more distant relationships between noncoding regions more sophisticated analysis is necessary to discern and quantify homologies. A very useful technique is that of generating dot matrices (337, 338). Two sequences to be compared are analyzed by computer for homologous segments. The user is able to alter the stringency of the comparison so as to pick up important homologies without having the comparison destroyed by too much random matching. The computer generates a two-dimensional graph of the comparison with one sequence plotted along each axis. Areas of homology then appear as diagonal lines. One of the beauties of this method is that insertions or deletions within one of the sequences do not destroy the homology, but simply result in a shifting of the location of the diagonal line. The dot-matrix method has been enormously helpful in defining genes that are orthologous but distantly related.

Since much of the initial sequence data from various species was of the coding regions of genes, and selection pressures reduce the rate of divergence of coding regions so that distant relationships are easier to discern, considerable efforts have been devoted to developing quantitative methods of converting coding-region sequence comparisons between orthologous or paralogous genes into an estimated time of divergence. The concept that the proportion of the differing sites between two genes is directly proportional to their time of divergence, commonly known as the evolutionary clock hypothesis, has both strong proponents and strong detractors (336, 339). The first comparisons of globin gene sequences suggested that the rate of divergence of "replacement-site" nucleotides, which result in amino acid changes, is about 1% in 10 million years; the rate for silent-site mutations is considerably more rapid, about 1% every 1.4 million years (63, 340). These are useful approximate guidelines and give divergence times for the β-globin gene that are consistent with the fossil record. However, there are many assumptions implicit in such a model, some of which are not strictly valid.

The rate of replacement-site changes is the product of the mutation rate of an organism times the probability of fixation of a mutation and, hence, its persistence. Both of the terms in this product can vary. There is a suggestion that the mutation rate in man has decreased owing to a longer generation time or to better DNA repair mechanisms (341). The probability of fixing a mutation depends somewhat on the flexibility of the given protein. For example, fibrinogen apparently tolerates amino acid replacements quite well and has a high fixation rate, whereas a highly conserved protein such as cytochrome c tolerates very few replacements. Hemoglobin is roughly in an intermediate range. Even within the globin system, there is a detectable difference in the accumulation of replacement substitutions between the various genes, the epsilon-derived genes in general having the slowest accumulation of replacement sites (136). This probably reflects the quite different environments in which embryonic and adult globins must function.

The divergence of silent sites also seems to follow a molecular clock, but somewhat less well. Whether such sites are truly silent is open to question; it is interesting that in the goat, which has a triplicated globin locus, the silent-site differences between the $\psi\beta^X$ and $\psi\beta^Z$ genes are about four times greater than the silent-site differences between the β^A and the β^C genes, although the time since divergence of these pairs should be the same (27). This suggests that there are constraints on silent sites in functional genes that do not apply in pseudogenes, so that silent sites of a coding region may actually diverge at a rate intermediate between coding sites and intergenic DNA. This selection pressure on silent sites might be attributable to the availability of various tRNAs as well as the need to avoid creating or disrupting splice junctions or unrecognized control signals.

When the evolutionary-clock method is used to compute the duplication time of paralogous genes based on their coding regions, there is a danger that gene conversion, which as we shall see below is widespread in evolutionary time, will lead to a calculation of the most recent conversion rather than the actual time of duplication. For example, this sort of analysis was used to suggest that the δ and β genes of man arose by duplication only 40 million years ago (63). More recent information, taking into account noncoding and flanking regions, makes it clear that the δ gene is much more ancient than this but has recently been converted (323, 342–343a).

A means of avoiding the possibility of variation in the rate at which the evolutionary clock runs is the "maximum parsimony method" of constructing evolutionary trees (332, 343b). This method makes no assumptions about rates of mutation. Basically, the relationship between several species is determined by constructing a tree that gives the least number of independent mutations connecting the various branches of the tree. If sufficient sequence information is available, it may be possible to select one tree relating a set of organisms that is consistent with the observations and to reject other possible trees as statistically unlikely. A detailed phylogenetic tree using this approach has been constructed for both α- and β-like chains (4, 332). However, this analysis is also subject to the objection that gene conversions will alter the conclusions. The problem with all these methods is that no reference gene-bank of mammalian ancestors is available, but that is a problem for which no solution is likely to be found.

Given this background, we now propose to summarize a possible history of the β-globin system, beginning with a single ancestral β gene about 500 million years (My) ago (Fig. 18). Duplication of the ancestral β gene into a two-gene cluster, the 5' member of which is the ancestor of the ε and γ genes and the 3' member of which is the ancestor of the δ and β genes, presumably occurred around 200 My ago (323). The ε and γ genes separated about 120 My ago and the δ and β genes at about the same time, giving rise to an ancestral four-gene system that appears to be the β-globin complex precursor in all mammalian species. Following the suggestion of Hardies et al. (342), these genes are denoted as proto-ε, proto-γ, proto-δ, and proto-β in Fig. 18. The time of divergence between the δ and β genes is thus considerably earlier than that based on the coding-region similarity of the δ and β genes, which placed their divergence at 40 My ago (63). More recent analyses of the 5' flanking region, the large intervening

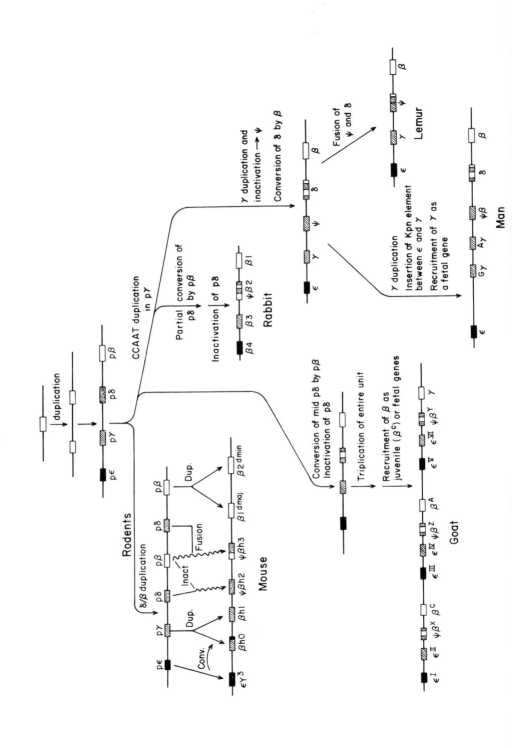

sequence, the silent site nucleotides in exon 3, and the 3' flanking region, have shown that 40 My ago actually is the time of a gene conversion event that converted the 5' part of the δ gene to the large intervening sequence (*323*, *342*). The large intervening sequence and 3' flanking regions thus remain as presumably unaltered descendants of the proto-δ gene. It has also been suggested that the 3' end of the proto-ε gene was converted by the corresponding region of the β gene at about the time of the mammalian radiation (*323*).

The rather complex series of events that apparently has led to the present mouse globin complex is depicted in Fig. 18 (see Addendum 2). The divergence of rodents prior to the general mammalian radiation 85 My ago shown in the figure is intentional and is supported by maximum parsimony analysis (*332*, *342*). A duplication of the δ and β genes in tandem occurred. The descendant of the proto-γ gene went on to duplicate and the 3' portion of one proto-γ gene was converted by the embryonic gene to give rise to the present-day βh0. Its closely related paralog βh1 did not undergo a conversion. The ψβh2 gene is suggested by dot-matrix analysis to be a descendant of proto-δ, but has developed a number of inactivating mutations. The ψβh3 gene, also inactive, contains sequences in its 5' end that are β-like, but has sequences at the 3' end that are δ-like, suggesting that it is a Lepore-like fusion product of an unequal crossing-over, as shown in the figure. Finally, the present-day $\beta 1^{dmaj}$ and $\beta 2^{dmin}$ genes appear to be the result of a relatively recent duplication of the adult β-globin gene. These relationships, which previously have been obscure, have been care-

FIG. 18. Evolution of the β-globin cluster from an ancestral β-like gene in mammals. This is a summation of results obtained by homology comparisons in the mouse (*342*), goat (*25*, *27*, *27a*, *136*, *342*), rabbit (*323*), and primates (*32*, *343*, *342*). Undoubtedly, this represents an oversimplification of the complex series of events responsible for the present-day clusters. We have followed the convention of Hardies et al. (*342*) and designated the ancestral four genes as proto-ε (filled bar), proto-γ (hatched bar), proto-δ (stippled bar), and proto-β (open bar). Descendants of the ancestral genes can be followed by the shading used. Note that the blocks are intended to include the 5' and 3' flanking regions, not just the transcriptional unit, in this figure. Thus the human δ gene in its present form has a 5' flanking region extending upstream from −80, which is proto-δ like, but the segment from the CCAAT box to the 5' end of IVS-2 has been converted by β, as has the 3' flanking region. Exon 3 remains proto-δ like. Not shown is a possible conversion of the 3' untranslated region of proto-ε by proto-β prior to the mammalian radiation (*323*).

The scheme shown for the goat may be an oversimplification, as ε^{II} is only slightly more related to proto-γ than proto-ε sequences (*136*). We have depicted the 3' cluster of goat genes containing ε^{V}, ε^{VI}, $\psi\beta^{Y}$, and γ as it would be expected to appear, based on the apparent triplication of the locus, but this region has not yet been unequivocally mapped (see Fig. 1). Further details are discussed in Section V,B.

fully dissected by a dot-matrix approach and separate analyses of the 5' flanking regions, large intervening sequences, and 3' flanking region in order to account for events such as gene conversion and unequal crossing-over (342).

The globin region in the goat also has an interesting evolutionary history. In its present form, the cluster appears to consist of a triplicated four-gene system, although complete coverage of the locus with overlapping clones has not yet been achieved. Each of the four geneclusters consists of two embryonic genes, a pseudogene, and a β-like gene that may be fetal in its expression (γ), juvenile (β^C), or adult (β^A). The $\psi\beta^X$ and $\psi\beta^Z$ genes have been subjected to careful dot-matrix analysis; the 5' flanking region of these genes is δ-like, but the intervening sequence is β-like, suggesting that a conversion event of proto-δ and proto-β occurred prior to triplication of the unit (25, 27, 27a, 136, 342, 344). The observation that β^A and β^C are more related to each other than either is to the γ gene suggests that the initial duplication gave rise to a β cluster and a γ cluster, and the β cluster reduplicated to give rise to a β^A and β^C cluster. It is apparent from consideration of this model that the gene that functions as a fetal gene in goats is directly derived from the proto-β gene, which is in marked contrast to the situation in man. This may in fact be a good example of convergent evolution, where different genes have been independently recruited for the same purpose in different organisms.

The close evolutionary relationship of the goat γ gene to the goat β^A gene naturally leads to a consideration of sequence differences that might account for differential expression. Comparison of the goat γ and goat β^A promoter regions from the cap site (Fig. 6) reveals only 6 point differences (25). Aligning this region with other globin genes does lead to one possibly significant observation. The four nucleotides just 5' to the ATAAAA box are usually GGGN for the β-globin genes, which are expressed in adult life (see Fig. 6). This includes the human β, rabbit β1, mouse β^{min}, goat β^A, and chicken β. However, genes expressed in fetal or embryonic life seem to have the sequence AAGA in the same location; this applies to human γ, rabbit β3, human ε, goat ε^I, and mouse βh0, βh1, and εY^3. Exceptions are rabbit β4, which has d-(GGGA) in this location, and goat ε^{II}, which has AGGA, though both are embryonic in function. Interestingly, however, 2 of the 6 point mutations in the goat γ gene relative to the goat β^A gene convert this sequence from GGGC to AAGC, which might have significance for control of gene expression.

Another interesting feature of the goat γ gene is the presence of a

247-bp insertion in the large intron that is flanked by a 13-bp direct repeat (25). This insertion contains a degenerated inverted repeat bearing some resemblance to the long-terminal-repeat (LTR) of Moloney murine leukemia virus. It is tempting to speculate that the insertion of this element, which is found to be repetitive, has altered the differential expression of the γ gene. A third hypothesis for the change in expression would be that the most 3' part of the β-globin locus abuts a chromosomal domain that is active during fetal life but becomes closed after birth. Hampered as we currently are by the absence of a system in which hemoglobin switching can be demonstrated with cloned genes, there is no good way to distinguish these possibilities at present.

Because the CCAAT box in the descendant of the proto-γ gene is duplicated in rabbit and in man, it seems likely that this duplication occurred during the mammalian radiation but after the divergence of goats, which previous molecular analyses suggest are not as closely related to man as are rabbits (332). The rabbit β system is relatively simple (323). The $\beta 4$ and $\beta 3$ genes, being the descendants of the proto-ε and proto-γ genes, respectively, both function as embryonic genes. It is noteworthy that the $\beta 3$ gene is actually expressed slightly earlier than the $\beta 4$ gene; similarly, in the mouse, the $\beta h0$ and $\beta h1$ genes are expressed prior to the εY^3 gene, so that it seems likely that the proto-γ gene functioned as an early embryonic gene (60), and the proto-ε gene as a slightly later embryonic gene, in the ancestor of all mammals. Thus, the rule of β-globin genes being arranged from the 5' to 3' direction in order of their activation, while true in man, seems not to be true in many other mammals. This observation and the intriguing fact that the goat adult-gene is actually in the middle of a cluster with the fetal and juvenile forms at either end makes any theory of hemoglobin switching that requires a processive 5' to 3' movement of open chromatin domains less than generally applicable.

The descendant of the proto-δ in rabbits, $\psi\beta 2$, has undergone a gene conversion, although a different one than in goat or man (343a). Why the descendants of the proto-δ are so susceptible to conversions and inactivations is unknown. This may reflect the fact that a second β-like gene is dispensable, so that alterations of this gene are better tolerated than others. However, this would not explain why it is always the descendants of the proto-δ gene, not the proto-β gene, that are affected. In rabbit, the 5' end of the $\psi\beta 2$ gene was apparently converted by $\beta 1$, with subsequent inactivation mutations, which include the loss of the CCAAT box (Fig. 18). The consistent location of pseudogenes toward the middle of the globin cluster has led to the suggestion (342) that the cluster is surrounded on both ends by *cis*-acting control elements that operate most effectively on nearby genes.

The primate evolutionary tree contains further surprises. In man, there is a $\psi\beta$ gene between the fetal and δ genes. Comparison of the 5′ flanking region of this pseudogene shows that it is most closely homologous with the γ gene, as is the coding region. However, the homology is quite weak, and a maximum parsimony analysis places the divergence of the $\psi\beta$ from the γ gene at approximately the same time as the γ-ε duplication (345). This is puzzling, since no descendant of $\psi\beta 1$ is found in rabbits or goats. The mapping of the globin cluster of the brown lemur, a prosimian (32, 134) reveals a simple four-gene cluster as depicted in Fig. 18, and the pseudogene between γ and β has been cloned and sequenced (134). The analysis shows that the lemur pseudogene is a Lepore-like fusion gene with its 5′ portion, up to the second intervening sequence, homologous to the $\psi\beta 1$ gene in man, whereas the 3′ portion resembles the human δ gene.

After divergence of the prosimians (represented by lemur), a duplication of the γ gene occurred approximately 35 My ago to give rise to the $^G\gamma$ and $^A\gamma$ genes. Of particular note is the acquisition of 5 to 6 kb of DNA between the embryonic and fetal genes in man relative to the positions of these genes in the lemur. In humans, this region is known to contain a representative of the *Kpn* repetitive element and, at the 3′ end of this, a sequence somewhat resembling a retrovirus LTR (290). It is likely that these elements were inserted into the genome after the divergence of prosimians (32). It is therefore of particular note that the γ gene in lemurs may *not* yet have been recruited as a fetal gene, since hemoglobin analyses of newborn and adult lemurs are identical (346, 347). It is possible that the recruitment of the proto-γ gene to function as a fetal gene coincided with the insertion of the *Kpn* repetitive element and pseudo-LTR between ε and γ. In this regard, the occurrence of an insertional element in the goat γ gene, which also contains an LTR-like sequence (25), is a fascinating observation. It will be important to study other primates such as New-World monkeys to see whether insertion of this *Kpn* element corresponds in an evolutionary sense with the recruitment of the γ gene to produce a fetal hemoglobin, or whether this represents a coincidence.

There is at least one puzzling detail in the history outlined above. As noted, the observation of a duplicated CCAAT box in both the $\beta 3$ gene of rabbits and the fetal gene of man suggests that this event occurred early. One would have expected the $\psi\beta$ gene of man to have also a duplicated CCAAT box because of its origins from the proto-γ gene. Superficially this is so, as the $\psi\beta$ gene has an almost perfect 38-bp repeat containing the CCAAT box (Fig. 6). However, its very perfectness is puzzling in view of the antiquity of the CCAAT box duplication and the divergence of the tandemly repeated copies in the rabbit β3 and the human γ genes, which suggest that most of this sequence is not subject to strong selection. Why then would the repeat have been so perfectly preserved in a pseudogene? To add to the confusion, the lemur pseudogene that, as we said, has its 5′ portion derived from the $\psi\beta$1 gene, does *not* have a duplicated CCAAT box. This last observation leads to the suggestion that the duplicated box was lost in the $\psi\beta$1 gene shortly after it arose. The unequal crossover that gave rise to the lemur pseudogene thus would result in a gene without a duplicated CCAAT. More recently, a reduplication in the human $\psi\beta$ gene may have occurred to give rise to the almost perfect repeat now seen. This hypothesis might suggest that something unusual about this particular promoter invites segmental duplication.

A final example of evolutionary analysis that provides an opportunity to follow gene silencing and reactivation in more recent times is the study of the δ gene in primates. Its current form arose approximately 40 My ago by a conversion of the promoter and the coding region of the gene by β, which spared IVS-2 (Fig. 18). Old-World monkeys, which diverged approximately 15–20 My ago, do not have δ transcripts in marrow (343); and cloning and sequencing of the δ globin gene from two representatives, the rhesus and colobus monkeys, shows that they share three nucleotide substitutions in the promoter region relative to man. Presumably one or a combination of these was responsible for resilencing of the gene in the ancestor of Old-World monkeys. This analysis is a beautiful example of how study of relatively closely related species can allow dissection of molecular events leading to inactivation of a gene.

Clearly, more information is needed on other species to understand more details of the evolution of the β-globin cluster. Though indirect, this sort of analysis may be quite useful in ascertaining how differential expression of these genes occurs. Just as the current wave of careful analysis of flanking sequences and the recognition of the possibility of gene conversion have led to major revisions in the proposed scheme of gene derivations, other surprises that will force revisions in Fig. 18 may be waiting in the wings. Thus the scheme presented here should be considered at present an educated guess.

VI. Naturally Occurring Mutations

A. General Comments

The contribution of mutations in the hemoglobin loci to morbidity and mortality in man is considerable, and the postulated basis for the advantage of the adult heterozygote for many of these conditions in combating malarial infections has presumably allowed for the selection and persistence of a large number of such mutations in the α- and β-globin loci. At the protein level, such mutations have been detectable for more than 30 years, through standard techniques of hemoglobin electrophoresis and peptide mapping. A large number of variant hemoglobins have been described and catalogued and have provided important insights into the normal function of the hemoglobin tetramer. Because of space limitations, we will not attempt to deal with any of these variant hemoglobins in spite of their enormous clinical and biological significance. The interested reader is referred to several comprehensive recent resource materials (4, 3, 348, 349).

However, because of their bearing on our understanding of gene expression, we will consider mutations that alter gene structure in ways other than simple amino-acid substitutions. Many of these conditions result in thalassemia; we will also have occasion to review a related condition known as hereditary persistence of fetal hemoglobin, an interesting and important example of altered control of gene expression. However, the separation of mutations into those causing amino-acid substitutions and those directly affecting gene expression can at times be somewhat blurred; in hemoglobin E, for example, a single base-substitution in the β-globin gene results in both an amino-acid change at the site of substitution and the creation of an alternate splice site that results in a quantitative reduction of mRNA (350).

B. The Thalassemias

In its broadest sense thalassemia can be defined as an imbalance of the level of production of α- and β-globin chains. The clinical and hematologic features vary greatly depending on the specific genetic abnormality, and the reader is referred to several excellent recent review articles (10, 351–357) and an authoritative textbook (2) that treat this topic in more detail. In addition, a recent thorough review of the α-thalassemias is available (358). Our goal here is to summarize the known mutations, particularly as they affect our understanding of normal gene expression.

1. α-Thalassemia

For many years the α-thalassemias represented a confusing group of disorders characterized by an extensive near-continuum of clinical severity, as well as the occurrence of many pedigrees difficult to analyze satisfactorily by standard Mendelian methods (2). However, with the advent of genomic blotting techniques, it became possible to look directly at the genotype, and the complexity of phenotypes became a resolvable problem. In denoting the genotype, it is useful to use the notation ab/cd to denote the function of the various α genes, where a and b are the α genes on one chromosome and c and d are the α genes on the other chromosome of the same individual (358). An individual with all four genes intact is denoted as αα/αα. An individual with one α gene inactivated or reduced in function has been previously given the designation α-thalassemia 2 and is denoted as -α/αα; such an individual has minimal hematologic abnormalities and is denoted as a "silent carrier." On the other hand, α-thalassemia 1 involves loss of two functioning α genes and results in mild anemia. Such an individual could be either -α/-α, the form usually seen in blacks (359–361) or, if both nonfunctional α genes are on the same chromosome, --/αα. If three of the four α genes are nonfunctional, the designation is --/-α, and the individual exhibits a moderate anemia and red cell inclusions that consist of tetramers of β-globin known as hemoglobin H (2). Such individuals are said to have hemoglobin H disease, which commonly results from the inheritance of an α-thalassemia-1 chromosome from one parent and an α-thalassemia-2 chromosome from the other. The most severe form of α-thalassemia is that in which there are *no* functional α genes; this gives rise to hydropic infants who are stillborn or die shortly after birth and who have hemoglobin Barts (a tetramer of γ-globin) as their primary hemoglobin. Since γ_4 has extremely high oxygen affinity, it can deliver almost no oxygen to the tissues, resulting in fetal hypoxia, hydrops, and death.

In α-thalassemia, most of the mutations so far described are deletions of relatively large (greater than 1 kb) segments of DNA, in contrast to the β-thalassemias, where most abnormalities are point mutations or frameshifts. Presumably the basis for this difference is the long duplication of the DNA embedding the α gene, with the resulting potential for unequal crossing-over. Illegitimate recombinations have also occurred in the α locus. Figure 19 shows the deletions that have been described. There are two deletions that result in the -α/ haplotype. The -$α^{3.7}$/ deletion (363–365, 370), also known as the "rightward deletion," seems to have arisen by an unequal crossing-

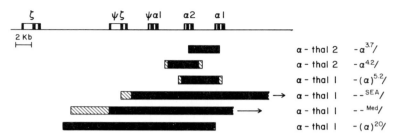

FIG. 19. Deletion forms of α-thalassemia. The α-globin locus and the six known deletions that result in α-thalassemia are shown. Those that leave one functioning α gene cause the phenotype of α-thalassemia 2, whereas those that destroy function of both α genes are denoted as α-thalassemia 1. The filled in areas represent portions known to be deleted, whereas the hatched regions represent uncertainties in the end points of the deletion in those mutations not yet sequenced. In those deletions where the total amount of DNA removed has been characterized, this appears as a superscript in the haplotype designation. For example, $-\alpha^{3.7}/$ represents a haplotype in which one α gene remains functioning and a total deletion of 3.7 kb has occurred. The Southeast Asian (SEA) and one variety of Mediterranean (Med) α-thalassemia 1 have yet to have their 3' end-points characterized, as indicated by the arrows in the figure. The deletion denoted as $-(\alpha)^{20}/$ in the figure was originally reported to involve a deletion of greater than 25 kb, but more recent data have led to a revision of this conclusion (362). References in which these deletions are described are: $-\alpha^{3.7}/$ (363–365); $-\alpha^{4.2}/$ (363); $-(\alpha)^{5.2}/$ (366); $--^{SEA}/$ (367, 368); $--^{Med}/$ (367); $-(\alpha)^{20}/$ (369).

over between misaligned α2 and α1 genes, with the crossover occurring somewhere in the so-called Z segment (see Fig. 12). The observation that this deletion has been found associated with several different polymorphic backgrounds (based on the variable length regions shown in Fig. 12) makes it likely that such a deletion has multiple origins (242, 358).

Sequence analysis of a Chinese patient with $-\alpha^{3.7}/$ (365) shows that the region just 5' to the Z block is clearly derived from α2, while the region from IVS-2 to the end is from α1, so presumably the site of crossing-over lies between these boundaries. Unique restriction sites in IVS-2 and the 3' untranslated portion of the gene, which distinguish α2 and α1, show that this 3' portion of the $-\alpha^{3.7}/$ gene is α1-like in several different ethnic groups in which this gene occurs. In the Chinese patient where the complete sequence was obtained, however, there is a surprising "patchwork" arrangement of the residues 5' to the α gene that distinguish α2 and α1; this cannot be accounted for by a simple crossover and suggests the possibility that the unequal crossover was associated with the formation of heteroduplex DNA and mismatch repair (see Section IV,C).

The other α-thal-2 is denoted -$α^{4.2}$/ because a 4.2-kb segment is deleted (363, 370). This is also known as the "leftward deletion" and most likely has arisen by unequal crossing-over between the X segments shown in Fig. 12. Because the insertions between the X and Z segments of the α2 copy are larger than in the α1 copy, this deletion, although arising similarly by unequal crossing-over, involves a slightly larger amount of the genome. These deletions can therefore be distinguished by blotting techniques. Crossovers in the Y segment are not known, but they would be difficult to distinguish from -$α^{3.7}$/ by blotting. Individuals with triplicated α genes, representing the other product of unequal crossing-over, are also known for both "rightward" and "leftward" crossovers (371–375a). Single or triplicated ζ genes, most likely arising by the same mechanism, have also been described (376).

Deletions leading to loss of function of both α genes on a given chromosome (α-thal-1), on the other hand, are more difficult to explain and apparently represent illegitimate recombination. Three such deletions have been described in the Mediterranean population (366, 367, 369) and the fourth in Southeast Asia (367, 368). As shown in Fig. 19 the 3' end-points of two of these deletions have not yet been precisely determined. The 5' end-point of one of the Mediterranean forms was initially reported to extend beyond the ζ gene in the 5' direction (367), but this was based on a restriction map of a clone containing the deletion end-points, and more recent data suggest that the actual end-point is between the ζ and ψζ genes, such that the total amount of deleted material is about 20 kb (362). We therefore denote this lesion -$(α)^{20}$/.

Before leaving the deletion types of α-thalassemia, it is appropriate to mention three patients who had the interesting combination of hemoglobin H disease and mental retardation (377, 378). In each case, the mother of the affected individual was found to have α-thalassemia 1 (-α/αα) and the father was genotypically normal (αα/αα). However, the proband in each case had the typical phenotype of hemoglobin H. In one instance, genomic blotting studies taking advantage of appropriate polymorphisms demonstrated that the proband had not inherited any α or ζ genes from his father, suggesting a *de novo* deletion of the entire α-globin locus. However, no karyotypic deletion was visible, even by prometaphase banding. A second patient with a similar phenotype and pedigree had solid evidence of intact α and ζ genes inherited from the father, but apparently both the paternal α genes were nonfunctional. This is a particularly interesting observation, as it is puzzling that a single nondeletion mutation seems to have inactivated both the α2 and α1 genes. It is tempting to speculate that the actual mutation in this patient might be not in the *cis*-acting regulatory factors, but in the *trans* controlling elements, with the result that the three intact α genes had an effective output of only one gene. It is obviously important to study this particular patient further. Weatherall *et al.* (*377, 378*) raise the interesting question whether the mental retardation and mild dysmorphic features seen in these children occur on the basis of deletion of other nearby genes on chromosome 16, or whether the loss of one ζ gene, which was the case in the first patient, results in embryonic hypoxia sufficient to interfere with normal development.

Nondeletion forms of α-thalassemia have also been described, although these are relatively few by comparison with the β-thalassemias. The only mutation, so far described, that alters a splice site is a 5-bp deletion that completely removes the donor splice site of the first intervening sequence of α2 and results in the activation of a cryptic splice site in the first exon as shown in Fig. 9 (*179, 379*). Two nondeletion mutations resulting in α-thalassemia have recently been uncovered in Saudi Arabian individuals (*380*). It has been known for some time that an unusual haplotype must be present in Saudis; when this is present in the homozygous state, the phenotype is hemoglobin H disease, suggesting that the α-globin production from four α genes is roughly equivalent to that from one normal gene (*358*). Molecular cloning and sequencing has revealed that the α1 gene from such individuals is completely nonproductive, owing to a frameshift mutation at codon 14. The α2 gene from the same chromosome contains a base change in the polyadenylation signal, altering the normal from AATAAA to AATAAG. Although some normal polyadenylated message is made in spite of this mutation, there is apparently inefficient production of mature messenger RNA such that the output from this α2 gene is reduced by about 50%.

Other nondeletion α-thalassemias appear to result from posttranscriptional defects and are not here dealt with extensively. A good example is hemoglobin Constant Spring, due to a point mutation of the translation termination codon and resulting in the appearance of an extra 31 amino-acid residues (*380a*). The thalassemic phenotype in Constant Spring apparently results from instability of the mRNA rather than from a reduced rate of transcription. Other α-thalassemia phenotypes, such as hemoglobin Quong Sze (125 Leu → Pro), are due to an amino-acid substitution that results in instability of the globin chain itself (*380b*).

Presumably the high frequency of α-thalassemia in some parts of the world, particularly Southeast Asia, relates to its heterozygote selective advantage. An alternative hypothesis is that α-thalassemia, by its beneficial influence on moderate to severe β-thalassemia (381–383) or sickle-cell disease (384, 384a), might allow the survival and reproductive fitness of such individuals in a population with a high rate of heterozygosity for these β-thalassemia mutations, and thus is coselected. This seems an unlikely explanation, however, as most homozygous β-thalassemics do not have their disease sufficiently moderated by concurrent α-thalassemia to achieve normal reproductive levels.

2. β-THALASSEMIA

The β-thalassemias are a heterogeneous group of disorders characterized by quantitative deficiency of β chains relative to α chains in erythroid cells. The result of this is precipitation of the excess α chains, leading to red-cell damage and ineffective erythropoiesis (2). Homozygotes for β-thalassemia are divided into β^0-thalassemia, in which there is virtually no detectable synthesis of β chains, and β^+-thalassemia, in which β chains are present but reduced. Heterozygotes for the β-thalassemias are generally clinically unaffected although they do have red-cell morphological abnormalities. Because of the great molecular diversity and etiology of the β-thalassemias, homozygotes for this condition who are not from inbred areas frequently turn out to be compound heterozygotes with two different β-gene mutations.

Homozygous β-thalassemia in its more severe form requires periodic transfusion therapy. However, the concomitant iron overloading, unless managed with an effective chelation program, leads to a cardiomyopathy that is usually progressive and fatal. Because of this serious prognosis and the progressive improvements in bone-marrow transplantation therapy, treatment of some children with β-thalassemia with marrow transplants from compatible donors is being investigated as a possible means of curing the disease (385).

Although it has been less than five years since the first mutation causing β-thalassemia in man was described in molecular terms, the description of such mutations has proceeded at a truly astonishing rate. In Table II no fewer than 29 mutations resulting in β-thalassemia are listed. This includes all the mutations described up to the time of writing this article except for a deletion type of β-thalassemia seen in Indians that involves the loss of 619 nucleotides and is therefore considered in Section VI,C (386). Figure 20 shows in diagrammatic form the location of these mutations in the β-globin gene, divided into

TABLE II
Mutations Causing β-Thalassemia

Mutation	Type	Ethnic origin	Reference
I. Chain terminator mutation			
A. Nonsense mutations			
Codon 15	β^0	Indian	387b
Codon 17	β^0	Chinese	388
Codon 39	β^0	Mediterranean	389–392
B. Frameshift mutations			
Codon 6 (−1 bp)	β^0	Mediterranean	393, 393a
Codon 8 (−2 bp)	β^0	Turkish	389
Codon 8 (+1 bp)	β^0	Indian	387b
Codon 16 (−1 bp)	β^0	Indian	387b
Codon 41/42 (−4 bp)	β^0	Indian and Taiwan	387b, 394
Codon 44 (−1 bp)	β^0	Kurdish	395
II. Defective promoter			
A. Distal element			
−87 C→G (ACAC*G*C)	β^+	Mediterranean	162, 161
−88 C→T (ACA*T*CC)	β^+	Black	395b
B. ATA Box			
−29 A→G (CAT*G*AAA)	β^+	Black	387a
−28 A→C (CATA*C*AA)	β^+	Kurdish	396
−28 A→G (CATA*G*AA)	β^+	Chinese	397
III. Defective RNA processing			
A. Splice junction alterations			
1. Donor Site:			
IVS-1 GT→AT	β^0	Mediterranean	162, 161
IVS-1 GT→TT	β^0	Indian	387b
IVS-2 GT→AT	β^0	Mediterranean	161, 180
IVS-1 position 5 G→C	β^+	Indian	161, 387b
IVS-1 position 6 T→C	β^+	Mediterranean	162, 151
2. Acceptor Site:			
IVS-1 25 bp 3′ end	β^0	Indian	183
IVS-2 AG→GG	β^0	Black	387a
B. Creation of new splice signal in IVS			
1. New acceptor			
IVS-1 position 110 G→A	β^+	Mediterranean	398, 399
2. New donor			
IVS-2 position 654 C→T	β^0	Chinese	387
IVS-2 position 705 T→G	β^0	Mediterranean	400, 181
IVS-2 position 745 C→G	β^+	Mediterranean	162, 161
C. Enhanced activity of cryptic splice site in exon			
1. Exon-1 cryptic donor			
Codon 24 T→A (silent)	β^+	Black	401
Codon 26 G→A (Glu→Lys)	β^E	S. E. Asian	350
Codon 27 G→T (Ala→Ser)	$\beta^{Knossos}$	Mediterranean	387, 387c
D. Poly(A) addition signal			
AATAAA→AACAAA	β^+	Black	395a

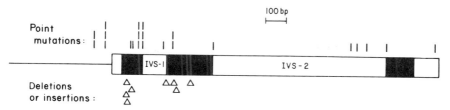

FIG. 20. Schematic representation of mutations known to cause β-thalassemia (as of October, 1983). The mutations described in Table II have been divided into those that are point mutations and those that represent small deletions or insertions, and the positions of these mutations are plotted on a diagram of the β gene. Open regions of the gene represent untranslated sequences, whereas filled-in areas are coding regions. The clustering of mutations around the first intervening sequence is apparent.

point mutations and deletions/insertions. The majority of the mutations occur in the 5' half of the gene, and there is a particular clustering around the 5' splice junction of the first intervening sequence. Table II divides the various mutations by the kind of alteration they impose upon the β-globin gene responsible for the thalassemic phenotype. Many of these mutations are self-evident, and several others have been mentioned in the section on normal splicing (Section III,D), since these mutations have contributed much to our understanding of the normal consensus sequence for donor and acceptor sites. A mouse model of β-thalassemia in which the $β^{maj}$ gene is deleted has been described (401a).

In assessing the mechanism of the reduction in β-globin synthesis of these mutations, it is appropriate that we describe the assay systems used to study transcription from a cloned thalassemia gene. Some of this information has already been presented in Section II. Ideally, one would like to study mRNA from the bone marrow of an affected individual in order to draw conclusions about mechanism. However, most patients will not be homozygotes for a given mutation, and marrow is not always available from the few that are. Therefore, cultured cell systems have provided a very necessary link for our understanding of the mechanism of these mutations. One possibility is to use SV40 virions in which a portion of the SV40 genome has been replaced by the β gene under study (99). These virions are then used to infect cell lines along with a helper virus to supply the necessary but missing SV40 functions, and the mRNA produced is analyzed. This method was successful in demonstrating the splicing abnormality in the most common form of $β^+$ thalassemia in Mediterranean patients, which is a point mutation in the first intervening sequence that creates an alternate acceptor site preferred to the normal site (*402*). However, there are constraints on the size of the gene that can be introduced by this method.

Other studies, referred to as transient expression experiments, have used plasmid vectors containing the necessary replication origin and antibiotic resistance genes for their growth in bacteria, the gene whose expression is to be studied, and usually an SV40 origin of replication and enhancer sequences, so that the plasmid is efficiently expressed in a eukaryotic cell recipient (*161, 351, 403*). Use of this vector system is accomplished by introducing the resultant plasmid into eukaryotic cells through standard calcium phosphate precipitation methods, followed by purification and analysis of the mRNA produced after a 30- to 48-hour period. The recipient cells used have most often been monkey kidney cells containing a stably incorporated SV40 genome (COS cells) (*158*) or the human carcinoma line called HeLa (*155, 161, 404*). For quantitation of the amount of message produced from the introduced β gene, an internal standard, often a normal α gene, is cotransfected.

Quantitation of mRNA can be done by "Northern blotting," although a complementary approach uses S1 nuclease digestion to give both quantitative and qualitative information about mRNA (*351, 404a*). In this approach, a single-stranded labeled-DNA probe representing the "nonsense" strand of a genomic clone of the gene of interest is annealed to mRNA and the resulting DNA · RNA hybrid is then subjected to gentle S1 nuclease digestion (*180*). Since the DNA probe will anneal to mRNA only in the regions that encode exons, electrophoresis of the resulting fragments followed by autoradiography allows determination of where the globin gene was spliced as well as quantitation of the amount, based on the intensity of the bands that appear. Another useful and also complementary technique is to anneal the mRNA with a single-stranded DNA primer of defined sequence, and extend the primer with reverse transcriptase (*180*). The length of the resulting extension product will depend on the precise site of transcription initiation as well as the splicing pattern that has been followed. It is then possible to purify the primer extension product by gel electrophoresis and perform a standard Maxam–Gilbert sequence-analysis to demonstrate the exact nucleotide sequence of the message corresponding to this band. With this repertory of methods, determination of the molecular pathology of the defects in most β-thalassemia mutations has been possible (*180, 356*).

Most of the mutations in Table II are self-explanatory, but there are a number of interesting features that we have not previously touched on, and that deserve emphasis. The first β mutation described causing β^0 thalassemia results from an A to T mutation at codon 17. This converts the normal lysine codon to a premature terminator. Kan and co-workers (405) constructed a human suppressor tRNA gene from a lysine tRNA by mutation of the anticodon. Using a *Xenopus* oocyte system, they demonstrated synthesis of normal β-globin chains if both the β^0-thalassemia gene and the suppressor tRNA were injected. The normal termination codon for the β-globin gene is UAA, so this was not interfered with.

In the past, the relative amounts of α- and β-globin mRNA in erythroid cells have been estimated for several point mutations that result in accumulation of low levels of a mutant globin chain. Reduced levels of mutant mRNA were found, and it was assumed that this was a consequence of cytoplasmic degradation of the mRNA. Weatherall and colleagues (380a) have examined this problem for the hemoglobin Constant Spring α-chain mutant using a sensitive assay for α- and β-globin mRNA. (Constant Spring is one of four mutants in the termination codon for the α chain, all of which lead to very low levels of mutant globin chain production.) These studies showed a more nearly normal $\alpha : \beta$ mRNA ratio in nuclei of marrow cells than in reticulocyte cytoplasm, suggesting the deficient mRNA was a consequence of a posttranscriptional event.

Initial studies in β^+-thalassemia suggested that a more complex situation might exist. Pulse-labeling studies showed a normal $\beta : \alpha$ RNA ratio in nuclei of marrow cells, but even after relatively brief labeling the ratio in the cytoplasm of marrow cells was reduced to levels more nearly resembling those of accumulated reticulocyte RNA. This suggests that most of the degradation of abnormally spliced RNA occurs in the nucleus, or in conjunction with export to the cytoplasm (402).

Recent observations in codon-39 β^0-thalassemia are relevant to this question (406, 407). The relative deficiency of β mRNA in mutations associated with premature termination of translation has usually been attributed to instability of mRNA not covered by polyribosomes on its 3' end. Using the $\beta^0 39$ gene in a plasmid expression vector system transfected into monkey kidney (COS) cells, the surprising finding was that both cytoplasmic and nuclear mRNA were reduced by 5- to 10-fold relative to a normal control. Using an actinomycin-D chase, the small amount of cytoplasmic message did *not* appear to have an increased turnover rate, but was as stable as a normal control. Furthermore, use of a cotransfected tyrosine suppressor tRNA restored both cytoplasmic and nuclear mRNA levels. These observations are not consistent with previous postulations of cytoplasmic mRNA instability, and they suggest some sort of coupling of the splicing and translation processes, perhaps at the nuclear membrane. It remains somewhat puzzling why a previous analysis of β^039 message in HeLa cells did not detect a quantitative difference between normal and β^039 message (390). Intranuclear incorporation of uridine in the marrow of a patient with the β^039 mutation was decreased, however, suggesting that nuclear message *is* reduced in amount *in vivo* in this mutation.

The promoter mutations listed in Table II also provide important insight about the function of the various promoter elements in expression systems that have been described. The finding of a β-thalassemia with a base change in the ACACCCC sequence (161, 162), which has been defined (see Section III,C) in rabbit β-globin as essential for efficient transcription (153, 408), is a convincing argument for the importance of this element *in vivo*; by functional assay a β-globin gene with this base change at -87 produces about 10% of the normal amount of message. The three mutations that have been described in the ATA box provide further evidence of its importance. The mutation at -29 results in about 25% of the normal amount of message (387), and the A to G mutation at -28 results in about 20% of normal. The A to C mutation at -28 (396) is being analyzed. Thus far no mutations in the CCAAT box have been defined, but it is likely that such will be found.

The mutations that result in defective RNA splicing represent slightly more than half of all the mutations so far described, and many of these have already been touched on (Section III,D). A particularly intriguing trio of mutations are those at IVS-2 positions 654, 705, and 745, all of which create sequences looking very much like the consensus sequence for a splice junction donor (see Fig. 8). All these mutations result in activation of a cryptic acceptor splice site at position 579 of the IVS, which matches the consensus sequence closely (Fig. 9). The position 745 mutation (161, 162) still allows some normally spliced message to be produced, but no normal message is seen in the 654 (387) or 705 (181, 400) mutations. In the position 654 mutation, all of the message produced has a splice from the normal 5' donor site to position 579 and a splice from position 654 to the normal 3' acceptor site, so that a region of 75 nucleotides from IVS-2 appears in the mature message.

The position-705 mutation is particularly intriguing; by S1-nuclease analysis no normal transcripts are formed. One group of transcripts corresponds to the same situation seen in the position 654 mutation where an extra exon between position 579 and 705 is incorporated in the mature message. The second class of transcript splices from the normal donor site to position 579 but contains no further splices 3' to this, so that the entire sequence from 579 to the poly(A) signal is represented in processed message. In this kind of message, therefore, the presence of a potential donor site at 705 results in the use of a cryptic splice site at 579, even though the potential splice site created by the mutation is not used. This observation has led to the postulate that a competitive binding situation might apply (*181*). In such a model, the potential splice donor at 705 and the normal acceptor site might be reversibly bound by the RNA processing apparatus. With these sites unavailable, the normal donor site would then be forced to splice into position 579. Then, if the reversibly bound complex between 705 and the normal acceptor came apart, the observed mRNA would result. It is clear that this particular mutation cannot be explained by a 5' to 3' scanning model of splicing.

Undoubtedly the table of thalassemic mutations presented will need continuing updating over the next few years, though it is doubtful whether the roster of mutations causing β-thalassemia will ever rival in length the list of variant hemoglobins, since mechanistically it would seem that there are fewer possible ways to reduce the amount of β message than there are to alter the protein for which it codes.

Before leaving this section, it is also appropriate to note that a few patients with low or absent levels of hemoglobin A_2 have been found; β-globin production is normal, and no detectable deletion of the β-globin complex has been apparent by genomic blotting (*409–411*). These individuals are classed as having δ^+ thalassemia if some A_2 is detectable, or δ^0-thalassemia if not (*410*). Mutations responsible for these phenotypes have not yet been identified. A very puzzling observation arose in the cloning of the δ gene from a homozygote for δ^0 thalassemia and the determination of its complete sequence (*411*). An interesting series of polymorphisms that differed between Caucasian (*45*) and Japanese individuals was detected, but there was no difference in sequence between the normal Japanese δ gene and that from the δ^0-thalassemia patient in the region from 300 bases 5' to the cap site to 130 nucleotides 3' to the poly(A) signal. The authors suggested that a more distally placed control region might be affected, though this would be quite startling. The only other example of a normal gene that fails to function because of a distant alteration is the $\gamma\delta\beta$-thalassemia (*412*) described in the following section; that patient had an enormous deletion extending within 2 kb 5' from the β gene, so that it is somewhat easier to conceptualize an alteration in the chromatin domain being responsible for failure of β-gene expression. The Japanese patient, on the other hand, had no such deletion detectable by genomic blotting or restriction analysis of cloned DNA. Use of this δ gene in an expression vector would be an important step to be certain that some sequence difference was not overlooked.

C. Deletions in the β-Globin Complex

Even prior to the advent of recombinant DNA techniques, deletions in the β-globin cluster were suspected (2). In particular, the observation that Lepore hemoglobin consisted of the amino terminus of δ and the carboxyl terminus of β suggested that an unequal crossing-over had occurred between adjacent δ and β genes. A similar argument was formulated to explain the occurrence of hemoglobin Kenya, which is a fusion hemoglobin of $^A\gamma$ and β. These assumptions led to the correct postulation of the order of the γ, δ, and β genes on the chromosome and have subsequently been borne out by studies of genomic DNA. With the advent of genomic blotting, it became possible to characterize many more such deletions. In some of these the end-points of the deletion have been cloned and characterized by DNA sequencing.

At least 15 such large deletions in the β complex are now known and are shown in Fig. 21. These range in size from a 619-bp deletion causing a form of $β^0$-thalassemia in Indians (386, 433) to a γδβ-thalassemia deletion that removes the entire β-globin cluster and involves at least 105 kb of contiguous DNA (431). Although many of these deletions are quite large, in none of them has a visible chromosomal deletion been reported on the short arm of chromosome 11. It is also worthy of note that homozygotes or double heterozygotes for several of the large deletions of at least 60 kb 3′ to the β gene have no extrahematopoietic phenotypic effects. This suggests that no other genes that are crucial for normal development or function reside in this 3′ area.

In analysis of the phenotype of these various deletions, it is useful to draw a distinction between the phenotype of hereditary persistence of fetal hemoglobin (HPFH) and that of δβ-thalassemia (2, 434). Hereditary persistence of fetal hemoglobin is a benign condition characterized by an increased synthesis of fetal globin in the adult, usually accompanied by decreased or absent synthesis of β-globin from the same chromosome. The total output of globin chains from the affected chromosome in such an individual is sufficient, however, to balance or nearly balance α-chain production, so that no clinical effects result. In δβ-thalassemia, on the other hand, there is decreased or absent syn-

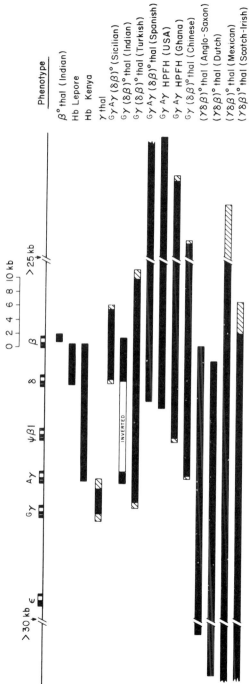

FIG. 21. Summary of deletions in the β-globin complex. This figure summarizes all the known large deletions in the β-globin complex. The phenotype and, in most instances, the ethnic origin in the patient in which the deletion was first described, are also shown. The designation (δβ)⁰thal indicates that no δ and β chains are made cis to the mutation; (γδβ)⁰thal mutations display no synthesis of any of the β-globin cluster genes cis to the mutation. The ᴳγ and ᴬγ designations in the phenotype indicate which fetal globin is made; for example, in ᶜγ(δβ)⁰thal (Chinese), ᶜγ chains are made, but ᴬγ, δ, and β are not. Hemoglobin Lepore has a thalassemic phenotype, whereas hemoglobin Kenya has the phenotype of hereditary persistence of fetal hemoglobin (HPFH). The broken line 3' to β is used to indicate that greater than 25 kb of DNA has been omitted from the diagram at this point; just how much sequence resides in this interval remains to be determined. Similarly, the broken lines 5' to ε are used to show that more than 30 kb of DNA separates the region in which the 5' end-points of the Anglo-Saxon and Dutch (γδβ)⁰thal are found from the β-globin cluster. Note that the total amount of deleted DNA in the ᴳγᴬγ HPFH (USA), ᴳγᴬγ HPFH (Ghana), and ᶜγ(δβ)⁰thal (Chinese) is very nearly the same; the 3' end-point of the ᴳγᴬγ(δβ)⁰thal (Spanish) has not yet been mapped. Also, note that the Anglo-Saxon and Dutch (γδβ)⁰thal mutations are of approximately the same length, but with the location of the deletion shifted by a few kilobases. References describing these mutations are: β⁰thal (Indian) (386); Hb Lepore (413, 414, 416–417a); Hb Kenya (418); γ thal (419); ᴳγᴬγ(δβ)⁰thal (Sicilian) (420a); ᶜγ(δβ)⁰thal (Indian) (421); ᶜγ(δβ)⁰thal (Turkish) (422); ᶜγ(δβ)⁰thal (Spanish) (423, 424); ᶜγᴬγ HPFH (USA) (425–427); ᶜγᴬγ HPFH (Ghana) (425, 426); ᶜγ(δβ)⁰thal (Chinese) (428); (γδβ)⁰thal (Anglo-Saxon) (429); (γδβ)⁰thal (Dutch) (412, 415, 430); (γδβ)⁰thal (Mexican) (431); and (γδβ)⁰thal (Scotch-Irish) (432) (see Addendum 2).

thesis of both δ and β from the affected chromosome, and the increase in fetal globin synthesis, although usually present, is insufficient to prevent the development of a thalassemia phenotype. A further characterization of these conditions that is often useful is whether fetal hemoglobin is present in all red cells in the peripheral blood (pancellular), or is seen in only a proportion of these cells (heterocellular). In δβ-thalassemia the pattern is invariably heterocellular, while both heterocellular and pancellular HPFH types are known. In many instances, however, the pancellular distribution seems to be associated with the highest levels of hemoglobin F synthesis, and it has been suggested that this may be a distinction that is more apparent than real (2, 434).

As one can see from Fig. 21, the deletions that lead to δβ-thalassemia are quite variable in size. Those characterized by synthesis of fetal hemoglobin containing both $^G\gamma$ and $^A\gamma$ genes are denoted $^G\gamma^A\gamma(\delta\beta)^0$thal, whereas those in which only $^G\gamma$ chains are made are denoted as $^G\gamma(\delta\beta)^0$thal. It is important to note that hemoglobin Lepore has a thalassemia phenotype, and that the amount of fetal hemoglobin synthesized in this condition is actually somewhat less (1–5% in heterozygotes) than in the δβ-thalassemias (5–15%) (434), a fact that has so far eluded any adequate explanation. Most of the deletions resulting in δβ-thalassemia are simple deletions, but a complex rearrangement in Indians is characterized by both a deletion and an inversion (421).

Fewer deletions resulting in HPFH have been described (425). In the most common type, seen in American blacks (U.S. HPFH), a deletion extends from an end-point in between the ψβ gene and the δ gene at least 60 kb in the 3' direction. The deletion end-points in this mutant have been cloned; the 5' end of the deletion begins *precisely* in the middle of an *Alu* sequence, but the sequence left after the deletion is not *Alu*-like and therefore this represents an "illegitimate" recombination (427). Another type of HPFH deletion seen in Ghana has a 5' end-point beginning about 5 kb upstream from that of the U.S. form; interestingly, the 3' end of this deletion is offset by almost exactly the same amount relative to the U.S. HPFH (415, 426). Hemoglobin Kenya is also properly included in this category. As mentioned above, this is a deletion presumably resulting from an unequal crossover between the $^A\gamma$ and β genes, resulting in a fusion hemoglobin produced in relatively large amounts in the adult as is the unaffected $^G\gamma$ gene, so that the phenotype is that of HPFH (418).

While the Mexican (431) and Scotch-Irish (432) γδβ-thalassemias are characterized by large deletions that completely remove all of the β-globin complex, the Anglo-Saxon γδβ-thal (*429*) has its 3' end-point within the β gene, and the Dutch form (*412*) actually ends 2 kb 5' to the β gene, but this gene is not expressed *cis* to the mutation. This very interesting observation has led to the cloning of the β gene from this chromosome to see if it contains some second mutation that might be responsible for the failure of expression (*430*). Sequencing of most of the β gene has shown it to be normal, and when this gene is put into a transient expression system, the gene functions normally. *In vivo*, the β gene on the affected chromosome is present in an inactive configuration, as shown by DNase sensitivity and methylation analysis (*430*).

It is worthy of note that the U.S. HPFH, Ghana HPFH, and Chinese $^G\gamma$ δβ-thal are characterized by deletions greater than 60 kb, but the total length of the deleted DNA must be almost the same in all three mutations (*415, 426*). In Dutch and Anglo-Saxon γδβ-thal, the total length of the deletion is also almost the same, although the actual location of the deletion is offset by a few kilobases (*415*). This correspondence is not quite exact, but its occurrence seems to be more than a coincidence. A possible explanation would be that the DNA strand is anchored to the nuclear matrix at certain points spaced a fixed distance apart. The DNA might be capable of reeling through these fixed points while still maintaining a fixed distance between them. If breakage occurred, an entire loop might be lost; this would result in deletions of a fixed or nearly fixed total amount of DNA, but displaced from one another depending on the actual location of the fixed matrix points at the time of the deletion (*415*). Whether this in fact is the mechanism for such deletions remains conjecture, although it is an attractive hypothesis. There is some evidence for such fixed points in the *Drosophila* nucleus (*435*).

The existence of deletion mutations that result in HPFH in some instances and in δβ-thalassemia in others has given rise to several hypotheses regarding the control sequences necessary for normal hemoglobin switching. Certainly there is good evidence that such *cis*-acting regulatory sequences must exist. One such observation is the fact that heterozygotes for $^G\gamma^A\gamma$ HPFH who are also heterozygous for a variant fetal hemoglobin on the opposite chromosome do not synthesize increased amounts of the fetal hemoglobin in *trans*, but only in *cis* (*436*).

Further evidence for *cis*-acting elements can be inferred from the Dutch (γδβ)^{0}thal; heterozygotes for this condition synthesize β-globin normally from their normal chromosome but fail to synthesize any from the affected chromosome in spite of the fact that it contains an intact β gene (*430*). Thus *cis*-acting factors must be important for normal β expression, since any *trans* effect should be satisfactorily present in a heterozygote who has a normal β-globin complex on the opposite chromosome.

At least three hypotheses regarding such *cis*-acting elements have been proposed. As long ago as 1974, it was proposed that a regulatory sequence between γ and δ might be responsible for shutting off fetal hemoglobin synthesis at about the time of birth (*437*). According to this proposal, deletion of this regulatory sequence when associated with deletion of δ and β genes would result in HPFH, whereas deletions that removed δ and β but did not remove the regulatory sequence would result in δβ-thalassemia. Genomic blotting studies demonstrating the nature of the deletions in Sicilian δβ-thal and U.S. HPFH (*261, 425*) suggested that if such a control element exists it must lie within 5 kb of the 5' end of the δ gene. More recently, cloning of the deletion end-point of the U.S. HPFH and of the Spanish (δβ)⁰thal (*424*) shows that these end-points are only about 1 kb apart and occur in a region 5' to the δ gene in which two *Alu* repetitive sequences occur. The U.S. HPFH deletion ends at the exact midpoint of the 5' member of this pair of *Alus* (Fig. 22), whereas the Spanish (δβ)⁰thal ends at the upstream end of the 3' member. This then narrows down the area of such a possible control element to a 1-kb region. The DNA sequence of this region in normal individuals has been determined and seems to contain no particularly unusual sequences other than the *Alu* elements themselves. We have cloned the DNA in between the two *Alu* sequences and have determined that the sequence is single-copy. Sequencing of this region from several different individuals demonstrated an interesting dimorphism, in which there appear to be two alleles differing by 16 point-mutations and two deletions (*133, 263*). However, these alleles do not correlate with the amount of fetal hemoglobin synthesized, and their specific relationship to hemoglobin switching seems unlikely.

In addition, two other deletions shown in Fig. 21 do not agree with the predictions of this idea. The Turkish $^G\gamma(\delta\beta)^0$ thalassemia (*422*) and Chinese $^G\gamma(\delta\beta)^0$-thalassemia (*428*) both remove the putative control element but result in $^G\gamma(\delta\beta)^0$-thalassemia, not in HPFH. However, one could argue that the inactivation of the $^A\gamma$ gene in both of these conditions in some important way alters the ability of complete compensation by increased fetal globin production; in fact, in the Turkish $^G\gamma(\delta\beta)^0$thal, the $^G\gamma$ gene is 2–3 times more active than in Sicilian $^G\gamma(\delta\beta)^0$thal (*434*). In support of the control element theory is the observation that hemoglobin Lepore has a thalassemia phenotype, whereas hemoglobin Kenya, in which the putative control element is deleted, has an HPFH phenotype.

A second proposal (262) to explain the observed phenotypes is that there are two chromatin domains, one surrounding the fetal genes with distinct 5' and 3' borders, and another flanking the α and β genes with similarly distinct borders. In this proposal, fetal hemoglobin switching occurs when the fetal domain closes and the $\delta\beta$ adult domain opens. Mutations that alter the 5' border of the $\delta\beta$ domain still allow some switching to occur and result in a thalassemia phenotype, whereas mutations that affect both the 5' and 3' $\delta\beta$ domains result in persistent open expression of the fetal domain and HPFH. Mutations that delete the adult domain but also affect the 3' border of the fetal domain, such as the Turkish and Chinese $^G\gamma(\delta\beta)^0$thal, result in a thalassemia phenotype because of their alteration of the chromatin structure in the region of the fetal genes. Hemoglobin Kenya is explained in this proposal by the deletion of the 3' border of the fetal domain and the 5' border of the adult domain, resulting in a single domain that remains open throughout. If correct, this hypothesis predicts that the small region in between the end-points of the Spanish $\delta\beta$-thalassemia and the U.S. HPFH (Fig. 22) must be the 5' border of the adult chromatin domain, and in this sense this hypothesis is an extension of the 1974 proposal (437) described above. A somewhat related idea (438) involves switching the origin of DNA replication. All of these propositions fail to account for the nondeletion forms of HPFH (see below).

A third possible explanation (426) suggests that what is important is not so much the region of DNA deleted, but the sequences brought into apposition to the β-globin complex by the deletion. It is suggested that in U.S. and Ghana HPFH, sequences that act as *cis* enhancers are brought in from 3' to the β-globin complex, thus allowing expression of the fetal genes into adulthood. Similarly, this construction would propose that in Dutch ($\gamma\delta\beta$) thalassemia, the reason why the β gene is not expressed is that some sequence brought in from distantly 5' results in a closed chromatin configuration and inability to express the β gene. A possible counter example to this idea, however, is the Chinese $^G\gamma(\delta\beta)$ thalassemia. Since the total amount of deleted DNA is very similar in this condition to the U.S. and Ghana HPFH, the 3' sequences brought into apposition with the $^G\gamma$ gene will be at approximately the same distance from this gene as they are in the two types of deletion HPFH. Therefore, the differences in phenotypes do not appear to be explicable under the assumptions of this proposal, although it is worth noting that the total amount of deleted DNA is not quite the same, and the additional 3' sequence present in the Chinese

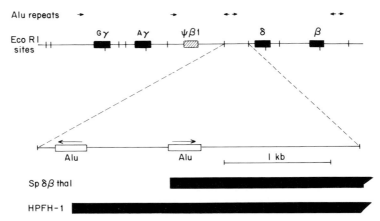

FIG. 22. Closer view of the region 5' to the δ gene possibly implicated as a controlling element in the fetal to adult globin switch. The upper part of the diagram depicts the β-globin cluster with the *Alu* repeats shown as arrows. The 3.1-kb *Eco*RI fragment 5' to δ is expanded in the lower part of the figure. The end-points of the Spanish $(\delta\beta)^0$ thalassemia and the $^G\gamma^A\gamma$ HPFH (U.S.), here denoted as HPFH-1, are shown. The 3' end-points of these deletions extend many kilobases 3' to the β gene (Fig. 21). The 5' end-points, however, are seen to be only about 1 kb apart. If a *cis*-acting controlling element that normally switches off fetal globin production at about the time of birth is located in this general area, these two deletions suggest that such an element should be located between their two end-points.

$^G\gamma(\delta\beta)$thal could be hypothesized to interfere with fetal globin gene activation.

At present, it is impossible to distinguish between these proposals, or even to state whether they are mutually exclusive. The possibility is quite real that more than one mechanism exists for *cis* regulation of fetal hemoglobin switching. It is interesting that no individuals have yet been described in whom there is marked overproduction of fetal hemoglobin from *both* chromosomes, as might be expected for a mutation in the *trans*-acting factors, which would presumably be recessively inherited. The failure to observe such mutations, in spite of the considerable evidence that *trans* effectors must exist, might lead one to believe that the pertinent *trans* effectors have functions in addition to fetal to adult globin switching, so that mutations in these are lethal.

There are also a number of interesting conditions in which an elevated synthesis of fetal hemoglobin occurs in adult life, but in which no detectable deletion is present by careful genomic blotting techniques. We refer to these conditions as nondeletion forms of HPFH, and they are conveniently separated into those in which the fetal hemoglobin is distributed pancellularly, and those in which the distribution is heterocellular, although as noted above this distinction can, at times, be blurred. The detailed hematologic characteristics of such patients have been thoroughly reviewed (434). The best-known forms of pancellular nondeletion HPFH are the so-called Greek and the $^G\gamma\beta^+$ HPFH. The Greek form (439), which is common on the island of Serifos but rare elsewhere, is characterized by the production of 10–20% hemoglobin F in heterozygotes; no homozygotes are known. The fetal hemoglobin made is approximately 90% $^A\gamma$, so that the $^G\gamma$ production is elevated minimally, if at all. There is some suggestion that β chains are also made in *cis* to the mutation, since individuals who are doubly heterozygous for Greek HPFH and β-thalassemia produce more hemoglobin A than would be expected from their thalassemia allele (440). Careful genomic blotting studies, using probes covering the entire β-globin complex, have failed to reveal any evidence of a significant deletion or rearrangement (262, 425, 441, 442), and cloning of the β-globin complex from two heterozygous individuals in this laboratory further confirms that no apparent rearrangements are present (443) (see Addendum 2). No large pedigrees have been studied to prove linkage between the HPFH phenotype and the β-globin complex.

Somewhat similar to the Greek HPFH is that in a recently described family (444), which differed in that the amount of fetal hemoglobin produced in a heterozygote was slightly less and not quite so pancellularly distributed. This form of HPFH, denoted as the Chinese variety, may represent an intermediate form between the Greek and the British (see below) HPFH.

Another interesting form of pancellular HPFH is the so-called $^G\gamma\beta^+$ variety, which is seen in American and African blacks. Again, there are no homozygotes known, but the heterozygotes produce 15–25% hemoglobin F (434). Double heterozygotes for this condition and hemoglobin S clearly demonstrate that there is β-chain synthesis *cis* to the HPFH mutation. Curiously, however, the amount of β production from the HPFH chromosome is somewhat reduced, so that the total output of non-α chains from this chromosome is approximately normal. Again, genomic blotting studies on several patients (444, 445) have failed to show any evidence of abnormalities, and we have confirmed this by cloning the β-globin complex from such a patient and analyzing it with multiple restriction enzymes (446). We have, however, identified a point mutation 202 bp 5' to the cap site of the $^G\gamma$ gene. This mutation occurs in a (G+C)-rich palindrome and creates a sequence that in some ways resembles elements of the herpes thymidine kinase (*142, 143*) and SV40 21-bp repeat (*139, 140, 447*), promoter elements known to be necessary for efficient *in vivo* transcription (Fig. 23). Recent studies of the expression of transfected insulin and chymotrypsin genes (*447a*) have uncovered cell-specific promoter sequences in the -160 to -300 region, and it is possible the -202 mutation up-regulates such an element in the $^G\gamma$ gene, resulting in its increased transcription into adult life. It is interesting that in this situation persistent expression of the $^G\gamma$ gene has apparently not affected the normal shutoff of $^A\gamma$.

Since the C→G mutation at -202 alters normal restriction sites at this locus, it has been possible to search for this mutation in genomic DNA. DNA from over 100 black individuals without HPFH has not shown the mutation (*448*), and 13 different individuals with $^G\gamma\beta^+$ HPFH not known to be related to out patient have the same restriction site alteration (*446*). Therefore this is either the mutation responsible for the phenotype or a closely linked relatively rare polymorphism. Distinguishing these possibilities will require expression systems capable of reproducing the phenotype.

Similar to the Greek form of HPFH in the production primarily of $^A\gamma$ chains, but differing in that the distribution is heterocellular and the amount of hemoglobin F in heterozygotes is somewhat less (4–13%), is a condition known as British HPFH (*121, 434, 441, 449*). Since known homozygotes have approximately 20% hemoglobin F and 80% hemoglobin (A+A$_2$), there must be β and δ synthesis in *cis*. A large pedigree study has shown tight linkage to the β-globin locus

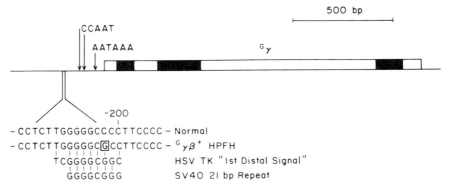

FIG. 23. Point mutation in the 5' flanking region of the $^G\gamma$ gene from a patient with $^G\gamma\beta^+$ HPFH. The C → G mutation at −202 relative to the $^G\gamma$ cap site, confirmed by genomic blotting, is shown. The sequence created has some similarities to promoter elements of the Herpes simplex thymidine kinase and SV40 early genes *in vivo* transcription (see Section V,C).

(*450*), but the mutation responsible for the phenotype remains at present unknown.

Finally, the inheritance of small amounts of hemoglobin F (1–4%) is usually denoted as the Swiss type of HPFH (*434*) and can, at times, present difficulties in diagnosis because of overlap with the upper range of the normal distribution. In two different families with this condition, evidence for recombination events between the locus causing Swiss HPFH and the β-globin locus has been presented (*450, 451*), suggesting that the mutation responsible for this phenotype may be at some distance from the β-globin complex (*451a*).

With the possible exception of the $^G\gamma\beta^+$ form, therefore, the nondeletion forms of HPFH remain unexplained, and presumably represent more subtle mutations of the hemoglobin switching apparatus (see Addendum 2). A nondeletion $\delta\beta$-thal has also been reported (*423*). Uncovering these mutations may have far-reaching implications for our understanding of this process and warrants the considerable efforts being expended in their pursuit.

Before leaving the section on deletions, it is also interesting to note the occurrence of a type of γ-thalassemia resulting from unequal crossing-over between the $^G\gamma$ and the $^A\gamma$ gene (*419*) leading to a 5-kb deletion (see Fig. 21). This is the fetal analog of the event that results in Lepore hemoglobin when the δ and β genes are involved. Just as there is an anti-Lepore from the other possible product of such an unequal δ-β crossing-over (*2, 452*), so there also appears to be a condition characterized by the presence of three fetal genes ($^G\gamma$-$^G\gamma$-$^A\gamma$), which may be relatively common in Vanuatu and appears to have no serious clinical consequences (*453*).

D. Polymorphisms without Known Functional Correlates

In this section we consider polymorphic differences in the β-globin cluster that appear not to have functional consequences but are highly interesting in their own right. In the α-globin cluster, few such point mutations responsible for loss or gain of a restriction site have been identified; the polymorphisms seen in the α cluster are instead primarily those of variable length, as previously discussed. In the β-globin cluster, however, many polymorphic restriction sites have now been described. A recent sequence comparison (*258*) suggests a minimum of 0.5% sequence variation in the $\delta\beta$ region. A variable-length polymorphism also occurs 1.3 kb 5' to the β gene where the sequence (ATTTT)$_n$ is found and n may be 4, 5, or 6 (*454, 455*). Variation in the copy number of this 5-bp repeat presumably could arise by slippage and unequal crossing-over. However, caution in interpreting these results is appropriate, as we have observed the same phenomena during propagation of subclones in *Escherichia coli*.

Polymorphic sites that alter a restriction site are highly useful for following the inheritance of a particular globin allele, whether on an evolutionary time scale or for purposes of prenatal diagnosis. Figure 24 is a summary of the polymorphic sites in the β-globin cluster that have been described in the literature since the first *Hpa*I polymorphic site was described in 1978 (462). Restriction sites drawn below the gene map have been intensively studied (162, 356, 458), whereas those sites drawn above the gene map have been reported as polymorphic in at least one situation, but less information about their prevalence and usefulness is available. References for these less-common sites can be found in the legend to Fig. 24. Although most of these sites were discovered by genomic blotting using the enzyme in question, at least three of them, notably the *Taq* site 5' to δ (133, 263), the *Hgi*A site at codon 2 of the β gene (162), and the *Hinf*I site 5' to β (455) were uncovered by noting sequence differences from different individuals.

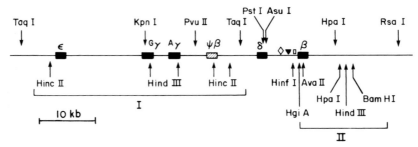

FIG. 24. Polymorphic sites thus far described in the β-globin cluster. The arrangement of the β-globin cluster is shown, and polymorphic restriction sites thus far described are marked with arrows. Sites below the line are those that have been intensively studied (162, 356, 426, 455–464). Sites within the group marked I are in linkage disequilibrium, as are sites within the group marked II, but there is random association between these two groups. The *Hinf*I site in between (455, 464) is not linked to either cluster. Three regions of DNA sequence that might have bearing on the linkage information are also shown: ◊ represents a $(TG)_n$ simple repeat; ▼ represents the sequence $(ATTTT)_n$, where n may be 4, 5, or 6 (454, 455); and □ shows the location of a 52-bp alternating purine and pyrimidine sequence that functions as a strong S1-nuclease hypersensitive site in supercoiled plasmids (465). The sites above the line have been described as polymorphic in some situations, but are less well studied. The primary references for description of these sites are *Taq*I and *Rsa*I at the extremes of the cluster (356), *Kpn*I (466), *Pvu*II (467, 468), *Taq*I between ψβ and δ (133, 263), *Pst*I (46), *Asu*I (469), and the relatively rare *Hpa*I site leading to the 7.0-kb β-globin fragment (462).

It is helpful to think of the combination of alleles at these polymorphic sites as defining a haplotype of the β-globin cluster for each chromosome of an individual. At each position marked with an arrow on the figure, the restriction site may be present (+) or absent (−), and the combination of all of these sites provides a rich background of markers inherited in a Mendelian fashion.

The collection of a large amount of data on the haplotypes in Mediterraneans, blacks, and Southeast Asians, has revealed some interesting findings. First, the alleles at these various polymorphic sites are not randomly associated (458). Considering the γ HindIII and $\psi\beta$ HincII sites in group I, for example, haplotypes − − − −, + + − +, + − + +, + + − −, and + − + − are found in Mediterraneans, but the other eleven possible combinations are seen rarely or not at all (162). Statistical tests confirm that the association of alleles at these restriction sites is very nonrandom; another way of stating this is to describe sites as being in linkage disequilibrium, as the frequencies of the haplotypes are significantly different from what would be expected by chance, given the frequencies of the alleles at the various sites.

The same nonrandom association of sites occurs in the group of restriction polymorphisms denoted by II in the figure. However, random association *does* occur between the clusters. A surprising finding is that the $Hinf$I site just 5′ to β is randomly associated with *both* cluster I and cluster II (356).

The fact that alleles at the polymorphic sites within the clusters have been maintained in linkage disequilibrium suggests that recombination within these areas has been relatively infrequent, whereas the random association between the two clusters suggests that crossing-over in this area is much more frequent. This cannot be simply accounted for by the distances involved, as the width of cluster I is greater than the distance between clusters I and II. Is there a "hot spot" for recombination around the δ gene or in the $\delta\beta$ intergenic area? The observation that the $Hinf$I site does not correlate with either cluster suggests that such may be present on both sides of the $Hinf$I site.

In this regard, it may be of particular significance that there are four examples where a mutant β-globin gene that normally resides on a specific haplotype has been found residing on the same haplotype in another individual, except that the $Hinf$I site is discordant (470). Possible explanations for this include a double crossover or independent origin of the mutation, but an alternate possibility would be that the region of the $Hinf$I site is particularly prone to interallelic gene conversion, which could alter the $Hinf$I site without affecting the rest of the haplotype (see Fig. 16). This hypothesis has the added attractiveness of providing a possible explanation for the high rate of recombination in this area, since in yeast, where gene conversion has been best studied, interallelic meiotic gene conversion is often associated with reciprocal crossover between flanking markers (308). As described in the section on gene conversion, this process is thought to be initiated by a single-stranded break, so that any DNA sequence that could contribute to a localized area of single-strandedness might induce such a phenomenon. In the fetal globin genes, for example, a region of $d(T-G)_n$ seems to have functioned as the border of a gene conversion event between $^G\gamma$ and $^A\gamma$ (43, 44). Interestingly, the sequence $(T-G)_{16}$ appears in the region between δ and β. Alternatively, the $(ATTTT)_n$ sequence mentioned above, which is 400 bases 5′ from the HinfI site, might be prone to localized single-strand formation both on the basis of slippage and A · T pairing weakness. Finally, there is a 52-bp stretch of alternating purine and pyrimidine residues 410 bp 3′ to the HinfI site that acts as a strong nuclease-S1-sensitive site in supercoiled plasmids (465). DNA sequences with this property often function also as hypersensitive regions in intact chromatin (see Section III,E), again raising the possibility of single-stranded breaks that could initiate gene conversion with or without recombination. Testing of these hypotheses might become feasible if other polymorphic markers in this region could be found (see Addendum 2).

The β genes themselves can be classified according to polymorphic markers. Detailed sequencing has revealed that there are five polymorphic sites that can be used to define β-gene frameworks, as shown in Fig. 25 (162). Two of these sites are recognized by restric-

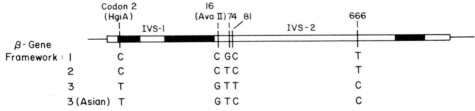

FIG. 25. Polymorphisms defining specific β-gene frameworks in man. The results of sequencing many different β genes has revealed five sites in the gene that are polymorphic and that can be used to track a particular β gene in a population (162). One of these sites occurs in codon 2 and affects an HgiA site. The other four sites are in the large intervening sequence; one of these at position 16 affects an AvaII site, but the other three do not alter known restriction sites. These frameworks have been particularly useful in deducing the evolutionary history of particular β-gene mutations (see Section VI,D).

tion enzymes *Hgi*A and *Ava*II, whereas the other three sites in the large intervening sequence do not alter known restriction sites. These frameworks, which represent a subset of cluster II, are presumably of ancient origin.

The availability of these polymorphic haplotypes has made it possible to ask certain questions about mutant globin genes. Are certain haplotypes seen only in association with an abnormal β-globin? Does the combination of a clinical phenotype and a specific haplotype define the genetic lesion? Have specific β-globin mutations arisen more than once? In analysis of the sickle mutations, the initial report of the *Hpa*I polymorphism suggested that the presence of a 13-kb *Hpa*I fragment hybridizing to the β-globin gene might be almost specific for $β^S$ (462). A subsequent study, however, showed that somewhat less than 60% of $β^S$ genes in a different American black population occurred on the 13-kb fragment, and that occasional individuals have a $β^A$ gene on a 13-kb fragment (471). Subsequent analysis using multiple restriction sites has shown that two haplotypes account for 140 out of 170 $β^S$ alleles (459). The most common of these is occasionally seen in association with $β^A$, but the next most common haplotype thus far has been seen only in association with $β^S$. However, this probably reflects the origin of the $β^S$ mutation on a haplotype that is uncommon in American and Jamaican blacks today, but may have been more common in the African population where the mutation arose.

The situation is quite different in the β-thalassemias. Analysis of haplotypes in $β^A$ chromosomes and unselected β-thal chromosomes has in general showed a roughly similar distribution. However, the exciting observation (162) that the presence of a given haplotype on a β-thalassemia chromosome is highly predictive of the specific mutation has been most clearly described thus far in the Mediterranean population. The majority of individuals with a given haplotype and β-thalassemia turn out to have the same mutation, with a few important exceptions as noted below. This observation has permitted a strategy for searching for new β-thalassemia mutations; of the nine common haplotypes found in the Mediterranean population, eight different thalassemias were quickly identified (162). This correlation suggests that the thalassemia mutations are relatively recent with respect to the polymorphic sites in the β-globin cluster.

There are some notable exceptions to simple coupling of haplotypes with specific defects. It is perhaps not surprising that some of the common haplotypes can be associated with more than one thalassemic mutation, since there is no reason why more than one thalassemic mutation should not have arisen from a common haplotype background. An example is the most common Mediterranean haplotype, denoted as I (*162*); most individuals with this haplotype and β-thalassemia are found to have the IVS-1 position-110 mutation, but the codon-6 frameshift mutation and the hemoglobin Knossos mutation are also found in this setting (*356*). A more puzzling exception is the situation where a given β-gene mutation has been found in more than one haplotype. In most such instances, the β mutation remains in the same β-gene framework as defined in Fig. 25, and it is reasonable to conclude that the presence of a different haplotype can be attributed to a crossing-over event between clusters I and II. However, analysis of 170 β^S chromosomes has shown that 16 different haplotypes occur (*459*), and these can be divided into four groups differing from each other by at least two crossing-over events. The most impressive haplotype is a single individual in which the β^S mutation appears in a framework-3 β gene instead of a framework-1 gene, where all the other mutations are found. Similar results were found in an analysis of Jamaican β^S chromosomes (*459a*).

There are at least three possible explanations for this variety of haplotype backgrounds in β^S. The occurrence of multiple origins of the same mutation must be given strong consideration, since selective pressure on the heterozygote exists. A second possibility would be a double crossover in the framework-3 individual mentioned above; because of flanking polymorphic markers, however, there are only 11 bases on the 5' side of the β^S mutation and 441 on the 3' side in which the crossover must have occurred, and a double crossover in such a narrow interval seems unlikely. Finally, the possibility of interallelic gene conversion has to be considered, as there are well-described examples in other gene systems of conversions of very short stretches of DNA. At present, the relative probability of these explanations is difficult to assess.

Similar data on the β^E mutation, which has been found in both framework-2 and framework-3, suggest the possibility of multiple independent origins of this mutation or interallelic gene conversion (*460, 461*). The possibility of intrachromosomal rather than interallelic gene conversion as the source of such a recurring mutation for β^S or β^E was intriguing, but the appropriate donor is not present at homologous codons of the other genes of the β-globin cluster.

E. Prenatal Diagnosis

Prenatal diagnosis of the hemoglobinopathies was initially achieved by fetal blood sampling and analysis of globin-chain synthesis by radioactive labeling (*472, 473*). This procedure, while remarkably successful in providing accurate diagnosis, is technically demanding and poses a relatively high risk (5–10%) of miscarriage (474). It was natural, therefore, to seek ways of direct DNA analysis once specific β-gene mutations were described. An excellent review of prenatal diagnosis by fetal DNA analysis has recently appeared (475).

Some hemoglobin abnormalities can be demonstrated directly by fetal DNA analysis, since the mutation itself produces a recognizable abnormality in restriction fragment length. An important example is that of the sickle (β^S) mutation. The normal nucleotide sequence of codons 5, 6, and 7 is CCTGAGGAG. In β^S this becomes CCTGTGGAG. The enzyme DdeI, which recognizes the sequence CTNAG (where N is any nucleoside) thus cuts the normal sequence, but not the sickle sequence. This observation has been used (476, 477) to distinguish normal from sickle genes, but the technique is technically difficult because of the small size of the DdeI fragments (less than 400 bp) that must be detected. More recently, a much simpler analysis has been possible with the introduction of the enzyme MstII, which recognizes the sequence CCTNAGG (478–480). Digestion of genomic DNA with MstII followed by transfer to nitrocellulose and hybridization with a probe containing the 5' flanking region of the β gene yields a band of 1.2 kb for the β^A gene and a band of 1.4 kb for a β^S. These two fragments are readily resolvable, and this analysis has been successfully applied to prenatal diagnosis (478, 479, 481). Some care must be taken in applying this technique widely, as other mutations in the MstII recognition site would give the same result (an example is the codon-6 frameshift mutation that causes β-thalassemia; see Table II). However, no silent substitutions were found in this sequence, in a small number of DNA samples from different ethnic groups (478).

A few β-thalassemias also create directly a difference in restriction-fragment length (162), but these are in the minority. Many of the α-thalassemias, on the other hand, can be directly assayed because they involve large deletions (482). Similarly, $\delta\beta$-thalassemia and hemoglobin Lepore can be directly assayed because of the presence of a deletion (483).

The majority of the β-thalassemias, then, cannot be ascertained directly by DNA analysis. In Sardinia, where almost all β-thalassemia is the codon-39 nonsense mutation, the linkage disequilibrium of this mutation with a BamHI polymorphism has been used for prenatal diagnosis (463). Specifically, the β-thal mutation occurs on a 9.3-kb BamHI fragment, which is also seen with β^A. However, a 22-kb BamHI fragment occurs with β^A but not β^039, so that the demonstration of the 22-kb fragment in a fetus at risk has been used as an indication that the fetus is not homozygous for β^0-thalassemia. This is a special case, however, since in most situations the possible thalassemia mutations to be considered are quite heterogeneous and such linkage disequilibrium cannot be relied on. Therefore, in most situations a family study is needed, preferably with an affected sibling available, to try to establish the haplotypes of the parents and the linkage between the thalassemic mutations and the haplotypes in that family (*484, 485*). Diagnosis may not be possible if one or both parents are homozygous for a particular haplotype. This limitation reportedly prevents successful diagnosis in approximately 50% of families in the United Kingdom (*475, 485a*), but diagnosis is possible in as many as 85% of families in the United States (*470*), with the difference presumably attributable to the greater population heterogeneity.

A promising recent development for the specific diagnosis of point mutations in the β-globin gene has been the demonstration that synthetic oligonucleotides can distinguish a single-base change under proper conditions of hybridization (*486–488*). In this technique, a 19-base synthetic oligoncleotide is constructed that is exactly homologous to the normal sequence in the region of interest, and another oligonucleotide is made homologous to the mutated gene. Under careful conditions of hybridization and using a restriction enzyme that places the fragment in question well away from the majority of the nonspecific background hybridization, it is possible to demonstrate a band when the probe is a perfect match, but not when there is a one-base mismatch. This approach has been used successfully for the sickle mutation (*486*), the IVS-1 position-110 β^+-thalassemia mutation (*487*), and the codon 39 β^0-thalassemia mutation (*488*). A battery of such probes would make it possible to demonstrate the exact thalassemia mutation in a given situation, and would allow accurate prenatal diagnosis without the need for informative polymorphisms.

The source of fetal DNA for analyses such as this can be amniotic fluid cells, either prepared directly or cultured to increase the amount of DNA recovered (*485, 489*). However, since amniocentesis cannot be done until week 16 of pregnancy, a diagnosis cannot be achieved until the mid-trimester. Accordingly, much effort is currently under way to develop the technique of chorionic villus biopsy in the first trimester of pregnancy to allow earlier diagnosis. In this technique (*490, 491*), a small sample of chorionic tissue is obtained through a catheter advanced through the cervix. The villi can be removed free of maternal tissue under a microscope, and DNA amounts from 5 to 50 μg can be reliably prepared. This amount is sufficient for genomic blotting and has already been used successfully for prenatal diagnosis of thalassemia *(467)* and sickle-cell disease *(481)*. Appropriate caution must being taken in the introduction of this method, however, since the risk of inducing miscarriage by the procedure remains largely unknown.

VII. Summary and Prospects

Intensive study of the structure and expression of globin genes in the last several years has led to major advances. As we have seen, however, there remain many major questions with only partial answers, and it seems that the globin gene system will continue to be a fertile ground for generating and exploring concepts in animal-cell molecular genetics.

The complete DNA sequences of the α- and β-globin clusters, an unimaginable achievement only a few years ago, have nearly been assembled. Analyses of the promoter regions, the splice junctions, and the 3' flanking regions have yielded important insights into the *cis*-acting sequences that are responsible for normal transcription and mRNA processing. The correlation between conservation of particular sequence elements in comparisons between genes, and the subsequent demonstration of the importance of these sequences in various transcriptional assays, have been impressive. However, at a somewhat different level, our understanding of other *cis*-acting sequences, such as those responsible for limitation of expression of globin genes to erythroid cells and those responsible for the differential expression of globin genes (particularly the fetal to adult globin switch in man) remains inadequate. Naturally occurring mutations resulting in hereditary persistence of fetal hemoglobin provide some clues, but the complete picture is far from being in focus. It will be important to continue to search for new HPFH mutations, both deletion and nondeletion, in an attempt to uncover the mechanisms involved. A major deficiency in our current experimental approach is a system in which switching can be demonstrated in cloned genes. The inducible synthesis of β, and constitutive production of γ, from globin cosmids transfected into mouse erythroleukemia (MEL) cells is exciting but probably not truly a faithful representation of the switching phenomenon. The sequences responsible for induction of β in this system, therefore, are not likely to be the same ones responsible for *cis* regulation of the γ to β switch *in vivo*. Perhaps one of the problems is the need to use MEL cells for these studies, since mice do not have a fetal hemoglobin and may lack the *trans*-acting factors needed for demonstration of the switch. Further efforts to develop a primate cell line that can be induced to synthesize adult globin are certainly warranted, as is the investigation of the elusive *trans*-acting factors affecting differential expression of globin genes.

Some DNA sequence signals recognized by the splicing apparatus are now fairly well understood. However, there are still important issues in mRNA processing that the globin system is well suited to address. One such issue is the suggestion, based on observations in codon 39 β^0-thalassemia, that mRNA transcription and translation may in some way be coupled in eukaryotes. Another is the complex area of RNA splicing, for which a complete set of rules, a purified *in vitro* splicing system, and a well supported mechanistic model remain to be defined.

A considerable amount of revealing information about the evolution of the mammalian β-globin cluster has been obtained and has revealed a complexity and fluidity of interactions not necessarily anticipated. Further comparisons, particularly between more closely related species, may further sharpen the focus in this area. The phenomenon of gene conversion in this cluster occurring on an evolutionary scale seems firmly established and is an attractive area for further experimental investigation in somatic cells.

Of the areas covered in this review, one of the most striking achievements of the past five years has been the rapid definition at the DNA level of the heterogeneous collection of mutations that lead to α- and β-thalassemia, which have themselves provided crucial information about the control of gene expression *in vivo*. The recognition of linked polymorphisms, the development of mutation-specific oligonucleotide probes, and the possibility of first-trimester chrionic villus biopsy have made prenatal diagnosis of sickle-cell anemia and the β-thalassemias of immediate applicability, and appropriate resources must be marshaled to offer this technology to the areas of the world where it is most needed. The challenge to the future is the application of this detailed knowledge not just to diagnosis, but to therapy of these devastating disorders. Two totally different therapies, bone marrow transplantation and 5-azacytidine, already show promise in this regard, but both carry significant real or potential morbidity. Therefore, actual transfection of a cloned gene into marrow cells to provide the needed protein product continues to merit intense attention.

Much has been written (*492–496*) about the coming of the age of gene therapy, though practical scenarios for doing this are relatively few. In some ways the hemoglobinopathies are ideal candidates for such attempts, since the tissue to be treated (bone marrow) can be removed, manipulated, and reinfused relatively easily. In other respects, however, these disorders may prove to be difficult: there is no intrinsic selection available for erythroid cells that have acquired the transfected gene, the success in obtaining globin-gene expression in DNA transfer experiments is limited, and the erythroid stem cells one would like to transfect constitute only a small fraction of marrow cells. This last difficulty points out the need for a means of stem cell purification or else the use of gene-transfer systems (such as retrovirus vectors) that allow a very high efficiency of gene integration into recipient cells. Finally, unlike the situation in most enzyme deficiencies, expression of the transfected globin gene must be regulated within a relatively tight range, or a thalassemic phenotype will result. In spite of these obstacles, however, it would be hard to find a person working in the field who is not at least guardedly optimistic about the prospects for successful gene therapy in the next one or two decades. One of the original criteria suggested for the application of such therapy is that it should have no deleterious effects (492). More realistically, however, gene therapy should be judged on the same basis as any new medical advance: Does the benefit of the proposed therapy exceed the associated risks, and does the new approach provide advantages over previously existing therapeutic options? The successful negotiation of these difficult issues is a challenge to scientists, physicians, and lay people alike.

Acknowledgments

We would like to thank our many colleagues for preprints and generous discussion of their work prior to publication. Discussions with Stephen Hardies and Marshall Edgell were particularly helpful in developing the section on evolution. We also benefited from numerous conversations with Bernard Forget, Christian Stoeckert, and Peter Rogan. Numerous helpful comments on the manuscript were contributed by Y. W. Kan, Haig Kazazian, Ken Kidd, and Stuart Orkin. The patient and tireless assistance of Ann M. Mulvey in preparing the manuscript was invaluable. F. C. acknowledges with gratitude the support of fellowships from the Cooley's Anemia Foundation, Inc., and the Charles E. Culpepper Foundation. S. M. W. acknowledges support from NIH Grants GM32156 and AM28376.

References

1. T. Maniatis, E. F. Fritsch, J. Lauer, and R. M. Lawn, *ARGen* **14**, 145 (1980).
2. D. J. Weatherall and J. B. Clegg, "The Thalassaemia Syndromes," 3rd ed. Blackwell, Oxford, 1981.
3. H. F. Bunn, B. G. Forget, and H. M. Ranney, "Human Hemoglobins." Saunders, Philadelphia, 1977.
4. R. E. Dickerson and I. Geis, "Hemoglobin: Structure, Function, Evolution, and Pathology." Benjamin/Cummings, Reading, Massachusetts, 1983.
5. B. G. Forget, *Ann. Int. Med.* **91**, 605 (1979).

6. B. G. Forget, *Recent Prog. Horm. Res.* **38**, 257 (1982).
7. A. Bank, J. G. Mears, and F. Ramirez, *Science (Washington, D. C.)* **207**, 486 (1980).
8. A. Bank, *Prog. Hematol.* **12**, 25 (1981).
9. D. J. Weatherall and J. B. Clegg, *Cell* **16**, 467 (1979).
10. R. A. Spritz and B. G. Forget, *Am. J. Hum. Genet.* **35**, 333 (1983).
11. N. J. Proudfoot, M. H. M. Shander, J. L. Manley, M. L. Gefter, and T. Maniatis, *Science (Washington, D. C.)* **209**, 1329 (1980).
12. F. Ramirez, J. G. Mears, and A. Bank, *MCBchem* **31**, 133 (1980).
13. A. W. Nienhuis and E. J. Benz, *N. Engl. J. Med.* **297**, 1318 (1977).
14. A. W. Nienhuis and E. J. Benz, *N. Engl. J. Med.* **297**, 1430 (1977).
15. G. Stamatoyannopoulos, T. Papayannopoulou, M. Brice, S. Kurachi, B. Nakamoto, G. Lim, and M. Farquhar, in "Hemoglobins in Development and Differentiation" (G. Stamatoyannopoulos and A. Nienhuis, eds.), p. 287, Liss, New York, 1981.
15a. T. Papayannopoulou and G. Stamatoyannopoulos, *J. Cell. Physiol. Suppl.* **1**, 145 (1982).
16. T. Papayannopoulou, S. Kurachi, B. Nakamoto, E. D. Sanjani, and G. Stamatoyannopoulos, *PNAS* **79**, 6579 (1982).
16a. G. Stamatoyannopoulos, B. Nakamoto, S. Kurachi, and T. Papayannopoulou, *PNAS* **80**, 5650 (1983).
17. T. Shibuya, Y. Niho, and T. Mak, *J. Exp. Med.* **156**, 398 (1982).
18. T. Shibuya and T. W. Mak, *J. Cell. Physiol. Suppl.* **1**, 185 (1982).
19. T. Shibuya and T. W. Mak, *Cell* **31**, 483 (1982).
20. T. Shibuya and T. W. Mak, *Nature (London)* **296**, 577 (1982).
21. T. Shibuya and T. W. Mak, *PNAS* **80**, 3721 (1983).
22. T. Shibuya and T. W. Mak, *J. Cell. Physiol.* (in Press).
24. T. Graf and H. Beug, *Cell* **34**, 7 (1983).
25. E. A. Schon, M. L. Cleary, J. R. Haynes, and J. B. Lingrel, *Cell* **27**, 359 (1981).
26. E. A. Schon, S. M. Wernke, and J. B. Lingrel, *JBC* **257**, 6825 (1982).
27. J. B. Lingrel, S. G. Shapiro, T. Townes, S. Wernke, and E. A. Schon, in "Recombinant DNA Technology" (S. Woo, ed.), Vol. 2. In press.
27a. T. M. Townes, S. G. Shapiro, S. M. Wernke, and J. B. Lingrel, *JBC* **259**, 1896 (1984).
28. E. Lacy, R. C. Hardison, D. Quon, and T. Maniatis, *Cell* **18**, 1273 (1979).
29. R. C. Hardison, E. T. Butler, E. Lacy, T. Maniatis, N. Rosenthal, and A. Efstratiadis, *Cell* **18**, 1285 (1979).
30. E. F. Fritsch, R. M. Lawn, and T. Mantiatis, *Cell* **19**, 959 (1980).
31. J. Lauer, C. J. Shen, and T. Maniatis, *Cell* **20**, 119 (1980).
32. P. A. Barrie, A. J. Jeffreys, and A. F. Scott, *JMB* **149**, 319 (1981).
33. C. L. Jahn, C. A. Hutchison, S. J. Phillips, S. Weaver, N. L. Haigwood, C. F. Voliva, and M. H. Edgell, *Cell* **21**, 159 (1980).
34. S. Weaver, M. B. Comer, C. L. Jahn, C. A. Hutchison, and M. H. Edgell, *Cell* **24**, 403 (1981).
35. Y. Nishioka and P. Leder, *Cell* **18**, 875 (1979).
36. R. A. Popp, P. A. Lalley, J. B. Whitney, and W. F. Anderson, *PNAS* **78**, 6362 (1981).
37. M. Dolan, B. J. Sugarman, J. B. Dodgson, and J. D. Engel, *Cell* **24**, 669 (1981).
38. B. Villeponteau, G. M. Landes, M. J. Pankratz, and H. G. Martinson, *JBC* **257**, 11023 (1982).
39. J. D. Engel and J. B. Dodgson, *PNAS* **77**, 2596 (1980).
40. H. A. Hosbach, T. Wyler, and R. Weber, *Cell* **32**, 45 (1983).
41. J. S. Lee, G. G. Brown, and D. P. S. Verma, *NARes* **11**, 5541 (1983).
42. F. E. Baralle, C. C. Shoulders, and N. J. Proudfoot, *Cell* **21**, 621 (1980).

43. J. L. Slightom, A. E. Blechl, and O. Smithies, *Cell* **21**, 627 (1980).
44. S. Shen, J. L. Slightom, and O. Smithies, *Cell* **26**, 191 (1981).
45. R. A. Spritz, J. K. Deriel, B. G. Forget, and S. M. Weissman, *Cell* **21**, 639 (1980).
46. R. M. Lawn, A. Efstratiadis, C. O'Connell, and T. Maniatis, *Cell* **21**, 647 (1980).
47. N. J. Proudfoot, A. Gil, and T. Maniatis, *Cell* **31**, 553 (1982).
48. A. M. Michelson and S. H. Orkin, *Cell* **22**, 371 (1980).
49. S. A. Liebhaber, M. J. Goossens, and Y. W. Kan, *PNAS* **77**, 7054 (1980).
50. A. Blanchetot, V. Wilson, D. Wood, and A. J. Jeffreys, *Nature (London)* **301**, 732 (1983).
51. M. Go, *Nature (London)* **291**, 90 (1981).
52. E. O. Jensen, K. Paludan, J. J. Hyldig-Kielsen, P. Jorgensen, and K. A. Marcker, *Nature (London)* **291**, 677 (1981).
53. J. J. Hyldig-Nielsen, E. O. Jensen, K. Paludan, O. Wiborg, R. Garrett, P. Jorgensen, and K. A. Marcker, *NARes* **10**, 689 (1982).
53a. O. Wiborg, J. J. Hyldig-Nielsen, E. O. Jensen, K. Pakulan, and K. A. Marcker, *NARes* **10**, 3487 (1982).
54. J. M. Adams, D. J. Kemp, O. Bernard, N. Gough, E. Webb, B. Tyler, S. Gerondakis, and S. Cory, *Immunol. Rev.* **59**, 5 (1981).
55. L. Hood, M. Steinmetz, and B. Malissen, *Annu. Rev. Immunol.* **1**, 529 (1983).
56. M. Malissen, B. Malissen, and B. R. Jordan, *PNAS* **79**, 893 (1982).
57. P. A. Biro, J. Pan, A. K. Sood, R. Kole, V. B. Reddy, and S. M. Weissman, *CSHSQB* **47**, 1079 (1983).
58. W. Gilbert, *Nature (London)* **271**, 501 (1978).
59. S. Tonegawa, A. M. Maxam, R. Tizard, O. Bernard, and W. Gilbert, *PNAS* **75**, 1485 (1978).
59a. W. A. Eaton, *Nature (London)* **284**, 183 (1980).
60. A. Hill, S. C. Hardies, S. J. Phillips, M. G. Davis, C. A. Hutchison, and M. H. Edgell, *JBC* **259**, 3739 (1984).
61. F. Sanger, A. R. Coulson, G. F. Hong, D. F. Hill, and G. B. Peterson, *JMB* **162**, 729 (1982).
62. G. Modiano, G. Battistuzzi, and A. G. Motulsky, *PNAS* **78**, 1110 (1981).
63. A. Efstratiadis, J. W. Posakony, T. Maniatis, R. M. Lawn, C. O'Connell, R. A. Spritz, J. K. DeRiel, B. G. Forget, S. M. Weissman, J. L. Slightom, A. E. Blechl, O. Smithies, F. E. Baralle, C. C. Shoulders, and N. J. Proudfoot, *Cell* **21**, 653 (1980).
64. B. Erni and T. Staehelin, *FEBS Lett.* **148**, 79 (1982).
65. O. Hagenbuchle, M. Santer, J. A. Steitz, and R. J. Mans, *Cell* **13**, 551 (1978).
66. G. W. Both, *FEBS Lett.* **101**, 220 (1979).
67. S. L. Martin, E. A. Zimmer, W. S. Davidson, A. C. Wilson, and Y. W. Kan, *Cell* **25**, 737 (1981).
67a. C. Friend, *J. Exp. Med.* **105**, 307 (1957).
68. P. A. Marks and R. A. Rifkind, *ARB* **47**, 419 (1978).
69. B. A. Brown, R. W. Padgett, S. C. Hardies, C. A. Hutchison, and M. H. Edgell, *PNAS* **79**, 2753 (1982).
70. H. R. Profous-Juchelka, R. C. Reuben, P. A. Marks, and R. A. Rifkind, *MCBiol* **3**, 229 (1983).
71. D. M. Miller, P. Turner, A. W. Nienhuis, D. E. Axelrod, and T. V. Gopalakrishnan, *Cell* **14**, 511 (1978).
72. T. Shibuya and T. W. Mak, submitted.
73. D. Mager, M. E. MacDonald, I. B. Robson, T. W. Mak, and A. Bernstein, *MCBiol* **1**, 721 (1981).

74. A. Deisseroth, R. Burk, D. Picciano, J. Minna, W. F. Anderson, and A. Nienhuis, *PNAS* **72**, 1102 (1978).
75. A. Deisseroth, J. Barker, W. F. Anderson, and A. W. Nienhuis, *PNAS* **72**, 2682 (1978).
76. A. Deisseroth and D. Hendrick, *Cell* **15**, 55 (1978).
77. J. Pyati, R. S. Kucherlapati, and A. Skoultchi, *PNAS* **77**, 3435 (1980).
78. M. C. Willing, A. Nienhuis, and W. F. Anderson, *Nature (London)* **277**, 534 (1979).
79. P. J. Zavodny, R. S. Roginski, and A. I. Skoultchi, *in* "Globin Gene Expression and Hematopoietic Differentiation (G. Stamatoyannopoulos and A. Nienhuis, eds.), p. 53. Liss, New York, 1983.
80. C. B. Lozzio and B. B. Lozzio, *Blood* **45**, 321 (1975).
81. P. Martin and T. Papayannopoulou, *Science (Washington, D. C.)* **216**, 1233 (1982).
82. R. F. Mueller, J. C. Murray, R. Gelinas, and M. Farquhar, *Hemoglobin* **7**, 245 (1983).
83. U. Testa, W. Vainchienker, Y. Beuzard, P. Rouyer-Fessard, A. Guerrasio, M. Titeux, P. Lapotre, J. Bouguet, J. Breton-Gorius, and J. Rosa, *EJB* **121**, 649 (1982).
85. C. W. Miller, K. Young, D. Dumenil, B. P. Alter, J. M. Schofield, and A. Bank, *Blood* (in press).
86. P. T. Rowley, B. M. Ohlsson-Wilhelm, N. S. Rudolph, B. A. Farley, B. Kosciolek, and S. LaBella, *Blood* **59**, 1098 (1982).
87. A. Dean, T. J. Ley, R. K. Humphries, M. Fordis, and A. N. Schechter, *PNAS* **80**, 5515 (1983).
88. P. Charnay and T. Maniatis, *Science (Washington, D. C.)* **220**, 1281 (1983).
89. N. Hsiung, R. S. Roginski, P. Henthorn, O. Smithies, R. Kucherlapati, and A. I. Skoultchi, *MCBiol* **2**, 401 (1982).
89a. R. S. Roginski, A. I. Skoultchi, P. Henthorn, O. Smithies, N. Hsiung, and R. Kucherlapati, *Cell* **35**, 149 (1983).
90. M. Perucho, D. Hanahan, and M. Wigler, *Cell* **22**, 309 (1980).
91. F. Colbere-Garakin, A. Garakin, and P. Kourilsky, *Curr. Top. Microbiol. Immunol.* **96**, 145 (1982).
92. R. C. Mulligan and P. Berg, *Science* **209**, 1422 (1980).
93. E. Neumann, M. Schaeffer-Ridder, Y. Wang, and P. H. Hofneider, *EMBO J.* **1**, 841 (1982).
93a. C. W. Lo, *MCBiol* **3**, 1803 (1983).
94. J. Paul and D. A. Spandidos, *Adv. Exp. Med. Biol.* **158**, 89 (1982).
95. S. Wright, E. deBoer, F. G. Grosveld, and R. A. Flavell, *Nature (London)* **305**, 333 (1983).
96. D. A. Spandidos and J. Paul, *EMBO J.* **1**, 15 (1982).
97. M. V. Chao, P. Mellon, P. Charnay, T. Maniatis, and R. Axel, *Cell* **32**, 483 (1983).
98. G. Khoury and P. Gruss, *Cell* **33**, 313 (1983).
98a. M. R. Green, R. Treisman, and T. Maniatis, *Cell* **35**, 137 (1983).
98b. R. Treisman, M. R. Green, and T. Maniatis, *PNAS* **80**, 7428 (1983).
99. J. T. Elder, R. A. Spritz, and S. M. Weissman, *ARGen* **15**, 295 (1981).
100. P. W. J. Rigby, *in* "Genetic Engineering" (R. Williamson, ed.), Vol. 3, p. 83. Academic Press, New York, 1982.
100a. A. S. Perkins, P. J. Kirshmeier, S. Guttoni-Ulli, and I. B. Weinstein, *MCBiol* **3**, 1123 (1983).
100b. R. Mann, R. C. Mulligan, and D. Baltimore, *Cell* **33**, 153 (1983).
100c. C. M. Wei, M. Gibson, P. G. Spear, and E. M. Scolnick, *J. Virol.* **39**, 935 (1981).
100d. C. Tabin, J. Hoffman, S. Groff, and R. Weinberg, *MCBiol* **2**, 426 (1982).

100e. K. Shimotohno and H. M. Temin, *Nature (London)* **299**, 265 (1982).
100f. K. Shimotohno and H. M. Temin, *Cell* **26**, 67 (1981).
100g. A. D. Miller, D. J. Jolly, T. Friedmann, and I. M. Verma, *PNAS* **80**, 4709 (1983).
100g. N. P. Sarver, P. Gruss, M. F. Law, G. Khoury, and P. M. Howley, *MCBiol* **1**, 486 (1981).
101. Y. Lau and Y. W. Kan, *PNAS* **80**, 5225 (1983).
102. W. F. Anderson, L. Killos, L. Sanders-Haigh, P. J. Kretschmer, and E. G. Diacumakos, *PNAS* **77**, 5399 (1980).
103. T. E. Wagner, P. C. Hoppe, J. D. Jollick, D. R. Scholl, R. L. Hodinka, and J. B. Gault, *PNAS* **78**, 6376 (1981).
104. E. F. Wagner, T. A. Stewart, and B. Mintz, *PNAS* **78**, 5016 (1981).
105. R. L. Brinster, H. Y. Chen, and M. Trumbauer, *Cell* **27**, 223 (1981).
106. R. L. Brinster, H. Y. Chen, R. Warren, A. Sarthy, and R. D. Palmiter, *Nature (London)* **296**, 39 (1982).
107. R. D. Palmiter, H. Y. Chen, and R. L. Brinster, *Cell* **29**, 701 (1982).
108. R. D. Palmiter, R. L. Brinster, R. E. Hammer, M. E. Trumbauer, M. G. Rosenfeld, N. C. Birnberg, and R. M. Evans, *Nature (London)* **300**, 611 (1982).
108a. R. D. Palmiter, G. Norstedt, R. E. Gelinas, R. E. Hammer, and R. L. Brinster, *Science (Washington, D. C.)* **222**, 809 (1983).
109. G. S. McKnight, R. E. Hammer, E. A. Kuenzel, and R. L. Brinster, *Cell* **34**, 335 (1983).
109a. R. L. Brinster, K. A. Ritchie, R. E. Hammer, R. L. O'Brien, B. Arp, and U. Storb, *Nature (London)* **306**, 332 (1983).
110. A. Pellicer, E. F. Wagner, A. El Kareh, M. J. Dewey, A. J. Reuser, S. Silverstein, R. Axel, and B. Mintz, *PNAS* **77**, 2098 (1980).
111. F. Costantini and E. Lacy, *Nature (London)* **294**, 92 (1981).
112. E. Lacy, S. Roberts, E. P. Evans, M. D. Burtenshaw, and F. D. Costantini, *Cell* **34**, 343 (1983).
113. E. F. Wagner and B. Mintz, *MCBiol* **2**, 190 (1982).
114. T. A. Stewart, E. F. Wagner, and B. Mintz, *Science (Washington, D. C.)* **217**, 1046 (1982).
115. B. P. Alter, *Blood* **545**, 1158 (1979).
116. M. Jensen, H. Attenberger, C. Schneider, and J. U. Walther, *Eur. J. Pediatr.* **138**, 311 (1982).
117. T. H. J. Huisman, H. Harris, and M. Gravely, *MCBchem* **17**, 45 (1977).
118. T. H. J. Huisman and C. Altay, *Blood* **58**, 491 (1981).
119. P. Powers, C. Altay, T. H. J. Huisman, and O. S. Smithies, personal communication.
120. P. J. Kretschmer, H. C. Coon, A. Davis, M. Harrison, and A. W. Nienhuis, *JBC* **256**, 1975 (1981).
121. W. G. Wood, D. J. Weatherall, J. B. Clegg, T. J. Hamblin, J. H. Edwards, and A. M. Barlow, *Br. J. Haematol.* **36**, 461 (1977).
122. R. P. Perrine, *Br. J. Haematol.* **40**, 415 (1978).
123. E. Hofer, R. Hofer-Warbinek, and J. E. Darnell, *Cell* **29**, 887 (1982).
123a. J. E. Darnell, personal communication.
124. M. Salditt-Georgieff and J. E. Darnell, *PNAS* **80**, 4694 (1983).
124a. J. E. Darnell, *Nature (London)* **297**, 365 (1982).
125. T. Shenk, in "Methods of DNA and RNA Sequencing" (S. M. Weissman, ed.), p. 349. Praeger, New York, 1983.
126. S. McKnight, *NARes* **8**, 949 (1980).

127. K. N. Subramanian, R. Dhar, and S. M. Weissman, *J. Biol. Chem.* **252**, 355 (1977).
128. A. van Ooyen, J. van den Berg, N. Mantei, and C. Weissmann, *Science* **206**, 337 (1979).
129. J. R. Haynes, P. Rosteck, and J. B. Lingrel, *PNAS* **77**, 7127 (1980).
130. R. C. Hardison, *JBC* **258**, 8739 (1983).
131. D. A. Konkel, J. V. Maizel, and P. Leder, *Cell* **18**, 865 (1979).
132. E. Lacy and T. Maniatis, *Cell* **21**, 545 (1980).
133. Y. Fukumaki, F. Collins, R. Kole, C. J. Stoeckert, P. Jagadeeswaran, C. H. Duncan, S. M. Weissman, J. Pan, B. G. Forget, P. A. Biro, A. K. Sood, and V. B. Reddy, *CSHSQB* **47**, 1079 (1983).
134. A. J. Jeffreys, P. A. Barrie, S. Harris, D. H. Fawcett, Z. J. Nugent, and A. C. Boyd, *JMB* **156**, 487 (1982).
135. R. C. Hardison, *JBC* **256**, 11780 (1981).
136. S. G. Shapiro, E. A. Schon, T. M. Townes, and J. B. Lingrel, *JMB* **169**, 31 (1983).
137. F. E. Baralle, C. C. Shoulders, S. Goodbourn, A. Jeffreys, and N. J. Proudfoot, *NARes* **8**, 4393 (1980).
138. J. N. Hansen, D. A. Konkel, and P. Leder, *JBC* **257**, 1048 (1982).
139. R. D. Everett, D. Baty, and P. Chambon, *NARes* **11**, 2447 (1983).
140. M. Fromm and P. Berg, *J. Mol. Appl. Genet.* **1**, 457 (1982).
141. S. W. Hartzell, J. Yamaguchi, and K. N. Subramanian, *NARes* **11**, 1601 (1983).
142. S. L. McKnight, *Cell* **31**, 355 (1982).
143. S. L. McKnight and R. Kingsbury, *Science (Washington, D. C.)* **217**, 316 (1982).
143a. W. S. Dynan and R. Tjian, *Cell* **35**, 79 (1983).
144. M. R. Capecchi, *Cell* **22**, 479 (1980).
145. J. Banerji, S. Rusconi, and W. Schaffner, *Cell* **27**, 299 (1981).
146. R. Grosschedl and M. L. Birnstiel, *PNAS* **77**, 1432 (1980).
147. P. A. Luciw, J. M. Bishop, H. E. Varmus, and M. R. Capecchi, *Cell* **33**, 705 (1983).
148. J. Banerji, L. Olson, and W. Schaffner, *Cell* **33**, 729 (1983).
149. S. D. Gillies, S. L. Morrison, V. T. Oi, and S. Tonegawa, *Cell* **33**, 717 (1983).
150. C. Queen and D. Baltimore, *Cell* **33**, 741 (1983).
151. D. DiMaio, R. Treisman, and T. Maniatis, *PNAS* **79**, 4030 (1982).
152. S. P. Gregory and P. H. W. Butterworth, *NARes* **11**, 5317 (1983).
153. P. Dierks, A. van Ooyen, M. D. Cochran, C. Dobkin, J. Reiser, and C. Weissmann, *Cell* **32**, 695 (1983).
154. G. C. Grosveld, C. K. Shewmaker, P. Jat, and R. A. Flavell, *Cell* **25**, 215 (1981).
155. G. C. Grosveld, E. deBoer, C. K. Shewmaker, and R. A. Flavell, *Nature (London)* **295**, 120 (1982).
156. C. A. Talkington and P. Leder, *Nature (London)* **298**, 192 (1982).
158. R. K. Humphries, T. Ley, P. Turner, A. D. Moulton, and A. W. Nienhuis, *Cell* **30**, 173 (1982).
159. G. Modiano and G. Pepe, *Mol. Biol. Med.* **1**, 157 (1983).
159a. J. G. Adams, W. T. Morrison, and M. H. Steinberg, *Science (Washington, D. C.)* **218**, 291 (1982).
160. J. Ross and A. Pizarro, *JMB* **167**, 607 (1983).
161. R. Treisman, S. H. Orkin, and T. Maniatis, *Nature (London)* **302**, 591 (1983).
162. S. H. Orkin, H. H. Kazazian, S. E. Antonarakis, S. C. Goff, C. D. Boehm, J. P. Sexton, P. G. Waber, and P. J. V. Giardina, *Nature (London)* **296**, 627 (1982).
164. J. D. Engel, D. J. Rusling, K. C. McCune, and J. B. Dodgson, *PNAS* **80**, 1392 (1983).
165. J. Niessing, C. Erbil, and V. Neubauer, *Gene* **18**, 187 (1982).
166. T. J. Ley and A. W. Nienhuis, *BBRC* **112**, 1041 (1983).

166a. D. P. Carlson and J. Ross, *Cell* **34**, 857 (1983).
167. M. Allan, G. Grindlay, L. Stefani, and J. Paul, *NARes* **10**, 5133 (1982).
168. M. Allan, W. G. Lanyon, and J. Paul, *Cell* **35**, 187 (1983).
169. D. H. Hamer and P. Leder, *Cell* **17**, 737 (1979).
170. P. Berg and co-workers, quoted in reference 99.
171. S. Mount and J. Steitz, *Nature (London)* **303**, 380 (1983).
172. D. S. Greenspan and S. M. Weissman, unpublished observations.
173. R. Breathnach and P. Chambon, *ARB* **50**, 349 (1981).
174. S. M. Mount and J. A. Steitz, in "Methods of DNA and RNA Sequencing" (S. M. Weissman, ed.), p. 399. Praeger, New York, 1983.
175. J. B. Dodgson and J. D. Engel, *JBC* **258**, 4623 (1983).
177. B. Wieringa, F. Meyer, J. Reiser, and C. Weissmann, *Nature (London)* **301**, 38 (1983).
178. B. Wieringa, F. Meyer, J. Reiser, and C. Weissmann, in "Gene Regulation," (B. W. O'Malley, ed.), p. 65. Academic Press, New York, 1982.
178a. T. Kuhne, B. Wieringa, J. Reiser, and C. Weissmann, *EMBO J.* **2**, 727 (1983).
179. B. K. Felber, S. H. Orkin, and D. H. Hamer, *Cell* **29**, 895 (1982).
180. R. Treisman, N. J. Proudfoot, M. Shander, and T. Maniatis, *Cell* **29**, 903 (1982).
181. C. Dobkin, R. G. Pergolizzi, P. Bahre, and A. Bank, *PNAS* **80**, 1184 (1983).
182. K. M. Lang and R. A. Spritz, *Science (Washington, D. C.)* **220**, 1351 (1983).
183. S. H. Orkin, J. P. Sexton, S. C. Goff, and H. H. Kazazian, *JBC* **258**, 7249 (1983).
184. C. Wu, *Nature (London)* **286**, 854 (1980).
185. J. D. McGhee, W. I. Wood, M. Dolan, J. D. Engel, and G. Felsenfeld, *Cell* **27**, 45 (1981).
186. A. Larsen and H. Weintraub, *Cell* **29**, 609 (1982).
186a. J. M. Nickol and G. Felsenfeld, *Cell* **35**, 467 (1983).
186b. E. Schon, T. Evans, J. Welsh, and A. Efstratiadis, *Cell* **35**, 837 (1983).
187. J. Stalder, A. Larson, J. D. Engel, M. Dolan, M. Groudine, and H. Weintraub, *Cell* **20**, 451 (1980).
188. M. Dolan, J. B. Dodgson, and J. D. Engel, *JBC* **258**, 3983 (1983).
188a. M. Groudine and H. Weintraub, *Cell* **24**, 393 (1981).
189. T. Kohwi-Shigematsu, R. Gelinas, and H. Weintraub, *PNAS* **80**, 4389 (1983).
190. M. Sheffery, R. A. Rifkind, and P. A. Marks, *PNAS* **79**, 1180 (1982).
191. M. Groudine, T. Kohwi-Shigematsu, R. Gelinas, G. Stamatoyannopoulos, T. Papayannopoulou, *PNAS* **80**, 7551 (1983).
193. H. M. Lachman and J. G. Mears, *NARes* **11**, 6065 (1983).
194. M. Groudine and H. Weintraub, *Cell* **30**, 131 (1982).
194a. S. Sargosti, S. Cereghini, and M. Yaniv, *JMB* **160**, 133 (1982).
195. C. Cremisi, *NARes* **9**, 5949 (1981).
196. S. Saragosti, G. Moyne, and M. Yaniv, *Cell* **20**, 65 (1980).
196a. E. B. Jakobovits, S. Bratosin, and Y. Aloni, *Nature (London)* **285**, 263 (1980).
197. R. D. Gerard, M. Woodworth-Gutai, and W. A. Scott, *MCBiol* **2**, 782 (1982).
197a. M. Fromm and P. Berg, *MCBiol* **3**, 991 (1983).
198. A. Nordheim and A. Rich, *Nature (London)* **303**, 674 (1983).
199. H. Weintraub, *Cell* **32**, 1191 (1983).
199a. C. J. Shen, *NARes* **11**, 7899 (1983).
199b. C. J. Stoeckert, unpublished observation.
200. E. W. Johns, "The High Mobility Group of Chromosomal Proteins." Academic Press, New York, 1982.
201. S. Weisbrod and H. Weintraub, *Cell* **23**, 391 (1981).

202. S. Weisbrod, *Nature (London)* **297**, 289 (1982).
202a. P. Triadou, J. C. Lelong, F. Gross, and M. Crepin, *BBRC* **101**, 45 (1981).
202b. P. Triadou, M. Crepin, F. Gross, and J. C. Lelong, *Bchem* **21**, 6060 (1982).
203. C. M. Kane, P. F. Cheng, J. B. E. Burch, and H. Weintraub, *PNAS* **79**, 6265 (1982).
203a. B. M. Emerson and G. Felsenfeld, *PNAS* **81**, 95 (1984).
204. W. Doerfler, *ARB* **52**, 93 (1983).
204a. J. W. Gautsch and M. C. Wilson, *Nature (London)* **301**, 32 (1983).
204b. H. Noguchi, G. P. veer Reddy, and A. B. Pardee, *Cell* **32**, 443 (1983).
204c. T. H. Bestor and V. M. Ingram, *PNAS* **80**, 5559 (1983).
205. M. Busslinger, E. deBoer, S. Wright, F. G. Grosveld, and R. A. Flavell, *NARes* **11**, 3559 (1983).
205a. E. Keshet and H. Cedar, *NARes* **11**, 3571 (1983).
206. S. Jentsch, U. Gunthert, and T. A. Trautner, *NARes* **9**, 2753 (1981).
207. C. Waalwijk and R. A. Flavell, *NARes* **5**, 3231 (1978).
208. C. J. Shen, in "DNA Methylation" (A. Razin, H. Cedar, and A. Riggs, eds.), Springer-Verlag, Berlin and New York, 1984. (In press.)
209. M. Sheffery, R. A. Rifkind, and P. A. Marks, submitted.
210. Y. Shaul, I. Ginzburg, and H. Aviv, *EJB* **114**, 591 (1981).
211. M. Sheffery, R. A. Rifkind, and P. A. Marks, *PNAS* (in press).
212. H. Weintraub, H. Beug, M. Groudine, and T. Graf, *Cell* **28**, 931 (1981).
212a. K. Tanaka, E. Appella, and G. Jay, *Cell* **35**, 457 (1983).
213. L. S. Haigh, B. B. Owens, S. Hellewell, and V. M. Ingram, *PNAS* **79**, 5332 (1982).
214. L. S. Haigh, B. B. Owens, and V. M. Ingram, in "Globin Gene Expression and Hematopoietic Differentiation" (G. Stamatoyannopoulos and A. W. Nienhuis, eds.), p. 205. Liss, New York, 1981.
215. L. S. Haigh, S. Hellewell, I. B. Roninson, B. B. Owens, and V. M. Ingram, in "Cell Function and Differentiation," p. 35. Liss, New York, 1982.
216. F. Mavilio, A. Giampaolo, A. Care, G. Migliaccio, M. Calandrini, G. Russo, G. L. Pagliardi, G. Mastroberardino, M. Marinucci, and C. Peschle, *PNAS* **80**, 6907 (1983).
216a. L. H. van der Ploeg and R. A. Flavell, *Cell* **19**, 947 (1980).
217. R. Stein, Y. Gruenbaum, Y. Pollack, A. Razin, and H. Cedar, *PNAS* **79**, 61 (1982).
218. M. Busslinger, J. Hurst, and R. A. Flavell, *Cell* **34**, 197 (1983).
218a. A. M. Maxam and W. Gilbert, *Methods Enzymol.* **65**, 499 (1980).
219. C. Wu and W. Gilbert, *PNAS* **78**, 1577 (1981).
220. F. Creusot, G. Acs, and J. K. Christman, *JBC* **257**, 2041 (1982).
220a. P. A. Jones and S. M. Taylor, *JMB* **162**, 679 (1982).
221. P. A. Jones and S. M. Taylor, *Cell* **20**, 85 (1980).
222. J. DeSimone, S. Biel, and P. Heller, *PNAS* **75**, 2937 (1978).
223. J. DeSimone, P. Heller, J. Amsel, and M. Usman, *J. Clin. Invest.* **65**, 224, (1980).
224. J. DeSimone, P. Heller, L. Hall, and D. Zwiers, in "Regulation of Hemoglobin Biosynthesis" (E. Goldwasser, ed.), p. 351. Elsevier, Amsterdam, 1983.
225. J. DeSimone, P. Heller, L. Hall, and D. Zwiers, *PNAS* **79**, 4428 (1982).
226. J. DeSimone, P. Heller, J. C. Schimenti, and C. H. Duncan, in "Globin Gene Expression and Hematopoietic Differentiation" (G. Stamatoyannopoulos and A. W. Nienhuis, eds.), p. 489. Liss, New York, 1983.
227. T. Ley, J. DeSimone, N. Anagnou, G. Keller, R. Humphries, P. Turner, N. Young, P. Heller, and A. Nienhuis, *N. Engl. J. Med.* **307**, 1469 (1982).
228. T. Ley, J. DeSimone, C. Noguchi, P. Turner, A. Schechter, P. Heller, and A. Nienhuis, *Blood* **62**, 370 (1983).

229. S. Charache, G. Dover, K. Smith, C. Talbot, M. Moyer, and S. Boyer, PNAS 80, 4842 (1983).
229a. G. Kolata, Science (Washington, D. C.) 223, 470 (1984).
230. E. Benz, N. Engl. J. Med. 307, 1515 (1982).
231. S. Orkin, Nature (London) 301, 108 (1983).
232. P. Barton, S. Malcolm, C. Murphy, and M. A. Fergusson-Smith, JMB 156, 269 (1982).
233. P. F. R. Little, Cell 28, 683 (1982).
234. N. J. Proudfoot and T. Maniatis, Cell 21, 537 (1980).
234a. E. Whitelaw and N. J. Proudfoot, NARes 11, 7717 (1983).
235. I. Sawada, M. P. Beal, C. J. Shen, B. Chapman, A. C. Wilson, and C. W. Schmid, NARes 11, 8087 (1983).
236. S. A. Liebhaber, M. Goossens, and Y. W. Kan, Nature (London) 290, 26 (1981).
237. S. A. Liebhaber and Y. W. Kan, J. Clin. Invest. 68, 439 (1981).
238. S. H. Orkin and S. C. Goff, Cell 24, 345 (1981).
238a. S. A. Liebhaber and Y. W. Kan, JBC 257, 11852 (1982).
239. J. F. Hess, M. Fox, C. Schmid, and C. J. Shen, PNAS 80, 5970 (1983).
239a. S. A. Liebhaber and K. A. Begley, NARes 11, 8915 (1983).
240. D. R. Higgs, S. E. Y. Goodbourn, J. S. Wainscoat, J. B. Clegg, and D. J. Weatherall, NARes 9, 4213 (1981).
241. B. S. Chapman, K. A. Vincent, and A. C. Wilson, submitted.
242. S. E. Y. Goodbourn, D. R. Higgs, J. B. Clegg, and D. J. Weatherall, PNAS 80, 5022 (1983).
243. G. I. Bell, M. J. Selby, and W. J. Rutter, Nature (London) 295 31 (1981).
244. R. V. Lebo, A. Chakravarti, K. H. Buetow, M. Cheung, H. Cann, B. Cordell, and H. Goodman, PNAS 80, 4808 (1983).
245. C. J. Shen and T. Maniatis, J. Mol. Appl. Genet. 1, 343 (1982).
246. J. Gusella, A. Varsanyi-Breiner, F. T. Kao, C. Jones, T. T. Puck, C. Keys, S. Orkin, and D. Housman, PNAS 76, 5239 (1979).
247. J. Gusella, PNAS 77, 2829 (1980).
248. J. Gusella, C. Jones, F. T. Kao, D. Housman, and T. T. Puck, PNAS 79, 7804 (1982).
249. B. deMartinville and U. Francke, Nature (London) 305, 641 (1983).
249a. S. E. Antonarakis, J. A. Phillips, R. L. Mallonee, H. H. Kazazian, E. R. Fearon, P. G. Waber, H. M. Kronenberg, A. Ullrich, and D. A. Meyers, PNAS 80, 6615 (1983).
249b. C. C. Morton, I. R. Kirsch, R. Taub, S. M. Orkin, and J. A. Brown, Am. J. Hum. Genet. 36, 576 (1984).
250. R. Bernards, P. F. Little, G. Annison, R. Williamson, and R. A. Flavell, PNAS 76, 4827 (1979).
251. P. F. Little, R. A. Flavell, J. M. Kooter, G. Annison, and R. Williamson, Nature (London) 278, 227 (1979).
252. D. Tuan, P. A. Biro, J. K. deRiel, H. Lazarus, and B. G. Forget, NARes 6, 2519 (1979).
253. F. G. Grosveld, H. H. Dahl, E. deBoer, and R. A. Flavell, Gene 13, 227 (1981).
254. S. Shen and O. Smithies, NARes 10, 7809 (1982).
255. G. DiSegni, G. Carrara, G. R. Tocchini-Valentini, C. C. Shoulders, and F. E. Baralle, NARes 9, 6709 (1981).
255a. P. Rogan, J. Pan, and S. M. Weissman, unpublished data.
256. P. Rogan, C. Stoeckert, F. Collins, J. Pan, P. Jagadeeswaran, B. Forget, and S. M. Weissman, unpublished data.
257. J. Pan, C. Stoeckert, Y. Fukumaki, and F. Collins, unpublished data.

258. M. Poncz, E. Schwartz, M. Ballantine, and S. Surrey, *JBC* **258**, 11599 (1983).
259. R. E. Kaufman, P. J. Kretschmer, J. W. Adams, H. C. Coon, W. F. Anderson, and A. W. Nienhuis, *PNAS* **77**, 4229 (1980).
260. F. G. Grosveld, T. Lund, E. J. Murray, A. L. Mellor, H. H. M. Dahl, and R. A. Flavell, *NARes* **10**, 6715 (1982).
261. E. F. Fritsch, R. M. Lawn, and T. Maniatis, *Nature (London)* **279**, 598 (1979).
262. R. Bernards and R. A. Flavell, *NARes* **8**, 1521 (1980).
263. N. Maeda, J. B. Bliska, and O. Smithies, *PNAS* **80**, 5012 (1983).
265. G. P. Smith, *Science (Washington, D. C.)* **191**, 528 (1976).
266. P. Soriano, M. Meunier-Rotival, and G. Bernardi, *PNAS* **80**, 1816 (1983).
267. M. F. Singer, *Cell* **28**, 433 (1982).
268. W. F. Doolittle and C. Sapienza, *Nature (London)* **284**, 601 (1980).
269. L. E. Orgel and F. H. C. Crick, *Nature (London)* **284**, 604 (1980).
270. C. M. Houck, F. P. Rinehart, and C. W. Schmid, *JMB* **132**, 289 (1979).
271. W. Jelinek, T. P. Toomey, L. Leinwand, C. H. Duncan, P. A. Biro, P. V. Choudary, S. M. Weissman, C. M. Rubin, C. M. Houck, P. L. Deininger, and C. W. Schmid, *PNAS* **77**, 1398 (1980).
271a. P. L. Deininger, D. J. Jolly, C. M. Rubin, T. Friedman, and C. W. Schmid, *JMB* **151**, 17 (1981).
272. V. B. Reddy, S. S. Tevethia, M. J. Tevethia, and S. M. Weissman, *PNAS* **79**, 2064 (1982).
273. J. T. Elder, J. Pan, and S. M. Weissman, *NARes* **9**, (1981).
274. C. Duncan, P. A. Biro, P. V. Choudary, J. T. Elder, R. R. Wang, B. G. Forget, J. K. deRiel, and S. M. Weissman, *PNAS* **76**, 5095 (1979).
275. S. A. Fuhrman, P. L. Deininger, P. LaPorte, T. Friedmann, and E. P. Geiduschek, *NARes* **9**, 6439 (1981).
276. C. H. Duncan, P. Jagadeeswaran, R. Wang, and S. Weissman, *Gene* **3**, 185 (1981).
277. P. Jagadeeswaran, B. G. Forget, and S. M. Weissman, *Cell* **26**, 141 (1981).
278. P. A. Sharp, *Nature (London)* **301**, 471 (1983).
279. G. I. Bell, R. Pictet, and W. J. Rutter, *NARes* **8**, 4091 (1980).
280. G. Grimaldi and M. F. Singer, *PNAS* **79**, 1497 (1982).
281. J. T. Elder, Ph.D. Thesis, Yale University, 1982.
281a. G. R. Daniels and P. L. Deininger, *NARes* **11**, 7595 (1983).
282. L. W. Coggins, G. J. Grindlay, J. K. Vass, A. A. Slater, P. Montague, M. A. Stinson, and J. Paul, *NARes* **8**, 3319 (1980).
283. L. W. Coggins, W. G. Lanyon, A. A. Slater, G. J. Grindlay, and J. Paul, *Biosci. Rep.* **1**, 309 (1981).
284. L. W. Coggins, J. K. Vass, M. A. Stinson, W. G. Lanyon, and J. Paul, *Gene* **17**, 113 (1982).
285. C. W. Schmid and W. E. Jelinek, *Science (Washington, D. C.)* **216**, 1065, (1982).
286. J. J. Maio, F. L. Brown, W. G. McKenna, and P. R. Musich, *Chromosoma* **83**, 127 (1981).
287. D. Gillespie, J. W. Adams, C. Costanzi, and M. J. Caranfa, *Gene* **20**, 409 (1982).
287a. B. Shafit-Zagardo, J. J. Maio, and F. L. Brown, *NARes* **10**, 3175 (1982).
288. P. R. Musich, F. L. Brown, and J. J. Maio, *Chromosoma* **80**, 331 (1980).
289. J. W. Adams, R. E. Kaufman, P. J. Kretschmer, M. Harrison, and A. W. Nienhuis, *NARes* **8**, 6113 (1980).
290. B. G. Forget, D. Tuan, P. A. Biro, P. Jagadeeswaran, and S. M. Weissman, *Trans. Am. Assoc. Phys.* **94**, 204 (1981).
291. L. Manuelidis and P. A. Biro, *NARes* **10**, 3221 (1982).

291a. H. Takahashi, Y. Hakamata, Y. Watanabe, R. Kikuno, T. Miyata, and S. Numa, *NARes* **11**, 6847 (1983).
291b. S. Potter, *PNAS* **81**, 1012 (1984).
291c. D. Murphy, P. M. Brickell, D. S. Latchman, K. Willison, and P. W. J. Rigby, *Cell* **35**, 865 (1983).
291d. P. Brulet, M. Kaghad, Y. Xu, O. Croissant, and F. Jacob, *PNAS* **80**, 5641 (1983).
291e. J. G. Sutcliff, R. J. Milner, F. E. Bloom, and R. A. Lerner, *PNAS* **79**, 4242 (1982).
291f. J. Chamberlain and T. Stanner, personal communication.
291g. S. Potter and R. Jones, *NARes* **11**, 3137 (1983).
292. M. I. Lerman, R. E. Thayer, and M. F. Singer, *PNAS* **80**, 3966 (1983).
292a. P. Rogan, personal communication.
293. L. Manuelidis, *NARes* **10**, 3211 (1982).
294. R. E. Thayer and M. F. Singer, *MCBiol* **3**, 967 (1983).
295. T. G. Fanning, *NARes* **10**, 5003 (1982).
296. R. Wilson and U. Storb, *NARes* **11**, 1803 (1983).
297. M. F. Singer, R. E. Thayer, G. Grimaldi, M. I. Lerman, and T. G. Fanning, *NARes* **11**, 5739 (1983).
298. J. D. M. Brown and M. Piechaczyk, *JMB* **165**, 249 (1983).
298a. L. Donehower, A. Huang, and G. Hager, *J. Virol.* **37**, 226 (1981).
299. T. Tsukada, Y. Watanabe, Y. Nakai, H. Imura, S. Nakanishi, and S. Numa, *NARes* **10**, 1471 (1982).
301. L. DiGiovanni, S. R. Haynes, R. Misra, and W. R. Jelinek, *PNAS* **80**, 6533 (1983).
302. G. Grimaldi and M. F. Singer, *NARes* **11**, 321 (1983).
303. L. B. Kole, S. R. Haynes, and W. R. Jelinek, *JMB* **165**, 257 (1983).
303a. P. Soriano, M. Meunier-Rotival, and G. Bernardi, *PNAS* **80**, 1816 (1983).
304. B. Shafit-Zagardo, F. L. Brown, J. J. Maio, and J. W. Adams, *Gene* **20**, 397 (1982).
304a. W. Gebhard and H. G. Zachau, *JMB* **170**, 255 (1983).
304b. W. Gebhard, T. Meitinger, J. Hochtl, and H. G. Zachau, *JMB* **157**, 453 (1982).
304c. J. Rogers, *Nature* **306**, 113 (1983).
305. P. Jagadeeswaran, P. A. Biro, D. Tuan, J. Pan, B. G. Forget, and S. M. Weissman, in "Human Genetics, Part A: The Unfolding Genome," p. 29, Liss, New York, 1982.
307. R. Egel, *Nature (London)* **290**, 191 (1981).
308. T. Petes and G. R. Fink, *Nature (London)* **300**, 216 (1982).
309. P. Kourilsky, *Biochimie* **65**, 85 (1983).
310. D. Baltimore, *Cell* **24**, 592 (1981).
311. H. L. Klein, and T. D. Petes, *Nature (London)* **289**, 144 (1981).
312. J. A. Jackson and G. R. Fink, *Nature (London)* **292**, 306 (1981).
313. P. Munz, H. Amstutz, J. Kohli, and U. Leupold, *Nature (London)* **300**, 225 (1982).
314. M. S. Meselson and C. M. Radding, *PNAS* **72**, 358 (1975).
315. J. W. Szostak, T. L. Orr-Weaver, R. J. Rothstein, and F. W. Stahl, *Cell* **33**, 25 (1983).
316. O. Smithies, W. R. Engels, J. R. Devereux, J. L. Slightom, and S. Shen, *Cell* **26**, 345 (1981).
317. C. J. Stoeckert, F. S. Collins, and S. M. Weissman, *NARes*, in press (1984).
318. J. Rogers, *Nature (London)* **305**, 101 (1983).
319. S. Neidle, *Nature (London)* **302**, 574 (1983).
320. D. B. Haniford and D. E. Pulleyblank, *Nature (London)* **302**, 632 (1983).
321. A. Nordheim and A. Rich, *PNAS* **80**, 1821 (1983).
321a. M. W. Kilpatrick *et al.*, *JBC* **259**, 7268 (1984).
322. R. Miesfeld, M. Krystal, and N. Arnheim, *NARes* **9**, 5931 (1981).

323. R. C. Hardison, *Mol. Biol. Evol.* (in press).
324. T. D. Petes, *Am. J. Hum. Genet.* **34**, 820 (1982).
325. R. Nairn and K. Yamaga, *ARGen* **14**, 241 (1980).
326. E. H. Weiss, A. Mellor, L. Golden, K. Fahrner, E. Simpson, J. Hurst, and R. A. Flavell, *Nature (London)* **301**, 671 (1983).
327. D. H. Schulze, L. R. Pease, S. S. Geier, A. A. Reyes, L. A. Sarmiento, R. B. Wallace, and S. G. Nathenson, *PNAS* **80**, 2007 (1983).
327a. A. L. Mellor, E. H. Weiss, K. Ramachandran, and R. A. Flavell, *Nature (London)* **306**, 792 (1983).
328. R. Ollo and F. Rougeon, *Cell* **32**, 515 (1983).
329. E. A. Zimmer, S. L. Martin, S. M. Beverley, Y. W. Kan, and A. C. Wilson, *PNAS* **77**, 2158 (1980).
330. R. M. Liskay and J. L. Stachelek, *Cell* **35**, 157 (1983).
331. M. Goodman and G. W. Moore, *Nature (London)* **253**, 603 (1975).
332. J. Czelusniak, M. Goodman, D. Hewett-Emmett, M. L. Weiss, P. J. Venta, and R. E. Tashian, *Nature (London)* **298**, 297 (1982).
332a. A. J. Jeffreys, V. Wilson, D. Wood, J. P. Simons, R. M. Kay, and J. G. Williams, *Cell* **21**, 555 (1980).
333. I. B. Robinson and V. M. Ingram, *PNAS* **78**, 4782 (1981).
334. I. B. Robinson and V. M. Ingram, *Cell* **28**, 515 (1982).
335. I. B. Robinson and V. M. Ingram, *JBC* **258**, 802 (1983).
336. A. C. Wilson, S. S. Carlson, and T. J. White, *ARB* **46**, 573 (1977).
337. P. Jagadeeswaran and P. M. McGuire, *NARes* **10**, 433 (1982).
338. R. Staden, *NARes* **10**, 2951 (1982).
339. G. Dover, *Nature (London)* **299**, 111 (1982).
340. F. Perler, A. Efstratiadis, P. Lomedico, W. Gilbert, R. Kolodner, and J. Dodgson, *Cell* **20**, 555 (1980).
341. M. Goodman, *J. Mol. Evol.* **17**, 114 (1981).
342. S. C. Hardies, M. H. Edgell, and C. A. Hutchison, *JBC* **259**, 3748 (1984).
343. S. L. Martin, K. A. Vincent, and A. C. Wilson, *JMB* **164**, 513 (1983).
343a. R. Hardison and J. Margot, *Mol. Biol. Evol.* **1**, 302 (1984).
343b. A. R. Templeton, *Evolution* **37**, 221 (1983).
344. J. B. Lingrel, E. A. Schon, M. L. Clearly, and S. G. Shapiro, *in* "Regulation of Hemoglobin Biosynthesis" (E. Goldwasser, ed.), p. 89. Elsevier, Amsterdam, 1983.
345. S. Hardies and M. Edgell, personal communication.
346. J. Buettner-Janusch, V. Buettner-Janusch, and D. Coppenhaver, *Folia Primatol.* **17**, 177 (1972).
347. D. H. Coppenhaver, J. D. Dixon, and L. K. Duffy, *Hemoglobin* **7**, 1 (1983).
348. International Hemoglobin Information Center Tabulation, Comprehensive Sickle Cell Center, Augusta, Georgia, 1981.
349. V. A. McKusick, "Mendelian Inheritance in Man," 6th ed. Johns Hopkins Press, Baltimore, Maryland, 1983.
350. S. H. Orkin, H. H. Kazazian Jr., S. E. Antonarakis, H. Ostrer, S. C. Goff, and J. P. Sexton, *Nature (London)* **300**, 768 (1982).
351. R. Treisman, S. H. Orkin, T. Maniatis, Structural and Functional Defects in β-Thalassemia, *in* "Globin Gene Expression and Hematopoietic Differentiation" (A. W. Nienhuis and G. Stamatoyannopoulos, eds.). Liss, New York, 1983.
352. E. J. Benz and B. G. Forget, *Annu. Rev. Med.* **33**, 363 (1982).
353. E. J. Benz and B. G. Forget, *Pathobiology Annual* **10**, 1 (1980).
354. D. J. Weatherall and J. B. Clegg, *Cell* **29**, 7 (1982).

355. M. H. Steinberg and J. G. Adams, *Am. J. Pathol.* 113, 396 (1983).
356. S. H. Orkin and S. E. Antonarakis, and H. H. Kazazian, *Prog. Hematol.* 13, 49 (1983).
357. S. H. Orkin and D. G. Nathan, *Adv. Hum. Genet.* 11, 233 (1981).
358. D. R. Higgs and D. J. Weatherall, *Curr. Top. Hematol.* 4, 37 (1983).
359. A. M. Dozy, Y. W. Kan, S. H. Embury, W. C. Mentzer, W. C. Wang, B. Lubin, J. R. Davis, and H. M. Koenig, *Nature (London)* 280, 605 (1979).
360. K. Ohene-Frempong, E. Rappaport, J. Atwater, E. Schwartz, and S. Surrey, *Blood* 56, 931 (1980).
361. D. R. Higgs, L. Pressley, J. M. Old, D. M. Hunt, J. B. Clegg, D. J. Weatherall, and G. R. Serjeant, *Lancet* 2, 272 (1979).
362. D. J. Weatherall, personal communication.
363. S. H. Embury, J. A. Miller, A. M. Dozy, Y. W. Kan, V. Chan, and D. Todd, *J. Clin. Invest.* 66, 1319 (1980).
364. J. A. Phillips, T. A. Vik, A. F. Scott, K. E. Young, H. H. Kazazian, K. D. Smith, V. F. Fairbanks, and H. M. Koenig, *Blood* 55, 1066 (1980).
365. A. M. Michelson and S. H. Orkin, *JBC* 258, 15245 (1983).
366. L. Pressley, D. R. Higgs, B. Aldridge, A. Metaxatou-Mavromati, J. B. Clegg, and D. J. Weatherall, *NARes* 8, 4889 (1980).
367. L. Pressley, D. R. Higgs, J. B. Clegg, and D. J. Weatherall, *PNAS* 77, 3586 (1980).
368. D. R. Higgs, L. Pressley, B. Aldridge, J. B. Clegg, D. J. Weatherall, A. Cao, M. G. Hadjiminas, C. Kattamis, A. Metaxatou-Mavromati, E. A. Rachmilewitz, and T. Sophocleous, *PNAS* 78, 5833 (1981).
369. S. H. Orkin and A. Michelson, *Nature (London)* 286, 538 (1980).
370. S. H. Embury, R. V. Lebo, A. M. Dozy, and Y. W. Kan, *J. Clin. Invest.* 63, 1307 (1979).
371. D. R. Higgs, J. M. Old, L. Pressley, J. B. Clegg, and D. J. Weatherall, *Nature (London)* 284, 632 (1980).
372. M. Goossens, A. M. Dozy, S. H. Embury, Z. Zachariades, M. G. Hadjiminas, G. Stamatoyannopoulos, and Y. W. Kan, *PNAS* 77, 518 (1980).
373. R. J. Trent, D. R. Higgs, J. B. Clegg, and D. J. Weatherall, *Br. J Haematol.* 49, 149 (1981).
374. L. E. Lie-Injo, A. R. Herrera, and Y. W. Kan, *NARes* 9, 3707 (1981).
375. D. J. Weatherall, *Br. J. Haematol.* 49, 149 (1981).
375a. S. L. Thein, I. Al-Hakim, and A. V. Moffbrand, *Br. J. Haematol.* 56, 333 (1984).
376. P. Winichagoon, D. R. Higgs, S. E. Y. Goodbourn, J. Lamb, J. B. Clegg, and D. J. Weatherall, *NARes* 10, 5853 (1982).
377. D. J. Weatherall, D. R. Higgs, C. Bunch, J. M. Old, D. M. Hunt, L. Pressley, J. B. Clegg, N. C. Bethlenfalvay, S. Sjolin, R. D. Koller, E. Magens, J. L. Francis, and D. Bebbington, *N. Engl. J. Med.* 305, 607 (1981).
378. D. J. Weatherall, D. R. Higgs, J. B. Clegg, and W. E. Wood, *Br. J. Haematol.* 52, 351 (1982).
379. S. H. Orkin, S. C. Goff, and R. L. Hechtman, *PNAS* 78, 5041 (1981).
380. D. R. Higgs, S. E. Y. Goodbourn, J. Lamb, J. B. Clegg, and D. J. Weatherall, *Nature (London)* 306, 398 (1983).
380a. D. M. Hunt, D. R. Higgs, P. Winichagoon, J. B. Clegg, and D. J. Weatherall, *Br. J. Haematol.* 51, 405 (1982).
380b. M. Goossens, K. Y. Lee, S. A. Liebhaber, and Y. W. Kan, *Nature (London)* 296, 864 (1982).
381. R. J. Trent, J. S. Wainscoat, E. R. Huehns, J. B. Clegg, and D. J. Weatherall, *Br. J. Haematol.* 52, 511 (1982).

382. J. S. Wainscoat, J. M. Old, D. J. Weatherall, and S. H. Orkin, *Lancet* **1**, 1235 (1983).
383. J. S. Wainscoat, J. I. Bell, J. M. Old, D. J. Weatherall, M. Furbetta, R. Galanello, and A. Cao, *Mol. Biol. Med.* **1**, 1 (1983).
384. D. R. Higgs, B. E. Aldridge, J. Lamb, J. B. Clegg, D. J. Weatherall, R. J. Hayes, Y. Grandison, Y. Lowrie, K. P. Mason, B. E. Serjeant, and G. R. Serjeant, *N. Engl. J. Med.* **306**, 1441 (1982).
384a. J. G. Mears, H. M. Loachman, D. Labie, and R. L. Nagel, *Blood* **62**, 286 (1983).
385. E. D. Thomas, C. D. Buckner, J. E. Sanders, T. Papayannopoulou, C. Borgna-Pignatti, P. DeStefano, K. M. Sullivan, R. A. Clift, and R. Storb, *Lancet* **2**, 227 (1982).
386. R. A. Spritz and S. H. Orkin, *NARes* **10**, 8025 (1982).
387. T. Cheng, S. H. Orkin, S. E. Antonarakis, M. J. Potter, J. P. Sexton, A. F. Markham, P. J. V. Giardina, A. Li, and H. H. Kazazian, *PNAS* **81**, 2821 (1984).
387a. S. E. Antonarakis, S. H. Orkin, T. Cheng, A. F. Scott, J. P. Sexton, S. Trusko, S. Charache, and H. H. Kazazian, *PNAS* **81**, 1154 (1980).
387b. H. H. Kazazian, S. H. Orkin, S. E. Antonarakis, J. P. Sexton, C. D. Boehm, S. C. Goff, and P. G. Waber, *EMBO J.* **3**, 593 (1984).
387c. S. M. Orkin, S. E. Antonarakis, and D. Loukopoulos, *Blood* **64**, 311 (1984).
388. J. C. Chang and Y. W. Kan, *PNAS* **76**, 2886 (1979).
389. S. H. Orkin and S. C. Goff, *JBC* **256**, 9782 (1981).
390. N. Moschonas, E. DeBoer, F. G. Grosveld, H. H. Dahl, S. Wright, C. K. Shewmaker, and R. A. Flavell, *NARes* **9**, 4391 (1981).
391. R. F. Trecartin, S. A. Liebhaber, J. C. Chang, K. Y. Lee, and Y. W. Kan, *J. Clin. Invest.* **68**, 1012 (1981).
392. R. Pergolizzi, R. A. Spritz, S. Spence, M. Goossens, Y. W. Kan, and A. Bank, *NARes* **9**, 7065 (1981).
393. H. H. Kazazian, S. H. Orkin, C. D. Boehm, J. P. Sexton, and S. E. Antonarakis, *Am. J. Hum. Genet.* **35**, 1028 (1983).
393a. J. C. Chang, A. Alberti, and Y. W. Kan, *NARes* **11**, 7789 (1983).
394. A. Kimura, E. Matsunaga, Y. Takihara, T. Nakamura, Y. Takagi, S. Lin, and H. Lee, *JBC* **258**, 2748 (1983).
395. A. J. Kinniburg, L. E. Maquat, T. Schedl, E. Rachmilewitz, and J. Ross, *NARes* **25**, 5421 (1982).
395a. S. H. Orkin, personal communication.
395b. S. M. Orkin, S. E. Antonarakis, and D. Loukopoulos, *JBC*, in press.
396. M. Poncz, M. Ballantine, D. Solowiejczyk, I. Barak, E. Schwartz, and S. Surrey, *JBC* **257**, 5994 (1982).
397. S. H. Orkin, J. P. Sexton, T. Cheng, S. C. Goff, P. J. V. Giardina, J. I. Lee, and H. H. Kazazian, *NARes* **11**, 4727 (1983).
398. R. A. Spritz, P. Jagadeeswaran, P. V. Choudary, P. A. Biro, J. T. Elder, J. K. Deriel, J. L. Manley, M. L. Gefter, B. G. Forget, and S. M. Weissman, *PNAS* **78**, 2455 (1981).
399. D. Westway and R. Williamson, *NARes* **9**, 1777 (1981).
400. S. E. Spence, R. G. Pergolizzi, M. Donovan-Peluso, K. A. Rosche, C. S. Dobkin, and A. Bank, *NARes* **10**, 1283 (1982).
401. M. E. Goldsmith, R. K. Humphries, T. Ley, A. Cline, J. A. Kantor, and A. W. Nienhuis, *PNAS* **80**, 2318 (1983).
401a. L. C. Skow, B. A. Burkhart, F. M. Johnson, R. A. Popp, D. M. Popp, S. Z. Goldberg, W. F. Anderson, L. B. Barnett, and S. E. Lewis, *Cell* **34**, 1043 (1983).
402. Y. Fukumaki, P. K. Ghosh, E. J. Benz, V. B. Reddy, P. Lebowitz, B. G. Forget, and S. M. Weissman, *Cell* **28**, 585 (1982).

403. J. Banerji, S. Rusconi, and W. Schaffner, *Cell* **27**, 299 (1981).
404. R. A. Flavell, F. G. Grosveld, G. C. Grosveld, S. Wright, M. Busslinger, E. deBoer, D. Kioussis, A. L. Mellor, L. Golden, E. Weiss, J. Hurst, H. Bud, H. Bullman, E. Simpson, R. James, A. R. M. Townsend, P. M. Taylor, W. Schmidt, J. Ferluga, L. Leben, M. Santamaria, G. Atfield, and H. Festenstein, *ICN–UCLA Symp. Mol. Cell. Biol.* (1982).
404a. T. J. Ley, N. P. Anagnou, G. Pepe, and A. W. Nienhuis, *PNAS* **79**, 4775 (1982).
405. G. F. Temple, A. M. Dozy, K. L. Roy, and Y. W. Kan, *Nature (London)* **296**, 537 (1982).
406. K. Takeshita, B. G. Forget, A. Scarpa, and E. J. Benz, *Blood* **64**, 13 (1984).
407. R. K. Humphries, T. J. Ley, N. P. Anagnou, A. W. Baur, and A. W. Nienhuis, *Blood* (in press).
408. P. Dierks, A. van Ooyen, N. Mantei, and C. Weissmann, *PNAS* **78**, 1411 (1981).
409. J. T. Wilson, L. B. Wilson, and Y. Ohta *BBRC* **99**, 1035 (1981).
410. M. Pirastu, R. Galanello, M. A. Melis, C. Brancati, A. Tagarelli, A. Cao, and Y. W. Kan, *Blood* **62**, 341 (1983).
411. A. Kimura, E. Matsunaga, Y. Ohta, T. Fujiyoshi, T. Matsuo, T. Nakamura, T. Imamura, T. Yanase, and Y. Takagi, *NARes* **10**, 5725 (1982).
412. L. H. T. van der Ploeg, A. Konings, M. Oort, D. Roos, L. Bernini, and R. A. Flavell, *Nature (London)* **283**, 637 (1980).
413. R. A. Flavell, J. M. Kooter, E. deBoer, P. F. Little, and R. Williamson, *Cell* **15**, 25 (1978).
414. S. Ottolenghi, B. Giglioni, P. Comi, A. M. Gianni, E. Polli, C. T. Acquaye, J. H. Oldham, and G. Masera, *Nature (London)* **278**, 654 (1979).
415. E. Vanin, P. Henthorn, D. Kioussis, F. Grosveld, and O. Smithies, *Cell* **35**, 701 (1983).
416. F. Ramirez, J. G. Mears, U. Nudel, A. Bank, L. Luzzatto, G. DiPrisco, R. D'Avino, G. Pepe, R. Gambino, R. Cimino, and N. Quattrin, *J. Clin. Invest.* **63**, 736 (1979).
417. M. Baird, H. Schreiner, C. Driscoll, and A. Bank, *J. Clin. Invest.* **68**, 560 (1981).
417a. F. Mavilio, A. Giampaolo, A. Care, N. M. Sposi, and M. Marinucci, *Blood* **62**, 230 (1983).
418. P. J. Ojwang, T. Nakatsuji, M. B. Gardiner, A. L. Reese, J. G. Gilman, and T. H. J. Huisman, *Hemoglobin* **7**, 115 (1983).
419. P. K. Sukumaran, T. Nakatsuji, M. B. Gardiner, A. L. Reese, J. G. Gilman, and T. H. J. Huisman, *NARes* **11**, 4635 (1983).
420. R. Bernards, J. M. Kooter, and R. A. Flavell, *Gene* **6**, 265 (1979).
421. R. W. Jones, J. M. Old, R. J. Trent, J. B. Clegg, and D. J. Weatherall, *Nature (London)* **291**, 39 (1981).
421a. S. L. Thein et al., *Br. J. Haematol.* **57**, 271 (1984).
422. S. H. Orkin, B. P. Alter, and C. Altay, *J. Clin. Invest.* **64**, 866 (1979).
423. S. Ottolenghi, B. Giglioni, R. Taramelli, P. Comi, U. Mazza, G. Saglio, C. Camaschella, P. Izzo, A. Cao, R. Galanello, E. Gimferrer, M. Baiget, and A. M. Gianni, *PNAS* **79**, 2347 (1982).
424. S. Ottolenghi and B. Giglioni, *Nature (London)* **300**, 770 (1982).
425. D. Tuan, M. J. Murnane, J. K. deRiel, and B. G. Forget, *Nature (London)* **258**, 335 (1980).
426. D. Tuan, E. Feingold, M. Newman, S. M. Weissman, and B. G. Forget, *PNAS* **80**, 6937 (1983).
427. P. Jagadeeswaran, D. Tuan, B. G. Forget, and S. M. Weissman, *Nature (London)* **296**, 469 (1982).
428. R. W. Jones, J. M. Old, J. B. Clegg, and D. J. Weatherall, *NARes* **9**, 6813 (1981).

429. S. H. Orkin, S. C. Goff, and D. G. Nathan, *J. Clin. Invest.* **67**, 878 (1981).
430. D. Kioussis, E. Vanin, T. deLange, R. A. Flavell, and F. G. Grosveld, *Nature (London)* **306**, 662 (1983).
431. E. R. Fearon, H. H. Kazazian, P. G. Waber, J. I. Lee, S. E. Antonarakis, S. H. Orkin, E. F. Vanin, P. S. Henthorn, F. G. Grosveld, A. F. Scott, and G. R. Buchanan, *Blood* **61**, 1269 (1983).
432. M. Pirastu, Y. W. Kan, C. C. Lin, R. M. Baine, and C. T. Holbrook, *J. Clin. Invest.* **72**, 602 (1983).
433. S. H. Orkin, J. M. Old, D. J. Weatherall, and D. G. Nathan, *PNAS* **76**, 2400 (1979).
434. W. G. Wood, J. B. Clegg, and D. J. Weatherall, *Br. J. Haematol.* **43**, 509 (1979).
435. D. Mathog, M. Hochstrasser, Y. Gruenbaum, H. Saumweber, and J. Sedat, submitted.
436. T. H. J. Huisman, M. E. Gravely, B. Webber, K. Okonjo, J. Henson, and A. L. Reese, *Blood* **58**, 62 (1981).
437. T. H. J. Huisman, W. A. Schroeder, G. D. Efremov, H. Duma, B. Mladenovskil, C. B. Hyman, E. A. Rachmilewitz, N. Bouver, A. Miller, A. Brodie, J. R. Shelton, J. B. Shelton, and G. Apell, *Ann. N.Y. Acad. Sci.* **232**, 107 (1974).
438. O. Smithies, *J. Cell. Physiol. Suppl.* **1**, 137 (1982).
439. P. Fessas and G. Stamatoyannopoulos, *Blood* **24**, 223 (1964).
440. K. Sofroniadou, W. G. Wood, P. E. Nute, and G. Stamatoyannopoulos, *Br. J. Haematol.* **29**, 137 (1975).
441. R. W. Jones, J. M. Old, W. G. Wood, J. B. Clegg, and D. J. Weatherall, *Br. J. Haematol.* **50**, 415 (1982).
442. T. Papayannopoulou, R. M. Lawn, G. Stamatoyannopoulos, and T. Maniatis, *Br. J. Haematol.* **50**, 387 (1982).
443. F. S. Collins, unpublished observation.
444. M. Farquhar, R. Gelinas, B. Tatsis, J. Murray, M. Yagi, R. Mueller, and G. Stamatoyannopoulos, *Am. J. Hum. Genet.* **35**, 611 (1983).
445. J. F. Balsley, E. Rappaport, E. Schwartz, and S. Surrey, *Blood* **59**, 828 (1982).
446. F. S. Collins, C. J. Stoeckert, G. Serjeant, B. Forget, S. M. Weissman, *PNAS* **81**, 4894 (1984).
447. C. Benoist and P. Chambon, *Nature (London)* **290**, 304 (1981).
447a. M. D. Walker, T. Edlund, A. M. Boulet, and W. J. Rutter, *Nature (London)* **306**, 557 (1983).
448. F. C. Collins *et al.*, submitted.
449. W. G. Wood, I. A. MacRae, P. D. Darbre, J. B. Clegg, and D. J. Weatherall, *Br. J. Haematol.* **50**, 401 (1982).
450. J. M. Old, H. Ayyub, W. G. Wood, J. B. Clegg, and D. J. Weatherall, *Science (Washington, D.C.)* **215**, 981 (1982).
451. A. M. Gianni, M. Bregni, M. C. Cappellini, G. Fiorelli, R. Taramelli, B. Giglioni, P. Comi, and S. Ottolenghi, *EMBO J.* **2**, 921 (1983).
451a. A. Giampaolo *et al.*, *Hum. Genet.* **66**, 151 (1984).
452. M. C. Driscoll, M. Baird, Y. Ohta, and A. Bank, *Blood* **58**, 539 (1981).
453. R. J. Trent, D. K. Bowden, J. M. Old, J. S. Wainscoat, J. B. Clegg, and D. J. Weatherall, *NARes* **9**, 6723 (1981).
454. R. A. Spritz, *NARes* **9**, 5037 (1981).
455. N. Moschonas, E. deBoer, and R. A. Flavell, *NARes* **10**, 2109 (1982).
456. A. J. Jeffreys, *Cell* **18**, 1 (1979).
457. P. F. Little, R. Williamson, G. Annison, R. A. Flavell, E. deBoer, L. F. Bernini, S. Ottolenghi, G. Saglio, and U. Mazza, *Nature (London)* **282**, 316 (1979).

458. S. E. Antonarakis, C. D. Boehm, P. J. V. Giardina, and H. H. Kazazian, PNAS **79**, 137 (1982).
459. S. E. Antonarakis, C. D. Boehm, G. R. Serjeant, C. E. Theisen, G. J. Dover, and H. H. Kazazian, PNAS **81**, 853 (1984).
459a. J. S. Wainscoat, J. I. Bell, S. L. Thein, D. R. Higgs, G. R. Serjeant, T. E. A. Peto, and D. J. Weatherall, Mol. Biol. Med. **1**, 191 (1983).
460. S. E. Antonarakis, S. H. Orkin, H. H. Kazazian, S. C. Goff, C. D. Boehm, P. G. Waber, J. P. Sexton, H. Ostrer, V. F. Fairbanks, and A. Chakravarti, PNAS **79**, 6608 (1982).
461. H. H. Kazazian, P. G. Waber, C. D. Boehm, J. I. Lee, S. E. Antonarakis, and V. F. Fairbanks, Am. J. Hum. Genet. **36**, 212 (1984).
462. Y. W. Kan and A. M. Dozy, PNAS **75**, 5631 (1978).
463. Y. W. Kan, K. Y. Lee, M. Furbetta, A. Angius, and A. Cao, N. Engl. J. Med. **302**, 185 (1980).
464. G. Kohen, N. Philippe, and J. Godet, Hum. Genet. **62**, 121 (1982).
465. P. N. Cockerill and G. H. Goodwin, BBRC **112**, 547 (1983).
466. F. Ramirez, A. L. Burns, J. G. Mears, S. Spence, D. Starkman, and A. Bank, NARes **7**, 1147 (1979).
467. J. M. Old, R. H. T. Ward, F. Karagozlu, M. Petrou, B. Modell, and D. J. Weatherall, Lancet **2**, 1413 (1982).
468. J. M. Old and J. S. Wainscoat, Br. J. Haematol. **58**, 337 (1983).
469. M. C. Driscoll, M. Baird, A. Bank, and E. A. Rachmilewitz, J. Clin. Invest. **68**, 915 (1981).
470. H. H. Kazazian, personal communication.
471. S. R. Panny, A. F. Scott, K. D. Smith, J. A. Phillips, H. H. Kazazian, C. C. Talbot, and C. D. Boehm, Am. J. Hum. Genet. **33**, 25 (1981).
472. J. C. Hobbins and M. J. Mahoney, N. Engl. J. Med. **290**, 1065 (1974).
473. H. Chang, J. C. Hobbins, G. Cividalli, F. D. Frigoletto, M. J. Mahoney, Y. W. Kan, and D. G. Nathan, N. Engl. J. Med. **290**, 1067 (1974).
474. B. P. Alter, Lancet **2**, 1151 (1981).
475. D. J. Weatherall and J. M. Old, Mol. Biol. Med. **1**, 151 (1983).
475a. C. D. Boehm, S. E. Antonarakis, J. A. Phillips, G. Stetten, and H. H. Kazazian, N. Engl. J. Med. **308**, 1054 (1983).
476. J. C. Chang and Y. W. Kan, Lancet **2**, 1127 (1981).
477. R. F. Geever, L. B. Wilson, F. S. Nallaseth, P. F. Milner, M. Bittner, and J. T. Wilson, PNAS **78**, 5081 (1981).
478. J. C. Chang and Y. W. Kan, N. Engl. J. Med. **307**, 30 (1982).
479. S. H. Orkin, P. F. R. Little, H. H. Kazazian, and C. D. Boehm, N. Engl. J. Med. **307**, 32, (1982).
480. J. T. Wilson, P. F. Milner, M. E. Summer, F. S. Nallaseth, H. E. Fadel, R. H. Reindollar, P. G. McDonough, and L. B. Wilson, PNAS **79**, 3628 (1982).
481. M. Goossens, Y. Dumez, L. Kaplan, M. Lupker, C. Chabret, R. Henrion, and J. Rosa, N. Engl. J. Med. **309**, 831 (1983).
482. A. M. Dozy, D. N. Abuelo, G. Barsel-Bowers, M. J. Mahoney, B. G. Forget, Y. W. Kan, and E. N. Forman, JAMA **241**, 1610 (1979).
483. S. H. Orkin, B. P. Alter, C. Altay, M. J. Mahoney, H. Lazarus, H. C. Hobbins, and D. G. Nathan, N. Engl. J. Med. **299**, 166 (1978).
484. P. F. R. Little, G. Annison, S. Darling, R. Williamson, L. Camba, and B. Modell, Nature (London) **285**, 144 (1980).

485. H. H. Kazazian, J. A. Phillips, C. D. Boehm, T. A. Vik, M. J. Mahoney, and A. K. Ritchey, *Blood* **56**, 926 (1980).
485a. J. M. Old, M. Petrou, B. Modell, and D. J. Weatherall, *Br. J. Haematol.* **57**, 255 (1984).
486. B. J. Conner, A. A. Reyes, C. Morin, K. Itakura, R. L. Teplitz, and R. B. Wallace, *PNAS* **80**, 278 (1983).
487. S. H. Orkin, A. F. Markham, and H. H. Kazazian, *J. Clin. Invest.* **71**, 775 (1983).
488. M. Pirastu, Y. W. Kan, A. Cao, B. J. Conner, R. L. Teplitz, and R. B. Wallace, *N. Engl. J. Med.* **309**, 284 (1983).
489. M. Goossens and Y. Y. Kan, *Methods Enzymol.* **76**, 805 (1981).
490. R. Williamson, J. Eskdale, D. V. Coleman, M. Niazi, F. E. Loeffler, and B. M. Modell, *Lancet* **2**, 1125 (1981).
491. R. G. Elles, R. Williamson, M. Niazi, D. V. Coleman, and D. Horwell, *N. Engl. J. Med.* **308**, 1433 (1983).
492. W. F. Anderson and J. C. Fletcher, *N. Engl. J. Med.* **303**, 1293 (1980).
493. K. W. Mercola and M. J. Cline, *N. Engl. J. Med.* **303**, 1297 (1980).
494. M. J. Cline, *J. Lab. Clin. Med.* **99**, 299 (1982).
495. R. Williamson, *Nature (London)* **298**, 416 (1982).
496. A. G. Motulsky, *Science (Washington, D.C.)* **219**, 135 (1983).

Addendum 1

See pp. 439–458.

```
TGTCACCACC TTTAAGGCAA ATGTTAAATG CGCTTTGGCT GAACTTTTTC CTATTTTGAG GAACTTTTTC CTATTTTGAG CTATTTGAG ATTTGTCCT TTATATGAGG CTTTCTTGA AAAGGAGAAT    100
GGGAGACATG GATATCATTT TGAAGAGTGA TGAAGAGGGT AAAAAAGGGT ACAAATGAA ATTTGTGTTG CAGATAGTAT GAGGAGCCAA CAAAAAGAG    200
CCTCAGGATC CAGCACACAT TATCACAAAC TTAGTGTCCA TCCATCACTG CTGACCCTCT CCCGACCTGA CTCCACCCCT GAGGACACAG GTCAGCCTTG    300
ACCAATGACT TTTAAGTACC ATGGAGAACA GGGGGCCAGA ACTTCGGCAG TAAAGAATAA AAGGCCAGAC AGAGAGGCAG CAGCACATAT CTGCTTCCGA    400
                                                                          Epsilon Gene
                                                                              Cap
CACAGTGCA ATCACTAGCA AGCTCTCAGG CCTGGCATCA TGGTGCATTT TACTGCTGAG GAGCAAGGCTG CCGTCACTAG CCTGTGGAGC AAGATGAATG    500
                                Init
TGGAAGAGGC TGGAGGTGAA GCCTTGGGCA GTAAGCATT GGTTCTCAAT GCATGGAAT GAAGGTGAA TATTACCCTA GCAAGTTGAT TGGGAAAGTC    600
                                         IVS 1
                                                      Exon 2
CTCAAGATTT TTTGATGTC TAATTTGTA TCTGATATGG TGTCATTTCA TAGACTCCTC GTGTTTACC CCTGGACCCA GAGATTTTTT GACAGCTTTG    700
GAAACCTGTC GTCTCCCTCT GCCATCCTGG GCAACCCCAA GGTCAAGGCC CATGGCAAGA AGTGCTGAC TTCCTTTGA GATGCTATTA AAAACATGGA    800
CAACCTCAAG CCCGGCTTTG CTAAGCTGAG TGTGACAAGC TGCATGTGAA TCCTGAGAAC TTCAAGGTGA GTTCAGGTGC TGGTGATGTG    900
ATTTTTTGGC TTTATATTTT GACATTAATT GAAGCTCATA ATCTTATTGG AAAGACCAAC AAAGATCTCA GAAATCATGG GTCGAGCTTG ATGTTAGAAC    1000
AGCAGACTTC TAGTGAGCAT AACCAAAACT TACATGATTC AGAACTAGTG AATTGTATTA TAGAATTGTA GACTTGTGAA AGAAGAATGA AATTTGCTT TTGGTAGATG AAGTCCATT    1100
TTGCCAGAAC TTGATGTGTT TATCCCAGAG AATTGTTATT GGGTCATGAT AATTGAGGTT TAGAAGAGAT TTTTGCAAAA AAATAAAAG ATTTGCTCAA AGAAAATAA    1200
TCAAGGAAAT AGAAATGCTT TATTTTATGT TTAAATTTCC CATCAGTATT GTGACCAAGT GAAGGCTTGT TTCCGAATTT GTTGGGGATT TTAAACTCCC GCTGAGAACT    1300
GACACATTTT CTAAAATATG TTAAATTTCC CATCAGTATT AATTAGACAA ATTGCTTAAG AAAAACAGGG AGAGAGGGAA CCCAATAATA CTGGTAAAAT GGGGAAGGGG    1400
CTTGCAGCAC TCACATTCTA CATTACAAA AATTAGACAA TTGCTTAAG AAAAACAGGG AGAGAGGGAA CCCAATAATA CTGGTAAAAT GGGGAAGGGG    1500
GTGAGGGTGT AGGTAGGTAG AATGTTGAAT GTAGGGCTCA TAGAATAAAA TTGAACCTAA GCTCATCTGA ATTTTTTGGG TGGGCACAAA CCTTGAACA    1600
GTTTGAGGTC AGGGTTGTCT AGGAATGTAG GTATAAAGCC GTTTTTGTTT ATTTGTTTGT TTTTTCATCA AGTTGTTTTC GAAACTTCT ACTCAACATG    1700
                                  Exon 3
CCTGTGTGTT ATTTTGTCTT TTGCCTAACA GCTCCTGGGT AACGTGATGG TGATTATTCT GGCTACTCAC TTTGGCAAGG AGTTCACCCC TGAAGTGCAG    1800
GCTGCCTGGC AGAAGCTGGT GTCTGCTGTC GCCATTGCCC TGGCCCATAA GCTGAGAGA AGCCTTCTG GTTCTCTTCC AGTTTGCAGG TGTTCCTGTG ACCCTGACAC    1900
                          Ter
CCTCCTTCTG CACATGGGGA CTGGGCTTGG CCTTGAGAGA AGCCTTCTG TACATTTTCT TCAGTAATCA AAATGCAA TTTTATCTTC    2000
                                                    Poly-A
TCCATCTTTT ACTCTCTGTGT TAAAAGGAAA AAGTGTTCAT GGGCTGAGGG ATGGAGAGAA ACATAGGAAG AACCAAGAGC TTCCTTAAGA AATGTATGGG    2100
GGCTTGTAAA ATTAATGTGG ATGTTATGGG AGAATTCCCA AGATTCCCAA GGAGGATGAT ATGATGGAGA AAAATCTTTA TCGGGGTGGG AAAATGTTA    2200
                             Eco RI
ATTAAGTGGC AGAGACTCCT AGGCAGTTTT TACTGCACCG GGGAAAGAAG GAGTGTTTAG TGTACCTGA GAAAGCAGAT TTGTGTACA TGTCACTTTT    2300
CATTAAAAAC AAAACAAAA CAAAACAAAA CTTCATAGAT ATCCAAGATA TAGGCTGAGA ATTACTATTT TAATTTACTC TTATTTACAT TTTGAAGTAG    2400
CTAGCTTGTC ACATGTTTTA TGAAATTGAT TTGGAGATAA GATGAGTGTG TATCAACAAT AGCCTGCTCT TTCCATGAAG GATTCCATTA TTTCATGGGT    2500
```

```
TAGCTGAAGC TAAGCACAT GATATCATTG TGCATTATCT TCTGATACAA TGTAACATGC ACTAAAATAA AGTTAGAGTT AGGACCTGAG TGGGAAAGTT  2600
TTTGGAGAGT GTGATGAAGA CTTTCCGTGG GAGATAGAAT ACTAATAAAG GCTTAAATTC TAAAACCAGC AAGCTAGGGC TTCGTGACTT GCATGAAACT  2700
GGCTCTCTGG AAGTAGAAGG GAGAGTAAGA CATACGTACA GGACTAGGAA AGACCAGATA GTACAGGGCC TGGCTACAAA AATACAAGCT TTTACTATGC  2800
TATTGCAATA CTAAACGATA AGCATTAGGA TGTTAAGTGA CTCAGGAAAT AAGATTTTGG GAAAAAGTAA TCTGCTTATG TGCACAAAAT GGATTCAAGT  2900
TTGCAGATAA AATAAAATAT GGATGATGAT TCAAGGGGAC AGATACAATG GTCAAACCCC AAGAGGAGCA GTGAGTCTGT GGAATTTTGA AGAATGGACA  3000
AAGTGGGGT GAGAAAGACA TAGTATTCGA CCTGACTGTG GGAGATGAGA AGGAAGAAGG AGGTGATAAA TGACTGAAAG CTCCCAGACT GGTGAAGATA  3100
ACAGGAGGAA ACCATGCACT TGACCCTGGT GACTCTCATG TGTGAAGGGT AGAGGATAT TAACAGATTT ACTTTTTAGG AAGTGCTAGA TTGGTCAGGG  3200
AGTTTTGACC TTCAGGTCTT GTGTCTTTCA TATCAAGAAA CCTTTGCATT ACTTTTGAAT TTCCAAGTTA GAGTGCCATA TTTTGGCAAA TATAACTTTA TTAGTAATTT  3300
TATAGTGCTC TCACATTGAT CAGACTTTTT CCTGTGAATT CCTGTGAATT ACTTTTGAAT TTGGCTGTAT ATATCCAGAA TATGGGAGAG AGACAAATAA TTATTGTAGT  3400
TGCAGGGTAT CAACAATACT GGTCTCTCTG AGCCTTATAA CCTTTCAATA TGCCCCATAA ACAGAGTAAA CAGGATTAT TCATGGCACT AAATATTTTC  3500
ACCTAGGTCA GTCAACAAAT GGAGGCAATG TGCATTTTTT GATACATATT TTTATATATT TATGTATTG AGAAAATTTC AAATCCTCAT TTCTGACCAT CATGTGACT  3600
GATTAAGTCT AGATATTTAG GATATCCATT ACTTTGAGCA TTTATCATTT CTATGTATTG AGAAAATTTC AAATCCTCAT TTCTGACCAT TTTGAAATAT  3700
ATAATAAATA GTAATTAACT ATAGTCACCC TACTCAAATA TCAACATTAT AAACTAACTA ATCCTCTTT CCACTTTTTT ACCAACCAC ATCTCTTAAA  3800
TCCCCTGCCA TACAACATAC ACATTTTCA GTCTGATAA CTATCATTCT ACTCTCATAC CACCATGAGA CCACTTTTTT AGTCCACAG ATGAATAAAA  3900
ACATGTGATA TTTGACTTTC TGTATCTGGC TTATTTATT AICTATCTCT TTGGCATACC AAGAGTTTGT TTTGTTCTG CTTCAGGGGT TTCAATTAAC  4000
ATAATGACCT CTGGTTCCAT CCATGTTGCT ACAAATGACA AGATTTCATT CTTTTTCATG GCAAAATAGT ACTGTGCAAA AAATACAATT TTTTAATCCG  4100
TTCATCTGTT GATAGACACT TAGGTTGATC CCAAACCTTA ACTATTGTGA ATAGGTGCTT CAATAAACAT GAGTGTAATG TGTCCATTGG ATATACTGAT  4200
TTCCTTCTT TTGGATAAAT AACCACTAGT GAGATTGCTG GATTGTATGA TAGTTCTGTT TTTAGTTTAT TGAGAAATCT TCATACTGTT TCCATAATG  4300
GTTGTACTAT TTTACATTCC CACCAACAGT GTGTAAGAAA GAGTTCCCTT CTTCCCATAT CCTCACAAGG ATCTGTTATT TGAGATTTTT TCATATGTTT TTTGTTAATA  4400
GCATTTAAC TAGAGTAAGT AGATATCTCA TTGTAGTTTT GATTGGCATT CTACTTTTAT TCCCTGATCA TAGTGATGT TGAGATTTTT TCATATGTTT GTTGGTCATT  4500
TGTATATCTT TTTCTGAGAT TGTCGTTCA GTATAGATTG TGAAGATTTT CTCCTCTGTG GGTTGTCTGT TTATTCTGCA GACTCTTCCT TTTGCCATGC AAAAGCTCTT  4600
ATATTATTCT TTTGTCAGAT GTATAGATTG TGAAGATTTT CTCCTCTGTG GGTTGTCTGT TTATTCTGCA GACTCTTCCT TTTGCCATGC AAAAGCTCTT  4700
TAGTTTAATT TAGTCCCAGA TATTTCTTT GTTTATGT GTTGCATTT GTGTCTTGT CATGAAATCC TTTCCTAAGC CAATGTGTAG AAGGTTTTT  4800
CCGATGTTAT TTCTAGAT TGTTACAGTT TCAGGCTTAG ATTTAAGTCC TTGATCCATC TTAAGTTGAT TTTGTATAA GGTGAGAGAT GAAGATCCAG  4900
TTCATTCTC CTACATGA CTTGCCAGCT ATCCCGACTC ATTTGTTGAA TAGGGTGCCC TTTCCCATTT ATGTTTTGT TTGCTTTGTC AAAGATCAGT  5000
```

```
TCGGATGTAA GTATTTGAGT TTATTCTGG  GTTCTCTATT CTGTTCCATT GGTCCGATGT GCCTATTTGT ACACCAGCAT CATGTGTGT TTTTGGTGAC  5100
TATGGCCTTA TTGTATAGTT TGAAATAGAG TAATGTAATG CCATTCAGAT TTGTTCTTT TTTAGACTT GCTTGTTTAT TGGCTCTTT TTTGGTTCCA  5200
TAAGAATTTT AGGATGTTT TTTCTAGTTC TGTGAAGGCT AATGGTGTA TTTATGGAA TTGCAATGCA ATTTGTAGGT TGCTTCTGGC ATTATGGCCA  5300
TTTTCACAAT ATTGATTCTA CCCATCTATG AGAATGGCAT GTGTTTCCAT CTTATATGAT TGTTTGTT TACTATCAGC CGTGTTTTGT AGTTTTCCTT  5400
GTAGATGTCT TTCACCTCCT TGGTTAGGTA TATATTCCTA AGTTTTGTT TTGTTTTGT TTGTTTTTG CAGCTATTGT AAAAGGGGT GAGTTATTGA  5500
TTTTATTCTC ATCTTGGTCA TTGCTGGTAT GTAAGAAAGC AACTCATTGG TGTACGTTAA TTTTGTATCC AGAAACTTTG CTGAATTAT TTATCAGTTC  5600
TAGGGGGTTT TGGAGGAGTC TTTAGAGTTT TCTACATACA CAATCATATC ATCAGCAAAC AGTGACAGTT TGACTTCTC TTTAACAAT TGGATGTGCT  5700
TTACTTGTTT CTCTTGTCTG ATTGCTCTG CTAGGACTTC CAGTAATATG TTAAAGAGAA GTGGTGAGAG TGGGTATCCT TGTCTCATTC CAGTTTTCAG  5800
ACAGAATGCT TTTAACTTT TCCCATTCAA TATAATGTTG GCTGTGTGTT TACCATAGCT GGCTTTATT ACATTGAGGT ATGTCCTTTG TAAACCGATT  5900
TTGCTGAGTT TTAGTCATAA AGTGATGTTG AATTTTGTTG AATGCAGTTT CTGTGGCTAT TGAGATAATC ACATGATTTT TGTTCCAAT TCTCTTTATG  6000
TTGTGTATCA CACTTATTGA CTTGCGTATG TTAAACCATC CGTGCATCCC TCGCATGAAA CCACTTGATC ATGGGTTTTG ATATGCCGTG TGGGATGCTA  6100
TTAGCTATAT TTTGTCAAGG ATGTTGGCAT CTATGTTCAT CAGGGATATT GATCTGTAGT GTTTTTTTT TTTGGTTATG TTCTTTCCCA GTTTTGGTAT  6200
TAAGGTGATA CTGGCTTCAT AGAATGATTT AGGGAGGATT CTCTCTTTCT CTATCTGTA GAATACTGTC AATAGGATTG GTATCAATTC TTCTTTGAAT  6300
   Eco RI
GTCTGGTAGA ATTCGAACGT CTCCTTTAGG TTTTCTAGTT TATTCATGTA AAGGTGTTCA TAGTAACCTT GGTTAATCTT GCTAATGGTC TATCAGTTTT ATTATCTTT  6400
TAATAGTATC TCCTGTTTG TTTCTAACTG AGTTTATTG CACTTCCTC TCTCTTTTCT GGTTAATCTT CATTTAGTTC TCCTCTATC TTAGTTATTC CCTTCTTTT  6500
TCAAAGAACC AGCTTTTTAT TTCATTTAGC TTTTGTATT CTCTAGTTC TTGGTGGTG ACCTTATATT GTCGTCCTC TTCAGACTC TTTGACATCG ACATTTAGG  6600
GCTGGGTTTT GGTTCTGTT GTTTTTGTT GTTTTTGTT CTCTAGTTTC TGTGGGTGTG GAGTTTTGA TAGGTGTGTC ACTATTGTCG GTCAGTTCAA GTAATTTTGT TGTTCTTATT  6700
CTGTGAACTT TCCTTTTAGC ACCATCTTG CTGTATCCTA GAGGTTTTGA TAGGTTACAT GTGCCATGGT GGTTGCTGCT CCCATCAACC ACATTTAGGG  6800
ATACTTTAAG TTCTGGGATA CATGTGCAGA ATGTGCAGT TTGTTACATA GGTATAGATG TGCCATGGTG GTTTGCTGCT TCCCCTCCC TGTCATCTAC TGTTCTCATT  6900
ATTAGTTATT TCTTTTAATG TTATCCCTCT CCTAACCCG TCACCCCCG ACAGGCCCTG GTGTGTGATG TTCCCCTCCC TGTGTCCATG TGTTCTCATT  7000
GTTCAACTCC CACTTATGAG TGAGAACGTG TGGTGTTTGG TTTCTCGTT CCTGTGTTAG ATGATGTTC CACCTTCA-C CATGTCCCTG  7100
CAAAGACATG AACTCATCAT TTTATGGCTG CATATATTC ATGGTGTATA TGTGCCACAT TTTCTTTATC CATTATATCG CTGATGGCCA TTTGGGTTGG  7200
TTCCAAGTCT TTGGTATTGT GAATAGTGCC GCAATAAACA TACGTGTGCA CATGTCTTTA TAGTAGAATG ATTTCTAATT CTTTGGTAT ATACCCAGTA  7300
ATGGGATTGC TGGGTCAAAC AGTATTTCTG GTTCTAGATC CTTGAGGAAT TGCCACACTG TCTTCCACAA TGGTTGAACT AATTACACA CCCATCAACA  7400
GTGTAAAATT TTTCCTATTC TTCCACATCC TCTCCAGCAC CTTTTGTTTC CTGACTTTT AATAATTGCC ATTCTAACTG GCATGAGATG GTATCTCATT  7500
```

441

```
GTGGTTTTGA TTTGCATTTC TCTAATGACC AGTGATGATG AGCTTCTTTT CATGTGTTTC TTGGCCACAT AAATGACTTC TTTAGAGAAG CATCTGTTCA    7600
TATCCTTTGT CCACTTTTTG ATGGGTCGT TAGGTTTTT CTTGTAAATT TGTTGAAGTT CTTGTAGAT TTTGATGTT AGCCCTTTGT CAGATGGATA    7700
GATTGGCAAA AATTTTCTCC CATTCTGTAG GTTGCCTGTT CACTCTGATG ATAGTCTTTT GCTGTGCAGA AGCTCTTTAG TTTAATTAGA TCCCATATGT    7800
CAATTTTGGC CTTTGTTGTC ATTGCTTTTG ATGTTAGTC ATGTTGGTG GTGGAATTTT GCCCATGCCT AGTCCTGAA TGGTATTGCC TAGTTATCT TCTAGGATTT    7900
TTATGGTTTT AGGTTGCACA TTTAAGTCTT TAATCCACCT TGAGTTAATT TTTGTATAAG GTGTAAGGAA GGGGTACAGT TTCAGTTTTA TGCATATTGC    8000
TAGCCAGTTT TTCCAGCACC ATTTATTAAA TAGGGAATTC TTTCTCCATT GCTTTTGTGA TGTTTGTCAA AGATCAGATG GTCGTAGATG TGTGGCATTA    8100
                                  Eco RI
TTTCTGAGGC TTCTGTTCTG TTCCACTGGT CTATATATCT GTTTTGGTAC CAGTACCATG CTGTTTTTGT TACTGTAGCC TTGTAGTATA GCTTGAAGTC    8200
AGGTAGCATC ATGCCTCCAG CTTTGTTCTT TTTGTTTAGG ATTGTCTTGG CTATATGGGC TCTTTTTTGA TTCCATATGA CATTTAAAGT AGTTTTTCT    8300
AATTCTTTGA AAAAAGTCAG TGGTAGCTTG ATGGGGATAG CATTGAATCT ATAAATTACT TTGGGCAGTA TGGCCATTTT AAAGATATTG ATTCTTTCTA    8400
TCTATGACGA TGGAATGTTT TTCCATTTGT TTGTCCTC TCTTATTTCC TTGAGCAGTG AGTGGTTTGT AGCTCTCCTT GAAGAGGTTC TTCACATCCC    8500
TTATAAGTTG TATTTCTAGG TATTTTATTT TATTCTCTTT GCAGCAATTG TGAATGGGAG TTCACCCATG ATTTGGCTCT CTGCTTGTCT ATATTGGTG    8600
TATAGGAATG CTTGTGATTT TTGCACACTG ATTTTGTATC TTGAGACTTT GCTGAAGCTG TTTATCAGCT TAAGATTTTG GGCTGAGATG ACAGGGTCTT    8700
CTAAATATAC AATCATGTCA TCTGCAAACA GAGACAATTT GACTTCCTCT CTTCCTATTT GAATATGCTT TATTCTTTC TCTTGCCTGA TTGTCCTGGC    8800
GAGAACTTCC AATACTATGT TGAGTAAGAG TGGCGAGAGG GCATCCTTGT CTTGTGCCGG TTTTCAAAGC AAATGATTTT TAAATTCCG TCTTGATTTC    8900
ATTGTTGACC CAATGATCAT TCAGGAGCAG GTTATTTAAT TTCCCTGTAT TTGCATGGTT GAGGCTTGTT TGAAGGTTC CTTTTGTAGT TGATTTCAA TTTTATTCTA    9000
CTGTGTCTG AGAGAGTGCT TGATATAATT TCAATTTTA AAAATTTATT GAGGCTTTTC TAAATATCTG TCATATGGCC TATCTGTAGG AAAGTTCCAT    9100
GTGCTGATGA ATAGAATGTG TATTCTGCAG TTGTTGGGTA GAATGTCCTG GAATGTCCTG TAAATATCTG ACTATTATTA TGTTGTCTGT GAAAATGTCAT TGTTTATAA ATTTGGATC    9200
ACTGTCTTGA TGACCTGCCT AGTGCAGTCA TATTAAGAAT TGTAATATTC TCCCATTGGA CAAGGGCTTT TATCATTATA TGATGTCCCT CTTTGTCTTT TTTAACTGCT    9300
TCCAGTATTA GATGCATATA TATTAAGAAT TGTAATATTC TCCCATTGGA CAAGGGCTTT TATCATTATA TGATGTCCCT CTTTGTCTTT TTTAACTGCT    9400
GTTCTTTAA AGTTTGTTTT GTCTGACATA AGAATAGCTG CTTTGGCTCG GCGGCAGATA ACTGGTTGGT GAATTCTATT CATTCTGCAA TTCTGTATCT TTTAAGTGAA    9500
                                                                           Eco RI
GTTTATGTGA GTCCTTATGT GTTAGGTGAG TCTCCTGAAG GCGGCAGATA ACTGGTTGGT GAATTCTATT CATTCTGCAA TTCTGTATCT TTTAAGTGAA    9600
GCATTAGTC CATTACATT CAACATCAGT ATTGAGGTGT GAGGTGACTA TTCCATTCTT CCTGTATTT GTTGCCTGTG TATCTTTTA TCTGTATTTT    9700
TGTTGTATAT GTCCTATGGG ATTTATGCTT TAAAGACGTT CTGTTTTGAT GTGCTTCCAG GGTTTATTTC AAGATTTAGA GCTCCTTTTA TCATTCTGT    9800
                   Eco RI
AGTGTTGGCT TGGTAGTGCC GAATTCTC AGCATTTGTT TTCTGAAAA ACACTGTGTA TTTCTTCAT TTGTGAAGCT TAGTTTCACT GGATATAAAA    9900
TTCTTGGCT ATAATGTTT TGTTTAAGAA GGCTGAAGAT AGGGCCATAT TCACTTCTAG CTTTTACGGT TTCTGCTGAG AAATCTGCTG TTAATCTGAT  10000
```

```
AGGTTTTCTT TCATAGGTTA CCTGTAGTT TCACCTCACA GCTCTTAAGA TTCTCTTGT CTTTAGATAA CTTTGGATAC TCTGATGACA ATGTACCTAG    10100
GCAATGATAT TTTTGCAATG AATTTCCCAG GTGTTTATTG AGCTTCTTTG TATTTGGATA TCTAGGTCTC TAGCAAGGAG GGGGAAGTTT TCCTTGATTA    10200
TTTCCATGGA CAAGTTTTCC AAACTTTTAG ATTTCTCTTC TTTCTCAGGA ATGCTGATTA TCCTTAGGTT TGATTGTTTA ACATAATCCC AGATTTCTTG    10300
GAGGCTTTGT TCATATTTTC TTATTCTTTT TTCTTTGTCT TTGTTGGATT GGGTAATTCA AAAACTTTGT CTTCAAGCTC TGAATTTCTT CTGCTTGAT    10400
TCTATTGCTG AGACTTTCTA GAGCATTTTG CATTTCTATA AGTGCATCCA TTCATCCATT GTTTCCTGAA GTTTTGAATG TTTTTTATTT ATGCTATCTC    10500
TTTAACTGAA GATTTCTCCC CTCATTTCTT GTATCATATT TTTGGTTTTT TTAAAATTGG ACTTCACCTT CCTCCGATGC CTCCTTGATT AGCTTAATAA    10600
CTGACCTTCT GAATTATTTT TCAGTAAAT CAGGGATTTC TTCTTGGTTT GGATGCATTG CTGGTGAGCT AGTATGATTT TTTGGGGGT GTTAAAGAAC    10700
CTTGTTTTTC ATATTACCAG AGTTAGTTTT CTGGTTCCTT CTCACTTGGG TAGGCTCTGT CAGAGGGAAA GTCTAGGCCT CAAGGCTGAG ACTTTTGTCC    10800
CAGCAGGTGT TCCCTTGATG TAGCACAGTC CCCCTTTTCC TAGGACGTGG GGCTTCCTGA GAGCCGAACT GTAGTGATTG TTATCTCTCT TCTGATCTA    10900
GCCACCCATC AGTCTACCA GACTCCAGGC TGGTACTGGG GTTTGTCTGC ACAGAGTCTT GTGACGTGAA CCATCGTGG GTCTCTCAGC CATAGATACA    11000
ACCACCTGCT CCAATGGAGG TGGTAGAGGA TGAAATGAAC TCTGTGAGGG TCCTTACTTT TGGTTGTTCA ATGCACTATC TTTTTGTGCT GGTTGGCCTC    11100
CTGCCAGAGG GTGGCACTTT CTAGAAAGCA TCAGCAGAGG CAGTCAGTG GTGGTGGCTG GGGGGGCTGG GGCACTAGAA CTCCCAAGAA TATATGCCCT    11200
TTGTCTTCAG CTACTAGGGT GAGTAAGGAA GGACCATCAG GTGGGGCAG GACTAGTCGT GTCTGAGCTC AGAGTCTCCT TGGGCAGGTC TTTCTGTGCC    11300
TACTGTGGGA GGATGGGGGT GTAGTTCCA GGTCAATGAA TTTATGTTCC TAGGACAATT ATGGCTGCCT CTGCTGTGTC ATGCAGGTCA TCAGAAAAGT    11400
GGGGAAAGC AAGCAGTCAC GTGACTTGCC CAGCTCCCAT GCAACTCAAA AGTTGGTCT CACTTCCAGC GTGCACCCTC CCCGCAACA GCTCCGAATC    11500
TGTTTCCATG CAGTCAGTGA GCAAGGCTGA GAACTTGCCC AGGCTACCAG CTGCGAAACC AAGTAGGGCT GTCCTACTTC CCTGCCAGTG GAGTCTGCAC    11600
ACCAAATCA TGTCCCCCCA CCAACCCCCC CACTGCCCAG CCCCCTAGATC TGGCCAGGTG GAGATTTCT TTTTCCTGTC TCTTTTCCCA GTTCCTCTGG    11700
CAGCCCTCCC AAATGACCCC TGTGAGGCAA GGCAGAAATG GCTTCCTAGG GACCCAGAG AGCCACAGG GCTTTTCCG CTGCTTCCTC TACCCCTGTA    11800
TTTTGCTTGG CCCTCTAAAT TGACTCAGCT CCAGGTAAGG TCAGAATCTT CTCCTGTGGT CTAGATCTTC AGGTTCCCAG TGAGGATGTG TGTTTGGGGG    11900
TAGACGGTCC CCCTTTTCCA CTTCCACAGT TTGGGCACTC ATGCAGTGG TTCTGCAAAA AAAATTCCTG ATGGGAGACT TCACATGCTG CTCTGTGCAT AGAGGGTGTG    12000
TGCGTTCTCT CAGCTTTCT GAATTTATTT CTGCAGGTGG TTCTGCAAAA TCTAATTTGT CATTTTAAT CAATCTTTT TTCTCTCT CTCTTTTCTT    12100
GCTGCAATGT ACTTCTGCTG CCACCCATCT GCCATCACCC TCTAATTTGT CGTAATATG GGCCGTGAT GTTGAGGGT GGCAGTGGA TACACTCTTT ACCCCTTAGG GAGCATATCT    12200
CTCCCCAAA ACTATACTGC CCTTTGATAT CAAGGAATCA AGGCCGTGAT GTTGAGGGT GGCAGTGGA TACACTCTTT ACCCCTTAGG GAGCATATCT    12300
AGATTTAGAT ATTGCCAATT CAAGATAACT TAATTGAAAG CAAATTCATA ATGAATACAC ACACACACAC ACATATCTGC ATGACAAGAT TTTTAATAGT    12400
TGAAAGAATA ACTAATAATT GTCCACAGGC AATAAGGGCT TTTTAAGCAA AACAGTTGTG ATAAACAGG TCATTCTTAG AATAGTAATC CAGCCAATAG    12500
```

```
TACAGGTTGC TTAGAGATTA TGACATTACC AGAGTTAAAA TTCAATAAATG GTTCTCACT CCTACCACT GAGGACAAGT TTATGTCCTT AGGTTTATGC  12600
TTCCCTGAAA CAATACCACC TGCTATTCTC CACTTACAT ATCAACGGCA CTGGTTCTTT ATCTAACTCT CTGGCACAGC AGGAGTTTGT TTTCTTCTGC  12700
                                                                                                    Region
TTCAGAGCTT TGAATTTACT ATTTCAGCTT CTAAACTTTA TTGCAATGC CTTCCCATGG CAGACTCCTT CGTCATTTT GCCTCTGTTC GAAACTTTC   12800
of DNA sequence uncertainty
CCCTTAATTT CATTCTTAGT TAAAAAAATC TGAAATTATT TTGTTGTTTA ACATAATTAT TAAGTATGT ATGTTCTACC TAGATATAT CTTCTAGGGG  12900
ATTGTTTTAT TCTCTGACGT TATTTAACTT AAATGCCCAC TACCTTTAAA AATTATGACA TTTATTTAAC AGATATTTGC TGAACAAAGT GTTTGAAAAT 13000
ACATGGGAAA GAATGCTTGA AAACACTTGA AATTGCTTGT GTAAAGAAAC AGTTTTATCA GTTAGGATTT AATCAATGTC AGAAGCAATG ATATAGGAAA 13100
AATCGAGGAA TAAGACAGTT ATGGATAAGG AGAAATACAA AAACTCTTAA AGATATTGC CTCAAAAGCA TAAGAGGAAA TAAGGGTTTA TACATGACTT 13200
TTAGAACACT GCCTGGGTTT TTGGATAAAT GGGGAAGTTG TTGAAAACA GGAGGGATCC TAGATATTCC TTAGTCTGAG GAGGAGCAAT TAAGATTCAC 13300
                     Alu element begins
TTGTTTAGAG GCTGGGAGTG GTGGCTCACG CCTGTAATCC CAGAATTTTG GGAGGCCAAG GCAGGCAGAT CACCTGAGGT CAAGAGTTCA AGACCAACCT 13400
GGCCAACATG GTGAAATCCC ATCTCTACAA AAATACAAAA ATTAGACAGG CATGATGGCA AGTGCCTGTA ATCCCAGCTA CTTGGGAGGC TGAGGAAGGA 13500
GAATTGCTTG AACCTGGAAG GCAGGAGTTG CAGTGAGCCG AGATCATACC AGCCTGGGTG ACAGAACAAG ACTCTGTCTC AAAAAAAAAA 13600
         ends
AAGAGAGATT CAAAAGATTC ACTTGTTTAG GCCTTAGCGG GCTTAGACAC CAGTCTCTGA CACATTCTTA AAGGTCAGGC TCTACAAATG GAACCCAACC 13700
AGACTCTCAG ATATGCCAA AGATCTATAC CACACCATCT TTGAAGAAGG CTATCTTATTA GAGACCCTAA TTTGGGTTCA CCTCAGTCTC TATAATCTGT 13800
ACCAGCATAC CAATAAAAT CTTTCTCACC CATCCTTAGA TTGAGAGAAG TCACTTATTA TATGTGAGT AACTGGAAGA TACTGATAAG TTGACAAATC 13900
TTTTCTTTC CTTTCTTATT CAACTTTTAT TTTAACTTCC AAAGAACAAG TGCAATATGT GCAGCTTTGT TGCGCAGGTC AACATGTATC TTTCTGGTCT 14000
TTTAGCCGCC TAACACTTTG AGCAGATATA AGCCTTACAC AGGATTATGA AGTCTGAAAG GATTCCACCA ATATTATTAT AATTCCTATC AACCTGATAA 14100
GTTAGGGGAA GGTAGAGCTC TCCTCCAATA AGCCAGATTT CCAGAGTTTC TGAGCTCATA ATCTACCAAG GTCATGATC GAGTTCAGAG AAAAACAAA  14200
AGCAAAACCA AACTTACCAA AAATAAAA TCCCAAAGAA AAATAAAGA AAAAACAGC ATGAATACTT CCTGCCATGT TAAGTGGCCA ATATGTCAGA 14300
AACAGCACTG AGTTACAGAT AAAGATGCT AAACTACAGT GACATCCCAG CTGTCACAGT GTGTGGACTA TTAGTCAATA AAACAGTCCC TGCCTCTTAA 14400
GAGTTGTTTT CCATGCAAAT ACATGTCTTA TGTCTTAGAA TAAGATTCCC TAAGAAGTGA ACCTAGCATT TATACAAGAT AATTAATTCT AATCCATAGT 14500
ATCTGGTAAA GAGCATTCTA CCATCATCTT TACCGAGCAT TGAATATAA AAAATACTT TTGCTGAGAT CACCAAAACC CTGGGTCATC CAGTGATAAA 14600
TACACATCAT CGGGTGCCTA CATCATACC TGAATATATA ATATTACATA ACATTAATCT ATTCCTGCAC TGAAACTGTT GCTTTATAGG ATTTTTCACT CGGATAAGTA 14700
GATATTGAAG TAAGGATTCA GTCTTATATT ATATTACATA GAGATATAT TCAAGAATAA GTATAGCACT TCTTATTTGG AAACCAATGC TTACTAAATG ACACTAATGA 14800
GAACTTAAGA GATAATGGCC TAAAACCACA GAGATATAT TCAAGAATAA GTATAGCACT TCTTATTTGG AAACCAATGC TTACTAAATG AGACTAAGAC 14900
GTGTCCCATC AAAATCCTG GACCTATGCC TAAAACACAT TTCACAATCC CTGAACTTTT CAAAAATTGG TACATGCTTT AACTTTAAAC TACAGGCCTC 15000
```

```
ACTGGAGCTA CAGACAAGAA GGTGAAAAAC GGCTGACAAA AGAAGTCCTG GTATCTTCTA TGGTGGGAGA AGAAAACTAG CTAAAGGAA GAATAAATTA   16100
GAGAAAAATT GGAATGACTG AATCGGAACA AGGCAAAGGC TATAAAAAAA ATTAAGCAGC AGTATCCTCT TGGGGCCCC TTCCCCACAC TATCTCAATG   16200
CAAATATCTG TCTGAAACGG TTCCTGGCTA AACTCCACCC ATGGGTTGGC CAGCCTTGCC TTGACCAATA GCCTTGACAA GCCAAACTTG ACCAATAGTC   15300
                                                                          G-Gamma Gene   Cap
TTAGAGTATC CAGTGAGGGC AGGGGCCGGC AGGTGGCTAG GGATGAAGAA TAAAGGAAG CACCCTTCAG CAGTTCCACA CACTCGCTTC TGGAACGTCT   15400
                                                Init
GAGGTTATCA ATAAGTCCT AGTCCAGACG CCATGGGTCA TTTCACAGAG GAGGACAAGG CTACTATCAC AAGCCTGTGG GGCAAGGTGA ATGTGAAGA   15500
                    IVS 1
TGCTGGAGGA GAAACCCTGG GAAGTAGGC TCTGGTGACC AGGACAAGGG AGGGAAGGAA GGACCCGTG CCTGGCAAAA GTCCAGGTCG CTTCTCAGGA   15600
                                        Exon 2
TTTGTGGCAC CTTCTGACTG TCAAACTGTT CTTGTCAATC TCACAGGCTC CTGGTTGTCT ACCCATGGAC CCAGAGGTTC TTTGACAGCT TTGGCAACCT   15700
GTCCTCTGCC TCTGCCATCA TGGGCAACCC CAAAGTCAAG GCACATGGCA AGAAGGTGCT GACTTCCTTG GGAGATGCCA TAAAGCACCT GGATGATCTC   15800
AAGGGCACCT TTGCCCAGCT GAGTGAACTG CACTGTGACA AGCTGCATGT GGATCCTGAG AACTTCAAGG TGAGTCCAGG AGATGTTTCA GCACTGTTGC   15900
                                                                                          IVS 2
CTTTAGTCTC GAGGCAACTT AGACAACTGA GTATTGATCT GTTTTAGGGC TGTGTGCAGC TGTTTGAAGA TACTGGGGTT GGGAGTGAAG AAACTGCAGA   16000
GGACTAACTG GGCTGGCAAT CAGTGGCAAT GTTTATTAGAT TTCGGTAGAA AGAACTTTCA CCCTTCCCCT TCTAGATGGA CAACTTTGAC TTTGTTTTAA ACATCTATC TGGAGGCAGG   16100
AATGAGGAAA ATGACTTTTC TTTATTAGAT TTCGGTAGAA AGAACTTTCA ATTGGCTCAG TCAAAGTGGG GAACTTTGT GGCCAAACAT ACATTGCTAA GGCTATTCCT   16200
ACAAGTATGG TCGTTAAAAA GATGCAGGCA GAAGGCATAT AATGCTTCAT TACAAACTTA TATCCTTTAA TGTGTGTGTG GGGCAAAGTA TGTCCAGGGG TGAGGAACAA   16300
ATATCAGCTG GACACATATA AAATGCTGCT AATGCTTCAT TACAAACTTA TATCCTTTAA TGTGTGTGTG TGCGCCGTG TGTTTGTGT GGTGTGAGAG CGTGTGTTTC   16400
TTGAAACATT TGGGCTGGAG TAGATTTTGA AAGTCAGCTC GTTCATGGTG GCAAGAAGAT AACAAGATTT AATTATGCC CAGTGACTAG TCCTGCAAGA AGAACAACTA   16500
TTTTAACGTT TTCAGCCTAC AGCATACAGG GTTCATGGTG GCAAGAAGAT AACAAGATTT AATTATGCC CAGTGACTAG TCCTGCAAGA AGAACAACTA   16600
CCTGCATTTA ATGGAAAGC AAAATCTCAG GCTTTGAGGG AAGTTAACAT AGGCTTGATT CTGGGTGAA GCTTGGTGTG TAGTTATCTG GAGGCCAGGC   16700
                                                              Exon 3
TGGAGCTCTC AGTCACTAT GGGTTCATCT TTATTGTCTC CTTTCATCTC AACAGCTCCT AGTGGCCAGT GCCCTGTCCT CCAGATACCA TGCCATGAT   16800
               Eco RI                                                                           Ter
AAGAATTCA CCCCTGAGGT GCAGGCTTCC TGGCAGAAGA TGGTGACTGG AGTGGCCAGT GCCCTGTCCT CCAGATACCA CACATGGTTG TCTTCAGTTC TTTTTTTAT   16900
GCAGAGCTTT CAAGGATAGG CTTTATTCTG CAAGCAATAC AAATAATAA TCTATTCTGC TAAGAGATCA CACATGGTTG TATGTGTT TGTGTCATGT GCACACTCCA   17000
                                            Poly-A
GTCTTTTTAA ATATATGAGC CACAAAGGGT TTTATGTTGA GGATGTGTT TATGTGTATT TATACATGCC AGAAACTTAT ACATGGAGC GTCTCCAAGT GGGAGTAAAA   17100
CACTTTTTG TTTACGTTAG ATGTGGGTTT TGATGACCAA ATAAAAGAAC TAGGCAATAA TCAGAAACAG ATGTTTTGA AGAGATGGGG AAAGGTTCAG TGAAGGGGGGC   17200
GGTGCAGAG AAATCTGGTT GGAAGAAAGA CCTCTATAGG ACAGGACTCC TCAGAAACAG ATGTTTTTGA AGAGATGGGG AAAGGTTCAG TGAAGGGGGGC   17300
TGAACCCCCT TCCCTGATT GCAGCACAGC AGCGAGGAAG GGGCTCAACG AAGAAAAAGT GTTCCAAGCT TTAGGAAGTC AAGGTTTAGG CAGGGATAGC   17400
CATTCTATTT TATTAGGGGC AATACTATTT CCAACGGCAT CTGGCTTTTC TCACCCCTTG TCAGGCTCTA CGGGGAGGTT GAGGTGTTAG AGATCAGAGC   17500
```

```
AGGAAACAGG TTTTCTTTC CACGGTAACT ACAATGAAGT GATCCTTACT TTACTAAGGA ACTTTTTCAT TTTAAGTGTT GACGCATGCC TAAAGAGGTG    17600
AAATTAATCC CATACCCTTA AGTCTACAGA CTGTCACAG CATTTCAAGG AGAGACCTC ATTGTAAGCT TCTAGGGAGG TGGGGACCTA GGTGAAGGAA    17700
ATGACCCAGC AGAAGCTCAC AAGTCAGCAT CAGCGTGTCA TGTCTCAGCA GCAGAACAGC ACGGTCAGAT GAAAATATAG TGTGAAGAAT TTGTATAACA    17800
TTAATTGAGA AGGCAGATTC ACTGGAGTTC TTATATAATT GAAAGTTAAT GCACGTTAAT AAGCAAGAGT TTAGTTTAAT GTGATGGTGT TATGAACTTA    17900
ACGCTTGTGT CTCCAGAAAA TTCACATGCT GAATCCCCAA CTCCCAATTG GCTCCATTTG TGGGGAGGC TTTGGAAAAG TAATCAGGTT TAGAGGAGCT    18000
CATGAGAGCA GATCCCCATC ATAGAATTAT TTTCCTCATC AGAAGCAGAG AGATTAGCCA AAAAATCCT GTTGTTGAAG TTCTGGTGAG GACACAGTGG GAAGTCAGCC    18100
ACCTGCAACC CAGGAAGAGA GCCCTGACCA GGAACCAGCA GAAAAGTGAG AAAAATCCT GTTGTTGAAG TCACCCAGTC TATGCTATTT TGTTATAGCA    18200
CCTTGCACTA AGTAAGGCAG ATGAAGAAAG AGAAAAAAT AAGCTTCGGT GTTCAGTGGA TTAGAAACCA TGTTTATCTC AGGTTTACAA ATCTCCACTT    18300
                                                                                              Eco RI
GTCCTCGTG TTTCAGAATA AAATACCAAC TCTACTACTC TCATCTGTAA GATGCAAATA GTAAGCCTGA TCCCTTCTGT CTAACTTCGA ATTCTATTT    18400
TTCTTCAACG TACTTTAGGC TTGTAATGTG TTTATATACA GTGAAATGTC AAGTTCTTTC TTTATATTTC TTTCTTTCTT TTTTTCCTC AGCCTCAGAG    18500
TTTTCCACAT GCCCTTCCTA CCTTCAGGAA CTTCTTTCTC CAAACGTCTT TTTTTTTT CTGCCTGGCC TCCATTCAAA TCATAAAGA CCCACTTCAA ATGCCATCAC    18600
TCACTACCAT TTCACAATTC GCACTTTCTT TCTTTGTCCT TTTTTTTT AGTAAAACAA GTTATAAAA AATTGAAGAA ATAAATGAAT GGCTACTTCA    18700
TAGGCAGAGT AGACACAAGG GCTACTGGTT GCCCATTTTT ATTGTTATTT TTCAATAGTA TGCTAAACAA GGGGTAGATT ATTTATGCTG CCCATTTTTA    18800
GACCATAAAA GATAACTTCC TGATGTTGCC ATGGCATTTT TTTTCTTTT AATTTTATTT CATTTCATTT TAATTTCGAA GGTACATGTG CAGGATGTGC    18900
AGGCTTGTTA CATGGGTAAA TGTGTGTCTT TCTGGCCTTT TAGCCATCTG TATCAATGAG CAGATATAAG CTTTACACAG GATCATGAAG GATGAAAGAA    19000
                                                                                              Eco RI
TTTCACCAAT ATTATAATAA TTTCAATCAA CCTGATAGCT TAGGGGATAA ACTAATTTGA AGATACAGCT TGCCTCCGAT AAGCCAGAAT TCCAGAGCTT    19100
CTGGCATTAT AATCTAGCAA GGTTAGAGAT CATGATCAC TTTCAGAGAA AACAAAAAC AACTAACCA AAAGCAAAAC AGAACCAAAA AACCTCCATA    19200
AATACTTCCT ACCCAGTTAA TGGTCCAATA TGTCAGAAAC AGCACTGTGT TAGAAATAAA GCTGTCTAAA GTACACTAAT ATTCCAGTTA TAATAGTGTG    19300
TGGACTATTA GTCAATAAAA ACAACCCTTG CCTCTTTAGA GTTGTTTCC ATGTACACGC ACATCTTATG TCTTAGAGTA AGATTCCCTG AGAAGTGAAC    19400
CTAGCATTTA TACAAGATAA TTAATTCTAA TCCACAGTAG CTGCCAAAGA ACATTCTACC ATCATCTTTA CTGAGCATAG AAGAGCTACG CCAAAACCCT    19500
GGGTCATCAG CCAGCACACA CACTTATCCA GTGGTAAATA CACATCATCT GGTGTATACA TACATACCTG AATATGGAAT CAAATATTTT TCTAAGATGA    19600
AACAGTCATG ATTTATTCA AATAGGTACG GATAAGTAGA TATTGAGGTA AGCATTAGGT CTTATATTAT GTAACACTAA TCTATTACTG CGGCTGAAACT    19700
GTGGTCTTTA TGAAAATTGT TTTCACTACA CTATTGAGAA ATTAAGAGAT AATGCAAAA GTCACAAAGA GTATATTCAA AAAGAAGTAT AGCACTTTTT    19800
CCTTAGAAAC CACTGCTAAC TGAAAGAGAC TAAGATTTGT CCCGTCAAAA ATCCTGACC TATGCCTAAA ACACATTTCA CAATCCCTGA ACTTTTCAAA    19900
AATTGGTACA TGCTTAGCT TTAAACTACA GGCCTCACTG GAGCTACAGA CAAGAAGTA AAAAACGGCT GACAAAGAA GTCCTGGTAT CCTCTATGAT    20000
```

```
GGGAGAAGA AACTAGCTAA AGGGAAGAAT AAATTAGAGA AAAACTGGAA TGACTGAATC GAACAAGGC AAGGCTATA AAAAAATTA AGCAGCAGTA   20100
TCCTCTTGGG GGCCCCTTCC CCACACTATC TCAATGCAAA TATCTGTCTG AAACGGTCCC TGGCTAAACT CCACCCATGG GTTGGCCAGC CTTGCCTTGA   20200
                                                                                                A-Gamma Gene
CCAATAGCCT TGACAAGGCA AACTTGACCA ATAGTCTTAG AGTATCCAGT GAGGCCAGGG GCCGGCGGCT GGCTAGGGAT GAAGAATAAA AGGAAGCACC   20300
                                                                               Cap
CTTCAGCAGT TCCACACACT CGCTTCTGAA ACGTCTGAGA TTATCAATAA GCTCCTAGTC CAGACGCCAT GGTCATTTC ACAGAGGAGG ACAAGGCTAC   20400
                                                                               Init
TATCACAAGC CTGTGGGCA AGTGAATGT GGAAGATGCT GGAGGAGAAA CCCTGGGAAG GTAGGCTCTG GTGACCAGGA CAAGGAGGG AAGAAGGAC   20500
                                                                               IVS 1
CCTGCCCTG GCAAAGTCC AGGTCGCTTC TCAGGATTTG TGGCACCTTC TGACTGTCAA ACTGTTCTTG TCAATCTCAC AGGCTCCTGG TTGTCTACCC   20600
                                                                                                Exon 2
ATGGACCCAG AGGTTCTTTG ACAGCTTTGG CAACCTGTCC TCTGCCTCTG CCATCATGGG CAACCCCAAA GTCAAGGCAC ATGGCAAGAA GGTGCTGACT   20700
TCCTTGGGAG ATGCCATAAA GCACCTGGAT GATCTCAAGG GCACCTTTGC CCAGCTGAGT CAACTTAGAC AACTGAGTAT TGATCTGAGC ACAGCAGGGT CCTGAGAACT   20800
   IVS 2
TCAAGGTGAG TCCAGGAGAT GTTTCAGCAC TGTTGCCTTT AGTCTCGAGG CAACTTAGAC AACTGAGTAT TGATCTGAGC ACAGCAGGGT GTGAGCTGTT   20900
TGAAGATACT GGGGTTGGA GTGAAGAAAC TGCAGAGGAC TAACTGGGCT GAGACCCAGT GGCAATGTTT TAGGGCCTAA GGAGTGCCTC TGAAAATCTA   21000
GATGGACAAC TTTGACTTTG AGAAAAGAGA GGTGGAAATG AGGAAAATGA GTTTCTTTA TTAGATTCCG GTAGAAAGAA CTTTCACCTT TCCCCTATTT   21100
TTGTTATTCG TTTTAAAACA TCTATCTGAA GGCAGGACAA GTATGGTCGT TAAAAGATG CAGGCAGAAG GCATATATTG GCTCAGTCAA AGTGGGAAC   21200
TTTGGTGGCC AAACATACAT TGCTAAGGCT ATTCCTATAT CAGCTGGACA CATATAAAAT GTGAGTAGA TTTTGAAAGT CAGCTCTGTG TGTGTGTGTG TGTGTGTG   21300
AGATGGGGC AAAGTATGTC CAGGGGTGAG GAACAATTGA ACATTTGGG CTGAGTAGA ATGGTGGAA AGATTTAAAT TATGGCCAGT GACTAGTGCT   21400
TGTCAGCGTG TGTTTCTTT AACGTCTTCA CATTTAATGG GAAGGCAAAA TCTCAGGCTT TCATCTTTAT TGTCTCCTTT GAAGTAGCA AGATTTAAAT TATGGCCAGT GACTAGTGCT   21500
TGAAGGGAA CAACTACCTG CCAGGCTGGA GCTCTTCAGT CACTATGGGT CACTATGGGT CACTATGGGT CATCTCAACA GCTCCTGGGA AATGTGCTGG TGACCGTTTT   21600
                                                                               Exon 3
TATCTGGAGG CCAGGCTGGA GCTCTTCAGT CACTATGGGT CACTATGGGT CACTATGGGT CATCTCAACA GCTCCTGGGA AATGTGCTGG TGACCGTTTT   21700
                                                                                                Ter
GGCAATCCAT TTCGGCAAAG AATTCACCCC TGAGGTGCAG GCTTCCTGCA AGAMATGGT GACTGCAGTG GCCAGTGCCC TGTCCTCCAG ATACCACTGA   21800
                                                                Poly-A
GCCTCTTGCC CATGATTCAG AGCTTTCAAG GATAGGCTTT ATTCTCCAAG CAATACAAAT AATAAATCTA TTCTGCTGAG AGATCACACA TGATTTCTT   21900
CAGCTCTTTT TTTTACATCT TTTTAAATAT ATTTTTATAT GATGAGCCACA TTTTGATGAG CAATAAAAG GTGTGTATGT GTATTTCTGC ATGCCTGTTT GTGTTTGTGG   22000
TGTGTGCATG CTCCTCATTT ATTTTTATAT GAGATGTGCA TTTTGATGAG CAATAAAAG CAGTAAAGAC ACTTGTACAC GGGAGTTCTG CAAGTGGGAG   22100
TAAATGGTGT TGGAGAAATC CGGTGGGAAG AAAGACCTCT ATAGGACAGG ACTTCTCAGA AACAGATGTT TTGAAGAGA TGGGAAAAGG TTCAGTGAAG   22200
                                                                               Eco RI
ACCTGGGGGC TGGATTGATT GCAGCTGAGT AGCAAGGATG GTTCTTAATG AGGGAAAGT GTTCCAAGCT TTAGAATTC AAGGTTTAGT CAGTGTAGC   22300
AATTCTATTT TATTAGGAGG AATACTATTT CTAATGCAC TTAGCTTTTC ACAGCCCTTG TGGATGCCTA AGAAAGTGAA ATTAATCCCA TGCCCTCAAG   22400
TGTGCAGATT GGTCACAGCA TTTCAAGGGA GAGACCTCAT TGTAAGACTC TGGGGGAGGT GGGGACTTAG GGTAAGAAA TGAATCAGCA GAGCCTCACA   22500
```

```
AGTCAGCATG AGCATGTTAT GTCTGAGAAA CAGACCAGCA CTGTGAGATC AAATGTAGT GGGAAGAATT TGTACAACAT TAATTGGAAG GTTTACTTAA  22600
TGGAATTTTT GTATAGTTGG ATGTTAGTGC ATCTCTATAA GTAAGAGTTT AATATGATGG TGTTACGGAC CTGGTGTTTG TGTCTCCTCA AAATTCACAT  22700
GCTGAATCCC CAACTCCCAA CTGACCTTAT CTGTGGGGGA GGCTTTTGAA AAGTAATTAG GTTTAGCTGA GCTCATAAGA GCAGATCCCC ATCATAAAAT  22800
TATTTTCCTT ATCAGAAGCA GAGAGACAAG CCATTTCTCT TTCCTCCCGG TGAGGACACA GTGAAGTC CGCCATCGC AATCCAGGAA GAGAACCCTG  22900
ACCACGAGTC AGCCTTCAGA AATGTGAGAA AAAACTCTGT TGTTGAAGCC ACCCAGTCTT TGTATTTTG TTAGACACC TTACACTGAG TAAGGCAGAT  23000
GAAGAAGGAG AAAAAAATAA GCTTGGGTTT TGAGTGAACT ACAGACCATG TTATCTCAGG TTTGCAAAGC TCCCCTCGTC CCCTATGTTT CAGCATAAAA  23100
TACCTACTCT ACTACTCTCA TCTATAAGAC CCAAATAATA AGCCTGCGCC CTTCTCTCTA ACTTTGATTT CTCCTATTTT TACTTCAACA TGCTTTACTC  23200
TAGCCTTGTA ATGTCTTTAC ATACAGTGAA ATGTAAAGTT CTTTATTCTT TTTTTCTTTC TTTCTTTTTT CTCCTCAGCC TCAGAATTTG GCACATGCCC  23300
TTCCTTCTTT CAGGAACTTC TCCAACATCT CTGCCTGGCT CCATCATATC ATAAAGGTCC CACTTCAAAT GCAGTCACTA CCGTTTCAGG ATATGCACTT  23400
TCTTTCTTTT TTGTTTTTTG TTTTTTTTAA GTCAAAGCAA ATTTCTTGAG AGAGTAAAGA AATAAACGAA TGACTACTGC ATAGGCAGAG CAGCCCGAG  23500
GGCCGCTGGT TGTTCCTTTT ATGGTTATTT CTTGATGATA TGTTAAACAA GTTTTGGATT ATTTATGCCT TCTCTTTTTA GGCCATATAG GGTAACTTTC  23600
TGACATTGCC ATGGCATGTT TCTTTTAATT TAATTTACTG TTACCTTAAA TTCAGGGGTA CACGTACAGG ATATGCAGGT TTGTTTTATA GGTAAAAGTG  23700
TGCCATGGTT TTAATGGGTT TTTTTTTTCT TGTAAAGTTG TTTAAGTTTC TGTTTACTC TGATATTGG CCTTTGTCAG AAGAATAGAT TGGAAAATCT  23800
TTTTCCCATT CTGTAGATTG TCTTTCGCTC TGATGGTAGT TTCTTTTGCT GAGCAGGAGC TCTTTAGTTT AATTAGATTC CATTGGTCAA TTTTTGCTTT  23900
TGCTGCAAAT GCTTTTCACG CTTTCATCAT GAAATCTGTG CCCGTGTTTA TATCATGAAT AGTATTGCCT TGATTTTTTT CTAGGCTTTT TATAGTTTGG  24000
GGTTTTTCAT TTAAGTCTCT AATCCATCCG GAGTTAATTT CCCCATTGCT TGCTTTTGTC AGTTTCTAA ATGCTGGTCT AAGACAGATG GTTGTAGGTA CAATATGAG GCCAGTTCTC  24100
CCCCATCATT TATTAAATG AAAATCCTTT CCCCATTGCT TGCTTTTGTC AGTTTCTAA ATGCTGGTCT AAGACAGATG GTTGTAGGTA CAATATGAG GCCAGTTCTC  24200
TCATATAATA CCATGAAA TCTCTTATTA ATTCATTTCT TTAGTATGT ATGCTGGTCT CCTCTGCTCA CTATAGTGAG GGCACCATTA TTCTTCAAG  24300
CTGTCTGTCT AGTTCATGTA AGATTCTCAG AATTAAGAAA AATGGATGGC ATATGAATGA AACTTCATGG ATGACATATG GAATCTAATG TGTATTTGTT  24400
GAATTAATGC ATAAGATGCA ACAAGGGAAA GTTGACAAC TGCAGTGATA ACCTGGTATT GATGATAAA GAGTCTATAG ATCACAGTAG AAGCAATAAT  24500
CATGGAAAAC AATTGGAAAT GGGGACAGC CACAAACAAG AAAGAATCAA TACTACCAGG AAAGTGACTG CAGGTCACTT TTCCTGGAGC GGGTGAGAGA  24600
            EcoRI
AAAGTGGAAG TTGCAGTAAC TGCCGAATTC CTGGTTGGCT GATGGAAAGA TGGGCCAACT GTTCACTGGT ACGCAGGGTT TTAGATGTAT GTACCTAAGG  24700
ATATGAGGTA TGGCAATGAA CAGAAATTCT TTGGAATGA GTTCTAGGGC CATTAAAGGA CATGACCTGA AGTTTCCTCT GAGGCCAGTC CCCACAACTC  24800
AATATAAATG TGTTTCCTGC ATATAGTCAA AGTTGCCACT TCTTTTTCTT CATATCATCG ATCCTGCTC TTAAAGATAA TCTTGGTTTT GCCTCAAACT  24900
GTTTGTCACT ACAAACTTTC CCCATGTTCC TAAGTAAAAC AGGTAACTGC CTCTCAACTA TATCAACTAG ACTAAAATAT TGTGTCTCTA ATATCAGAAA  25000
```

```
TTCAGCTTTA ATATATTGGG TTCAACTTCTT TGAAATTTAG AGTCTCCTTG AAATACACAT GGGGGTGATT TCCTAAACTT TATTTCTTGT AAGGATTTAT  26100
CTCAGGGTA ACACAAAC CAGCATCCTG AACCTCTAAG TATGAGACA GTAAGCCTTA AGAATATAAA ATAAACTGTT CTTCTCTCTG CCGGTGGAAG   25200
TGTGCCCTGT CTATTCCTGA AATTGCTTGT TTGAGACGCA TGAGACGTGC AGCACATGAG ACGTGCAGCA CATGAGACAC GTGCAGCAGC CTGTGAATA  25300
TTGTCAGTGA AGAATGTCTT TGCCTGATTA GATATAAAGA CAAGTTAAAC ACAGCATTAG ACTATAGATC AAGCCTGTGC CAGACACAAA TGACCTAATG  25400
CCCAGCACGG GCCACGGAAT CTCCTATCCT CTTGCTTGAA CAGCAGCA CACTTCTCCC CCAACACTAT TAGATGTTCT GGCATAATTT TGTAGATATG  25500
TAGGATTTGA CATGGACTAT TGTTCAATGA TTCAGAGGAA ATCTCCTTTG TTCAGATAAG TACACTGACT ACTAAATGA TTAAAAACA CAGTAATAAA  25600
ACCAGTTTT CCCCTTACTT CCCTAGTTTG TTTCTTATTC TGCTTTCTTC CAAGTTGATG CTGGATAGAG GTGTTTATTT CTATTCTAAA AAGTGATGAA  25700
         Alu begins
ATTGCCCGGG CGCGGTGGCT CACACCTGTA ATCCCAGCAC TTTGGGAGGC TGAGGTGGGC GGATCACGAG GTCAGGAGAT CAAGACCATC CTGGCTAACA  25800
TGGTGAAACC CCATCTCTAC TAAAAATACA AAAAATTAGC CAGAGACAGT GGGGGTGCC TGTAGTCCCA GCTACTCGGG AGGCTGAGGC AGGAGAATGG  25900
                       Alu ends
CGTGAACCTG GGAGGCAGAG CTGTAGTGAG CAGAGATCGC GCCACTGCAC GCCAGCCT GGGTGACAAA GCGAGACTCC ATCTCAAAAA AAAAAAAAA   26000
AAAAAAAGA AAGAAAGAAA GAAAAAAAA GTGATGAAAT TGTGTATTCA TGTCTCACAA CTCCTGATCA AAGAATTG AAACCAAGA AAGCTGTGG CTTCTTCCAC  26100
ATAAGCCTG GATGAATAAC AGGATAACAC GTTGTTACAT TGTCTCACAA CTCCTGATCA CTCCTGATCA GGAATTGATG GCTAAGATAT TCGTAATTCT TATCCTTCTC  26200
AGTTGTAACT TATTTCCATT TGTCAGCTTC AGGTATTGCG GCTGCTGGCG AAGTCCTTGA GAAACAAACT GCACACTGA TGGTGGGGGT AGTGTAGAA   26300
AATGGAGGGG AAGGAAGTAA AGTTTCAAAT TAAGCCTGAA CAGGAAAGTT CCCCTGAGAA GGCCACCTGG ATCTATCAG AAACTCGAAT GTCCATCTG  26400
CAAAACTTCC TTGCCCAAAC CCCACCCCTTG GAGTCACAAC CCACCCCCTG CCAATAGATT CATTTCACTG GGAGAGGCAA AGGGCTGGTC AATAGATTCA  26500
                                              Pseudo-Beta Gene  "Cap"
TTTCACTGGG AGAGGCAAAG GGCTGGGGGC CAGAGAGGAG AAGTAAAAAG CCACACATGA AGCAGCAATG CAGGCATGCT TCTGGCTCAT CTGTGATCAC  26600
CAGGAAACTC CCAGATCTGA CACTGTAGTG CATTTCACTG CTGACAAGAA GGCTGCTGCC ACCAGCCTGT GAAGCAAGGT TAAGGTGAGA AGGCTGGAGG  26700
         IVS 1  "Init"
TGAGATTCTG GCCAGGTAGG TACTGGAAGC CGGGACAAGG TGCAGAAAGG CAGAAAGTGT TTCTGAAAGA GGGATTAGCC CGTTGTCTTA CATAGTCTGA  26800
         Exon 2
CTTTGCACCT GCTCTGTGAT TATGACTATC CCACAGTCTC CTGGTTGTCT ACCCATGGAC TTGAAAGTT TTGGATATCT GGGCTCTGAC           26900
TGTGCAATAA TGGGCAACCC CAAAGTCCAG GCACATGGCA AGAAGGTGCT GATCTCCTTC GAAAAGCTG TTATGCTCAC GGATGACCTC AAAGCACCT   27000
                                                    IVS 2
TTGCTACACT GAGTGACCTG CACTGTAACA AGCTGCACGT GGACCCTGAG AACTTCCTGG TGAGTAGTAA GTACACTCAC GCTTCTTCT TTACCCTTAG  27100
ATATTTGCAC TATGGGTACT TTTGAAAGCA GAGGTGGCTT TCTCTTGTGT TATGAGTCAG CTATGGATA TGATATTCA GCAGTGGAT TTTGAGAGTT  27200
ATGTTGCTGT AAATAACATA ACTAAAATTT GTAGAGCAA GGACTATGAA TAATGAAGG CCACTTACCA AAATAACAT TTTGATAGCT CTGAAAAACA CATCTTATAA  27300
AAAATTCTGG CCAAAATCAA ACTGAGTGTT TTGGATGAGG GAACAGAAGT TGAGATAGAG AAAATAACAT CTTTCCTTTG GTCAGCGAAA TTTTCTATAA  27400
AAATTAATAG TCACTTTTCT GCATAGTCCT GGAGGTTAGA AAAGATCAA CTGAACAAAG TAGTGGGAAG CTGTTAAAAG AGGATTGTTT CCCTCCGAAT  27500
```

449

```
GATGATGTA TACTTTTGTA CGCATGGTAC AGGATTCTTT GTTATGAGTG TTTGGAAAA TTGTATGTAT GTATGTATGT ATGTGATGAC TGGGACTTA      27600
TCCTATCCAT TACTGTTCCT TGAAGTACTA TTATCCTACT TTTTAAAAGG ACGAAGTCTC TAAAAAAAA ATGAAACAT CACAATATGT TGGGTAGTG      27700
AGTTGGCATA GCAAGTAAGA GAAGGATAGG ACAATGGGAG GTGCAGGGCT GCCAGTCATA TTGAAGCTGA TATCTAGCCC ATAATGGTGA GAGTTGCTCA   27800
AACTCTGGTC AAAAGGATG TAAGTGTTAT ATCTATTTAC TGCAAGTCA GCTTGAGGCC TTCTATTCAC TATGTACCAT TTTCTTTTTT ATCTTCACTC     27900
          Exon 3
CCTCCCCAGC TCTTAGGCAA CCTGATATTG ATTGTTTGGA CAACCCACTT CAGCGAGGAT TTTACCCTAC AGATACAGGC TTCCTGGCAG TAACTAACAA   28000
ATGCTGTGGT TAATGCTGTA GCCCACAAGA CCACTGAGTT TGCCTGCTGA TTCAGTTCCT GCATGATAAA CCCTGTCCAC TATGTTTGTA CCTATGTCCC  AAAATCTCAT CTCCTTTAGA TGGGGAGGT  28100
             Ter                                                    Poly-A
TGGGGAGAAG AGCAGTATCC TGCCTGCTGA TTCAGTTCCT GCATGATAAA CTATTTCTGA TGGAAATGAG AATGTTGGAG AATGGGAGTT TAAGGACAGA GAAGATACTT          28200
CTGTCTTTAT ATTTTACCCT GATTCAGCCA AAAGACGCA CTATTTCTGA TGGAAATGAG AATGTTGGAG AATGGGAGTT TAAGGACAGA GAAGATACTT                      28300
TCTTGCAATC CTGCAAGAAA AGAGAGAACT CGTGGGTGGA TTTAGTGGGG TAGTTACTCC TAGGAAGGGG AAATGCTCTC TAGAATAAGA CAATGTTTT                        28400
ACAGAAAGGG AGTCAATGG AGGTACTCTT TGGAGGTGTA AGAGGATTGT TGGTAGTGTG ACTTCTACTT ATTTAAACA ACATATTTT TATGATTTAT AATGAAGTGG              28500
CTATTATTTG TATGAAACTC AGATATAGC CAATCAAGGG TCATTTGGTG ACTGCAGTTC AACTTTCTCT TAACGTCTTC AATGGTATTA ACATATTTAT ATAGAGAATT            28600
GGATGGGGCT TCCTAGAGAC CAATCAAGGG CCAAACCTTG TGCTACCTCT GTGACCTGAA ACATATTTAT AATTCCATTA AGCTGTGCAT ATGATAGATT                       28700
GGCTCTCTG GTTTCATCT GTACTTCATC TGTAACTA ATTGAATTGA TACCTGTAAA GTCATTTATC ACACTACCCA ATGAAATAT ACACACATGT GTGCATTCAT                28800
ATTTCCTTA AAGGATTTT GTAAGACTA ATTGAATTGA TTTTAGTGGT AGTGATTTTA TTCTCTTCT ATATATAC GTGAGCAAAC AGCAGATTAA AAGCTGAGA                  28900
GTTTCTATAA ATATGTACAA GTTTTTATG TTTTAGTGGT AGTGATTTTA TTCTCTTCT ATATATAC GTGAGCAAAC AGCAGATTAA AAGCTGAGA TTTAGGAAAC                29000
AATTTTTATG AATAAAAAT TATTAGCAAT CAATATTGAA AACCACTGAT TTTGTTTAT GTGAGCAAAC AGCAGATTAA AAGCTGAGA TTTAGGAAAC                          29100
AGCAGCTTAA GTCAAGTTGA TAGAGGAGAA TATGGACATT TAAAAGAGGC AGGATGATAT AAAATTAGGG AAACTGGATG CAGAGACCAG ATGAAGTAG                         29200
AAAATAGCT ATCGTTTTGA GCAAAATCA CTGAAAGTTT CTGCATATGA GAGTGACATA ATAAATAGGG AAACTGGATG CAGAGACCAG ATGAAGTAG                          29300
TATATATAGA ACTGATTAGA CAAGTCTAA CTTGGGTATA GTCAGAGAG CTTGCTCTAA TTATATTGAG GTGATGATA AAAGAACTGA AGTTGATGGA                          29400
AACAATGAAG TTAAGAAAAA AAATGGAGTA AGAGACCATT GTGGCAGTGA TTGCACAGAA CTGAAAACA TTGTGAAACA GAGAGTCAGA GATGACAGCT                        29500
AAAATCCCTG TCTGTGAATG AAAGAAGGA AATTTATTGA CAGAACAGCA AATGCCTACA AGCCCCCTGT TTGGATCTGG CAATGAACGT AGCCATTCTG                        29600
TGGCAATCAC TTCAAACTCC TGTACCCAAG ACCCTTAGGA AGTATGTAGC ACCCTCAAAC CTAAAACCTC AAAGAAAGAG GTTTTAGAAG ATATAACC                         29700
CTTTCTTCTC CAGTTTCATT AATCCCAAAA CCTCTTCTC AAAGTATTTC CTCTATGTGT CCACCCAAA GAGCTCACCT CACCATATCT CTTGAGTGGG                         29800
AGCTCATAGA TAGGCGGTGC TACCATCTAA CAGCTTCTGA AATTCCTTTG TCATATTTTT GAGTCCCCAC TCAATAACCC ACAAAGCAGA ATAAATACCA                       29900
GTTGCTCTGT ACATAACACC ATGTTGTCTT GTAGCATACA TTAATTAAGC ACATTCTTG AATAATTACT GTGTCCAAAC AATCACACTT TAAATCTCAC                        30000
```

450

```
ACTTGTGCTA TCCCTTGCCC TTCTGAATGT CACTCTGTAT TTTTAAATGA AGAGATGAGG GTTGAATTTC CTGTGTTACT TATTGTTCAT TTCTCGATGA  30100
GGAGTTTTCA CATTCACCTT TACTGAAAA CACATAAGTA CACATCTTAC AGGAAAAATA TACCAAACTG ACATGTAGCA TGAATGCTTG TGCATGTAGT  30200
CATATAAAAT CTTGTAGCAA TGTAAACATT CTCTGATATA CACATACAGA TGTCTCACA ATTTCTTATG CTCCATGAAC AAACATTCCA           30300
TGCACACATA AGAACACACA CTGTTACAGA TGCATACTTG AGTGCATTGA CAAAATTACC CCAGTCAATC TAGAGAATTT GGATTTCTGC ATTTGACTCT  30400
GTTAGCTTTG TACATGCTGT TCATTTACTC TGGGTGATGT CTTCCCTCA TTTTGCCTTG TCTATCTTGT ACTCATACTT TAAGTCCTAA CTTATATGTT  30500
ATCTCAACTA AGAAGCTATT TTTTTTTAAT TTTAACTGGG CTTAAAGCCC TGTCTATAAA CTCTGCTACA ATTATGGGCT CTTTCTTATA ATATTTAGTG  30600
TTTTTCCTAC TAATGTACTT AATCTGCTCA TTGTATATTC CTACCACTAA ATTAGATGTC TCTTTTCTAC CTTTTATGGT AGAGACATTG TCTTGTAAAC TCTTATTCC  30700
CTAGTATTTG GAGATGAAAA AAAAGATTAA ATTATCCAAA ATTAGATGTC TCTTTTCTAC ATTATGAGTA TTACACTATC CATAGGGAAG TTTGTTTGAG  30800
ACCTAAACTG AGGAACCTTT GGTTCTAAAA TGACTATGTG ATATCTTAGT ATTTATGGT CATGAGGTTC CTTCCTTCTG CCTCTGCTAT AGTTGATTA  30900
GTCAGCAAGC ATGTGTCATG CATTTATTCA CATCAGAATT TCATACACTA ATAAGACATA GTATCAGAAG TCAGTTTATT AGTTATATCA GTTAGGGTCC  31000
ATCAAGGAAA GGACAAACCA TTATCAGTTA CTCAACCTAG AATTAAATAC AGCTCTTCCC GCCATTAGGC CCTTGTATTG GAAGAGCTAA AATATCAAAT  31100
AAAGGACAGT GCAGAAATCT AGATGTTAGT ACATCAGAAA ACCTCTTCC TGAGGTAGTG GGTGTCCTT GGAGAAGGGC AGAAGAGAGA AATGTTTATA CCACCAGAGT  31200
CCAGAACCAG AGCCCATAAC CAGAGGTCCA CTGGATTCAG TGAGGTAGTG ATCCGTTCAG AGAGAGAGAG AGACCAGAAA TAATCTTGCT TATGCTTTCC CTCAGCCAGT  31300
AGCCATAAAA AACCATAAAA AAGACTGTCT GCTGTAGGAG GAAACCTGGG AAATGTCAGT TCCTCAAATA CAGAGAACAC TGAGGAAGG ATGAGAATA  31400
GTTACCATT GCAGAATGTA CATGCGACTG TGGTAATTGA CAGAAGGAAA CTAGAGATGTG TCCAGTAAAT GAATAATTAC AGTGTGCAGT GATTATTGCA ATGATTAATG  31500
AATGGAAAG CAGACATGAA TGGTAATTGA CAGAAGGAAA CTAGAGATGTG TCCAGTAAAT GAATAATTAC AGTGTGCAGT GATTATTGCA ATGATTAATG  31600
TATTGATAAG ATAATATGAA AACACAGAAT TCAAACAGCA GTGAACTGAG ATTAGAATTG TGGAGAGCAC TGGCATTTAA GAATGCACA CTTAGAATGT  31700
                       Eco RI
GCTCTAGGC ATTGTTCTGT GCATATATCA TCTCAATATT CATTATCTGA AAATTATGAA TTAGGTACAA AGCTCAAATA ATTTATTTT TCAGGTTAGC  31800
           Alu element begins
AAGAACTTTT TTTTTTTTT TCTGAGATGG AGCATTGCTA TGGTTGCCCA GGCTGGAGTG CAATGGCATG ATCCAGGCTC ACTGCAACAT CTGCCTCCCA  31900
GGTCAAGCG ATTCTCCTGC CTCAGCCTCC CAAGTACAGG GCACTACAGG CATGCCCAC CACCATGCCT GGCTAATTTT CTATTTTTAG TAGATAGGGG  32000
                                                                                                      Alu ends
GTTTCACCAT GTTGGTCAGG CTGATCTCGA ACTCCTAACA TCAGTGATC CACCCTCTC GGCCTCTGAA AGTGCTGGGA TCACAGGCGT GAGCCACCAC  32100
ACCCAGCCAA CTTCCATTCT AACCCACATT GGCATTACAC TAATTAAAAT TCTAAAAATC GATACTGAG ATCGGGATT TGGGGACTA TGTCTTACTT  32200
ACATACATT CCCATCTTTC ACCCTACCTT TTCCTTTTTG TTTCAGCTTT TCACTGTGTC AAATCTAGA ACCTTATCTC CTACCTGTC TGAAACCAAC  32300
AGCAAGTTGA CTTCCATTCT AACCCACATT GGCATTACAC TAATTAAAAT TCTAAAAATC GATACTGAG ATCGGGATT TGGGGACTA TGTCTTACTT  32400
CATACTTCCT TGAGATTTCA CATTAAATGT TGGTGTTCAT TAAAGTCCT TCATTTAACT TGTATTCAT CACACCTTG GATTCACAGT TATATCTAA  32500
```

```
CTCTTAAATA CAGCCTGTAT AATCCCAATT CCCAACTCTG ATTTCTAACC TCTGACCTCA ACCTCAGTG CCAAACCCAT ATATCAAACA ATGTACTGGG   32600
CTTATTTATA TAGATGTCCT ATAGGCACCT CAGACTCAGC ATGGGTATTT CACTTGTTAT ACTAAAACTG TTTCTCTTCC AGTGTTTTCC ATTTTAGTCA   32700
TTAGATAGCT ACTTGCCCAT TCACCAAGGT CACAGATTAA AATCATTTCC CTACTCTAA TCAACAGTTC GATTCTGCTT CAATTTGTC CTATCTATA    32800
                                                                                           Alu element begins
ATCACCACTC TTACTGCCCA GTCAGGTCCT CATTGTTTCC TGAACAAGAG TAGATGCTAT TCTTTCCACT TTTAGACCTT ATCCTGGCTG GATGCGGTGG   32900
CTCAGGCTTG TAAACCCAGC ACTTTGGGAG GCCAAGGCAG GCAGATCACT TGAGGTCAGG AGTTCAAGAC CAGCCTGACC AACATGTGA AACCCCATCT    33000
CTACTAAAAA TACAAAATCA GCCGGGCGTG TGGTGCATGC CTGTAGTCCC AGCTATTCAG GTGGCTGAGG CAGGAGAATT GCTTGAACCC AGGAGGCAGA   33100
GGTTGCGGTG AGCCTAGATT GCACCATTGC ACTCTAGCTT GGGCAATAGG GATGAAACTC CATCCAGAA GAGAAAAGAA AAAAGACCT TATTCTGTTA    33200
                                                                                           Alu ends
TACAAATCCT CTCAATGCAA TCCATATAGA ATAAACATGT AACCAGATCT CCCAATGTGT AAATCATTT CAGGTAGAAC AGAATTAAAG TGAAAAGCCA    33300
AGTCTTTGGA ATAACAGAC AAGATCAAA TAACAGTCCT CATGGCCTTA AGAATTTACC TAACATTTTT TTAGAATCA ATTTCTTAT ATATGAATTG     33400
GAAACATAAT TCCTCCCTCA CAAACACATT CTAAGATTTT AAGGAGATAT TGATGAAGTA CATCATCTGT CATTTTTAAC AGGTAGTGGT AGTGATTCAC   33500
ACAGCACATT ATGATCTGTT CTTGTATGTT GCATACAATA CAATGTATCC AAGACTGTAT TTCTGATTTT GACCTGGTTG TATTCTTTCT GAGCTCCAGA   33600
                                                                                                TCCATATATC TAAGTACATC
TTTTTGCATT TTACAAGAGT GCATACACATA CAATGTATCC AAGACTGTAT TTCTGATTTT ATCGTACCAC TAAACTCACA AATGTGGCCC TATTCTTGTG   33700
TTCACGACTG ACATCACCGT CATGGTCCAA GTCTGATAAT AGAAATGGCA TTGTCACTTT CTTCCCTACT GCAACAGAAG CCCAGCTATT TGTCTCCCAT   33800
TTTCTCTACT TCTAAAATAC ATTTCTTCAC TAAGTGAGAA TAATCTTTTA AAGACACAAA TCAAACCATG CCACCACCTT TCTTGAATTA TTCAATATCT   33900
TTCGTTGGCT TCCAGGTTAC AGAAAAATAA CTTGTAACAA AGTTTAAAGG TCATTCATGG CTCCTCTCTA CCCTCTGGTT ACTTTTAATC AACCAAATGC   34000
                                                                                                      CCTTGTGATC
AGAATCTCAG GCACATCATC CATCTTTCTA TATACAAATA AGTCATATA ATTTCTGTGT GATTCTTTTC ATATATTCAT TCTTTGACTA TACCGTAATT   34100
                                                                                                      TATCAATTCT ACTGTTGTA
ATTTGTATCG CTACGTGTTA AGCAGTAGTT TTTGAGGTTT TTATGATTAT TGCTGTCATA AGCATTTCTA TGGATATACT TGGATACACA CATGCATGTG   34200
                                                                                                      TTTCTGAATA
AGCATTTAAG TGGCTACCCG TTTGAGGTTT TTATGATTAT TATCAAGCAT CCAGCATTTG TGGATATACT TAAAGGTTTT CCAAAGGGGT TATACTATTG   34300
TACAGTGTCA
TCTAAAAATG TAATTGCTAG GTAATAGACT TATCAAGCAT ATCACCACCA AAATTTGAAC TGTCAGTCTT TGTCTCTT TTGTCTCTTT TTTTTCCTTC    34400
                                                                                Poly-pyrimidine run begins
CCAACAGAGT TTGAGTTTCT ATTGATCCAT ATCACCACCA AAATTTGAAC TGTCAGTCTT TCCCTTCCCT TCTCTTTCT CTTCCCTAT CCTTCTCCTC    34500
CCTTCCCCTC TCTTCGTTTC TTTTCTCTC TCTTTCTCTT ATTTTCCCTT CTTCTCCTCC ATCCCTTCCA TCCCTCTCT TCCCCTCTTC CTTCTCCTCC   34600
TCCTTTTTTC TCCTCTCCTC TCCATTATTT ATTTTCCTT CTTCTCCTCC ATCCCTTCCA TCCCTCTCT TCCCCTCTTC CTTCTCCTCC TTCTCCATTT   34700
ends
CTTCCTCCTC TTTCCTTCAA TCCTTCCTTT TGGATATGCT CATGGGTGTG TATTTGTCTG CCATTGTGGC ATATTTGAA TTCAGAAAAG AGTGAAAAAC   34800
                                                                                     EcoRI
TACTGGATC TTCATTCTGG GTCTAATTCC ACATTTTTT TTAAGAACAC ACTCTGTAA AATGTTCTGT ACTAGATAT TCCCAGGAAC TTCGTTAAAT     34900
TTAATCTGGC TGAATATGGT AAATCTACTT TGCACTTGC ATCTTTCTT TAGTCATACC ATAATTTAA ACATTCAAA TATTTGTATA TAATATTTGA     35000
```

```
TTTTATCTGT CATTAAATG TTAACCTTAA AATTCATGTT TCCAGAACCT ATTTCAATAA CTGGTAAATA AACACTATTC ATTTTTTAAA TATTCTTTTA  36100
ATGGATATTT ATTTCAATAT AATAAAAAT TAGAGTTTTA TTATAGGAAG AATTTACCAA AAGAAGGAGG AAGCAAGCAA GTTAAACTG CAGCAATAGT  36200
TGTCCATTCC AACCTCTCAA AATTCCCTTG GAGACAAAAT CTCTAGAGGC AAAGAAGAAC TTTATATTGA GTCAACTTGT TAAAACATCT GCTTTTAGAT  36300
AAGTTTTCTT AGTATAAAGT GACAGAAACA AATAAGTTAA ACTCTAAGAT ACATTCCACT ATATTAGCCT AAAACACTTC TGCAAAAATG AAACTAGGAG  36400
GATATTTTTA GAAACAACTG CTGAAAGAGA TGCCGTGGGG AGATATGCAG AGGAGAACAG GGTTTCTGAG TCAAGACACA CATGACAGAA CAGCCAATCT  36500
CAGGGCAAGT TAAGGAATGA GTGGAATGAA GGTTCATTTT TCATTCTCAC AAACTAATGA AACCCTGCTT ATCTTAAACC AACCTGCTCA CTGGAGCAGG  36600
                                            Delta Gene                                                  Init
GAGGACAGGA CCAGCATAAA AGCCAGGGCA GAGTCGACTG TTGCTTACAC TTTCTTTCTGA CATAACAGTG TTCACTAGCA ACCTCAAACA GACACCATGG  36700
                                            Cap
TGCATCTGAC TCCTGAGGAG AAGACTGCTG TCAATGCCCT GTGGGCAAA GTGAACGTGG ATGCAGTTGG TGGTGAGGCC CTGGGCAGT TGGTATCAAG  36800
                                                                                   IVS 1
GTTATAAGAG AGGCTCAAGG AGCAAATGG AAACTGGGCA TGTAGACA GAGAGACTC TTTGAGTCCT TTGGGTTTCT GATAGGCACT GTCCTCTCCT CCCTTGGGCT  36900
          Exon 2
GTTTTCCTAC CCTCAGATTA CTGGTGGTCT AGTGGCCTTT AGTGATGGCC CAGAGGTTC TGGCTCACCT GGACAACCTC AAGGGCACTT TTTCTCAGT GAGTGAGCTG  36100
CACTGTGACA AGCTGCACGT GGATCCTGAG AACTTCAGGG TAGTCCAGG AGATGCTTCA CTTTTCTCTT TTACTTTCT TGTTAATTTT ATTCTGATT  36200
                                                                        IVS 2
TTACCTACCT GCTCTTCTCC CACATTTTTG TCATTTTACT ATATTTAAT ATCCTGTCTT TCTCCCCAA TGTCTCTCCA CATGGTATG GGAGAGCTC AACTCAAAG ATGAGAGGCA  36300
TTATGCATAC CAGCTCTCAC CTGCTAATTC TGCACTTAGA ATAATCCTTT TGTCACTTGA ATAATCCTTT GAATTTTATA GAATAATTG TAAATGGAAT GGAAAGGAAA GTGAATATTT  36400
TAGAATACTG TTTTAGAGGC TATAAATCAT TTTACAATAA GGAATAATTG AATTCTGTAG AATTCTGTAG TAGGAGACAG CCCATCATCA CACTGATTAA TCAATTAATT TGTATCTATT  36500
GATTATGAAA GACTAGGCAG TTACACTGGA GGTGGGCAG AAGTCGTTGC TAGGAGACAG AATCTGTGGC TAGGAGACAG CCCATCATCA CACTGATTAA TCAATTAATT TGTATCTATT  36600
AATCTGTTTA TAGTAATTAA TTTGTATATG CTATATACAC ATAATTTAAT AGAGGATGTG TGTGTATAGA AAAATGGGTG TTGGCTCAGT TCCTCAGAAG CCAGTCTTTA TTTCTCTGTT  36700
ATATATGTAC ATATATAGAC TACATGCTAG TTAAGTACAT AGAGGATGTG TGTGTATAGA AAAATGGGTG TTGGCTCAGT TCCTCAGAAG CCAGTCTTTA TTTCTCTGTT  36800
                                                                                                   Eco RI
GCTGATGGGA ATAACCTGGG GATCAGTTTT GTCTAAGATT TGGGCAGAAA TGCTAAGATT TGGCAGAAT TGCTGGTGTG TGCTGGTGCC CGCAACTTTG GCAAGGATT CACCCCACAA  36900
                                                      Exon 3
AACCATATGG ATGTATCTGC CTACCTCTTC TCCGCAGCTC TTGGGCAATG ATGCCCTGGC TCACAAGTAC CATTGAGATC CTGGACTGTT TCCTGATAAC CATAAGAGA  37000
ATGCAGGCTG CCTATCAGAA GGTGGTGGCT GGTGTGGCTA ATGCCCTGGC TCACAAGTAC CATTGAGATC CTGGACTGTT TCCTGATAAC CATAAGAGA  37100
                                                                        Poly-A
CCCTATTCC CTAGATTCTA TTTTCTGAAC TTGGAACAC AATGCCTACT TCAAGGGTAT GGCTTCTGCC TAATAAAGAA ACTTGTTAT TTACAAAGA GTACATGGA  37200
TTAATTCAC TTATTTCATT TTTTGTCCA GGTGTGTAAG AAGGTTCCTG AGGCTCTACA GATAGGGAGC ACTTGTTAT TTACAAAGA GTACATGGA  37300
AAGAGAAAA GCAAGGAAC CGTACAAGGC ATTAATGGGT GACACTTCTA CCTCCAAAGA GCAGAAATTA TCAAGAACTC TTGATACAAA GATAATACTG  37400
```

453

```
GCACTGCAGA GGTTCTAGGG AAGACCTCAA CCCTAAGACA TAGCCTGAAG GTAATAGCT AGATTAAAC TCCAACATT ACTGAGAAAA TAATGTGCTC   37600
AATTAAAGGC ATAATGATTA CTCAAGACAA TGTTATGTTG TCTTTCTTCC TCCTTCCTTT GCCTGCACAT TGTAGCCCAT AATACTATAC CCCATCAAGT   37700
GTTCCTGCTC CAAGAAATAG CTTCCTCCTC TTACTTGCCC CAGAACATCT CTGTAAAGAA TTTCCTCTTA TCTTCCCATA TTTCAGTCAA GATTCATTGC   37800
TCACGTATTA CTTGTGACCT CTCTTGACCC CAGCCACAAT AAACTTCTCT ATACTACCCA AAAAATCTTT CCAAACCCTC CCCGACACCA TATTTTATA   37900
TTTTCTTAT TTATTTCATG CACACACACA CACTCCGTGC TTTATAAGCA ATTCTGCCTA TTCTCTACCT TCTTACAATG CCTACTGTGC CTCATATTAA   38000
ATTCATCAAT GGGCAGAAAG AAATATTTA TTCAAGAAAA CAGTGAATGA ATGAACGAAT GAGTAAATGA GTAAATGAAG GAATGATTAT TCCTTGCTTT   38100
AGAACTTCTG GAATTAGAGG ACAATATTAA TAATACCATC GCACAGTGT TCTTTGTTGT TCTTTGTTGT ACATACAAAG AGGAAGCATG CAGTAAACAA   38200
CCGAACAGTT ATTTCCTTTC TGATCATAGG AGTAATATTT TTTTCCTTGA GCACATTTTT GCCATAGTTA AATTAGAAG GATTTTTAGA ACTTTCTCAG   38300
TTGTATACAT TTTTAAAAAT CTGTATTATA TGCATGTTGA TTAATTTAA ACTTACTTGA ATACCTAAAC AGAATCTGTT GTTTCCTTGT GTTTGAAAGT   38400
GCTTTCACAG TAACTCTGTC TGTACTGCCA GAATATACTG ACAATGTGTT ATAGTTAACT GTTTTGATCA CAACATTTTG AATTGACTGG CAGCAGAAGC   38500
TCTTTTTATA TCCATGTGTT TTCCTTAAGT CATTATACAT AGTAGGCATG AGACTCTTTA TACTGAATAA GATATTTAGG AACCACTGGT TTACATATCA   38600
GAAGCAGAGC TACTCAGGGC ATTTGGGGA AGATCACTTT CACATTCCTG AGCATAGGGA AGTTCTCATA AGAGTAAGAT ATTAAAAGGA GATACTTGTG   38700
TGGTATTCGA AAGACAGTAA GAGAGATTGT AGACCTTATG ATCTTGATAG GGAAAACAAA CTACATTCCT TTCTCCAAAA GTCAAAAAAA AAGAGCAAAT   38800
                                                                                                 Eco RI
ATAGCTTACT ATACCTTCTA TTCCTACACC ATTAGAAGTA GTCAGTGAGT CTAGGCAAGA TGTTGGCCCT CTTCATATTG AAAATCCAA ATACCAGAGA ATTCATGAGA   38900
ACATCACCTG GATGGACAT GTGCCGAGCA ACACAATTAC TATATGCTAG GCATTGCTAT CTTCATATTG AAGATGAGGA GGTCAAGAGA TGAAAAAGA   39000
CTTGGCACCT TGTTGTTATA TTAAAATTAT TTGTTAGAGT AGAGTTTTG TAAGAGTCTA GGAGTGTGGG AGCTAAATGA TGATACACAT GGACACAAAG   39100
AATAGATCAA CAGACACCCA GGCCTACTTG AGGGTTGAGG GTGGAAAGAG GGAGACGATG AAAAGAACC TATTGGGTAT TAAGTTCATC ACTGAGTGAT   39200
GAAATAATCT GTACATCAAG ACCCAGTGAT ATGCAATTTA CCTATATAAC TTGTACATGT AATGGTTGTT TTAAATAAA GTTAAAACAA AGTATAGGAA   39300
TGGAATTAAT TCCTCAAGAT TTGGCTTTAA TTTTATTGA TAATTTATCA ATGAAATCT GAGCCCAGTG GAGGAAATAT TAATGAACAA GTGCAGACT GAAATATAAA   39400
AGTATGTCTG AATGAAAGGG TGTGTGTGTG TGTGAAAGAG AGGGAGAGAG GAAGGAAGA GAGGACGTAA TAATGTGAAT TTGAGTTGAT GAAAATTTT   39500
CAATAAAATA ATTTATGTC AGGAGAATTA AGCTAATAG TCTCCTAAAT CATCCATCTC TTGAGCTTCA GAGCAGTCCT CTGAATTAAT GCCTACATGT   39600
TTGTAAAGGG TGTTCAGACT GAAGCCAAGA TTCTACCTCT AAAGAGATGC AATCTCAAAT TATCTGAAG ACTGTACCTC TGCTCTCCAT AATTGACAC   39700
CATGGCCCAC TTAATGAGGT TAAAAAAAG CTAATTCTGA ATGAAAATCT GAGCCCAGTG GAGAAATAT TAATGAACAA GTGCAGACT GAAATATAAA   39800
TTTTCTGTAA TAATTATGCA TATACTTTAG CAAAGTTCTG TCTATGTTGA CTTTATTGCT TTTGGTAAGA AATACAACTT TTTAAAGTGA ACTAAACTAT   39900
CCTATTCCA AACTATTTTG TGTGTGTGCG GTTTGTTTCT ATGGGTTCTG GTTTTCTTGG AGCATTTTTA TTTCATTTA ATTAATTAAT TCTGAGAGCT   40000
```

```
GCTGAGTTGT GTTACTGAG AGATTGTGTA TCTGCGAGAG AAGTCTGTAG CAAGTAGCTA GACTGTCTT GACCTAGGAA CATATACAGT AGATTGCTAA   40100
AATGTCTCAC TTGGGAATT TTAAGACTAA CAGTAGAGCA TGTATAAAAA TACTCTAGTC AAGTGCTGCT TTTGAAACAA ATGATAAAAC CACACTCCCA   40200
TAGATGAGTG TCATGATTTT CATGGAGGAA GTTAATATTC ATCCTCTAAG TATACCCAGA CTAGGGCCAT TCTGTATATA AACATTAGGA CTTAAGAAAG   40300
                                                                            Poly (GT)
ATTAATAGAC TGGAGTAAAG GAAATGACC TCTGTCTCTC TCGCTGTCTC TTTTTTGAGG ACTGTGTGT GTGTGTGTGT GTGTGTGT GTGTGTGTG        40400
GTCAGTGGGG CTGGAATAAA AGTAGAATAG ACCTGCACCT GCTGTGGCAT CCATTCACAG AGTAGAAGCA AGCTCACAAT AGTGAAGATG TCAGTAAGCT   40500
TGAATAGTTT TTCAGGAACT TTGAATGCTG ATTTAGATTT GAAACTGAGG CTCTGACCAT ACCAAATTT GCACTATTA TTGCTTCTTG AAACTTATTT     40600
GCCTGGTATG CCTGGGCTTT TGATGGTCTT AGTATAGCTT GCAGCCTTGT CCCTGCAGGG TATTATGGGT AATAGAAAGA AAAGTCTGCG TTACACTCTA   40700
GTCACACTAA GTAACTACCA TTGGAAAAGC AACCCCTGCC TTGAAGCCAG GATGATGGTA TCTGCAGCAG TTGCCAACAC AAGAGAAGA TCCATAGTTC    40800
ATCATTTAAA AAAGAAAACA AATAGAAAA AGGAAAACTA TTTCTGAGCA TAAGGAGTTG TAGGTAAGT CTTTAAGAAG GTGACAATTT CTGCCAATCA    40900
GGATTTCAAA GCTCTTGCTT TGACAATTTT GGTCTTTCAG AATACTATAA ATATAACCTA TATTATATT TCATAAAGTC TGTGCATTTT CTTTGACCCA    41000
GGATATTTGC AAAAGACATA TTCAAACTTC CCCAGAACAC TTTATTTCAC ATATACATGC CTCCTATATC AGGGATGTGA AACAGGGTCT TGAAAACTGT   41100
CTAAATCTAA AACAATGCTA ATGCAGGTTT AAATTTAATA AATAAAATC CAAAATCTAA CAGCCAAGTC AAATCTGTAT GTTTAACAT TTAAAATATT    41200
TTAAAGACGT CTTTTCCCAG GATTCAACAT GTGAAATCTT TCTTACCTCT ATAATCATAC ATAGGCATAA TTTTTTAACC TAGGCTCCAG ATAGCCATAG AAGAACCAAA 41300
TATAAAAAGA AAATACTTAA ATTTTATCCC TCTTACCTCT ATAATCATAC ATAGGCATAA TTTTTTAACC TAGGCTCCAG ATAGCCATAG AAGAACCAAA  41400
CACTTCTGC GTGTGTGAGA ATAATCAGAG TGAGATTTTT TCACAAGTAC CTGATGAGGG TTGAGACAGG TAGAAAAGT GAGAGATCTC TATTTATTTA    41500
GCAATAATAG AGAAAGCATT TAAGAGAATA AAGCAATGGA AATAAGAAAT TTGTAAATTT CCTTCTGATA ACTAGAAATA GAGGATCCAG TTTCTTTTGG  41600
  Repeating (ATTTT)
TTAACCTAAA TTTTATTCA TTTTATTGTT TTATTTTATT TTATTTTGTG TAATGCTAGT TTCAGAGTGT TAGAGCTGAA AGGAAGAAGT              41700
AGGAGAAACA TGCAAAGTAA AAGTATAACA CTTTCCTTAC TAAACCGACT GGGTTTCCAG GTAGGGCAG GATTCAGGAT GACTGACAGG GCCCTTAGGG   41800
AACACTGAGA CCCTACGCTG ACCTCATAAA TGCTTGCTAC AGGGAAAAA GTACAGGGGG ATGGGAGAAA GGCATCACG TTGGGAAGCT ATAGAAAAG TTCAAAAGTT 41900
TTTCCTCACC TGAGGAGTTA ATTTAGTACA AGGGAAAAA GTACAGGGGG ATGGGAGAAA GGCATCACG TTGGGAAGCT ATAGAAAAG AAGAGTAAAT    42000
TTTAGTAAAG GAGGTTTAAA CAAACAAAT ATAAAGAGAA ATAGGAACTT GAATCAAGGA AATGATTTTA AAACCAGTA TTCTTAGTGG ACTAGAGGAA    42100
AAAATAATC TGAGCCAAGT AGAAGACCTT TTCCCCTCCT ACCCCTACTT TCTAAGTCAC AGAGGGTTTT TGTTCCCCCA GACACTCTTG CAGATTAGTC    42200
CAGGCAGAAA CAGTTAGATG TCCCCAGTTA ACCTCCTATT TGACACCACT GATTACCCCA TTGATAGTCA CACTTTGGGT TGTAAGTGAC TTTTTATTTA   42300
TTTGTATTTT TGACTGCATT AAGAGGTCTC TAGTTTTTTA TCTCTGTTTT CCCAAAACCT AATAAGTAAC TAATGCACAG AGCACATTGA TTTGTATTTA  42400
                                                                              Alternating Purine-Pyrimidine Run
TTCTATTTTT AGACATAATT TATTAGCATG CATGAGCAAA TTAAGAAAAA CAACAACAAA TGAATGCATA TATAGCATA TGTATGTGTG TATATATACA   42500
```

```
CATATATATA TATATTTTT TTCTTTTCTT ACCAGAAGGT TTTAATCCAA ATAAGGAGAA GATATGCTTA GAACTGAGGT AGAGTTTTCA TCCATTCTGT  42600
CCTGTAAGTA TTTTGCATAT TCTGGAGACG CAGGAAGAGA TCCATCTACA TATCCAAAG CTGACAAAA CTCTTCCACT TTTAGTGCAT              42700
CAATTTCTTA TTTGTGTAAT AGAAAAATTG GGAAAACGAT CTTCAATATG CTTACCAAGC TGTGATTCCA AATATTACGT AATACACTT GCAAAGGAGG 42800
ATGTTTTTAG TAGCAATTTG TACTGATGGT ATGGGGCCAA GAGATATATC TTAGAGGGAG GGCTGAGGGT TTGAAGTCCA ACTCCTAAGC CAGTGCCAGA 42900
AGAGCCAAGG ACAGTACGG CTGTCATCAC TTAGACCTCA CCCTGTGGAG CCACACCCTA GGGTTGGCCA ATCTACTCCC AGGAGCAGGG AGGGACAGGAG 43000
                                              Cap                                                     Init
CCAGGGCTGC GCATATAAGT CAGGGCAGAG CCATCTATTG CTTACATTTG CTTCTGACAC AACTGTGTTC ACTAGCAACC TCAAACAGAC ACCATGGTGC 43100
                     Beta Gene                                                          ^^^           ^^^
ACCTGACTCC TGAGGAGAAG TCTGCCGTTA CTGCCCTGTG GGGCAAGGTG AACGTGGATG AAGTTGGTGG TGAGGCCCTG GGCAGGTTGG TATCAAGGTT 43200
                                                                                         IVS 1
ACAAGACAGG TTTAAGGAGA CCAATAGAAA CTGGGCATGT GGAGACAGAG AAGACTCTTG GGTTTCTGAT AGGCACTGAC TCTCTCTGCC TATTGGTCTA 43300
           Exon 2
TTTTCCCACC CTTAGGCTGC TGGTGGTCTA CCCTTGGACC CAGAGGTTCT TTGAGTCCTT TGGGGATCTG TCCACTCCTG ATGCTGTTAT GGGCAACCCT 43400
AAGGTGAAGG CTCATGGCAA GAAAGTGCTC GGTGCCTTTA GTGATGCCTT GGCTCACCTG GACAACCTCA AGGGCACCTT TGCCACACTG AGTGAGCTGC 43500
ACTGTGACAA GCTGCACGTG GATCCTGAGA ACTTCAGGGT GAGTCTATGG GACCCTTGAT GTTTTCTTTC CCCTTCTTTT CTATGGTTAA GTTCATGTCA 43600
                                                IVS 2
TAGGAAGGGG AGAAGTAACA GGGTACAGTT TAGAATGGGA AACAGACGAA TGATTGCATC AGTGTGGAAG TCTCAGGATC GTTTTAGTTT CTTTTATTTG 43700
CTGTTCATAA CAATTGTTTT CTTTTGTTTA ATCTTGCTT TCTTTTTT TCTCTCCGC AATTTTTACT ATATACTTA ATGCCTTAAC ATTGTGTATA       43800
ACAAAAGGAA ATATCTCTGA GATACATTAA GTAACTTAAA AAAAACTTT ACACAGTCTG CCTAGTACAT TACTATTTGG AATATATGTG TGCTTATTTG  43900
CATATTCATA ATCTCCCTAC TTTATTTTCT TTTATTTTA ATTGATACAT AATCATTATA GTTAAAGTG TAATGTTTTA CTTATTTCTA ATACTTTCCC   44000
ACATATTGAC CAAATCAGGG TAATTTTGCA TTTGTAATTT TAAAAATGC TTTCTTTTTT CACCATTCTA AAGAATAACA GTGATAATTT CTGGGTTAAG GCAATAGCAA 44100
TAATCTCTTT CTTTCAGGGC AATAATGATA TCTGCATAIA TGCCTCTTTG CACCATTCTA AAGAATAACA GTGATAATTT CTGGGTTAAG GCAATAGCAA 44200
TATTTCTGCA TATAAATATT TCTGCATATA AATTGTAACT GATGTAAGAG GTTTCATATT GCTAATGCA GCTAATAGCA GCTACAATCC AGTCTTTAT   44300
                                                                                                   Exon 3
TTTATGGTTG GGATAAGGCT GGATTATTCT GAGTCCAAGC TAGGCCCTTT TGCTAATCAT GTTCATACCT CTTATCTTCC TCCCACAGCT CCTGGGCAAC 44400
GTGCTGGTCT GTGTGCTGGC CCATCACTTT GGCAAAGAAT TCACCCCACC AGTGCAGGCT GCCTATCAGA AAGTGGTGGC TGGTGTGGCT AATGCCCTGG 44500
                                Eco RI
CCCACAAGTA TCACTAAGCT CGGTTCTTG CTGTCCAATT TCTATTAAAG GTTCCTTTGT TCCCTAAGTC CAACTACTAA ACTGGGGGAT ATTATGAAGG 44600
                                   Poly-A
GCCTTGAGCA TCTGATTCT GCCTAATAAA AAACATTTAT TTTCATTGCA ATGATGTATT TAAATTATTT CTGAATATTT TACTAAAAAG GGAATGTGGG 44700
                    ^^^^^^^^
AGGTCAGTGC ATTTAAAACA TAAAGAAATG AAGAGCTAGT TCAAACCTTG GGAAAATACA CTATATCTTA AACTCCATGA AAGAAGGTGA GGCTGCAAAC 44800
AGTAATGCA CATTGGCAAC AGCCCTGATG CCTATGCTTT CAGAAAAGA TTCAAGTAGA GGTTGATTT GGAGGTTAAA GTTTTGCTAT               44900
GCTGTATTTT ACATTACTTA TTGTTTAGC TGTCCTCATG AATGTCTTT CACTACCCAT TTGCTTATCC CACTAGGGGT CTGTCCTGACT CCACTCAGTT  45000
```

```
CTCTTGCTTA GAGATACCAC CTTTCCCCTG AAGTGTTCCT TCCATGTTTT ACGGGCGAGAT GGTTTCTCCT CGCCTGGCCA CTCAGCCTTA GTTGTCTCTG  45100
TTGTCTTATA GAGGTCTACT TGAAGAGGA AAAACAGGGG GCATGGTTTG ACTGTCTCGT GAGCCCTTCT TCCCTGCCTC CCCCACTCAC AGTGACCCGG  45200
AATCTGCAGT GCTAGTCTCC CGGAACTATC ACTCTTTCAC AGTCTGCTTT GGAAGGACTG GGCTTAGTAT GAAAAGTTAG GACTGAGAAG AATTTGAAAG  45300
GGGGCTTTTT GTAGCTTGAT ATTCACTACT GTCTATTATAC CCTATCATAG GCCCACCCCA AATGGAAGTC CCATTCTTCC TCAGGATGTT TAAGATTAGC  45400
ATTCAGGAAG AGATCAGAGG TCTGCTGGCT CCCTTATCAT GTCCCTTATG GTGCTTCTGG CTCTGCAGTT ATTAGCATAG TGTTACCATC AACCACCTTA  45500
ACTTCATTTT TCTTATTCAA TACCTAGTA GGTAGATGCT AGATTCTGAA AATAAAATAT GAGTCTCAAG TGGTCCTTGT CCTCTCTCCC AGTCAAATTC  45600
TGAATCTAGT TGGCAAGATT CTGAAATCAA GGCATATAAT CAGTAATAAG TGATGATAGA AGGGTATATA GAAGAATTTT ATTATATGAG AGGGTGAAAC  45700
CTAAAATGAA ATGAAATCAG ACCCTTGTCT TACACCATAA ACAAAAATAA ATTTGAATGG GTTAAAGAAT TAAACTAAGA CCTAAAACCA TAAAATTTT  45800
TAAAGAAATC AAAGAAGAA AATTCTAATA TTCATGTTGC AGCCGTTTTT TGAATTTGAT ATGAGAAGCA AAGGCAACAA AAGGAAAAAT AAAGAAGTGA  45900
GGCTACATCA AACTAAAAA TTTCCACACA AAAAGAAAA CAATGAACAA ATGAAAGGTG AACCATGAAA TGGCATATTT GCAAACCAAA TATTTCTTAA  46000
ATATTTTGGT TAATATCCAA AATATATAAG AAACACAGAT GATTCAATAA CAAACAAAAA ATTAAAATA GGAAAATAAA AAAATTAAAA AGAAGAAAAT  46100
CCTGCCATTT ATGCGAGAAT TGATGAACCT GGAGGATGTA AAACTAAGAA AAATAAGCCT GACACAAAAA GACAAATACT ACACAACCTT GCTCATATGT  46200
GAAACATAAA AAAGTCACTC TCATGGAAAC AGACAGTAGA GGTATGTTT CCAGGGGTTG GGGGTGGGAG AATCAGGAAA CTATTACTCA AAGGGTATAA  46300
AATTTCAGTT ATGTGGGATG AATAAATTCT AGATATCTAA TGTACAGCAT CGTGACTGTA GTTAATTGTA CTGTAAGTAT ATTTAAATT TGCAAAGAGA  46400
      Alu                  element begins
GTAGATTTTT TTGTTTTTTT AGATGGAGTT TTGCTCTTGT TGTCCAGGCT GAAGTGCAAT GGCAAGATCT TGGCTCACTG CAACCTCCGC CTCCTGGGTT  46500
CAAGCAAATC TCCTGCCTCA GCCTCCCGAG GCTCCCCAG TAGCTGGGAT TACAGGCATG CCCAGCTAAT TTTGTATTTT TAGTAGAGAC GGGGTTTCTC  46600
CATGTTGGTC AGGCTGATCC GCCTCCTCGG CCACCAAAGG GCTGGGATTA CAGGCGTGAC CAGGGCGCCT CACCGGGCCT GCCCGAGT AGATCTTAAA AGCATTTACC  46700
                                                                           Alu ends
ACAAGAAAA GGTAACTATG TGAGATAATG GGTATGTTAA ATGATACCTC TAGCTTGAT TGTGGTAATC ATTTCACAAG GTATACATAT ATTAAACAT CATGTTGTAC  46800
ACCTTAAATA TATACAATTT TTATTTGTGA ATCTTGATT ATCAGATAC AATAAAGTTG AAGAATAATA AAAAGAATA GACATCACAT GAATTAAAAA ACTAAAAAT  46900
AAAAAAATGC ATCTTGATGA TTAGAATTGC CCCTTCCCAA TCTATAATT TTCAGATAC AATATCCAT TATTTTAGAT ATTTTGTATA GTTTACTCC ACTCTTTTCC AAACAATAC AATAAATTT  47000
AGCACTTTAT CTTCATTTC CCCTTCCCAA TCTATAATTT TATATATATA TATTTTAGAT ATTTTGTATA GTTTACTCC CTAGATTTC TAGTGTTATT  47100
ATTAAATAGT GAAGAAATGT TTACACTTAT GTACAAAATG TTTTGCATGC TTTTCTTCAT TTCTACATT CTCTCTAAGT CTCTCTAAGT TTAAATCATG TTATTCTTAAT  47200
TATCCTTAAT ATTATCTCTT TCTGCTGAA ATAGTGAAT ATATATTGTT ACTTTTGTT TATCTAAAAA TGGCTTCATT TTCTTCATTC TAAATCATG TTATTCTTAAT  47300
ACCACTCATG TGTAAGTAAG ATAGTGAAT AACAGTAATT AATATGAAAT CCAAAACTA AATCTCACA AATATAATAA TGTGATATAT AAAATATAG CTTTTAAAT  47400
TAGCTTGGAA ATAAAAAACA AACAGTAATT GAACAACTAT ACTTTTTGAA AAGAGTAAAG TGAAATGCTT AACTGCATAT ACCACAATCG ATTACACAAT  47500
```

```
TAGGTGTGAA GGTAAAATTC AGTCACGAAA AACTAGAAT AAAATATGG GAAGACATGT ATATAATCGT AGATAACA GTGTTATTTA ATTATCAACC  47600
                                                                              Alu element begins
CAAAGTAGAA ACTATCAAGG GAGAAATAAA TTCAGTCAAC AATAAAAGCA TTTAAGAAGT TATTCTAGGC TGGGAGCGGT GGCTCACACC TGCAATTGCA  47700
GCACTTTGGG AGGCCTAGAC AGGGGGATCA CGACGTCAGG AGTTCAAGAT CAGCCTGGCC ACATAGTGA AACCTCATCG CTACTAAAAA TATAAAACT   47800
TAGCCTGGCG TGGTGGCAGG CATGTGTAAT CCCAGCAATT TGGGAGGCTG AGCCAGGAGA ATCGCTTGAT CCTGGGAGGC AGAGGTTGCA GTGAGCCAAG  47900
                                                                     Alu ends
ATTGTGCCAC TGCATTCCAG CCCAGGTGAC AGCATGAGAC TCCCTCACAA AAAAAAAGA AAAAAAGGG GGGGGGGGC GGTGGAGCCA AGATGACCGA  48000
ATAGGAACAG CTCCAGTCTA TAGTCCCAT CGTGAGTGAC GCAGAAGACG GGTGATTTCT GCATTTCCAA CTGAGGTACC AGTTCATCT CACAGGGAAG  48100
TGCCAGGCAG
```

FIG. 14. Complete nucleotide sequence of the β-globin cluster from the ε gene through the β gene. This represents a compilation of previously published data from several other laboratories together with approximately 18 kb of intergenic DNA sequence determined in our laboratory and not previously published. References for the source of the sequences follow. (1) From the 5' end to the EcoRI site 3' to ε, reference 42. (2) From this point to the BamHI site 5' to Gγ, the sequence was derived in our laboratory, references 255a and 256. (3) From this point to the EcoRI site 3' to Aγ, the sequence is taken from reference 44. (4) The next 7.0-kb EcoRI fragment containing the ψβ gene was sequenced in our laboratory, reference 257. (5) The next 3.1-kb EcoRI fragment has been sequenced by ourselves (257) and two other groups (263, 258). A series of polymorphic differences is found in this segment (263, 133); the sequence presented here is referred to by Maeda et al. (263) as the T allele. (6) The final 14-kb sequence containing the δ and β genes is taken from reference 258, except that we have removed the −28 ATA box mutation that characterized this thalassemic allele (see Table II).

Published sequences were entered by hand into a computer file and proofread twice. Previously unpublished sequences from our laboratory were obtained partly by Maxam-Gilbert method and partly by M13 dideoxy sequencing; in most instances both strands were sequenced and the accuracy should be 98% or better. A possible exception is the region from 12789 to 12920, for which only one strand was sequenced and the possibility of errors is greater. We are rechecking this region; those interested are invited to contact us for any changes.

The EcoRI sites are marked, as are the CAP sites, initiation codons, splice junctions, termination codons, and polyadenylation signals of the expressed genes. There are six Alu repetitive elements present; the direct repeats flanking these elements are overlined, except for the most 3' Alu, which appears not to have a flanking repeat. Also marked are a long polypyrimidine stretch 5' to the δ gene and three interesting elements between δ and β which may be involved in recombination (see Fig. 24).

Addendum 2

Recently a number of important new developments have appeared. A brief survey of a few of these is presented here, though space precludes our presenting any of them in very much detail.

Perhaps the most far-reaching advance has been the development of an efficient cell-free system for mRNA splicing (1, 2). Two groups have now made rapid progress in elucidation of the splicing mechanism by using purified adenovirus (1) or β-globin (2) mRNA precursors, which can be produced in bacteria in relatively large amounts. Study of the products and intermediates of splicing appear to show that the steps are (i) a U1-dependent (3) cleavage of the donor site; (ii) looping around of the 5' end of the intron sequence probably to form an unusual branched 2'–5' bond (4) to an adenosine residue 30–40 bp 5' to the acceptor site (creating a "lariat" structure); (iii) cleavage of the acceptor site and ligation of the exons. Analysis of the sequence around the branched adenosine residue in human β-globin IVS-1 and comparison with globin genes from other species has revealed a previously undetected consensus sequence YNYTRAY, where the A residue is located 23–40 bp 5' to the acceptor site (2). This is reminiscent of the "TACTAAC" box located in a similar position in yeast introns, point mutations of which disturb splicing (4a). Now that such an *in vitro* splicing system has been developed, the mechanism of some of the more puzzling splicing mutations causing β-thalassemia (such as IVS-2 position 705) can be elucidated.

Another interesting set of experiments has sought to identify the sequences responsible for Me_2SO induction of β-globin synthesis in MEL cells. Transfected human β-globin genes show marked increase in mRNA production after Me_2SO treatment; α-globin genes produce at a high level even before induction. A transfected fusion gene containing α sequences 5' to the cap site and β sequences 3' to this will induce readily with Me_2SO (5). So also will similar hybrid genes with either the 5' or the 3' portion contributed by β, and the remainder by either γ or a mouse H-2 gene (which are not themselves Me_2SO inducible) (6). The conclusion suggested is that both 5' flanking and intragenic (or 3' flanking) sequences of the β-globin gene are capable of conferring Me_2SO inducibility.

In the realm of hemoglobin gene transfer, a heartening development is the report by Costantini (7) of appropriately erythroid-specific expression in a few transgenic mice of a hybrid mouse–human β gene injected into oocytes. Though the level of expression of the injected gene remains considerably less than that of the native β gene, this report opens the way for study of the cis-acting sequences responsible for erythroid-specific expression in intact animals.

Evolutionary analyses of the β-globin cluster have also progressed. The triplicated β locus of the goat has now been completely crossed with overlapping clones, and the postulated order depicted in Fig. 18 has been borne out (8). However, other features of the scheme in this figure turn out to be oversimplified. In particular, the goat ε^{II} gene has been found to be ortholgous to the human $\psi\beta$ gene, rather than to γ (9–11). Apparently a five-gene system existed prior to the general mammalian radiation, with the three 5' genes [ε, γ, and the ancestor of the ε^{II} and $\psi\beta$ which Jeffreys (9) has suggested should be denoted η] being embryonic in function, and the two 3' genes (δ and β) being adult. The η gene has apparently been lost in the rabbit, and the γ has been lost in the goat; man, however, retains descendents of all five genes.

Two new point mutations causing thalassemia have been described. Kan and co-workers (12) found a T to C mutation in the ATG initiation codon of an α gene, and we have found a T to G change at position 116 of IVS-1 in a β gene, which creates an abnormal acceptor site and β-thalassemia (13).

Several more large deletions in the β-globin cluster have been described since Fig. 21 was constructed. One is a large 5' deletion which ends in the $^G\gamma$ gene IVS-2 (14). Somewhat surprisingly, the phenotype is of β-thalassemia; even though the β gene is about 28 kb away from the endpoint of the deletion (15), it is not expressed. Another informative case is a 6.5 kb deletion which removes the 5' end of δ and the region of DNA including the *Alu* repetitive sequences 5' to δ, which as we have described has been hypothesized to bear a switching control sequence. A homozygote for this deletion makes no β chains and has a thalassemia intermediate phenotype; heterozygotes make only slightly increased HbF. It is difficult to completely reconcile this mutation with the control region theory, at least in its simplest form. A Dutch $β^0$ thalassemia has been described in which a 10 kb deletion removes the β gene (16); the δ gene is intact, but the 3' end of this deletion maps very close to that for the Sicilian $(δβ)^0$ thalassemia (see Fig. 21). Finally, a Malaysian case of $^G\gamma(δβ)^0$ thalassemia has been described with a phenotype and 5' deletion endpoint very similar to the Turkish $^G\gamma(δβ)^0$ thalassemia (see Fig. 21), but with a different 3' endpoint (17).

In the nondeletion HPFH syndromes, more evidence for the role of the −200 region of the fetal globin genes has been obtained by finding the −202 $^G\gamma$ mutation in several more families with $^G\gamma\beta^+$

HPFH (18), and by the report of a different point mutation at −196 in the $^A\gamma$ gene from a Southern Italian individual with a previously undescribed type of $^A\gamma$ HPFH (19). The $^A\gamma$ gene from individuals with the Greek variety of HPFH has also been sequenced, and a G to A point mutation found at −117, just 5′ to the distal CCAAT box (20, 21). This is a rather unexpected but intriguing position for a mutation resulting in overexpression. In a related finding, cultured erythroid cells (BFU-E's) from heterozygotes for Ghana $^G\gamma^A\gamma$, Greek $^A\gamma$, and $^G\gamma\beta^+$ HPFH, while under normal conditions overproducing γ chains relative to normal BFU-E's, can be made to switch to production of normal amounts of β-globin by the addition of a fetal sheep serum factor; the suggestion is made that this trans-acting influence is able to overcome the cis-acting mutation (22).

Two more useful polymorphic sites have been described. One is an RsaI site (23) located within the 52 bp alternating purine and pyrimidine sequence 5′ to β (marked in Fig. 14). Another is an NcoI site 3 kb 5′ to this (18). Preliminary data suggest that both of these behave like the HinfI site just 5′ to β; that is, all three sites seem to lie in the region of high-frequency recombination discussed in Section VI,D.

Finally, although there are still few data in print, exciting advances in the application of retrovirus vectors to the transfer and expression of β-globin genes in eukaryotic cells are well underway, and developments in this area may begin to appear rapidly (24, 25). A particularly intriguing observation using such vectors raises the possibility that the human β-globin gene carries sequences capable of repressing transcription from an adjacent promoter. Inserting a 3.2 kb genomic human β gene into a retroviral vector and infecting MEL cells has shown that not only β mRNA but viral gene mRNA production is repressed before Me_2SO induction, but appears in wild-type amounts after induction (26). This finding, as well as the nondeletion HPFH results noted above, suggest that globin genes may normally carry important *negative* controlling elements whose direct investigation may prove fruitful.

Addendum References

1. P. J. Grabowski, R. A. Padgett, and P. A. Sharp, *Cell* **37**, 415 (1984).
2. B. Ruskin, A. R. Krainer, T. Maniatis, and M. R. Green, *Cell* (in press, 1984).
3. R. A. Padgett, S. M. Mount, J. A. Steitz, and P. A. Sharp, *Cell* **35**, 101 (1983).
4. J. C. Wallace and M. Edmonds, *PNAS* **80**, 950 (1983).
4a. C. J. Langford, F. J. Klinz, C. Donath, and D. Gallwitz, *Cell* **36**, 645 (1984).
5. P. Charnay, R. Treisman, P. Mellon, M. Chao, R. Axel, and T. Maniatis, *Cell* (in press, 1984).

6. S. Wright, E. deBoer, A. Rosenthal, R. A. Flavell, and F. Grosveld, *Cell* (in press, 1984).
7. F. Costantini, personal communication.
8. T. M. Townes, M. C. Fitzgerald, and J. B. Lingrel, *PNAS* (in press, 1984).
9. S. Harris, P. A. Barrie, M. L. Weiss, and A. J. Jeffreys, submitted.
10. M. Goodman, B. F. Koop, J. Czelusniak, M. L. Weiss, and J. L. Slightom, submitted.
11. L.-Y. Chang and J. L. Slightom, submitted.
12. M. Pirastu, G. Saglio, A. Cao, and Y. W. Kan, *Clin. Res.* **32**, 550A (1984).
13. F. S. Collins, J. E. Metherall, J. Pan, S. M. Weissman, and B. G. Forget, in preparation.
14. M. Pirastu, P. Curtin, Y. W. Kan, J. A. Gobert-Jones, A. D. Stephens, and H. Lehmann, *Clin. Res.* **32**, 493A (1984).
15. J. S. Wainscoat, presented at Fifth Cooley's Anemia Symposium, May 31, 1984, New York City.
16. J. G. Gilman, T. H. J. Huisman, and J. Abels, *Br. J. Haematol.* **56**, 339 (1984).
17. R. J. Trent, R. W. Jones, J. B. Clegg, D. J. Weatherall, R. Davidson, and W. G. Wood, *Br. J. Haematol.* **57**, 279 (1984).
18. F. S. Collins, C. D. Boehm, P. G. Waber, C. J. Stoeckert, B. G. Forget, S. M. Weissman, and H. H. Kazazian, submitted.
19. S. Ottolenghi, presented at Fifth Cooley's Anemia Symposium, May 31, 1984, New York City.
20. F. S. Collins, J. E. Metherall, M. Yamakawa, J. Pan, S. M. Weissman, and B. G. Forget, submitted.
21. R. Gelinas, and G. Stamatoyannopoulos, submitted.
22. T. Papayannopoulou, B. Tatsis, S. Kurachi, B. Nakamoto, and G. Stamatoyannopoulos, *Nature* **309**, 71 (1984).
23. G. L. Semenza, P. Malladi, S. Surrey, K. Delgrosso, M. Poncz, and E. Schwartz, *JBC* **259**, 6045 (1984).
24. C. L. Cepko, B. E. Roberts, and R. C. Mulligan, *Cell* (in press, 1984).
25. R. D. Cone and R. C. Mulligan, *PNAS* (in press, 1984).
26. R. D. Cone, A. Weber, and R. C. Mulligan, in preparation.

Addendum to Article by Helga Kersten
Queuosine Modification in tRNA and Expression of Genes of Electron Transport Chains

In the article, "On the Biological Significance of tRNA Modification" I have presented the protist *Dictyostelium discoideum* and *Xiphophorine* fishes as suitable models to elucidate the function of queuine and of the Q-family of tRNAs in relation to differentiation, pteridine metabolism, and neoplastic transformation.

In the meantime, isogenic pairs of *E. coli* containing or lacking the tRNA-transglycosylase (tgt$^+$ or tgt$^-$, respectively) have been employed to study the function of Q in tRNA of eubacteria (1, 2). When grown on glucose, under anaerobic conditions, the tgt$^-$ mutants are defective with respect to the nitrate respiration system in contrast to the tgt$^+$ strains. The nitrate respiration system transports electrons from D(−)-lactate via menaquinone or ubiquinone and cytochrome $b_{556}^{NO_3^-}$ to nitrate. Low-temperature cytochrome spectra of the anaerobically grown tgt$^-$ mutants show a lowered amount of type *b* cytochromes involving the spectrum of cytochrome $b_{556}^{NO_3^-}$. When grown anaerobically on glucose, the cytoplasmic membrane of the tgt$^-$ mutants lacks two proteins. Their molecular weights correspond to those of two subunits of the nitrate reductase complex with molecular weights of about 60,000 and 20,000. The 140,000 subunit of the nitrate reductase is not present in the tgt$^-$ mutant grown on glucose either aerobically or anaerobically. In contrast to the tgt$^+$ strains, neither tgt$^-$ mutant can grow on lactate under anaerobic conditions with nitrate offered as electron acceptor, and NO_3^- is not reduced to NO_2^- (1, 2).

Furthermore, the tgt$^-$ mutants and the *E. coli* strain (MRE 600) treated with exogenous biopterin to inhibit the synthesis of Q in tRNA show increased levels of cytochrome *d* (2). These changes are exactly the same as those observed when ubiquinone or oxygen becomes limiting (3), suggesting an important role of queuine and the Q-family of tRNAs in the regulation of redox chains and lactate metabolism in *E. coli*.

Two structural genes required for the expression of the nitrate reductase complex in *E. coli* have been mapped at 27 min on the *E. coli* chromosome: the nitrate reductase subunit (Chl C) and the apocytochrome $b_{556}^{NO_3^-}$ (Chl I). Interestingly, two tandem tRNA^{Tyr} genes map also at 27 min (4), and tRNA^{Tyr} belongs to those tRNAs containing Q in the first position of the anticodon. The close linkage of genes of the nitrate reductase respiration system with the tandem tRNA^{Tyr} genes supports the view that the expression of these tRNA genes and the subsequent modification of G-34 to Q-34 in tRNA^{Tyr} are involved in the expression of genes for subunits of the nitrate reductase in this region of the *E. coli* chromosome. In this respect, it is of interest that the glutamic tRNA and the COB gene, coding for the subunit of the cytochrome bc_1 complex of yeast mitochondria are apparently co-transcribed (5). Assuming that the nitrate reductase subunits are co-transcribed with one of the tandem tRNA^{Tyr} genes, the other tRNAs of the Q-family might regulate the expression of other proteins genes.

Mechanisms by which tRNA modifying enzymes and alterations in tRNA modification might be involved in metabolic regulation of gene expression, dependent on the redox state of the cell, are discussed in the article. Also discussed are similar mechanisms for other tRNA modifications that have recently been found to be functionally related to electron acceptor pathways in anaerobic an aerobic growth, e.g., the modification of io^{6}A and ms^{2}io^{6}A in tRNAs of *Salmonella typhimurium* (6, 7).

Nitrate-respiring bacteria are plausible ancestors of animal mitochondria (8), and queuine has been found in mitochondrial tRNA^{Asp} of mammals (9). Therefore alterations in the Q-content of tRNA might play a role in the expression of genes of mitochondrial DNA that code for components of the respiratory chain.

These observations and those described in the article indicate that alterations in Q-modification of specific tRNAs play a crucial role in the regulation of the synthesis of components of electron transport chains in bacteria as well as in higher eukaryotes. A close linkage between the inhibitory effect of pteridines, of Q-modification in tRNA and pteridine-dependent hydroxylations of aromatic ring systems, important for the biosynthesis of ubiquinone, is evident.

The available information shows that the accumulation of lactate in tumor cells and the alterations in respiration observed 58 years ago by Warburg (*10*) are closely related to changes in the Q-content of tRNA. A strict correlation between the Q-content of tRNA and the malignancy of human lymphomas has been demonstrated (*11*).

A novel isoenzyme of lactate dehydrogenase, LDH$_k$ has been detected in various cells, transformed by the Kirsten murine sarcoma virus (KiMSV) (ref. *12* and summarizing references therein). This is a highly oncogene helper-dependent virus, with a genome containing type C mouse leukemia virus sequences recombined with one or two sets of specific rat cell genetic information. Most of the rat genetic information in the virus is homologous to and apparently derived from 30-S RNA species found in normal rat cells. The normal rat sequences encode the transforming activity. The 30-S RNA is suggested to represent a number of closely related, but not identical RNA species. There are about 30 copies of this DNA per haploid rat genome. The DNA has properties suggesting it may be related to transposons.

LDH_k has been considered as being evolutionarily related to cyochrome b from yeast (12). Contrary to the yeast enzyme, LDH_k is reversibly inhibited by oxygen as is the nitrate-respiration system of *E. coli*. The LDH_k has a molecular weight of 57,000 and can be split into a 35,000 and 22,000 subunit. The nitrate reductase system of *E. coli* contains three subunits with molecular weights of about 60,000, 20,000, and 140,000. Recall that the 60,000 and 20,000 subunits become expressed only under anaerobic conditions in *E. coli* cells containing Q in tRNA. One of the Q tRNAs possibly represses the expression of those subunits under aerobic conditions by mechanisms discussed in the article. Similarly Q in eukaryotic tRNA could repress LDH_k, which might have evolved from a gene encoding one subunit of the nitrate reductase of facultative aerobes.

Finally these observations and those reported for specific changes in lactate dehydrogenase patterns and cytochrome b_{559} in *D. discoideum* in response to queuine show the importance of queuine and alterations in Q-modification of tRNA in the regulation of anaerobic and aerobic pathways. Hitherto unrelated alterations in LDH isoenzymes patterns, lactate and pyruvate metabolism, respiration, pteridine levels, and the oncogenic helper function of viruses are apparently linked by the modified nucleoside queuosine in tRNA. All of them seem to be involved together with the oncogene products in the development of tumors.

References

1. G. Jänel, U. Michelsen, H. Kersten, and S. Nishimura, *EMBO J.*, in press (1984).
2. H. Kersten and G. Jänel, *in* "Biochemical and Clinical Aspects of Pteridines" (H. Wachter, H.Ch. Curtius, and W. Pfleiderer, eds.). de Gruyter, Berlin, 1984, in press.
3. N. A. Newton, G. B. Cox, and F. Gibson, *BBA* **244**, 55 (1971).
4. B. J. Bachmann, *Microbiol. Rev.* **47**, 180 (1983).
5. Th. Christianson, J. C. Edwards, D. M. Mueller, and M. Rabinowitz, *PNAS* **80**, 5564 (1983).
6. B. N. Ames, T. H. Tsang, M. Buck, and M. F. Christman, *PNAS* **80**, 5240 (1983).
7. M. Buck and B. N. Ames, *Cell* **36**, 523 (1984).
8. B. J. Finlay, A. S. W. Span, and J. M. P. Harman, *Nature* **303**, 333 (1983).
9. E. Randerath, H. P. Agrawal, and K. Randerath, *Cancer Res.* **44**, 1167 (1984).
10. O. Warburg, *in* "Über den Stoffwechsel der Tumoren." Springer-Verlag, Berlin, 1926.
11. B. Emmerich, E. Zubrod, G. Dess, P. Maubach, and H. Kersten, *Exp. Hematol.* **11**, 170 (1983).
12. G. R. Anderson, W. P. Kovacik, and K. R. Marotti, *JBC* **256**, 10583 (1981).

Index

A

2-Acetamidofluorene, damaged DNA immunoassay and, 31–38
Aflatoxin B_1, damaged DNA immunoassay and, 41–43
Alkylation, damaged DNA immunoassay and, 25–31
Amino acid(s), metabolism, N^6-(2-isopentenyl)adenosine and, 87–88
Artemia
  polypeptide-chain initiation in
    comparison of Artemia and rabbit reticulocyte eIF2, 237–245
    comparison of eIF2 from dormant and developing embryos, 225–227
    eIF2 and initiation of protein synthesis, 224–225
    regulation of eIF2 activity, 227–236
    status of message in embryos, 245–246
    mRNA, 248–255
    mRNP particles, 246–248

B

Bacillus subtilis, sporulation, N^6-(2-isopentenyl)adenosine and, 86–87
Benzo[a]pyrene, damaged DNA immunoassay and, 38–41

C

Chromosome, location of rRNA genes and, 118
Cisplatin, damaged DNA immunoassay and, 43–45
Codon-anticodon, selectivity on ribosome, 197–200
Codon/anticodon recognition, N^6-(2-isopentyl)adenosine in tRNA and, 83–84
Cytokinin, N^6-(2-isopentenyl)adenosine and, 93–95

D

Deoxyribonuclease, globin gene hypersensitivity, 348–354
Deoxyribonucleic acid
  chemical adducts in, immunoassay and
    2-acetamidofluorene, 31–38
    aflatoxin B_1, 41–43
    alkylation damage, 25–31
    benzo[a]pyrene, 38–41
    cisplatin, 43–45
    Z-DNA, 45–46
  ionizing radiation-damaged, immunoassay of
    antinucleoside antibodies, 22–23
    early studies, 18–22
    radiation damaged nucleosides, 23–25
  methodology of immunoassay,
    antibody specificity and affinity, 6–7
    immunization schedule, 4–5
    immunogens, 2–4
    monoclonal antibodies, 5–6
  radiation-damaged, immunoassay of
    antibody specificity, 15–17
    early studies, 12–15
    quantitation and *in situ* detection, 17–18
  undamaged, immunoassay, 8–9
Z-Deoxyribonucleic acid, immunoassay of, 45–46
Development, regulation of globin gene and, 330–332
Dictyostelium discoideum, queuine in, 96–98
Differentiation, queuine and, 98–101

E

Endosymbiosis, 5-S ribosomal RNA and, 185–187
Error, feedback, increased accuracy and, 212–216
Evolution
  globin gene and
    early dispersion, 379–381
    mammalian comparisons, 381–389
  N^6-(2-isopentenyl)adenosine in tRNA and, 83
  modified nucleosides in tRNA and, 62–65
Evolutionary relationships, assessing by use of 5-S ribosomal RNA signatures, 184–185

G

Genes
  expression by ambiguous translation, 192–194
  ribosomal
    amplification and extrachromosomal location, 118–120
    chromosomal location, 118
    redundancy of, 117
  ribosomal, organization of
    arrangement of coding and noncoding regions, 120–124
    coding regions, 130–132
    nontranscribed spacers, 133–136
    ribosomal-insertion sequences, 136–138
    transcribed spacers, 132–133
    transcription-initiation regions, 124–127
    transcription-termination regions, 127–130
  ribosomal, transcription of, 138–139
    processing of pre-rRNA, 145–147
    regulation of, 147–150
    in vitro, 142–145
    in vivo, 139–142
Globin genes
  approaches for studying expression
    cellular systems, 325–326
    systems for introducing globin genes into cells, 326–330
  evolutionary considerations
    early dispersion of globin gene family, 379–381
    mammalian comparisons, 381–389
  naturally occurring mutations
    deletions in β-globin complex, 402–411
    general comments, 390
    polymorphisms without known functional correlates, 411–416
    prenatal diagnosis, 416–419
    thalassemias, 390–402
  physiological expression of
    developmental regulation, 330–332
    DNase hypersensitivity, 348–354
    methylation, 354–358
    promoter function, 333–343
    splicing, 343–348
    transcriptional unit, 332–333
  structure of
    codon utilization, 321–324
    5' and 3' untranslated regions, 324–325
    general, 317–319
    organization of exons and introns, 319–321
  structure of clusters
    evidence for gene conversion of repeated genes, 373–379
    α-globin cluster, 359–363
    β-globin cluster, 363–373
β-Globin gene, nucleotide sequences of, 439–458
Growth, exponential, translational accuracy and, 208–212

H

Hypoxanthine phosphoribosyltransferase, comparison of normal and mutant proteins, 298–300
Hypoxanthine phosphoribosyltransferase gene
  general aspects; enzyme and genetic properties, 296–298
  molecular cloning and analysis of sequences, 301–305
  normal and mutant structures, 305–309

INDEX

I

Immunoassay
  methodology
    antibody specificity and affinity, 6–7
    immunization schedule, 4–5
    immunogens, 2–4
    monoclonal antibodies, 5–6
  of undamaged nucleic acid components
    DNA and RNA, 8–9
    normal nucleosides, 7–8
    oligonucleotides, 11
    sequences versus conformation, 11–12
    synthetic polynucleotides, 9–11
Iron, metabolism, N^6-(2-isopentenyl)-adenosine and, 87–88
N^6-(2-Isopentenyl)adenosine
  and derivatives in tRNA
    codon/anticodon recognition, 83–84
    evolutionary aspects, 83
    transcription termination control and, 84–86
  and derivatives
    central role in growth and development
      cytokinin, 93–95
      link between A-37 modification and major routes of metabolism, 88–93
      relation to iron and amino acid metabolism, 87–88
      sporulation in *Bacillus subtilis*, 86–87

M

Melanophore system, of xiphophorine fish, queuine and, 101–103
Metabolism, major routes, modification at tRNA A-37 and, 88–93
Methylation, of globin gene, 354–358

N

Nucleosides
  modified, evolutionary aspects of, 62–65
  normal, immunoassay of, 7–8

Nucleotides, sequence of β-globin gene, 439–458

O

Oligonucleotides, immunoassay of, 11

P

Peptide chains, elongation, role of ribosylthymine in, 69–70
Phylogenetic trees, 5-S ribosomal RNAs and, 182–183
Polynucleotides, synthetic, immunoassay of, 9–11
Polypeptide-chain initiation
  in *Artemia*
    comparison of eIF2 with rabbit reticulocyte eIF2, 237–245
    comparison of eIF2 from dormant and developing embryos, 225–227
    eIF2 and initiation of protein synthesis, 224–225
    regulation of eIF2 activity, 227–236
  intermediates, effect of iRNA on, 283–290
Promoter function, of globin gene, 333–343
Proofreading
  and error regulation, inexpensive, kinetic options for, 212–216
  kinetic, 200–205
  problem solved by, 205–208
Protein, interaction with 5-S ribosomal RNA, 177–178
  eubacterial, 178–180
  eukaryotic, 180–181
  significance of, 181
Protein synthesis
  initiation, role of ribosylthymine in, 66–68
  tetrahydrofolate-independent initiation, ribosylthymine and, 73–74

R

Ribonucleic acid
  conformation derived from enzymatic,

chemical and physical studies of 5-S ribosomal
  eubacterial, 174–175
  eukaryotic, 175–177
endogenous and exogenous messengers, differential effects of iRNA and iRNP on mRNA translation, 281–283
eukaryotic, translation modulator activity and, 269–273
evolutionary aspects of 5-S ribosomal, 181–182
  assessing evolutionary relationships and, 184–185
  endosymbiosis and, 185–187
  phylogenetic tree and, 182–183
  taxonomic reclassification and, 183–184
generalized structures of 5-S ribosomal
  alternate structures, 166–167
  conserved structure, 167–169
  helix II, 165
  helix III, 165
  helix V, 165–166
  non-Watson-Crick base pairs in helices IV and V, 162–164
  tertiary interactions, 167
hot spots of insertions and deletions in 5-S ribosomal
  changes in helix III, 171
  insertions and deletions in general, 171–174
  major deletions in helix V, 171
  major insertions in helix IV, 171
messenger, in *Artemia* embryos, 248–255
organization of ribosomal genes
  arrangement of coding and noncoding regions, 120–124
  coding regions, 130–132
  nontranscribed spacers, 133–136
  ribosomal-insertion sequences, 136–138
  transcribed spacers, 132–133
  transcription-initiation region, 124–127
  transcription-termination region, 127–130
protein interaction with 5-S ribosomal, 177–178
  eubacterial, 178–180
  eukaryotic, 180–181
  significance of, 181
ribosomal genes
  amplification and extrachromosomal location, 118–120
  chromosomal location, 118
  redundancy of, 117–118
transcription of ribosomal genes, 138–139
  processing of pre-rRNA, 145–147
  regulation of, 147–150
  in vitro, 142–145
  in vivo, 139–142
transfer
  evolutionary aspects of modified nucleosides in, 62–65
  expression of genes and thymine biosynthesis, 82
  function of queuine in, 95–96
    in *Dictyostelium discoideum*, 96–98
    melanophore system of Xiphophorine fish, 101–103
    possible role in differentiation and neoplastic transformation, 98–101
  function of undermodified, unformylated, 75–79
  N^6-(2-isopentenyl)adenosine and derivatives in
    codon/anticodon recognition, 83–84
    evolutionary aspects, 83
    transcription termination control, 84–86
  role of ribosylthymine in function of, 65–66
    elongation of peptide chains, 69–70
    fidelity of translation, 70–71
    initiation of protein synthesis, 66–68
    ribosomal A-site interaction, 68–69
  with variable thymine content, utilization of, 80–82
  variations of TψC sequence, 79
Ribonucleic acid
  translation-inhibiting

effect on different intermediates of polypeptide chain initiation, 283–290
translation-inhibiting, secondary structure of, 277–278
translation of polycistronic messenger, ribosylthymine and, 74–75
undamaged, immunoassay of, 8–9
Ribonucleoprotein particles
 messenger, in *Artemia* embryos, 246–248
 translation-inhibiting
  of chick embryo muscle, 273–277
  dissociation and reassociation of, 278–280
Ribosome
 A-site interaction with tRNA, role of ribosylthymine in, 68–69
 codon-anticodon selectivity on, 197–200
Ribosylthymine
 in eukaryotic elongator tRNA
  expression of tRNA genes and biosynthesis of thymine, 82
  influence on translation *in vitro*, 79–80
  utilization of tRNAs with variable thymine content, 80–82
  variations of TψC sequence, 79
 involvement in regulatory mechanisms, 71–72
 function of undemodified, unformylated tRNA^{fmet} and, 75–79
 stringent response, 72–73
 tetrahydrofolate-independent initiation, 73–74
 translation of polycistronic mRNA and, 74–75
 role in tRNA function in eubacteria, 65–66
  elongation of peptide chains, 69–70
  fidelity of translation, 70–71
  initiation of protein synthesis, 66–68
  ribosomal A-site interaction, 68–69

S

Splicing, of globin gene, 343–348
Stringent response, ribosylthymine and, 72–73

T

Taxonomy, reclassifications suggested by 5-S ribosomal RNA comparisons, 183–184
Thalassemias, globin genes and, 390–402
Thymine, biosynthesis, expression of tRNA genes and, 82
Transcription, termination control, N^6-(2-isopentenyl)adenosine in tRNA and, 84–86
Transcriptional unit, of globin gene, 332–333
Transformation, neoplastic, queuine and, 98–101
Translation
 accuracy
  exponential growth and, 208–212
  kinetic proofreading and, 200–205
 ambiguous, gene expression by, 192–194
 correlation of speed and accuracy in, 194–196
 of endogenous and exogenous mRNA, differential effects of iRNA and iRNP on, 281–283
 fidelity, role of ribosylthymine in, 70–71
 modulation, eukaryotic RNAs and, 269–273
 in vitro, influence of ribosylthymine on, 79–80

X

Xiphophorine fish, melanophore system, queuine and, 101–103

Contents of Previous Volumes

Volume 1
"Primer" in DNA Polymerase Reactions—*F. J. Bollum*
The Biosynthesis of Ribonucleic Acid in Animal Systems—*R. M. S. Smellie*
The Role of DNA in RNA Synthesis—*Jerard Hurwitz and J. T. August*
Polynucleotide Phosphorylase—*M. Grunberg-Manago*
Messenger Ribonucleic Acid—*Fritz Lipmann*
The Recent Excitement in the Coding Problem—*F. H. C. Crick*
Some Thoughts on the Double-Stranded Model of Deoxyribonucleic Acid—*Aaron Bendich and Herbert S. Rosenkranz*
Denaturation and Renaturation of Deoxyribonucleic Acid—*J. Marmur, R. Rownd, and C. L. Schildkraut*
Some Problems Concerning the Macromolecular Structure of Ribonucleic Acids—*A. S. Spirin*
The Structure of DNA as Determined by X-Ray Scattering Techniques—*Vittoria Luzzati*
Molecular Mechanisms of Radiation Effects—*A. Wacker*

Volume 2
Nucleic Acids and Information Transfer—*Liebe F. Cavalieri and Barbara H. Rosenberg*
Nuclear Ribonucleic Acid—*Henry Harris*
Plant Virus Nucleic Acids—*Roy Markham*
The Nucleases of *Escherichia coli*—*I. R. Lehman*
Specificity of Chemical Mutagenesis—*David R. Krieg*
Column Chromatography of Oligonucleotides and Polynucleotides—*Matthys Staehelin*
Mechanism of Action and Application of Azapyrimidines—*J. Skoda*
The Function of the Pyrimidine Base in the Ribonuclease Reaction—*Herbert Witzel*
Preparation, Fractionation, and Properties of sRNA—*G. L. Brown*

Volume 3
Isolation and Fractionation of Nucleic Acids—*K. S. Kirby*
Cellular Sites of RNA Synthesis—*David M. Prescott*
Ribonucleases in Taka-Diastase: Properties, Chemical Nature, and Applications—*Fujio Egami, Kenji Takahashi, and Tsuneko Uchida*
Chemical Effects of Ionizing Radiations on Nucleic Acids and Related Compounds—*Joseph J. Weiss*
The Regulation of RNA Synthesis in Bacteria—*Frederick C. Neidhardt*
Actinomycin and Nucleic Acid Function—*E. Reich and I. H. Goldberg*
De Novo Protein in Synthesis *in Vitro*—*B. Nisman and J. Pelmont*
Free Nucleotides in Animal Tissues—*P. Mandel*

Volume 4
Fluorinated Pyrimidines—*Charles Heidelberger*
Genetic Recombination in Bacteriophage—*E. Volkin*
DNA Polymerases from Mammalian Cells—*H. M. Keir*
The Evolution of Base Sequences in Polynucleotides—*B. J. McCarthy*
Biosynthesis of Ribosomes in Bacterial Cells—*Syozo Osawa*
5-Hydroxymethylpyrimidines and Their Derivatives—*T. L. V. Ulbright*

Amino Acid Esters of RNA, Nucleotides, and Related Compounds—*H. G. Zachau and H. Feldmann*
Uptake of DNA by Living Cells—*L. Ledoux*

Volume 5
Introduction to the Biochemistry of 4-Arabinosyl Nucleosides—*Seymour S. Cohen*
Effects of Some Chemical Mutagens and Carcinogens on Nucleic Acids—*P. D. Lawley*
Nucleic Acids in Chloroplasts and Metabolic DNA—*Tatsuichi Iwamura*
Enzymatic Alteration of Macromolecular Structure—*P. R. Srinivasan and Ernest Borek*
Hormones and the Synthesis and Utilization of Ribonucleic Acids—*J. R. Tata*
Nucleoside Antibiotics—*Jack J. Fox, Kyoichi A. Watanabe, and Alexander Bloch*
Recombination of DNA Molecules—*Charles A. Thomas, Jr.*
   Appendix I. Recombination of a Pool of DNA Fragments with Complementary Single-Chain Ends—*G. S. Watson, W. K. Smith, and Charles A. Thomas, Jr.*
   Appendix II. Proof that Sequences of A, C, G, and T Can Be Assembled to Produce Chains of Ultimate Length, Avoiding Repetitions Everywhere—*A. S. Fraenkel and J. Gillis*
The Chemistry of Pseudouridine—*Robert Warner Chambers*
The Biochemistry of Pseudouridine—*Eugene Goldwasser and Robert L. Heinrikson*

Volume 6
Nucleic Acids and Mutability—*Stephen Zamenhof*
Specificity in the Structure of Transfer RNA—*Kin-ichiro Miura*
Synthetic Polynucleotides—*A. M. Michelson, J. Massoulié, and W. Guschbauer*
The DNA of Chloroplasts, Mitochondria, and Centrioles—*S. Granick and Aharon Gibor*
Behavior, Neural Function, and RNA—*H. Hydén*
The Nucleolus and the Synthesis of Ribosomes—*Robert P. Perry*
The Nature and Biosynthesis of Nuclear Ribonucleic Acids—*G. P. Georgiev*
Replication of Phage RNA—*Charles Weissmann and Severo Ochoa*

Volume 7
Autoradiographic Studies on DNA Replication in Normal and Leukemic Human Chromosomes—*Felice Gavosto*
Proteins of the Cell Nucleus—*Lubomir S. Hnilica*
The Present Status of the Genetic Code—*Carl R. Woese*
The Search for the Messenger RNA of Hemoglobin—*H. Chantrenne, A. Burny, and G. Marbaix*
Ribonucleic Acids and Information Transfer in Animal Cells—*A. A. Hadjiolov*
Transfer of Genetic Information during Embryogenesis—*Martin Nemer*
Enzymatic Reduction of Ribonucleotides—*Agne Larsson and Peter Reichard*
The Mutagenic Action of Hydroxylamine—*J. H. Phillips and D. M. Brown*
Mammalian Nucleolytic Enzymes and Their Localization—*David Shugar and Halina Sierakowska*

Volume 8
Nucleic Acids—The First Hundred Years—*J. N. Davidson*
Nucleic Acids and Protamine in Salmon Testes—*Gordon H. Dixon and Michael Smith*
Experimental Approaches to the Determination of the Nucleotide Sequences of Large Oligonucleotides and Small Nucleic Acids—*Robert W. Holley*
Alterations of DNA Base Composition in Bacteria—*G. F. Gause*
Chemistry of Guanine and Its Biologically Significant Derivatives—*Robert Shapiro*

Bacteriophage φX174 and Related Viruses—*Robert L. Sinsheimer*
The Preparation and Characterization of Large Oligonucleotides—*George W. Rushizky and Herbert A. Sober*
Purine N-Oxides and Cancer—*George Bosworth Brown*
The Photochemistry, Photobiology, and Repair of Polynucleotides—*R. B. Setlow*
What Really Is DNA? Remarks on the Changing Aspects of a Scientific Concept—*Erwin Chargaff*
Recent Nucleic Acid Research in China—*Tien-Hsi Cheng and Roy H. Doi*

Volume 9
The Role of Conformation in Chemical Mutagenesis—*B. Singer and H. Fraenkel-Conrat*
Polarographic Techniques in Nucleic Acid Research—*E. Paleček*
RNA Polymerase and the Control of RNA Synthesis—*John P. Richardson*
Radiation-Induced Alterations in the Structure of Deoxyribonucleic Acid and Their Biological Consequences—*D. T. Kanazir*
Optical Rotatory Dispersion and Circular Dichroism of Nucleic Acids—*Jen Tsi Yang and Tatsuya Samejima*
The Specificity of Molecular Hybridization in Relation to Studies on Higher Organisms—*P. M. B. Walker*
Quantum-Mechanical Investigations of the Electronic Structure of Nucleic Acids and Their Constituents—*Bernard Pullman and Alberte Pullman*
The Chemical Modification of Nucleic Acids—*N. K. Kochetkov and E. I. Budowsky*

Volume 10
Induced Activation of Amino Acid Activating Enzymes by Amino Acids and tRNA—*Alan H. Mehler*
Transfer RNA and Cell Differentiation—*Noboru Sueoka and Tamiko Kano-Sueoka*
N^6-(Δ^2-Isopentenyl)adenosine: Chemical Reactions, Biosynthesis, Metabolism, and Significance to the Structure and Function of tRNA—*Ross H. Hall*
Nucleotide Biosynthesis from Preformed Purines in Mammalian Cells: Regulatory Mechanisms and Biological Significance—*A. W. Murray, Daphne C. Elliott, and M. R. Atkinson*
Ribosome Specificity of Protein Synthesis in Vitro—*Orio Ciferri and Bruno Parisi*
Synthetic Nucleotide-peptides—*Zoe A. Shabarova*
The Crystal Structures of Purines, Pyrimidines and Their Intermolecular Complexes—*Donald Voet and Alexander Rich*

Volume 11
The Induction of Interferon by Natural and Synthetic Polynucleotides—*Clarence Colby, Jr.*
Ribonucleic Acid Maturation in Animal Cells—*R. H. Burdon*
Liporibonucleoprotein as an Integral Part of Animal Cell Membranes—*V. S. Shapot and S. Ya. Davidova*
Uptake of Nonviral Nucleic Acids by Mammalian Cells—*Pushpa M. Bhargava and G. Shanmugam*
The Relaxed Control Phenomenon—*Ann M. Ryan and Ernest Borek*
Molecular Aspects of Genetic Recombination—*Cedric I. Davern*
Principles and Practices of Nucleic Acid Hybridization—*David E. Kennell*
Recent Studies Concerning the Coding Mechanism—*Thomas H. Jukes and Lila Gatlin*
The Ribosomal RNA Cistrons—*M. L. Birnstiel, M. Chipchase, and J. Speirs*
Three-Dimensional Structure of tRNA—*Friedrich Cramer*
Current Thoughts on the Replication of DNA—*Andrew Becker and Jerard Hurwitz*
Reaction of Aminoacyl-tRNA Synthetases with Heterologous tRNA's—*K. Bruce Jacobson*
On the Recognition of tRNA by Its Aminoacyl-tRNA Ligase—*Robert W. Chambers*

Volume 12

Ultraviolet Photochemistry as a Probe of Polyribonucleotide Conformation—*A. J. Lomant and Jacques R. Fresco*
Some Recent Developments in DNA Enzymology—*Mehran Goulian*
Minor Components in Transfer RNA: Their Characterization, Location, and Function—*Susumu Nishimura*
The Mechanism of Aminoacylation of Transfer RNA—*Robert B. Loftfield*
Regulation of RNA Synthesis—*Ekkehard K. F. Bautz*
The Poly(dA-dT) of Crab—*M. Laskowski, Sr.*
The Chemical Synthesis and the Biochemical Properties of Peptidyl-tRNA—*Yehuda Lapidot and Nathan de Groot*

Volume 13

Reactions of Nucleic Acids and Nucleoproteins with Formaldehyde—*M. Ya. Feldman*
Synthesis and Functions of the -C-C-A Terminus of Transfer RNA—*Murray P. Deutscher*
Mammalian RNA Polymerases—*Samson T. Jacob*
Poly(adenosine diphosphate ribose)—*Takashi Sugimura*
The Stereochemistry of Actinomycin Binding to DNA and Its Implications in Molecular Biology—*Henry M. Sobell*
Resistance Factors and Their Ecological Importance to Bacteria and to Man—*M. H. Richmond*
Lysogenic Induction—*Ernest Borek and Ann Ryan*
Recognition in Nucleic Acids and the Anticodon Families—*Jacques Ninio*
Translation and Transcription of the Tryptophan Operon—*Fumio Imamoto*
Lymphoid Cell RNA's and Immunity—*A. Arthur Gottlieb*

Volume 14

DNA Modification and Restriction—*Werner Arber*
Mechanism of Bacterial Transformation and Transfection—*Nihal K. Notani and Jane K. Setlow*
DNA Polymerases II and III of *Escherichia coli*—*Malcolm L. Gefter*
The Primary Structure of DNA—*Kenneth Murray and Robert W. Old*
RNA-Directed DNA Polymerase—Properties and Functions in Oncogenic RNA Viruses and Cells—*Maurice Green and Gray F. Gerard*

Volume 15

Information Transfer in Cells Infected by RNA Tumor Viruses and Extension to Human Neoplasia—*D. Gillespie, W. C. Saxinger, and R. C. Gallo*
Mammalian DNA Polymerases—*F. J. Bollum*
Eukaryotic RNA Polymerases and the Factors That Control Them—*B. B. Biswas, A. Ganguly, and D. Das*
Structural and Energetic Consequences of Noncomplementary Base Oppositions in Nucleic Acid Helices—*A. J. Lomant and Jacques R. Fresco*
The Chemical Effects of Nucleic Acid Alkylation and Their Relation to Mutagenesis and Carcinogenesis—*B. Singer*
Effects of the Antibiotics Netropsin and Distamycin A on the Structure and Function of Nucleic Acids—*Christoph Zimmer*

Volume 16

Initiation of Enzymic Synthesis of Deoxyribonucleic Acid by Ribonucleic Acid Primers—*Erwin Chargaff*
Transcription and Processing of Transfer RNA Precursors—*John D. Smith*

Bisulfite Modification of Nucleic Acids and Their Constituents—*Hikoya Hayatsu*
The Mechanism of the Mutagenic Action of Hydroxylamines—*E. I. Budowsky*
Diethyl Pyrocarbonate in Nucleic Acid Research—*L. Ehrenberg, I. Fedorcsák, and F. Solymosy*

Volume 17
The Enzymic Mechanism of Guanosine 5′, 3′-Polyphosphate Synthesis—*Fritz Lipmann and Jose Sy*
Effects of Polyamines on the Structure and Reactivity of tRNA—*Ted T. Sakai and Seymour S. Cohen*
Information Transfer and Sperm Uptake by Mammalian Somatic Cells—*Aaron Bendich, Ellen Borenfreund, Steven S. Witkins, Delia Beju, and Paul J. Higgins*
Studies on the Ribosome and Its Components—*Pnina Spitnik-Elson and David Elson*
Classical and Postclassical Modes of Regulation of the Synthesis of Degradative Bacterial Enzymes—*Boris Magasanik*
Characteristics and Significance of the Polyadenylate Sequence in Mammalian Messenger RNA—*George Brawerman*
Polyadenylate Polymerases—*Mary Edmonds and Mary Ann Winters*
Three-Dimensional Structure of Transfer RNA—*Sung-Hou Kim*
Insights into Protein Biosynthesis and Ribosome Function through Inhibitors—*Sidney Pestka*
Interaction with Nucleic Acids of Carcinogenic and Mutagenic N-Nitroso Compounds—*W. Lijinsky*
Biochemistry and Physiology of Bacterial Ribonuclease—*Alok K. Datta and Salil K. Niyogi*

Volume 18
The Ribosome of *Escherichia coli*—*R. Brimacombe, K. H. Nierhaus, R. A. Garrett and H. G. Wittmann*
Structure and Function of 5 S and 5.8 S RNA—*Volker A. Erdmann*
High-Resolution Nuclear Magnetic Resonance Investigations of the Structure of tRNA in Solution—*David R. Kearns*
Premelting Changes in DNA Conformation—*E. Paleček*
Quantum-Mechanical Studies on the Conformation of Nucleic Acids and Their Constituents—*Bernard Pullman and Anil Saran*

Volume 19: Symposium on mRNA: The Relation of Structure and Function
I. The 5′-Terminal Sequence ("Cap") of mRNAs
Caps in Eukaryotic mRNAs: Mechanism of Formation of Reovirus mRNA 5′-Terminal m^{7}GpppGm-C—*Y. Furuichi, S. Muthukrishnan, J. Tomasz and A. J. Shatkin*
Nucleotide Methylation Patterns in Eukaryotic mRNA—*Fritz M. Rottman, Ronald C. Desrosiers and Karen Friderici*
Structural and Functional Studies on the "5′-Cap": A Survey Method of mRNA—*Harris Busch, Friedrich Hirsch, Kaushal Kumar Gupta, Manchanahalli Rao, William Spohn and Benjamin C. Wu*
Modification of the 5′-Terminals of mRNAs by Viral and Cellular Enzymes—*Bernard Moss, Scott A. Martin, Marcia J. Ensinger, Robert F. Boone and Cha-Mer Wei*
Blocked and Unblocked 5′ Termini in Vesicular Stomatitis Virus Product RNA *in Vitro*: Their Possible Role in mRNA Biosynthesis—*Richard J. Colonno, Gordon Abraham and Amiya K. Banerjee*
The Genome of Poliovirus Is an Exceptional Eukaryotic mRNA—*Yuan Fon Lee, Akio Nomoto and Eckard Wimmer*

II. Sequences and Conformations of mRNAs

Transcribed Oligonucleotide Sequences in Hela Cell hnRNA and mRNA—*Mary Edmonds, Hiroshi Nakazato, E. L. Korwek and S. Venkatesan*

Polyadenylylation of Stored mRNA in Cotton Seed Germination—*Barry Harris and Leon Dure III*

mRNAs Containing and Lacking Poly(A) Function as Separate and Distinct Classes during Embryonic Development—*Martin Nemer and Saul Surrey*

Sequence Analysis of Eukaryotic mRNA—*N. J. Proudfoot, C. C. Cheng and G. G. Brownlee*

The Structure and Function of Protamine mRNA from Developing Trout Testis—*P. L. Davies, G. H. Dixon, L. N. Ferrier, L. Gedamu and K. Iatrou*

The Primary Structure of Regions of SV40 DNA Encoding the Ends of mRNA—*Kiranur N. Subramanian, Prabhat K. Ghoshi, Ravi Dhar, Bayar Thimmappaya, Sayeeda B. Zain, Julian Pan and Sherman M. Weissman*

Nucleotide Sequence Analysis of Coding and Noncoding Regions of Human β-Globin mRNA—*Charles A. Marotta, Bernard G. Forget, Michael Cohen/Solal and Sherman M. Weissman*

Determination of Globin mRNA Sequences and Their Insertion into Bacterial Plasmids—*Winston Salser, Jeff Browne, Pat Clarke, Howard Heindell, Russell Higuchi, Gary Paddock, John Roberts, Gary Studnicka and Paul Zakar*

The Chromosomal Arrangement of Coding Sequences in a Family of Repeated Genes—*G. M. Rubin, D. J. Finnegan and D. S. Hogness*

Mutation Rates in Globin Genes: The Genetic Load and Haldane's Dilemma—*Winston Salser and Judith Strommer Isaacson*

Heterogeneity of the 3' Portion of Sequences Related to Immunoglobulin κ-Chain mRNA—*Ursula Storb*

Structural Studies on Intact and Deadenylylated Rabbit Globin mRNA—*John N. Vournakis, Marcia S. Flashner, MaryAnn Katopes, Gary A. Kitos, Nikos C. Vamvakopoulos, Matthew S. Sell and Regina M. Wurst*

Molecular Weight Distribution of RNA Fractionated on Aqueous and 70% Formamide Sucrose Gradients—*Helga Boedtker and Hans Lehrach*

III. Processing of mRNAs

Bacteriophages T7 and T3 as Model Systems for RNA Synthesis and Processing—*J. J. Dunn, C. W. Anderson, J. F. Atkins, D. C. Bartelt and W. C. Crockett*

The Relationship between hnRNA and mRNA—*Robert P. Perry, Enzo Bard, B. David Hames, Dawn E. Kelley and Ueli Schibler*

A Comparison of Nuclear and Cytoplasmic Viral RNAs Synthesized Early in Productive Infection with Adenovirus 2—*Heschel J. Raskas and Elizabeth A. Craig*

Biogenesis of Silk Fibroin mRNA: An Example of Very Rapid Processing?—*Paul M. Lizardi*

Visualization of the Silk Fibroin Transcription Unit and Nascent Silk Fibroin Molecules on Polyribosomes of *Bombyx mori*—*Steven L. McKnight, Nelda L. Sullivan and Oscar L. Miller, Jr.*

Production and Fate of Balbiani Ring Products—*B. Daneholt, S. T. Case, J. Hyde, L. Nelson and L. Wieslander*

Distribution of hnRNA and mRNA Sequences in Nuclear Ribonucleoprotein Complexes—*Alan J. Kinniburgh, Peter B. Billings, Thomas J. Quinlan and Terence E. Martin*

IV. Chromatin Structure and Template Activity

The Structure of Specific Genes in Chromatin—*Richard Axel*

The Structure of DNA in Native Chromatin as Determined by Ethidium Bromide Binding—*J. Paoletti, B. B. Magee and P. T. Magee*

Cellular Skeletons and RNA Messages—*Ronald Herman, Gary Zieve, Jeffrey Williams, Robert Lenk and Sheldon Penman*

The Mechanism of Steroid-Hormone Regulation of Transcription of Specific Eukaryotic Genes—*Bert W. O'Malley and Anthony R. Means*

Nonhistone Chromosomal Proteins and Histone Gene Transcription—*Gary Stein, Janet Stein, Lewis Kleinsmith, William Park, Robert Jansing and Judith Thomson*

Selective Transcription of DNA Mediated by Nonhistone Proteins—*Tung Y. Wang, Nina C. Kostraba and Ruth S. Newman*

V. Control of Translation

Structure and Function of the RNAs of Brome Mosaic Virus—*Paul Kaesberg*

Effect of 5'-Terminal Structures on the Binding of Ribopolymers to Eukaryotic Ribosomes—*S. Muthukrishnan, Y. Furuichi, G. W. Both and A. J. Shatkin*

Translational Control in Embryonic Muscle—*Stuart M. Heywood and Doris S. Kennedy*

Protein and mRNA Synthesis in Cultured Muscle Cells—*R. G. Whalen, M. E. Buckingham and F. Gros*

VI. Summary

mRNA Structure and Function—*James E. Darnell*

Volume 20

Correlation of Biological Activities with Structural Features of Transfer RNA—*B. F. C. Clark*

Bleomycin, an Antibiotic That Removes Thymine from Double-Stranded DNA—*Werner E. G. Müller and Rudolf K. Zahn*

Mammalian Nucleolytic Enzymes—*Halina Sierakowska and David Shugar*

Transfer RNA in RNA Tumor Viruses—*Larry C. Waters and Beth C. Mullin*

Integration versus Degradation of Exogenous DNA in Plants: An Open Question—*Paul F. Lurquin*

Initiation Mechanisms of Protein Synthesis—*Marianne Grunberg-Manago and François Gros*

Volume 21

Informosomes and Their Protein Components: The Present State of Knowledge—*A. A. Preobrazhensky and A. S. Spirin*

Energetics of the Ribosome—*A. S. Spirin*

Mechanisms in Polypeptide Chain Elongation on Ribosomes—*Engin Bermek*

Synthetic Oligodeoxynucleotides for Analysis of DNA Structure and Function—*Ray Wu, Chander P. Bahl, and Saran A. Narang*

The Transfer RNAs of Eukaryotic Organelles—*W. Edgar Barnett, S. D. Schwartzbach, and L. I. Hecker*

Regulation of the Biosynthesis of Aminoacid:tRNA Ligases and of tRNA—*Susan D. Morgan and Dieter Söll*

Volume 22

The —C—C—A End of tRNA and Its Role in Protein Biosynthesis—*Mathias Sprinzl and Friedrich Cramer*

The Mechanism of Action of Antitumor Platinum Compounds—*J. J. Roberts and A. J. Thomson*

DNA Glycosylases, Endonucleases for Apurinic/Apyrimidinic Sites, and Base Excision-Repair—*Thomas Lindahl*

Naturally Occurring Nucleoside and Nucleotide Antibiotics—*Robert J. Suhadolnik*

Genetically Controlled Variation in the Shapes of Enzymes—*George Johnson*

Transcription Units for mRNA Production in Eukaryotic Cells and Their DNA Viruses—*James E. Darnell, Jr.*

Volume 23
The Peptidyltransferase Center of Ribosomes—*Alexander A. Krayevsky and Marina K. Kukhanova*
Patterns of Nucleic Acid Synthesis in *Physarum polycephalum*—*Geoffrey Turnock*
Biochemical Effects of the Modification of Nucleic Acids by Certain Polycyclic Aromatic Carcinogens—*Dezider Grunberger and I. Bernard Weinstein*
Participation of Modified Nucleosides in Translation and Transcription—*B. Singer and M. Kröger*
The Accuracy of Translation—*Michael Yarus*
Structure, Function, and Evolution of Transfer RNAs (with Appendix Giving Complete Sequences of 178 tRNAs)—*Ram P. Singhal and Pamela A. M. Fallis*

Volume 24
Structure of Transcribing Chromatin—*Diane Mathis, Pierre Oudet, and Pierre Chambon*
Ligand-Induced Conformational Changes in Ribonucleic Acids—*Hans Günter Gassen*
Replicative DNA Polymerases and Mechanisms at a Replication Fork—*Robert K. Fujimura and Shishir K. Das*
Antibodies Specific for Modified Nucleosides: An Immunochemical Approach for the Isolation and Characterization of Nucleic Acids—*Theodore W. Munns and M. Kathryn Liszewski*
DNA Structure and Gene Replication—*R. D. Wells, T. C. Goodman, W. Hillen, G. T. Horn, R. D. Klein, J. E. Larson, U. R. Müller, S. K. Neuendorf, N. Panayotatos, and S. M. Stirdivant*

Volume 25
Splicing of Viral mRNAs—*Yosef Aloni*
DNA Methylation and Its Possible Biological Roles—*Aharon Razin and Joseph Friedman*
Mechanisms of DNA Replication and Mutagenesis in Ultraviolet-Irradiated Bacteria and Mammalian Cells—*Jennifer D. Hall and David W. Mount*
The Regulation of Initiation of Mammalian Protein Synthesis—*Rosemary Jagus, W. French Anderson, and Brian Safer*
Structure, Replication, and Transcription of the SV40 Genome—*Gokul C. Das and Salil K. Niyogi*

Volume 26: Symposium on DNA: Multiprotein Interactions
Introduction: DNA–Multiprotein Interactions in Transcription, Replication, and Repair—*R. K. Fujimura*
Replicative DNA Polymerase and Its Complex: Summary—*David Korn*
Enzyme Studies of ϕX174 DNA Replication—*Ken-ichi Arai, Naoko Arai, Joseph Shlomai, Joan Kobori, Laurien Polder, Robert Low, Ulrich Hübscher, LeRoy Bertsch, and Arthur Kornberg*
The DNA Replication Origin (ori) of *Escherichia coli*: Structure and Function of the ori-Containing DNA Fragment—*Yukinori Hirota, Masao Yamada, Akiko Hishimura, Atsuhiro Oka, Kazunori Sugimoto, Kiyozo Asada, and Mitsuru Takanami*
Replication of Linear Duplex DNA *in Vitro* with Bacteriophage T5 DNA Polymerase—*R. K. Fujimura, S. K. Das, D. P. Allison, and B. C. Roop*
Mechanisms of Catalysis of Human DNA Polymerases α and β—*David Korn, Paul A. Fisher, and Teresa S.-F. Wang*
Structural and Functional Properties of Calf Thymus DNA Polymerase δ—*Marietta Y. W. Tsang Lee, Cheng-Keat Tan, Kathleen M. Downey, and Antero G. So*
Mechanisms of Transcription: Summary—*R. K. Fujimura*

Regulatory Circuits of Bacteriophage Lambda—*S. L. Adhya, S. Garges, and D. F. Ward*
Chromatin Transcription and Replication: Summary—*Ronald L. Seale*
Site of Histone Assembly—*Ronald L. Seale*
Chromatin Replication in *Tetrahymena pyriformis*—*A. T. Annunziato and C. L. F. Woodcock*
Role of Chromatin Structure, Histone Acetylation, and the Primary Sequence of DNA in the Expression of SV40 and Polyoma in Normal or Teratocarcinoma Cells—*G. Moyne, M. Katinka, S. Saragosti, A. Chestier, and M. Yaniv*
Control of Transcription in Eukaryotes: Summary—*William J. Rutter*
Repair Replication Schemes in Bacteria and Human Cells—*Philip C. Hanawalt, Priscilla K. Cooper, and Charles Allen Smith*
Recent Developments in the Enzymology of Excision Repair of DNA—*Errol C. Friedberg, Corrie T. M. Anderson, Thomas Bonura, Richard Cone, Eric H. Radany, and Richard J. Reynolds*
Multiprotein Interaction in Strand Cleavage of DNA Damaged by UV and Chemicals—*Erling Seeberg*
In Vitro Packaging of Damaged Bacteriophage T7 DNA—*Warren E. Masker, Nancy B. Kuemmerle, and Lori A. Dodson*
The Inducible Repair of Alkylated DNA—*John Cairns, Peter Robins, Barbara Sedgwick, and Phillipa Talmud*
Functions Induced by Damaged DNA: Summary—*Evelyn M. Witkin*
Inducible Error-Prone Repair and Induction of Prophage Lambda in *Escherichia coli*—*Raymond Devoret*
DNA and Nucleoside Triphosphate Binding Properties of recA Protein from *Escherichia coli*—*K. McEntee, G. M. Weinstock, and I. R. Lehman*
Molecular Mechanism for the Induction of "SOS" Functions—*Michio Oishi, Robert M. Irbe, and Lee M. E. Morin*
Induction and Enhanced Reactivation of Mammalian Viruses by Light—*Larry E. Bockstahler*
Comparative Induction Studies—*Ernest C. Pollard, D. J. Fluke, and Deno Kazanis*
Concluding Remarks—*Ernest C. Pollard*

Volume 27

Poly(adenosine diphosphate ribose)—*Paul Mandel, Hideo Okazaki, and Claude Niedergang*
The Regulatory Function of Poly(A) and Adjacent 3′ Sequences in Translated RNA—*Uriel Z. Littauer and Hermona Soreq*
tRNA-like Structures in the Genomes of RNA Viruses—*Anne-Lise Haenni, Sadhna Joshi, and Francois Chapeville*
Mechanism of Interferon Action: Progress toward Its Understanding—*Ganes C. Sen*
RNA-Helix-Destabilizing Proteins—*John O. Thomas and Wlodzimierz Szer*
Nucleotide Cyclases—*Laurence S. Bradham and Wai Yiu Cheung*
Cyclic Nucleotide Control of Protein Kinases—*R. K. Sharma*

Volume 28

The Structure of Ribosomal RNA and Its Organization Relative to Ribosomal Protein—*Richard Brimacombe, Peter Maly, and Christian Zwieb*
Structure, Biosynthesis, and Function of Queuosine in Transfer RNA—*Susumu Nishimura*
Queuine: An Addendum—*Ram P. Singhal*
The Fidelity of Translation—*Abraham K. Abraham*
Structure and Functions of Ribosomal Protein S1—*Alap-Raman Subramanian*

The Yeast Cell Cycle: Coordination of Growth and Division Rates—*Steven G. Elliott and Calvin S. McLaughlin*

Prokaryotic and Eukaryotic 5 S RNAs: Primary Sequences and Proposed Secondary Structures—*Ram P. Singhal and Joni K. Shaw*

Structure of Transfer RNAs: Listing of 150 Additional Sequences—*Ram P. Singhal, Edda F. Roberts, and Vikram N. Vakharia*

Volume 29: Genetic Mechanisms in Carcinogenesis

Introduction—*W. K. Yang*

I. Genetic Factors in Cancer

Evolution of RNA Tumor Viruses: Analogy for Nonviral Carcinogenesis—*Howard M. Temin*

Model Hereditary Cancers of Man—*Alfred G. Knudson, Jr.*

Bacterial "Inserted Sequence" Elements and Their Influence on Genetic Stability and Evolution—*Werner Arber*

Significance of Specific Chromosomal Translocations and Trisomies for the Genesis of Murine and Human Tumors of the Lymphocyte–Plasmacyte Lineage—*George Klein (Summary prepared by W. W. Au)*

Short Communications

Quantitation of One Aspect of Karyotype Instability Associated with Neoplastic Transformation in Chinese Hamster Cells—*L. S. Cram, M. F. Bartholdi, F. A. Ray, G. L. Travis, J. H. Jett, and P. M. Kraemer*

Mechanism of Mutation at the Adenine Phosphoribosyltransferase Locus in CHO cells—*A. Simon and M. W. Taylor*

Development of a Transplantable Mouse Myeloid Leukemia Model System: A Preliminary Report—*W. W. Au, H. E. Luippold, and J. A. Otten*

II. Genetic Elements in Radiation and Chemical Carcinogenesis

Molecular Studies of the Radiation Leukemia Virus (RadLV) and Related Retroviruses of C57BL/Ka Mice—*R. A. Grymes, M. L. Scott, J. P. Kim, K. E. Fry, and Henry S. Kaplan*

Endogenous Retrovirus and Radiation-Induced Leukemia in the RFM Mouse—*Raymond W. Tennant, L. R. Boone, P. A. Lalley, and W. K. Yang*

Genetic and Probability Aspects of Cell Transformation by Chemical Carcinogens—*Charles Heidelberger, J. R. Landolph, R. E. K. Fournier, A. Fernandez, and A. R. Peterson*

Short Communications

Replication and Demethylation of O^6-Methylguanine in DNA—*E. T. Snow, R. S. Foote, and S. Mitra*

Quantitative Assay of Low Level of Benzo[a]pyrenediol Epoxide Bound to DNA by Acid-Induced Liberation of Tetraols followed by Chromatography and Fluorometric Detection—*R. O. Rahn, J. M. Holland, and L. R. Shugart*

Transfer of Phorbol Ester Promotability by Transfection of DNA from Promotable into Nonpromotable Cells—*N. H. Colburn, C. B. Talmadge, and T. D. Gindhart*

Specificity of Interaction between Carcinogenic Polynuclear Aromatic Hydrocarbons and Nuclear Proteins: Widespread Occurrence of a Restricted Pattern of Histone Binding in Intact Cells—*M. C. MacLeod, J. C. Pelling, T. J. Slaga, P. A. Noghrei-Nikbakht, B. K. Mansfield, and J. K. Selkirk*

III. Mechanism of Viral Carcinogenesis

Role of Polyoma T Antigens in Malignant Cell Transformation—*Walter Eckhart*

Avian Leukosis Viruses and Cancer: Genetics of Insertional Mutagenesis—*Harriet L. Robinson*

Short Communications

Use of a Viral Probe to Study Recombinational Exchanges in Mammalian Cells—*G. J. Duigou and S. G. Zimmer*

CONTENTS OF PREVIOUS VOLUMES

Comparative Tryptic Peptide Analysis of P85 gag–mos of Mo-MuSV ts-110 and the P38-P23 mos-Related Products of Wild-Type Virus—*E. C. Murphy, Jr. and R. B. Arlinghaus*

Genomic Complexity and Molecular Cloning of a Proviral DNA Specific for a Feral Rat Endogenous C-Type Virus, Originated from a 3-Methylcholanthrene-Induced Fibrosarcoma—*S. S. Yang and R. Modali*

An SV40 Mammalian Inductest for Putative Carcinogens—*S. P. Moore and T. P. Coohill*

Characterization of a Cellular Protein That Promotes SV40 Infection in Human Cells—*V. F. Righthand and J. C. Bagshaw*

IV. Regulatory Functions and Genetic Control

Cancer as a Problem in Intercellular Communication: Regulation by Growth-Inhibiting Factors (Chalones)—*Van R. Potter*

Restriction of Murine Leukemia Viruses by Fv-1: A Model for Studying Host Genetic Control of Retroviral Gene Movement and Leukemogenesis—*Wen K. Yang, L. R. Boone, R. W. Tennant, and A. Brown*

Expression of a Viral Oncogene under Control of the Mouse Mammary Tumor Virus Promoter: A New System for the Study of Glucocorticoid Regulation—*Gordon L. Hager*

Short Communications

Variation of Long-Terminal-Repeat Size in Molecular Clones of the BALB/c Endogenous Ecotropic Murine Leukemia Virus—*L. R. Boone, F. E. Myer, D. M. Yang, J. O. Kiggans, C. Koh, R. W. Tennant, and W. K. Yang*

Role of N^6-Methyladenosine in Expression of Rous Sarcoma Virus RNA: Analyses Utilizing Immunoglobulin Specific for N^6-Methyladenosine—*R. J. Resnick, D. Noreen, T. W. Munns, and M. L. Perdue*

V. Growth and Differentiation in Neoplastic Transformation

Role of Tyrosine Phosphorylation in Malignant Transformation by Viruses and in Cellular Growth Control—*Tony Hunter and J. A. Cooper*

Molecular Interaction of the src Gene Product with Cellular Adhesion Plaques—*Larry R. Rohrschneider, M. J. Rosok, and L. E. Gentry*

The Receptor for Epidermal Growth Factor Functions as a Tyrosyl-Specific Kinase—*Stanley Cohen*

Short Communications

An in Vitro Model of Epithelial Cell Neoplastic Progression: Growth Properties and Polypeptide Composition of Cell Lines—*K. D. Somers, M. M. Murphey, and D. G. Stark*

Chromosomal Protein Antigens Formed in Experimental Hepatocarcinogenesis by Azo Dyes—*W. N. Schmidt, B. J. Gronert, R. C. Briggs, D. L. Page, and L. S. Hnilica*

Cross-linking of Nuclear Antigens to DNA in HeLa Cells—*Z. M. Banjar, R. C. Briggs, L. S. Hnilica, J. Stein, and G. Stein*

Expression of Mutated Actin Gene Associated with Malignant Transformation—*H. Hamada, J. Leavitt, and T. Kakunaga*

Phenotypic Changes in Epithelial Cell Population Undergoing Neoplastic Transformation in Vitro—*G. R. Braslawsky, S. J. Kennel, and P. Nettesheim*

VI. The Search for Human Transforming Genes

The Oncogene of a Human Bladder Carcinoma—*L. F. Parada, C. Shih, M. Murray, and Robert A. Weinberg*

Transforming Genes of Neoplasms—*Geoffrey M. Cooper*

Short Communications

A Sequence Homologous to Rous Sarcoma Virus v-src Is on Human Chromosome 20—*A. Y. Sakaguchi, S. L. Naylor, and T. B. Shows*

Variable Differentiative Response of 6-Thioguanine-Resistant HL60 Sublines: Possible Relationship to Double-Minute Chromosomes—*R. E. Gallagher, A. C. Ferrari, A. W. Zulich, and J. R. Testa*

Volume 30

RNA Processing in a Unicellular Microorganism: Implications for Eukaryotic Cells—*David Apirion*

Nearest-Neighbor Effects in the Structure and Function of Nucleic Acids—*E. Bubienko, P. Cruz, J. F. Thomason, and P. N. Borer*

The Elongation Factor EF-Tu and Its Two Encoding Genes—*L. Bosch, B. Kraal, P. H. Van der Meide, F. J. Duisterwinkel, and J. M. Van Noort*

Small Nuclear RNAs and RNA Processing—*Ram Reddy and Harris Busch*

Ribosome Evolution: The Structural Bases of Protein Synthesis in Archaebacteria, Eubacteria, and Eukaryotes—*James A. Lake*

Analysis of the Expression of Genes Encoding Animal mRNA by *in Vitro* Techniques—*James L. Manley*

Synthesis, Processing, and Gene Structure of Vasopressin and Oxytocin—*Dietmar Richter*